AF688320

Sensory Profiling of Dairy Products

Society of Dairy Technology Series

The Society of Dairy Technology has joined with Wiley to produce a series of technical dairy-related handbooks providing an invaluable resource for all those involved in the dairy industry; from practitioners to technologists working in both traditional and modern large-scale dairy operations.

Sensory Profiling of Dairy Products, ISBN: 9781119619215
By John J. Tuohy (Editor)

Probiotic Dairy Products, 2nd Edition, ISBN 9781119214106
by Adnan Y. Tamime and Linda V. Thomas (Editors)

Microbial Toxins in Dairy Products, ISBN 9781118756430
by Adnan Y. Tamime (Editor)

Biofilms in the Dairy Industry, ISBN 9781118876213
by Koon Hoong Teh, Steve Flint, John Brooks, and Geoff Knight (Editors)

Milk and Dairy Products as Functional Foods, ISBN 9781444336832
by Ara Kanekanian (Editor)

Membrane Processing: Dairy and Beverage Applications, ISBN 9781444333374
by Adnan Y. Tamime (Editor)

Processed Cheese and Analogues, ISBN 9781405186421
by Adnan Y. Tamime (Editor)

Technology of Cheesemaking, 2nd Edition, ISBN 9781405182980
by Barry A. Law and Adnan Y. Tamime (Editors)

Dairy Fats and Related Products, ISBN 9781405150903
by Adnan Y. Tamime (Editor)

Dairy Powders and Concentrated Products, ISBN 9781405157643
by Adnan Y. Tamime (Editor)

Milk Processing and Quality Management, ISBN 9781405145305
by Adnan Y. Tamime (Editor)

Cleaning-in-Place: Dairy, Food and Beverage Operations, 3rd Edition, ISBN 9781405155038
by Adnan Y. Tamime (Editor)

Structure of Dairy Products, ISBN 9781405129756
by Adnan Y. Tamime (Editor)

Brined Cheeses, ISBN 9781405124607
by Adnan Y. Tamime (Editor)

Fermented Milks, ISBN 9780632064588
by Adnan Y. Tamime (Editor)

Probiotic Dairy Products, ISBN 9781405121248
By Adnan Y. Tamime (Editor)

Sensory Profiling of Dairy Products

Edited by

John J. Tuohy

2e Technical Development Ltd, Fermoy, Co. Cork, Ireland

WILEY Blackwell

SDT Society of Dairy Technology
Passion for Dairy

This edition first published 2023
© 2023 John Wiley & Sons Ltd

All rights reserved. No part of this publication may be reproduced, stored in a retrieval system, or transmitted, in any form or by any means, electronic, mechanical, photocopying, recording or otherwise, except as permitted by law. Advice on how to obtain permission to reuse material from this title is available at http://www.wiley.com/go/permissions.

The right of John J. Tuohy to be identified as the author of the editorial material in this work has been asserted in accordance with law.

Registered Offices
John Wiley & Sons, Inc., 111 River Street, Hoboken, NJ 07030, USA
John Wiley & Sons Ltd, The Atrium, Southern Gate, Chichester, West Sussex, PO19 8SQ, UK

For details of our global editorial offices, customer services, and more information about Wiley products visit us at www.wiley.com.

Wiley also publishes its books in a variety of electronic formats and by print-on-demand. Some content that appears in standard print versions of this book may not be available in other formats.

Trademarks: Wiley and the Wiley logo are trademarks or registered trademarks of John Wiley & Sons, Inc. and/or its affiliates in the United States and other countries and may not be used without written permission. All other trademarks are the property of their respective owners. John Wiley & Sons, Inc. is not associated with any product or vendor mentioned in this book.

Limit of Liability/Disclaimer of Warranty
While the publisher and authors have used their best efforts in preparing this work, they make no representations or warranties with respect to the accuracy or completeness of the contents of this work and specifically disclaim all warranties, including without limitation any implied warranties of merchantability or fitness for a particular purpose. No warranty may be created or extended by sales representatives, written sales materials or promotional statements for this work. The fact that an organization, website, or product is referred to in this work as a citation and/or potential source of further information does not mean that the publisher and authors endorse the information or services the organization, website, or product may provide or recommendations it may make. This work is sold with the understanding that the publisher is not engaged in rendering professional services. The advice and strategies contained herein may not be suitable for your situation. You should consult with a specialist where appropriate. Further, readers should be aware that websites listed in this work may have changed or disappeared between when this work was written and when it is read. Neither the publisher nor authors shall be liable for any loss of profit or any other commercial damages, including but not limited to special, incidental, consequential, or other damages.

Library of Congress Cataloging-in-Publication Data

Names: Tuohy, John J., editor.
Title: Sensory profiling of dairy products / Edited by John J. Tuohy.
Other titles: Society of Dairy Technology series.
Description: First edition. | Hoboken, NJ : Wiley, 2023. | Series: Society of dairy technology | Includes index.
Identifiers: LCCN 2022054072 (print) | LCCN 2022054073 (ebook) | ISBN 9781119619215 (cloth) | ISBN 9781119619314 (adobe pdf) | ISBN 9781119619338 (epub)
Subjects: LCSH: Dairy products. | Dairly products–Flavor and odor.
Classification: LCC TX556.M5 S46 2023 (print) | LCC TX556.M5 (ebook) | DDC 641.3/7–dc23/eng/20221207
LC record available at https://lccn.loc.gov/2022054072
LC ebook record available at https://lccn.loc.gov/2022054073

Cover Design: Wiley
Cover Images: Courtesy of Teagasc Food Research Centre, Moorepark, Ireland, © Juice Verve/Shutterstock, PanuShot/Shutterstock

Set in 10/12.5pt Times by Straive, Pondicherry, India

Contents

List of Contributors

Editor

Dr John J. Tuohy
2e Technical Development Ltd,
Hollybrook House, Corrin, Fermoy,
Co. Cork, P61 KR74, Ireland
Email: john@2etd.eu
URL: https://www.2etd.eu
Tel.: +353-87-2657706

Contributors

Prof Marzia Albenzio
Department of Agriculture, Food, Natural Resources and Engineering (DAFNE),
University of Foggia,
Via Napoli, 25-71121, Foggia, Italy
Email: marzia.albenzio@unifg.it
URL: www.unifg.it
Tel.: + 39-0881-589327

Prof Elenilson G. Alves Filho
Departamento de Engenharia de Alimentos,
Universidade Federal do Ceará, Campus do Pici,
Fortaleza-CE, 60020181, Brazil
Email: elenilson.godoy@yahoo.com.br
URL: www.ufc.br
Tel.: +55-85-33917388

Prof Lourdes Amigo
Instituto de Investigación en Ciencias de la Alimentación (CIAL)
University of Madrid,
Calle Nicolás Cabrera 9,
28049, Madrid, Spain
Email: lourdes.amigo@csic.es
URL: www.cial.uam-csic.es/pagperso/biopep/personal.html
Tel.: + 34-910017900

Dr Giordana Demaman Arend
Departamento de Engenharia Química e de Alimentos,
Universidade Federal de Santa Catarina (UFSC),
Florianópolis, Brazil
Email: giordana.darend@gmail.com
URL: www.ufsc.br
Tel.: +55-48-37216443

Miss Barbara Biró
Department of Postharvest, Supply Chain, Commerce and Sensory Science,
Institute of Food Science and Technology,
Hungarian University of Agriculture and Life Sciences,
Budapest, Hungary
Email: biro.barbara.4@phd.uni-mate.hu
URL: www.uni-mate.hu
Tel.: +36-13057351

Dr Şebnem Ö. Budak
Department of Dairy Technology,
Faculty of Agriculture,
Ankara University, Ankara, Turkey
Email: agricul@agri.ankara.edu.tr
URL: www.ankara.edu.tr
Tel.: +90-3125961000

Dr Stephanie Bull
Dept of Food and Nutritional Sciences,
Harry Nursten Building, University of Reading,
Pepper Lane, Whiteknights,
Reading, RG6 6DZ, United Kingdom
Email: stephanie.bull@reading.ac.uk
URL: https://www.reading.ac.uk/food/our-staff/dr-stephanie-bull
Tel.: +44-118-378-8593

Callebe Camelo-Silva
Departamento de Engenharia Química e de Alimentos,
Universidade Federal de Santa Catarina (UFSC),
R. do Biotério Central,
S/n - Córrego Grande,
Florianópolis SC, 88037-010
Brazil
URL: www.ufsc.br
Tel.: +55 48-37216440

Ms Maria H. Canella
Postgraduate Program in Food Engineering,
Technology Center, Federal University of Santa Catarina,
Rod. Admar Gonzaga 1346, Itacorubi, 88034-001,
Florianópolis, Brazil
Email: helenacanella@gmail.com
URL: www.ufsc.br
Tel.: +55-48-37215366

Prof Adriano G. da Cruz
Departamento de Alimentos,
Instituto Federal de Educação, Ciência and Tecnologia do Rio de Janeiro,
Rio de Janeiro, 2027002, Brazil
Email: adriano.cruz@ifrj.edu.br
URL: www.ifrj.edu.br
Tel.: +55-21-25667733

Dr Erick A. Esmerino
Departamento de Tecnologia de Alimentos,
Universidade Federal Fluminense,
Rua Vital Brazil Filho 64,
Niterói, 24230340, Brazil
Email: erick.almeida@hotmail.com
URL: www.uff.br
Tel.: +55-21-26299545

Dr Elson R. Filho
Department of Food,
Federal Institute of Education, Science & Technology of Rio de Janeiro,
Rio de Janeiro, 20270021, Brazil
Email: elsontavares@live.com
URL: www.iff.edu.br
Tel.: 55-19-982811469

Prof Mônica Q. Freitas
Veterinary College,
Federal University Fluminense,
Niterói, Rio de Janeiro, 24230340, Brazil
Email: mqueiroz@vm.uff.br
URL: www.uff.br
Tel.: +55-21-26299529

Marie-Therese Fröhlich-Wyder
Agroscope,
Schwarzenburgstrasse 161,
3003 Bern, Switzerland
Email: marie-therese.froehlich@agroscope.admin.ch
URL: www.agroscope.ch
Tel.: +41-58-46-38223

Dr Pascal Fuchsmann
Agroscope,
Schwarzenburgstrasse 161,
3003 Bern, Switzerland
Email: pascal.fuchsmann@agroscope.admin.ch
URL: www.agroscope.ch
Tel.: +41-58-46-38260

Prof Adriana Gambaro
Sensory Evaluation Area,
Department of Food Science and Technology,
School of Chemistry, Universidad de la República (UdelaR),
General Flores 2124, Montevideo, Uruguay
Email: agambaro@fq.edu.uy
URL: www.udelar.edu.uy
Tel.: +598-2929-0707

Dr Barbara Guggenbühl Gasser
Agroscope,
Schwarzenburgstrasse 161,
3003 Bern, Switzerland
Email: barbara.guggenbuehl@agroscope.admin.ch
URL: www.agroscope.ch
Tel.: +41-5846-38301

Prof Attila Gere
Department of Postharvest, Supply Chain, Commerce and Sensory Science,
Institute of Food Science and Technology,
Hungarian University of Agriculture and Life Sciences,
Budapest, Hungary
Email: gere.attila@uni-mate.hu
URL: www.uni-mate.hu
Tel.: +36-202786768

Prof Hamid Ghoddusi
Microbiology Research Unit,
London Metropolitan University,
London, United Kingdom
Email: h.ghoddusi@londonmet.ac.uk
URL: https://www.londonmet.ac.uk/profiles/staff/hamid-ghoddusi/
Tel.: +44 (0)20-7133-4196

Dr Monique Juna Lopes Leite
Departamento de Engenharia Química e de Alimentos,
Universidade Federal de Santa Catarina (UFSC),
Florianópolis, Brazil
Email: moniquejlleite@gmail.com
URL: www.ufsc.br
Tel.: +55-48-37216443

Assoc Prof Dr Stella Lignou
Dept of Food and Nutritional Sciences,
Harry Nursten Building, University of Reading,
Pepper Lane, Whiteknights,
Reading, RG6 6DZ, United Kingdom
Email: s.lignou@reading.ac.uk
URL: https://www.reading.ac.uk/food/our-staff/stella-lignou
Tel.: +44-118-378-8717

Dr Daniel Lozano-Ojalvo
Instituto de Investigación en Ciencias de la Alimentación (CIAL)
University of Madrid,
Calle Nicolás Cabrera 9, 28049 Madrid,
Spain
Email: daniel.lozano@csic.es
URL: www.alergia-alimentos.csic.es
Tel.: +34-910017900

Dr Marco Di Luccio
Departamento de Engenharia Química e de Alimentos,
Universidade Federal de Santa Catarina (UFSC),
Florianópolis, Brazil
Email: di.luccio@ufsc.br
URL: www.ufsc.br
Tel.: +55-48-37216443

Dr Matthew McSweeney
School of Nutrition and Dietetics,
Acadia University,
15 University Ave,
Wolfville, Nova Scotia, B4P 2R6, Canada
Email: matthew.mcsweeney@acadiau.ca
URL: https://nutrition.acadiau.ca/sensory-research-of-food.html
Tel: +1-902-585-1230

Prof Bhavbhuti M. Mehta
Dairy Chemistry Department,
SMC College of Dairy Science,
Kamdhenu University,
Anand – 388110, Gujarat, India
Email: bhavbhuti5@yahoo.co.in; bhavbhutimehta@gmail.com
URL: www.kamdhenuuni.edu.in/cds-anand-faculty
Tel.: +919825807454

Prof Lisa Methven
Dept. of Food and Nutritional Sciences,
Harry Nursten Building, University of Reading,
Pepper Lane, Whiteknights,
Reading, RG6 6DZ, United Kingdom
Email: l.methven@reading.ac.uk
URL: https://www.reading.ac.uk/food/about/staff/l-methven.aspx
Tel.: +44-118-378-8714

Dr Marcelo Miraballes
Sensory Evaluation Area,
Department of Food Science and Technology,
School of Chemistry, Universidad de la República (UdelaR),
General Flores 2124, Montevideo, Uruguay
Email: mmiraballes@fq.edu.uy
URL: www.udelar.edu.uy
Tel.: +598-2929-0707

Dr Elena Molina
Instituto de Investigación en Ciencias de la Alimentación (CIAL),
University of Madrid, Calle Nicolás Cabrera 9,
28049 Madrid, Spain
Email: e.molina@csic.es
URL: www.alergia-alimentos.csic.es
Tel.: + 34-910017938

Dr Howard Moskowitz
Mind Genomics Associates, Inc.,
White Plains, New York 10604, USA
Email: mjihrm@gmail.com
Tel.: +1-914-421-0479

Marice N. Oliveira
Department of Biochemical and Pharmaceutical Technology,
University of São Paulo,
Av. Prof Lineu Prestes, 580,
Bl. 16, 05508-900, São Paulo, Brazil
Email: monolive@usp.br
URL: www.5.usp.br
Tel.: +55-11-3091-3659

Assoc Prof Dr Maria Jose Oruna-Concha
Dept of Food and Nutritional Sciences,
Harry Nursten Building,
University of Reading,
Pepper Lane, Whiteknights,
Reading, RG6 6DZ, United Kingdom
Email: m.j.oruna-concha@reading.ac.uk
URL: https://www.reading.ac.uk/food/our-staff/maria-jose-oruna-concha
Tel.: +44-118-378-5364

Dr Maurice G. O'Sullivan
School of Food & Nutritional Sciences,
University College Cork,
Cork, T12 Y337, Ireland
Email: maurice.osullivan@ucc.ie
URL: https/www.ucc.ie/en/fns/about/people/mauriceosullivan
Tel.: +353-21-4902058

Prof Barbaros Özer
Department of Dairy Technology, Faculty of Agriculture,
Ankara University, Ankara, Turkey
Email: adabarbaros@gmail.com
URL: www.ankara.edu.tr
Tel.: +90 (312) 596-17-39

Dr Petraq Papajorgji
European University of Tirana,
Tirana, Albania
Email: petraq@gmail.com
URL: www.unitir.edu.al
Tel.: +355 (0) 672018121

Dr Tatiana C. Pimentel
Instituto Federal de Educação, Ciência & Tecnologia do Paraná,
Paranavaí-PA, 87703536 Brazil
Email: tatiana.pimentel@ifpr.edu.br
URL: www.ifpr.edu.br
Tel.: +55-44-34820110

Prof Ana C.M. Pinheiro
Departamento de Ciência de Alimentos,
Universidade Federal de Lavras,
Lavras, Minas Gerais, 37200000, Brazil
Email: anacarlamp@dca.ufla.br
URL: www.ufla.br
Tel.: +55-35-39291391

Dr Suneeta Pinto
Dairy Technology Department,
SMC College of Dairy Science, Kamdhenu University,
Anand – 388110, Gujarat, India
Email: suneeta_pinto@yahoo.co.in
URL: www.kamdhenuuni.edu.in/cds-anand
Tel.: +919099561760

Ms Amanda A. Prestes
Postgraduate Program in Food Engineering,
Technology Center, Federal University of Santa Catarina,
Rod. Admar Gonzaga 1346, Itacorubi, 88034-001,
Florianópolis, Brazil
Email: aprestes04@gmail.com
URL: www.ufsc.br
Tel.: +55-48-37215366

Prof Elane S. Prudencio
Department of Food Science and Technology,
Federal University of Santa Catarina,
Rod. Admar Gonzaga 1346, Itacorubi, 88034-001,
Florianópolis, Santa Catarina, Brazil
Email: elane.prudencio@ufsc.br
URL: www.ufsc.br
Tel.: +55-48-37215366

Dr Dalma Radványi
Department of Food Chemistry and Analytical Chemistry,
Institute of Food Science and Technology,
Hungarian University of Agriculture and Life Sciences,
Budapest, Hungary
Email: gerene.radvanyi.dalma@uni-mate.hu
URL: www.uni-mate.hu
Tel.: +36-205118891

Dr Michele N. Ribeiro
Departamento de Ciência de Alimentos,
Universidade Federal de Lavras,
Lavras, Minas Gerais, 37200000, Brazil
Email: michele.ribeiro01@gmail.com
URL: www.ufla.br
Tel.: + 55-35-39291391

Dr Jéssica F. Rodrigues
Instituto Federal de Educação, Ciência & Tecnologia de Minas Gerais,
Bombuí, 38900000, Brazil
Email: jessica.rodrigues@ifmg.edu.br
URL: www.ifmg.edu.br
Tel.: +55-37-3431-5451

Dr Sueli Rodrigues
Departamento de Engenharia de Alimentos,
Universidade Federal do Ceará,
Campus do Pici, Fortaleza-CE, 60020181, Brazil
Email: suelir@ufc.br
URL: www.ufc.br
Tel.: +55-85-33917388

Prof Antonella Santillo
Department of Agriculture, Food, Natural Resources and Engineering (DAFNE),
University of Foggia,
Via Napoli, 25-71121 Foggia, Italy
Email: antonella.santillo@unifg.it
URL: www.unifg.it
Tel.: + 39-0881-58932

Dr Saionara Sartor
Department of Food Science and Technology,
Federal Technological University of Paraná,
Francisco Beltrão, 85601970, Brazil
Email: saionara.sartor@posgrad.ufsc.br
URL: www.ufpr.br
Tel.: +55-46-35202600

Dr Lorena M.A. Silva
Embrapa Agroindústria Tropical,
rua Dra Sara Mesquita,
Pici, Fortaleza-CE, Brazil
Email: lorena.mara@embrapa.br
URL: www.embrapa.br
Tel.: +55-85-33917388

Dr Rosa Sullivan
Dept of Food and Nutritional Sciences,
Harry Nursten Building, University of Reading,
Pepper Lane, Whiteknights,
Reading, RG6 6DZ, United Kingdom
Email: r.c.sullivan@pgr.reading.ac.uk
URL: https://www.linkedin.com/in/rsullivan-insights

Prof Silvani Verruck
Department of Food Science and Technology,
Federal University of Santa Catarina,
Rod. Admar Gonzaga 1346, Itacorubi, 88034-001,
Florianópolis, Santa Catarina, Brazil
Email: silvani.verruck@ufsc.br
URL: www.ufsc.br
Tel.: +55-48-37215397

Prof R. Andrew Wilbey
Honorary Fellow,
Dept. of Food and Nutritional Sciences,
Harry Nursten Building, University of Reading,
Pepper Lane, Whiteknights,
Reading, RG6 6DZ, United Kingdom
Email: andrew.wilbey@btinternet.com
URL: https://www.reading.ac.uk/food/about/staff

Mr Ryan Zemel
Limbic Reviews,
Chicago, Illinois 60642, USA
Email: rzemel@icloud.com

Preface to the Technical Series

For more than 75 years, the Society of Dairy Technology (SDT) has sought to provide education and training in the dairy field, disseminating knowledge and fostering personal development through symposia, conferences, residential courses, publications and its journal, the *International Journal of Dairy Technology* (previously published as the *Journal of the Society of Dairy Technology*).

Recently, there have been significant advances in our understanding of milk systems, probably the most complex natural food available to man. At the same time, improvements in process technology have been accompanied by massive changes in the scale of many processing operations and the manufacture of a wide range of dairy and other related products.

Some years ago, the Society embarked on a project with Wiley to produce a Technical series of dairy-related books to provide an invaluable source of information for practising dairy scientists and technologists covering the range from small enterprises to modern large-scale operations. This latest, the sixteenth, volume in the series, on *Sensory Profiling of Dairy Products*, provides a link between the other titles in the series and covers an aspect of dairy science that is of great importance to all those who work in the industry.

Liz Whitley
Chairman of the Publications Committee, SDT
July 2022

Preface

Globalisation of the international food industry, more extensive foreign travel by the current generations of consumers and access to social media platforms have led to an unprecedented interest in and curiosity about dairy products and new dairy foods. In response, there has been significant diversification of the range of dairy products that consumers have access to, via the mainstream grocery trade, speciality or ethnic food stores, farmers' markets and increasingly online over the internet. The ambition is for this book to be a source of basic sensory information on some important traditional dairy products and new dairy foods from around the world, drawing on the latest scientific and technological knowledge and in a format that is accessible to dairy professionals from across the dairy supply chain including: manufacturing, product development, innovation and marketing, retailing, culinary as well as in regulatory affairs.

Among safe, natural foods, dairy products stand out for the diversity of characteristic sensory properties including: appearance, flavour, aroma, and texture and mouthfeel, found within the category. For centuries, these characteristic sensory attributes of dairy products have been the basis of quality assurance, product innovation and important criteria in the marketing and regulation of national and international trade in dairy products. Advances in biology, physiology, flavour chemistry, texture analysis sensory data analysis, process engineering, as well as a better understanding of the psychology of human behaviour, have all contributed to improvements in sensory methodology, the interpretation of sensory test data and a more objective understanding of the sensory preferences of consumers and strategies to address them.

After a brief historical synopsis of the sensory evaluation of dairy products, the opening chapter deals with a modern psychological approach to understanding consumer attitudes to dairy products and the interpretation of consumer feedback to inform commercial decision-making. Advances in the physiology of sensory perception are covered in the second chapter with particular reference to dairy product applications. Chapters 3–5 deal essentially with sensory methodology and the analysis of sensory data concerning dairy products.

The core of the book comprises Chapters 6–16, which profile the sensory attributes of products among the most important dairy categories (fluid milks, cream and milkfat products and frozen dairy desserts) but with particular emphasis on representative cheese varieties such as: Cheddar (Chapter 10), Swiss (Chapter 11) and hard Italian cheeses (Chapter 12) while also featuring both well-known and lesser-known indigenous cheese varieties from the Middle East (Chapter 8), Pan-America (Chapter 9) and Iberia (Chapter 13).

John J. Tuohy

1 Sensory Analysis and Consumer Mind-Sets and Emotions for Dairy Products

Attila Gere, Barbara Biró, Dalma Radványi, Ryan Zemel,
Petraq Papajorgji and Howard Moskowitz

1.1 Introduction

1.1.1 *History of the sensory analysis of dairy products*

Examination of food with human senses is contemporaneous with humanity, since its appearance, colour, odour, texture and flavour convey important information about its composition and quality. Human senses can, therefore, help avoid the consumption of unsafe materials, such as toxins, pathogenic microorganisms or other perceptible contaminations.

With the development of civilisation, people's demands were changing; in addition to survival, the enjoyment value began to matter more and more, and the production of high-quality, more expensive products had begun. Nevertheless, scientific research into the sensory properties of foods, consumer acceptance and liking did not begin until the twentieth century (Kemp et al. 2009).

From the beginning of the 1900's, food manufacturers started to use professionals for sensory evaluation, specialised in particular product categories, such as dairy products. The first quality standard on dairy products was developed by the International Dairy Federation (IDF), in 1903. The first grading standard about dairy products was established by the US Government in 1913, for butter. Shortly after, in 1916, the first national Collegiate Butter Judging Contest was held in the US, by the American Dairy Science Association (ADSA), which was expanded with the judgement of milk and cheddar cheese only a year later. In 1934, the first edition of *Judging Dairy Products* was published by Nelson and Trout.

After 1940, the scientific interest in sensory analysis seemed to grow even more. The Triangle test was developed by Bengtsson and Helm, while the US Army Quartermaster Food and Container Institute started research work on food acceptance and developed the hedonic scale.

As a result, the food industry discovered the potential of sensory evaluation quickly, and the field became an important part of the research and development processes and

Sensory Profiling of Dairy Products, First Edition. Edited by John J. Tuohy.
© 2023 John Wiley & Sons Ltd. Published 2023 by John Wiley & Sons Ltd.

laid the foundations for university-level research. In the following years, new methods were developed and books that formed the basis of the discipline were also published.

In 1968, the American Society for Testing and Materials (ASTM) published the first manual (STP433 – Basic Principles of Sensory Evaluation), which was followed by the first ISO standard (ISO 3591 – Sensory analysis protocol – standardised apparatus – a wine tasting glass design) 1977 (Clark et al. 2009).

Since the sensory attributes of food products affect willingness to pay greatly, newer evaluation methods are needed to be developed to help maintain consistent quality. According to several experts, the sensory evaluation of dairy products must be based on scientific principles, it requires experience and the acquisition of certain skills.

The first steps for sensory analysis of dairy products were the previously mentioned annual Collegial Dairy Products Evaluation Competitions in the early 1900s. These events brought the development of scorecards for evaluating different product types with numbers, which were further developed and used by United States Department of Agriculture (USDA), as a tool for education. The Competition is still held annually to this day (Website: http://dairyproductscontest.org) (Bodyfelt et al. 2008).

In Europe, during the second half of the twentieth century, the IDF played an important role in the conception of the Codex Alimentarius Commission (CAC) in 1961 The task of the CAC is to create worldwide standards related to food, including dairy products, but these do not contain recommendations and instructions on sensory evaluation (Tietje and Brouder 2009).

Due to the people's demands and the developments of the last century, the dairy industry has grown into one of the most quality-oriented industries. Nowadays, instrumental test methods (chromatography, electronic nose and tongue, mass spectrometry) and sensory evaluation are used together in quality management and quality assurance (Clark et al. 2009).

The currently used methods cover the entire field of sensory evaluation: analytical sensory tests (descriptive sensory analysis, discrimination tests, threshold tests), traditional judging/grading (by ADSA and USDA) and consumer tests. With the application of analytical sensory tests, taste profiles and aroma wheels can be created, which facilitate the performing of traditional judging and consumer tests (Drake 2004, 2007).

Since the 1940s, numerous new methods have been developed for consumer sensory testing, which are often performed with newly developed dairy products. The 9-point hedonic scale and its adaptations, the labelled affective magnitude (LAM) scale and the just-about-right (JAR) scale shall be highlighted, which were performed with, for example, probiotic yoghurts (Hekmat and Reid 2006), drinking yoghurts (Pohjanheimo and Sandell 2009) and flavoured kefirs (Gere et al. 2017). The hedonic scale-based preference mapping is widely used with dairy products, such as milk-fruit beverages or cheese (Fernández-Vázquez et al. 2018; Meals et al. 2020). Conjoint analysis is also a well-used method for testing dairy products and milk-based beverages (Jervis et al. 2012; Maurice et al. 2004). From the novel descripting consumer sensory methods in the 2000s, many tests are performed on different dairy products, such as flash profiling (Delarue and Sieffermann 2004), check-all-that-apply (CATA) questionnaires (Esmerino et al. 2017a), temporal-check-all-that-apply (TCATA), temporal dominance of sensation (TDS), progressive profile (PP) (Esmerino et al. 2017b) or rate-all-that-apply (RATA) questionnaires with emoji scales (Pinto et al. 2020).

Based on the experience of the last more than 100 years, sensory analysis has been an essential part of the development of the food industry, including the dairy industry. Moreover, based on the published literature, the dairy industry has contributed significantly to the development of the science of sensory evaluation.

1.1.2 *Changes in consumer habits*

Consumer habits and tastes constantly change. This continuous evolution forces the product world to follow. With each generation, new ways of living, eating, and working evolve, and so must foods and meals. The increasingly brisk transition period of one generation to the next is accompanied by changes in lifestyles, ideas and thinking styles, each evolving and demanding that the world of food production and merchandising shift to keep step. The evolution occurs in the world of the product, and in the less tangible but equally important world of the perceived nutrition world. It is obvious, then, that in order to meet the requirements of the new generation, dairy companies must identify these consumer needs.

Traditional efforts in the world of food product development have focused on product acceptance, as well as the fit of the product to an underlying concept, whether that concept be the 'product idea' or the basis for advertising to consumers why they should buy the product. The concepts are the product and the positioning, respectively.

Methods for testing the attractiveness of a concept vary from the simple evaluation of the ideas underlying the concept (so-called 'promise testing'), all the way the evaluation of complete paragraphs, of the type one might read in a magazine, as a full page or half page advertisement. Sometimes the ideas are presented without prices, sometime ideas are presented with prices and so on. In the end, the objective is to identify those ideas, those messages, which 'drive' an individual to say that she or he will buy the product.

Acceptance and usage occasion, as well as price, have been enriched by new areas of focus in the world of responses to food. The traditional study of products in terms of acceptance and use-occasions has been enriched in the last two decades by the introduction of a third dimension beyond acceptance and occasion. That third dimension is 'emotion', a dimension that would not normally occur to the businessperson interested in the product as something to be sold. Yet, researchers have made a convincing argument that emotions are intimately tied to food and worthy of study. To that end, a variety of papers have appeared which reveal the linkage between well-known emotions and certain kinds of food (Desmet and Schifferstein 2008; Rousset et al. 2005; Schouteten et al. 2015; Torben 2005).

During the past 20 years, the notion has become popular that through systematic experimentation with messages or ideas that the researcher can learn a great deal about how people react to ideas about food. The approach is generically known as conjoint measurement (Luce and Tukey 1964), a heritage of efforts in mathematical psychology to create better forms of measurement. Business professor Paul Green at Wharton School of Business, University of Pennsylvania, brought conjoint measurement to the business and applications literature (Green and Rao 1971; Green and Srinivasan 1990). Moskowitz and Gofman wrote a best-selling business book on the application of conjoint measurement to business issues (Moskowitz and Gofman 2007). Many papers have appeared detailing the method, a number referred to in Moskowitz (2012).

The fundamentals of conjoint measurement come from the world of experimental design (Box et al. 2005). Experimental design is a well-known field of statistical endeavour, allows the researcher to understand how different variables combine to generate a response. Experimental design of mixtures differs from the typical research approach of isolating one variable, exploring that variable in depth, and from that exploration done on several different variables.

The specific instantiation of experimental design applied here is known as Mind Genomics (MG), an offshoot of the approaches presented by Moskowitz over the past 20+ years. MG deals with the science of judgement of the everyday. Many scientific experiments are conducted with so-called 'model systems', namely stimuli that are pure, without cognitive meaning. In contrast, we experience life with stimuli filled with cognitive meaning. A science of perception and decision-making focusing on the experience of daily life thus provides a new way to understand how we perceive, and how we decide. We are not looking for aberrant situations to teach us interesting 'factoids' about people, but rather erecting a science of what we experience during the normal course of being and doing within a society which communicates, and which expects certain types of social behaviours.

1.2 How MG approaches the problem of understanding new versus traditional in cheese

The specific steps of an MG study have been laid out in a number of previous publications' go back virtually to more than three and a half decades, when the approach was laid out by Moskowitz and Martin (1993) at an ESOMAR conference in Copenhagen Denmark. The approach is illustrated showing the data as examples.

a. Define the topic and create the raw materials. The first step of an MG study design defines the topic is as much detail as possible, and then creates the raw materials. The raw materials for this study begin with four questions which are linked together to 'tell a story', and then four answers for each question. The answers are simple, once the questions have been created. The hard part is to think about the story, rather than simply to identify a variable, isolate it, measure and report the results. The researcher must pose the four questions and provide the 16 answers. Table 1.1 shows four questions, each having four answers. In this study, the questions and answers are specific to cheese.
b. Combine the answers into small vignettes or 'test concepts' using experimental design. The test stimuli comprise small, easy-to-read combinations of answers, vignettes. The vignettes, created in the format shown in Figure 1.1, comprise one or no answers from a question. The vignette tells a story. The story is incomplete, however, for some of the vignettes, a fact which does not hinder the respondent

Table 1.1 The four questions and the four answers for each question. The topic is cheese.

Question 1: What is the nutritional benefit of the cheese?	
A1	Calcium fortification of traditional cheese products helps to strengthen bones
A2	Reducing fat content of traditional cheeses helps lose weight
A3	Vitamin fortification of traditional cheese products keeps vitality
A4	Probiotics in traditional cheese products have health-preserving effects
Question 2: What is the origin of the cheese	
B1	Only organic ingredients allowed
B2	Local ingredients should be used for the most part
B3	Origin of ingredients not important when ingredients have SAME sensory characteristics
B4	Production should follow traditional recopies . . . history very important, no need for NEW
Question 3: What are the sensory characteristics of the cheese?	
C1	Traditional products with new, unusual flavourings . . . nothing wrong with that
C2	Flavour enhancers to enhance the special flavours of the traditional products
C3	Colourings make products more appealing
C4	Traditional products taste better compared to the commercial ones
Question 4: What are the economic effects?	
D1	Traditional cheese products support local labour market . . . job creation
D2	Innovation is key for traditional cheese makers to stay on the market
D3	Traditional cheese makers cannot compete with low-cost and large-scale production of cheese
D4	Rural life and national traditions are supported by purchasing traditional products

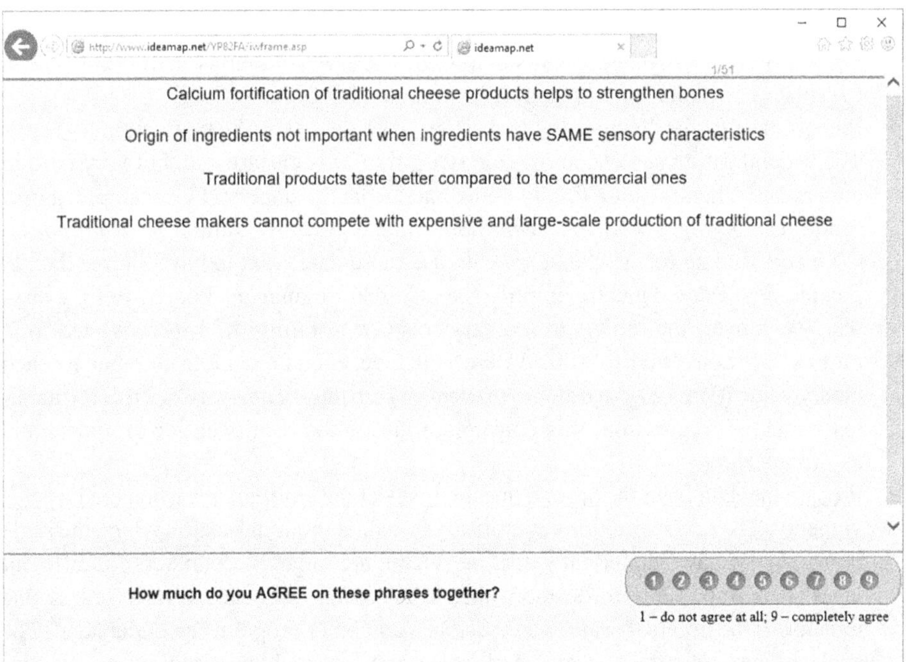

Figure 1.1 Example of a four-element vignette. The elements appear at the top, without connectives, making the vignette easy to read. Each vignette requires two answers, a rating and a selection of emotion. *Source:* Microsoft Corporation.

from evaluating the combination and assigning a rating. The participants were not shown the questions, but only the answers in certain combinations. The 16 answers were combined into a total of 24 vignettes, which were presented to paid US participants, recruited by an online panel provider. Although the combinations might seem 'random' to a respondent (an often-heard comment), the reality is that the combinations are carefully designed according to an underlying experiment design, which ensures that each answer appears equally often, and independent, in a statistical sense, of the other answers. Each respondent evaluated a *unique* set of 24 vignettes, constructed according to the same basic design, but with the elements permuted so that the elements, i.e. answers, remained the same, but the specific combinations varied across respondents. The experimental design ensures that a wide number of combinations are covered across the respondents, rather than a simple set of 24 combinations appearing again and again (Gofman and Moskowitz 2010).

c. Select the rating question(s). The respondent reads the presented elements and rates the elements on the vignette as a whole story or concept and answers the rating question (1 – do not agree at all; 9 – completely agree). The rating scale is phrased as follows: *How much do you AGREE on these phrases together?* To capture the emotions of the participants after reading the vignette, an additional question was asked. The question was phrased as follows: *How do you personally FEEL when you read the combination of these phrases together?* The options are assorted so that emotions are not placed in order of intensity (happy versus sad). These seven emotions are the following: 1 = curious, 2 = amused, 3 = don't care, 4 = disappointed, 5 = surprised, 6 = irritated, 7 = disinterested. The seven emotions are presented in an irregular order, interspersing positive, neutral and negative feelings.
d. Prepare ratings for analysis by converting from the Likert Scale to a binary scale. The 9-point rating scale/Likert Scale (Question 1) is easy to use, but very hard to understand when the data finally come back from the analyst. For example, a user of the results might well ask '*what does the 7.5 mean*' or similarly, '*what does a 4.5 mean*' and so forth. Managers who use these data want to have simple data to ponder, not scaled data. To simplify the results for managers, but really for everyone, we convert the ratings to a binary scale, with ratings of 1–6 converted to 0, ratings 7–9 converted to 100. As we add a very small random number to each binary value (0 or 100), in order to prevent a crash during analysis by OLS (ordinary least squares) regression, this conversion makes the results easier to understand and to interpret.
e. Prepare the data from the second question (select the emotion felt when reading this vignette). The seven emotions comprise a so-called 'nominal scale', where the scale points do not have numerical value, per se, but are rather placeholders for different ideas. It is impossible to do more than analyse the frequency of use, unless one transforms the nominal scale to a form that a computer program can understand. The analysis comprises the conversion of the scale of seven choices into seven new variables, each variable corresponding to one of the choices. For a given vignette, each of the seven variables begins with the value '0'. When the vignette is assigned a specific emotion, the initial assignment of '0' is converted to the value '100', but the

remaining six variables remain at '0'. Thus, the scale generates seven, presumably unrelated variables, but we 'know' the origin of these variables. For this analysis, we produce two new variables, the positive emotions and the negative emotions. We do so by summing the values of the variables corresponding to the positive emotions and summing the values of the variables corresponding to the negative emotions, respectively.

f. Create a model or linear equation relating the presence/absence of the 16 answers or elements to the binary response. The equation is expressed by the simple linear model, showing the contribution of each element to the binary rating: Rating = k_0 + k_1(A1) + k_2(A2) . . . k_{16}(D4). The elements are represented as A1, A2 . . . D4, the additive constant is k_0, and the 16 coefficients are denoted by k_i. Each parameter of the regression equation (additive constant, 16 coefficients, one per element of answer) has meaning, as will be shown in the explanation of the results. The 16 coefficients are particularly important because they reveal the contribution of the corresponding element to the rating.
g. Interpret the model in terms of 'meaning' for cheese. Table 1.2 shows the parameters of the model.

Table 1.2 Parameters for the relation between Question 1 (binary transformation of 'Agree') and the presence/absence of the individual elements. The table is sorted in descending order of coefficient.

	Statement	Coeff	T-Stat	P-Val
	Additive constant	55.05	7.23	0.00
A4	Probiotics in traditional cheese products have health-preserving effects	5.92	1.28	0.20
A1	Calcium fortification of traditional cheese products helps to strengthen bones	4.30	0.93	0.35
D4	Rural life and national traditions are supported by purchasing traditional products	3.00	0.65	0.52
D2	Innovation is key for traditional cheese makers to stay in the market	2.50	0.54	0.59
D1	Traditional cheese products support local labour market . . . job creation	0.63	0.14	0.89
B4	Production should follow traditional recopies . . . history very important, no need for NEW	−1.02	−0.22	0.83
C4	Traditional products taste better compared to the commercial ones	−1.76	−0.38	0.71
C2	Flavour enhancers to enhance the special flavours of the traditional products	−1.96	−0.42	0.67
A2	Reducing fat content of traditional cheeses helps lose weight	−2.88	−0.62	0.53
B2	Local ingredients should be used for the most part	−4.04	−0.87	0.39
A3	Vitamin fortification of traditional cheese products keeps vitality	−4.21	−0.91	0.37
C1	Traditional products with new, unusual flavourings . . . nothing wrong with that	−4.41	−0.95	0.35
C3	Colourings make products more appealing	−6.58	−1.42	0.16
D3	Traditional cheese makers cannot compete with low-cost and large-scale production of cheese	−8.29	−1.78	0.08
B1	Only organic ingredients allowed	−9.61	−2.04	0.04
B3	Origin of ingredients not important when ingredients have SAME sensory characteristics	−13.94	−2.98	0.00

1. The additive constant is the estimate percent of times that a rating would receive a 7–9 in the absence of elements. Clearly, all vignettes comprised a minimum of two elements, and a maximum of four elements, so the additive constant is a purely estimate value. Nonetheless, it has meaning. It is basic likelihood that the respondent will agree to the vignette, i.e. that the respondent will feel that the vignette 'makes sense', and that the respondent likes what is read. The additive constant is 55.05, meaning that they tend to start off positive, at least half the time or half the respondents. The rest will be the task of the elements to drive more 'agreement', i.e. more acceptance.
2. The coefficient for each element, which tells the additional percent of the responses that are likely to be 7–9 when the particular element is inserted into the vignette. When the coefficient is positive, the elements add a certain percent. Thus, element A4 (probiotics in traditional cheese products have health-preserving effects) has a coefficient of 5.92, meaning that when added to the vignette an extra 5.92% of the ratings will be 7–9. In contrast, some elements are negative, such as B3 (origin of ingredients not important when ingredients have SAME sensory characteristics.) This element has a strong negative effect, -13.94. When this element is inserted into the vignette, almost 14% fewer ratings of 7–9 are to be expected.
3. There are also two statistics worth noting, statistics which emerge automatically from today's off-the-shelf (so-called 'canned') statistical software. The first is the student t-statistic, a measure of the value of the coefficient divided by the standard error of the coefficient. The t-statistic is a measure of signal (the value of the coefficient) to noise (the expected variability of the coefficient if the study were to be run many times.) The t-statistic should be high, meaning that the value of the coefficient is more believable, the higher the t-statistic. Furthermore, a t-statistic closer to 0 means that upon replications of this same experiment, with the same conditions, we might find coefficients much closer to 0.
4. The second statistic of interest is the p-value, the likelihood or probability that the coefficient is different from 0. The lower the p-value, the more likely it is that the coefficient we observe is truly not 0.
5. It is important to note that despite the statistical measures associated with the coefficient (the student t-statistic and the p-value), our learning comes from the magnitude of the coefficient. Big positive numbers suggest a more likely positive response, that the respondent 'agrees' with the combination. Big negative numbers suggest a non-positive response, that the respondent does not 'agree' with the combination. We do not yet know whether the respondent disagrees with the combination, or simply does not agree.
6. Finally, the coefficients are additive. One can estimate the likely response to the vignette by combining the elements and estimating their sum of additive constant and individual coefficient.
7. The total panel data suggest that respondents diverge from the concept of using organic ingredients in traditional cheese, oppose the notion of importance of the origin of ingredients and disagree with the suggestion that traditional cheese makers cannot compete with larger companies. However, respondents from the

total panel agree with the ideas of probiotics and calcium fortification. Both concepts are suitable options for traditional cheese makers to increase the health benefits of their products, and to give some added value. From the results provided by the total panel we propose that for cheese, respondents are looking to purchase something that is innovative, and full of health benefits; something which moves beyond just traditional.

1.3 Looking at different groups of respondents

A key benefit of the MG approach is that it uses the structure of a 'within-subjects' design. That is, each respondent evaluates the 'right' set of combinations of messages so that one can build an individual-level model. That property has two major benefits:

1. When looking at key subgroups, one need not worry about 'balancing' the appearance of the different elements. Each respondent brings in the appropriate set of statistically independent elements.
2. When dividing the respondent population into basic groups, based upon the pattern of responses, one can use the pattern of coefficients generated by each respondent as the basis for the mind-set segmentation.

The focus of this study was to explore the total panel and to uncover any emergent mind-sets, these mind-sets comprising individuals who perceive the world in the same way, for the specific topic (innovation in cheese.). MG typically works with relatively small samples of 25–100, recognising that the experiments are exploratory. By exploratory, we mean that the focus is to demonstrate the existence of different mind-sets, but not to estimate the proportion of these mind-sets in a population. Such estimates can be made, and discussed later, but the bigger discovery is the nature of the mind-sets, should they exist.

Mind-sets based upon the pattern of responses emerge from the statistical method of k-means clustering (Hastie et al. 2001). In simple terms, each respondent generates a vector of 16 coefficients. Clustering divides these respondents into two or more groups, based upon a criterion defined ahead of time. The criterion is that the differences or variability of patterns within a cluster or mind-set is small, whereas the differences or variability of the averages of the mind-sets is large across the 16 elements (Dubes and Jain 1980).

Based upon thousands of studies over the past 35 years, a good rule of thumb has emerged. Regression coefficients above 8 are relevant 'positives', whereas regression coefficients below -8 are relevant 'negatives', again based on these thousands of previous studies. By relevant, we mean that the element pertains to something which occurs with the product in the 'outside' world. By looking at relevance rather than statistical significance, we move away from purely statistical considerations of 'difference' to 'meaningfulness'.

There are at least two interpretable groups, as revealed by the MG clustering. These mind-sets appear in Table 1.3. It is important to keep in mind that the clustering or segmentation is done as a purely mathematical/statistical procedure, and without

Table 1.3 Coefficients for total panel and two mind-set segments for question #1 (agree). The table shows the coefficients estimated for the top of the scale (actively agree) and for the bottom of the scale (actively disagree).

		Tot	Mind-Set1	Mind-Set2	Tot	Mind-Set1	Mind-Set2
Statement		Top3	Top3	Top3	Bot3	Bot3	Bot3
	Base size	53	29	24	53	29	24
	CONSTANT	55	60	47	18	14	23
A4	Probiotics in traditional cheese products have health-preserving effects	6	1	13	–4	2	–11
A1	Calcium fortification of traditional cheese products helps to strengthen bones	4	1	10	–3	0	–8
D2	Innovation of key for traditional cheese makers to stay on the market	3	–3	10	–6	1	–15
D4	Rural life and national traditions are supported by purchasing traditional products	3	1	6	–5	–5	–7
D1	Traditional cheese products support local labour market . . . job creation	1	–2	5	–6	–7	–5
B4	Production should follow traditional recopies . . . history very important, no need for NEW	–1	–6	5	0	–1	1
A2	Reducing fat content of traditional cheeses helps lose weight	–3	–4	0	0	2	–3
A3	Vitamin fortification of traditional cheese products keeps vitality	–4	–7	0	0	2	–3
C1	Traditional products with new, unusual flavourings . . . nothing wrong with that	–4	–9	0	3	3	3
C2	Flavour enhancers to enhance the special flavours of the traditional products	–2	–3	–1	0	1	–1
C4	Traditional products taste better compared to the commercial ones	–2	–2	–2	4	5	3
B2	Local ingredients should be used for the most part	–4	–5	–3	–2	–4	–1
D3	Traditional cheese makers cannot compete with expensive and large-scale production of traditional cheese	–8	–11	–4	–4	–2	–8
B1	Only organic ingredients allowed	–10	–13	–5	4	2	5
C3	Colourings make products more appealing	–7	–5	–9	6	5	7
B3	Origin of ingredients not important when ingredients have SAME sensory characteristics	–14	–12	–17	5	4	5

understanding WHAT the clusters really signify. That is, the clustering procedure is 'agnostic', doing its mathematical work without interpretation.

Table 1.3 shows the coefficients for the total panel and the two mind-sets. Table 1.3 divides into two sections. The columns marked **Top3** correspond to the analysis of strong positive responses, where ratings of 7–9 are converted to 100, whereas ratings of 1–6 are converted to 0. These are the degree to which the respondent 'agrees' with the element, when the element is in the vignette. The columns marked **Bot3** correspond to the analysis of the strong negative responses, where ratings of 1–3 are converted to 100 whereas ratings of 4–9 are converted to 0. These are the degree to which the respondent 'disagrees' with the element, when the element is in the vignette. It is important to stress here that when we look at a binary scale, the low end of the scale, converted to 0, does not mean 'disagree' or 'agree' but rather 'lack agreement (not disagree) or lack of disagreement (not agree)'.

Table 1.3 suggests that when it comes to cheese, there are two key mind-sets. Mind-Set1 tends to agree with the elements (additive constant 60 for Top3), tends not to disagree with the elements (additive constant 14 for Bot3), but really does not pay attention to the elements, Mind-Set1 is simply not really involved. In contrast, Mind-Set2 tends to agree less (additive constant 47 for Top3), tends to disagree more (additive constant 23 for Bot 3) and responds strongly about health and innovation. We might label Mind-Set1 'traditionalists', and Mind-Set2 'Innovators'.

What surprises is the low values of the coefficients for total panel, and for the mind-sets, except for three elements for Mind-Set2. Many previous studies with MG suggest that the mind-set segmentation is more profound. On the other hand, the typical segmentation generating strong large positive coefficients finds its crux in flavour-based messages, and not in health-oriented or heritage-oriented message.

Probiotics in traditional cheese products have health-preserving effects
Calcium fortification of traditional cheese products helps to strengthen bones
Innovation is key for traditional cheese makers to stay on the market

1.4 Linking emotions to messages

The second question required the respondent to select one of seven feelings/emotions. The feelings/emotions were presented in irregular order, to prevent the respondent from choosing a similar point on the scale for the two ratings. The analysis of the results by OLS regression proceeds in the same way as the analysis of the ratings from question #1. The only differences are the nature of the dependent variable and the nature of the regression model:

Two dependent variables: The seven emotions were reduced to the three positive emotions (curious, amused, surprise) and the three negative emotions (disappointed, irritated, disinterested). When a respondent selected a positive emotion, the new variable, PosEm was coded 0, and the other new variable, NegEm was coded 0. The opposite coding was done was then respondent selected a negative emotion. NegEm was coded 100, and PosEm was coded 0. Both PosEm and NegEm was coded 0 when the respondent selected 'Don't care'.

Table 1.4 shows that there are strong positive emotions, but these are not the emotions that drive agreement. Rather, the strong positive emotions linking to the elements tend to be more 'cause-related', and less to 'self-related'.

Rural life and national traditions are supported by purchasing traditional products
Only organic ingredients allowed
Traditional cheese products support local labour market . . . job creation

Both mind-sets, Mind-Set1 (Traditionalist) and Mind-Set2 (Innovator) respond with emotion, but the Innovator appears to have slightly greater emotional linkages to elements than the Traditionalist.

1.5 Finding mind-sets in the population for future communication, research and sales efforts

A continuing issue, or perhaps better a continuing opportunity, arises from the need to capitalise on the mind-sets for either social or business reasons. In terms of social reasons, working with Innovators allows a country, a company, a school or really any social group to change patterns of food consumption, to reduce neophobia and perhaps to eat better. The obesity epidemic, for example, is raging through the world. Finding Innovators, willing to try new foods, perhaps better-for-you foods, may be the first step.

How can one discover these new groups or mind-sets in the population and work with them? MG studies suggest both two mind-sets, at least when we talk about the health benefits. These mind-sets are spread through the population. In most MG studies, the data suggest that conventional subgroups of consumers may not allow one to assign a given individual to a mind-set unless the individual participates in a version of the study, devoted to the specific topic. That is, the mind-sets emerge as relevant to a topic, and not as general descriptions of a population. In this respect, the granularity of MG proceeds from the specific to the general. There is never the problem of discovering the 'proper message' to a member of a mind-set uncovered by MG, for it is from the reactions to the messaging that these mind-sets emerge for the particular topic.

Author Gere has created the PVI, the personal viewpoint identifier, a set of questions, directly related to the study. The pattern of the responses to these questions automatically assigns the person into the appropriate mind-set. The PVI works in either case to assign a new person to a mind-set, after which the new person can become part of the health movement or receive advertisements and coupon appropriate for her/his mind-set with respect to cheese.

Figure 1.2 presents the personal viewpoint identifier, which as of this writing resides at: https://www.pvi360.com/TypingToolPage.aspx?projectid=164&userid=2008. The top of Figure 1.2 presents the six questions. The bottom presents the feedback, simple web-based material that is automatically incorporated into the respondent's data file for a loyalty programme to guide further sales efforts or presented to the respondent as an information 'reward'.

Table 1.4 Linking between elements, emotions, and mind-sets.

Element		Tot	Mind-Set1	Mind-Set2	Tot	Mind-Set1	Mind-Set2
		PosEm	PosEm	PosEm	NegEm	NegEm	NegEm
D4	Rural life and national traditions are supported by purchasing traditional products	22	22	22	8	11	6
B1	Only organic ingredients allowed	20	23	18	14	12	16
D1	Traditional cheese products support local labour market . . . job creation	20	19	21	13	16	10
C1	Traditional products with new, unusual flavourings . . . nothing wrong with that	19	20	17	10	10	8
A1	Calcium fortification of traditional cheese products helps to strengthen bones	17	12	23	10	6	15
B4	Production should follow traditional recopies . . . history very important, no need for NEW	17	14	20	12	10	14
C4	Traditional products taste better compared to the commercial ones	17	23	11	9	9	8
A4	Probiotics in traditional cheese products have health-preserving effect	16	16	16	5	0	10
D2	Innovation of key for traditional cheese makers to stay on the market	16	13	19	12	15	9
A3	Vitamin fortification of traditional cheese products keeps vitality	14	9	20	9	6	13
B2	Local ingredients should be used for the most part	14	12	17	10	7	14
C2	Flavour enhancers to enhance the special flavours of the traditional products	14	12	16	10	9	11
D3	Traditional cheese makers cannot compete with expensive and large-scale production of traditional cheese	14	13	15	14	20	8
A2	Reducing fat content of traditional cheeses helps lose weight	13	8	19	7	2	13
B3	Origin of ingredients not important when ingredients have SAME sensory characteristics	12	8	16	16	16	17
C3	Colourings make products more appealing	10	11	8	14	16	10

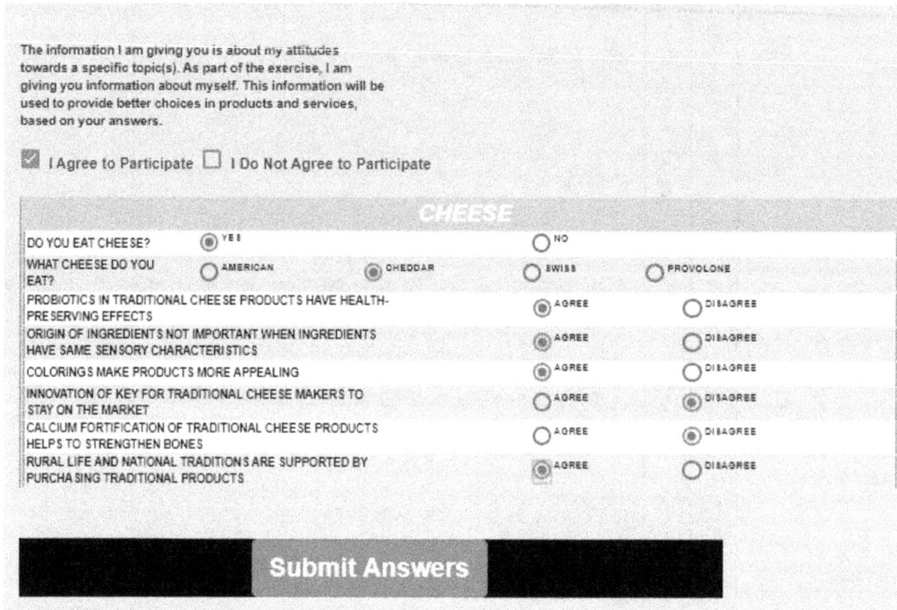
The information I am giving you is about my attitudes towards a specific topic(s). As part of the exercise, I am giving you information about myself. This information will be used to provide better choices in products and services, based on your answers.

I Agree to Participate I Do Not Agree to Participate

CHEESE

DO YOU EAT CHEESE?	YES		NO	
WHAT CHEESE DO YOU EAT?	AMERICAN	CHEDDAR	SWISS	PROVOLONE
PROBIOTICS IN TRADITIONAL CHEESE PRODUCTS HAVE HEALTH-PRESERVING EFFECTS			AGREE	DISAGREE
ORIGIN OF INGREDIENTS NOT IMPORTANT WHEN INGREDIENTS HAVE SAME SENSORY CHARACTERISTICS			AGREE	DISAGREE
COLORINGS MAKE PRODUCTS MORE APPEALING			AGREE	DISAGREE
INNOVATION OF KEY FOR TRADITIONAL CHEESE MAKERS TO STAY ON THE MARKET			AGREE	DISAGREE
CALCIUM FORTIFICATION OF TRADITIONAL CHEESE PRODUCTS HELPS TO STRENGTHEN BONES			AGREE	DISAGREE
RURAL LIFE AND NATIONAL TRADITIONS ARE SUPPORTED BY PURCHASING TRADITIONAL PRODUCTS			AGREE	DISAGREE

Submit Answers

Figure 1.2 The personal viewpoint identifier for cheese.

1.6 The multiple contributions of MG to scientific investigation

The focus of this chapter has been on the methodological advances in MG, and specifically its contribution to the world of product evaluation. Traditional research on the evaluation of products uses either product names or actual products, requiring the respondent to evaluate the product name in terms of a battery of attributes. The consequence of traditional work is that a dairy product such as 'cheese' often remains simply that, 'cheese'. The reality of experience, however, is that the experience of the product 'cheese' is much more detailed, much more filled with texture and nuance. Those involved in the description of cheese products know this as a fact of everyday life. Those involved in the study of cheese in the context of daily life, either observing behaviour or recording answers from surveys, often do not have the luxury to probe into the different aspects of cheese, be these aspects sensory, or as shown here, experiential and attitudinal. Further, the current science of consumer research often stops at the level of the general product, simply because the demands of the research and/or the proclivities of the research focus on the bigger picture, cheese as cheese in the large world of food-related behaviours.

There is another major contribution of MG to understanding how people react to messages about the ordinary aspects of their daily lives, such as dairy products. That contribution is the ability of MG to move beyond the mental editor of a person. Typically, when it comes to questionnaires, people like to please the interviewer, often changing what they say, and even believing what they say, in order to get approval from the interviewer.

MG as we have discussed here provides the researcher with a flexible microscope, magnifying glass or a telescope – or all three. When the research wants to look at cheese, focusing on a detail that is not product related such as the innovations we have here, we have a microscope to go into the details, or a magnifying glass to look at general areas in some detail. Yet, MG can work at the level of a telescope, looking at dairy-product-related messages in the constellation or galaxy of food behaviour on the one hand, or food innovation behaviour on the other.

1.7 The role of emotions

As noted in the introduction to this chapter, emotions are becoming of increasing interest to food researchers, which should not surprise. Experimental psychologists have known for a century or more, since Pavlov, that food is a 'reinforcer', the presence of which can increase the probability of a response, whether the response be salivation (classical conditioning) or bar pressing (operant conditioning). The key notion here is the ability of food to drive a response and reinforce that response. There is the subtle, unrecognised reality, that the process of ingestion is generally a positive one. It is only recently that this positive experience of eating food has been dimensionalised by researchers.

The key to MG and emotions is the ability to link emotions to specific messages, an ability that provides new direction for the current focus on dairy research, and indeed research in human experience. Most of the work involving emotions deals with the connection of emotions to facial expressions (Ekman et al. 1972), or even emotions in activities. The ability to parse compound stimuli by experimental design, and attach to each aspect of the compound stimulus a gradation of linkages to several emotions, as done here, provides another methodological advance for dairy science on the one hand, and psychological and sociological science on the other.

Acknowledgement

Barbara Biró thanks the support of the Doctoral School of Food Sciences, Szent István University. Attila Gere thanks the support of the Premium Postdoctoral Research Program of the Hungarian Academy of Sciences and the support of National Research, Development and Innovation Office of Hungary (OTKA, contracts No. K119269). The Project is supported by the European Union and co-financed by the European Social Fund (grant agreement no. EFOP-3.6.3-VEKOP-16-2017-00005). Barbara Biró is supported by the ÚNKP-19-3-I-SZIE-65 New National Excellence Program of the Ministry for Innovation and Technology.

References

Bodyfelt, F.W., Drake, M.A., and Rankin, S.A. (2008). Developments in dairy foods sensory science and education: from student contests to impact on product quality. *International Dairy Journal* 18 (7): 729–734. https://doi.org/10.1016/j.idairyj.2008.03.011.

Box, G., Hunter, S., and Hunter, W. (2005). *Statistics for Experimenters: Design, Innovation, and Discovery*, 2e. Hoboken, New Jersey, NY: Wiley.

Clark, S., Costello, M., Drake, M., and Bodyfelt, F. (2009). *The Sensory Evaluation of Dairy Products*, 2e. New York, NY: Springer https://doi.org/10.1007/978-0-387-77408-4.

Delarue, J. and Sieffermann, J.M. (2004). Sensory mapping using Flash profile. Comparison with a conventional descriptive method for the evaluation of the flavour of fruit dairy products. *Food Quality and Preference* 15 (4): 383–392. https://doi.org/10.1016/S0950-3293(03)00085-5.

Desmet, P.M.A. and Schifferstein, H.N.J. (2008). Sources of positive and negative emotions in food experience. *Appetite* 50 (2): 290–301. https://doi.org/10.1016/j.appet.2007.08.003.

Drake, M. (2004). ADSA Foundation Scholar Award: Defining dairy flavors. *Journal of Dairy Science* 87 (4): 777–784. https://doi.org/10.3168/jds.S0022-0302(04)73221-X.

Drake, M. (2007). Invited review: Sensory analysis of dairy foods. *Journal of Dairy Science* 90 (11): 4925–4937. https://doi.org/10.3168/jds.2007-0332.

Dubes, R. and Jain, A.K. (1980). Clustering methodologies in exploratory data analysis. *Advances in Computers* 19: 113–228. https://doi.org/10.1016/S0065-2458(08)60034-0.

Ekman, P., Friesen, W.V., and Ellsworth, P. (1972). Chapter X – Sampling emotions. In: *Emotion in the Human Face*, 1e (ed. P. Ekman, W.V. Friesen and P. Ellsworth). New York, NY: Pergamon Press Inc. https://doi.org/10.1016/B978-0-08-016643-8.50019-7.

Esmerino, E.A., Castura, J.C., Ferraz, J.P. et al. (2017a). Dynamic profiling of different ready-to-drink fermented dairy products: a comparative study using Temporal Check-All-That-Apply (TCATA), Temporal Dominance of Sensations (TDS) and Progressive Profile (PP). *Food Research International* 101: 249–258. https://doi.org/10.1016/j.foodres.2017.09.012.

Esmerino, E.A., Tavares Filho, E.R., Thomas Carr, B. et al. (2017b). Consumer-based product characterization using Pivot Profile, Projective Mapping and Check-all-that-apply (CATA): a comparative case with Greek yogurt samples. *Food Research International* 99 (May): 375–384. https://doi.org/10.1016/j.foodres.2017.06.001.

Fernández-Vázquez, R., Stinco, C.M., Hernanz Vila, D. et al. (2018). Internal preference mapping of milk–fruit beverages: influence of color and appearance on its acceptability. *Food Science and Nutrition* 6 (1): 27–35. https://doi.org/10.1002/fsn3.494.

Gere, A., Szabó, Z., Pásztor-Huszár, K. et al. (2017). Use of JAR-based analysis for improvement of product acceptance: a case study on flavored kefirs. *Journal of Food Science* 82 (5): 1200–1207. https://doi.org/10.1111/1750-3841.13701.

Gofman, A. and Moskowitz, H. (2010). Isomorphic permuted experimental designs and their application in conjoint analysis. *Journal of Sensory Studies* 25 (1): 127–145. https://doi.org/10.1111/j.1745-459X.2009.00258.x.

Green, P.E. and Rao, V.R. (1971). Conjoint measurement – for quantifying judgmental data. *Journal of Marketing Research* 8 (3): 355–363. https://doi.org/10.1177/002224377100800312.

Green, P.E. and Srinivasan, V. (1990). Conjoint analysis in marketing: new developments with implications for research and practice. *Journal of Marketing* 54 (4): 3–19. https://doi.org/10.1177/002224299005400402.

Hastie, T., Tibshirani, R., and Friedman, J. (2001). *The Elements of Statistical Learning*, 1e. New York, NY: Springer https://doi.org/10.1007/978-0-387-84858-7.

Hekmat, S. and Reid, G. (2006). Sensory properties of probiotic yogurt is comparable to standard yogurt. *Nutrition Research* 26 (4): 163–166. https://doi.org/10.1016/j.nutres.2006.04.004.

Jervis, S.M., Lopetcharat, K., and Drake, M.A. (2012). Application of ethnography and conjoint analysis to determine key consumer attributes for latte-style coffee beverages. *Journal of Sensory Studies* 27 (1): 48–58. https://doi.org/10.1111/j.1745-459X.2011.00366.x.

Kemp, S.E., Hollowood, T., and Hort, J. (2009). *Sensory Evaluation – A Practical Handbook*, 1e. Chichester, West Sussex: Wiley.

Luce, R.D. and Tukey, J.W. (1964). Simultaneous conjoint measurement: a new type of fundamental measurement. *Journal of Mathematical Psychology* 1 (1): 1–27. https://doi.org/10.1016/0022-2496(64)90015-X.

Murphy, M., Cowan, C., Meehan, H., and O'Reilly, S. (2004). A conjoint analysis of Irish consumer preferences for farmhouse cheese. *British Food Journal* 106 (4): 288–300. https://doi.org/10.1108/00070700410529555.

Meals, S.E., Schiano, A.N., and Drake, M.A. (2020). Drivers of liking for Cheddar cheese shreds. *Journal of Dairy Science* 103 (3): 2167–2185. https://doi.org/10.3168/jds.2019-16911.

Moskowitz, H. (2012). 'Mind genomics': the experimental, inductive science of the ordinary, and its application to aspects of food and feeding. *Physiology & Behavior* 107 (4): 606–613. https://doi.org/10.1016/j.physbeh.2012.04.009.

Moskowitz, H.R. and Gofman, A. (2007). *Selling Blue Elephants: How to Make Great Products that People Want Before they Even Know they Want Them*, 1e. New York, NY: Pearson Prentice Hall.

Moskowitz, H.R. and Martin, D.C. (1993). How computer aided design and presentation of concepts speeds up the product development process. *ESOMAR Congress 1993: The Many Faces of Quality Now And In Future*.

Pinto, V.R.A., Teixeira, C.G., Lima, T.S. et al. (2020). Health beliefs towards kefir correlate with emotion and attitude: a study using an emoji scale in Brazil. *Food Research International* 129: 108833. https://doi.org/10.1016/j.foodres.2019.108833.

Pohjanheimo, T. and Sandell, M. (2009). Explaining the liking for drinking yoghurt: the role of sensory quality, food choice motives, health concern and product information. *International Dairy Journal* 19 (8): 459–466. https://doi.org/10.1016/j.idairyj.2009.03.004.

Rousset, S., Deiss, V., Juillard, E. et al. (2005). Emotions generated by meat and other food products in women. *British Journal of Nutrition* 94 (4): 609–619. https://doi.org/10.1079/BJN20051538.

Schouteten, J.J., De Steur, H., De Pelsmaeker, S. et al. (2015). Impact of health labels on flavor perception and emotional profiling: a consumer study on cheese. *Nutrients* 7 (12): 10251–10268. https://doi.org/10.3390/nu7125533.

Tietje, C. and Brouder, A. (2009). *Handbook of Transnational Economic Governance Regimes*, 1e. Leiden, The Netherlands: BRILL.

Torben, H. (2005). Understanding consumer perception of food quality: the cases of shrimps and cheese. *British Food Journal* 107 (7): 500–525. https://doi.org/10.1108/00070700510606909.

2 Physiology of Sensory Perception of Flavour and Mouthfeel Stimuli Imparted by Dairy Products

Lisa Methven, Stella Lignou, Stephanie Bull, Maria Jose Oruna-Concha and Rosa Sullivan

2.1 Introduction

Our senses guide our selection, perception and enjoyment of foods. Vision enables us to distinguish between foods that appear good to eat and those that might be harmful, hence avoiding a rotten apple. Association with specific aromas are through learning and anticipation, we may learn to love the aroma of coffee but reject the aroma we have learned to relate to rancidity. Our sense of taste assists from birth, guiding us through sweet and savoury tastes to seek out carbohydrates and protein and to avoid bitter tasting foods that might be toxic. Therefore, the importance of learning cannot be ignored, humans learn as they develop to appreciate an extent of bitterness enabling them to like bitter vegetables, to appreciate hard and not only soft textures (Laureati et al. 2020; Nor et al. 2021). When we consider dairy products, we anticipate perception from all of our senses, whether it be the white appearance of milk that clings to the glass, the pungent aroma of mature cheese or the soft creamy texture of a yoghurt, our senses are stimulated. This chapter reviews the sensory perception of flavour and mouthfeel stimuli that can be imparted by dairy products.

2.2 Aroma perception

The aroma of dairy products is one of the most important factors that determine consumer acceptability but also their shelf life. Aroma from airborne volatile compounds is perceived by receptors located in the olfactory epithelium through two pathways: (i) from the external environment, via the nose during sniffing and generally referred to as 'smell' or 'orthonasal olfaction', or (ii) via the mouth, during eating or drinking where odorant compounds released during mastication reach the nasal cavity through the pharynx and this is known as 'retro-nasal olfaction' (Figure 2.1) (Negoias et al. 2008).

Sensory Profiling of Dairy Products, First Edition. Edited by John J. Tuohy.
© 2023 John Wiley & Sons Ltd. Published 2023 by John Wiley & Sons Ltd.

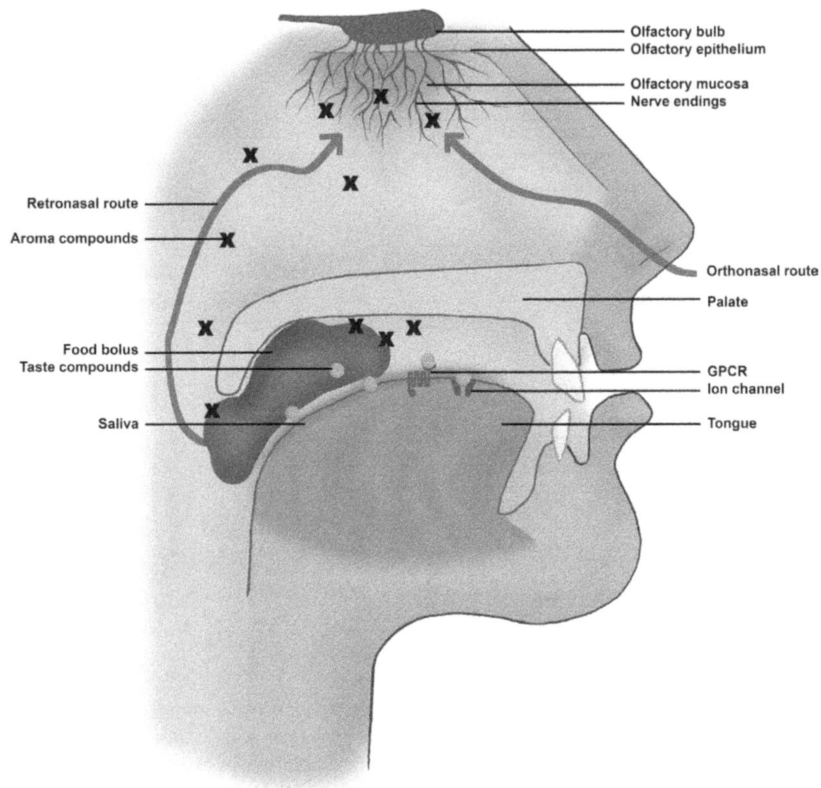

Figure 2.1 The route to flavour perception of foods: ortho- and retro-nasal routes by which aroma compounds reach the olfactory mucosa, and the route of tastants to taste receptors on the tongue (G-protein-coupled receptors, GPCR, and ion channels).

Volatile compounds including acids, aldehydes, ketones, alcohols, esters, and hydrocarbons can impart favourable flavours to the various dairy products. These compounds are either present in the raw material or can be formed during the different processes of pasteurisation, ultrahigh-temperature (UHT) treatments or fermentation.

2.2.1 *Physiology of aroma perception*

Aromas are sensed by the olfactory epithelium which is located within the nasal cavity and then processed by the olfactory bulb in the olfactory system. The olfactory epithelium has a $5\,cm^2$ area and is covered by a mucus layer that contains mucopolysaccharides, immunoglobulins, proteins and various enzymes. Within the mucus layer, nerve cells are projected in a form of very fine fibres (cilia) increasing the potential that olfactory receptors have. At the other end of the nerve, axons that project to the olfactory bulb pass through the cribriform plate of the skull and converge on junctions called glomeruli. Each glomerulus is extended to mitral cells which are the principal neurons in the olfactory bulb (Figure 2.2).

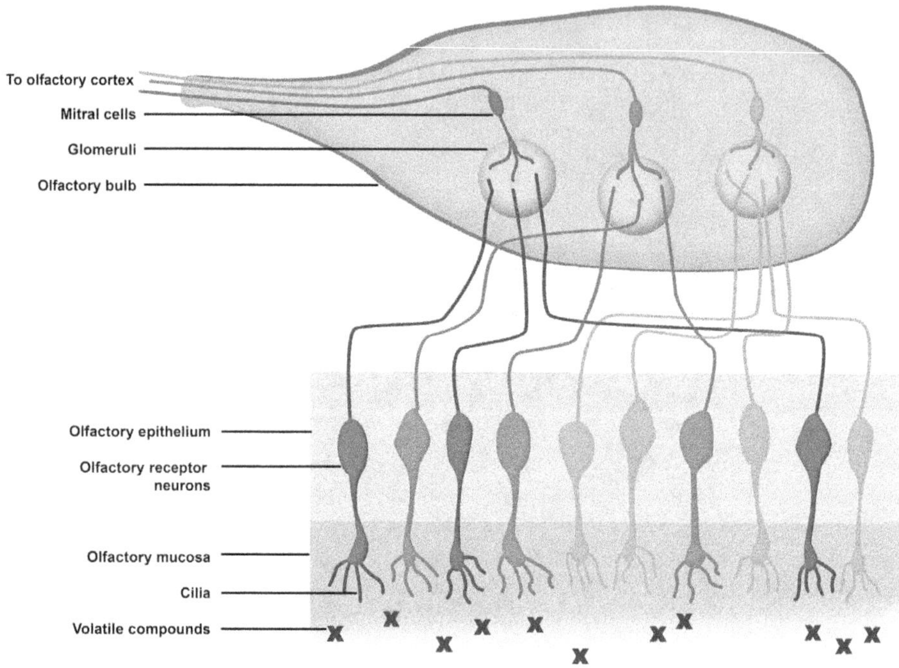

Figure 2.2 Olfactory system. On reaching the olfactory epithelium volatile compounds bind to a range of olfactory receptors (OR) within the cilia. Each OR neuron maintains odour-specific information (as denoted by the different colours) and the impulses converge at glomerulus in the olfactory bulb. Mitral cells convey information from the glomerulus to the olfactory cortex for final interpretation.

In 2004, Axel and Buck earned the Nobel Prize in Physiology or Medicine for the first molecular characterisation of the mammalian sense of olfaction (Buck and Axel 1991). They discovered a large gene family of about 1000 different genes (3% of our genes) that can give rise to an equivalent number of olfactory receptors. However, most of these are inactive pseudogenes and only around 350 code for functional olfactory receptors. Axel and Buck showed that the large family of olfactory receptors (OR) belong to the G-protein-coupled receptors (GPCR). When the receptor is activated by an odorant compound, it causes the activation of G-protein, which stimulates conversion of ATP to cyclic AMP (cAMP). The cAMP in turn opens cyclic nucleotide-gated ion channels which allow Ca^{2+} and Na^{+} ions to enter the cell, open Ca^{2+}-gated Cl^{-} channels that lead to Cl^{-} movement outside the cell, depolarisation of the olfactory sensory neuron, and an action potential transferred to the brain via the axons (Connor et al. 2018). Since ORs are proteins, the chain of amino acids that is fixed in the cell membrane will determine which odorant compounds will bind to them and activate them. This binding will create a potential in the olfactory cell that will generate stimulation in the olfactory nerve fibres (Buck and Axel 1991). Axel and Buck also reported that each single olfactory receptor cell expresses only one of the OR genes which means that we have as many types of olfactory receptor cells as ORs (Hayden and Teeling 2014). Each OR can bind to multiple odorant molecules but also each odorant can bind to multiple ORs thus leading to an 'odorant pattern' and activation of ORs

(Malnic et al. 1999). Since most of the odours are a combination of various volatile compounds, (e.g. the odour of coffee contains over 800 volatile compounds of which only 40 contribute to its characteristic aroma (Belitz et al. 2009)), each compound will activate several ORs and this can explain why humans are able to recognise and memorise approximately 10 000 different odours.

Many theories have been proposed to explain the mechanism of smelling odours; however, the exact mechanism by which an OR is activated is still not clear. The shape theory of olfaction suggests that the shape of the odorant molecule controls the olfactory cell to be stimulated by a classical lock-and-key mechanism, thus, the odour to be perceived (Amoore 1946; Pauling 1946; Zarzo 2007). The vibrational theory of olfaction originated by the work of Dyson in the 1930s suggested that a correlation between particular odours and vibrations frequencies of molecules exists (Dyson 1938). A vibrational explanation was later supported by Turin who claimed smell was a spectral sense; noting that molecules of identical structure but differing in vibration had different odours, and finding correlations between the electron tunnelling spectrum and odour sensation in structurally unrelated molecules (Turin 1996). It is possible that both shape and vibration have roles within the activation of ORs.

The olfactory function is also dependent on the genetic variation that exists in human OR genes. According to Olender et al. (2012), 66% of OR genes contain insertions/deletions, single nucleotide polymorphisms (SNPs), and copy number variants (CNVs). Consequently, genetic variation in ORs may lead to differences among individuals in their ability to smell and describe odours but also in their sensitivity to various odours (Hasin-Brumshtein et al. 2009; Hasin et al. 2008; Keller et al. 2007). For example, SNPs in *OR7D4* affected the degree of aversion to the smell of the steroid pheromone androstenone (boar taint), which smells unpleasant to many people (foul, sweaty) but pleasant to others (sweet, floral) (Keller et al. 2007; Lunde et al. 2012). Similarly, a SNP in *OR2J3* may alter the ability of detecting the grassy smell of cic-3-hexenol (McRae et al. 2012), and a SNP in *OR5A1* can affect the preference for floral scent of β-ionone (Jaeger et al. 2013).

2.3 Taste perception

Taste is a phenomenon that we perceive from receptors located predominantly on the surface of our tongue. The most common tastants fall into the five modalities of salty, sour, umami, sweet, and bitter. All these taste modalities have relevance to dairy products with arguably the former three having the most direct relevance if we consider the salty and umami perception from a hard cheese, the more dominant sour perception from fresh and soft cheese or from buttermilk or yoghurt. However, we cannot disregard the sweetness resulting from lactose, after all it is the daily addition of milk to coffee where the sweet taste of the lactose partly suppresses the bitter taste of resulting from the caffeine. However, the presence of short peptides resulting from proteolytic activity in dairy products, such as cheese, also results in bitter taste (Lemieux and Simard 1991, 1992; Toelstede and Hofmann 2008b). All the tastants responsible for these five taste sensations are water-soluble compounds which is of particular relevance to dairy products which are predominantly either water-in-oil or oil-in-water emulsions. The tastants will be solubilised in either the continuous water phase, for example in milk or yoghurt, or in the disperse phase, for example in cheese. However, there is a sixth taste of relevant to dairy products, that of oleogustus, or 'fat taste' for which the stimuli responsible are free fatty acids (FFA) (Running et al. 2015). This section considers the role of taste in the perception of dairy products.

2.3.1 *Physiology of taste perception*

Taste receptors in the oral cavity are predominantly located in the papilla that protrude from the surface of the tongue (Lindemann 2001). Such papillae give the tongue a rough surface and help to move food around the oral cavity. There are four types of papillae, three of which contain taste buds (fungiform, foliate, and circumvallate), whereas the fourth type (filiform) have no taste receptors but are involved in texture perception (Trivedi 2012) (Figure 2.3).

Although the distribution of different papilla types is unequal over the surface of the tongue, the taste buds within the papillae contain taste receptors for different taste qualities; so in fact sweet, bitter, umami, sour, salty, and oleogustus can occur across the tongue (Lindemann 1999, 2001; Trivedi 2012). There are various types of taste receptor, broadly fitting into two categories: GPCRs and ion channels. It is proposed that the transduction of ionic tastants occurs through ion channels, for example in the transduction of sodium chloride leading to the perception of salty taste; whereas the transduction of organic tastant molecules occurs predominantly through type II taste cells comprising GPCRs of either the type 1 (T1R) or type 2 (T2R) family (Bachmanov and Beauchamp 2007; Trivedi 2012).

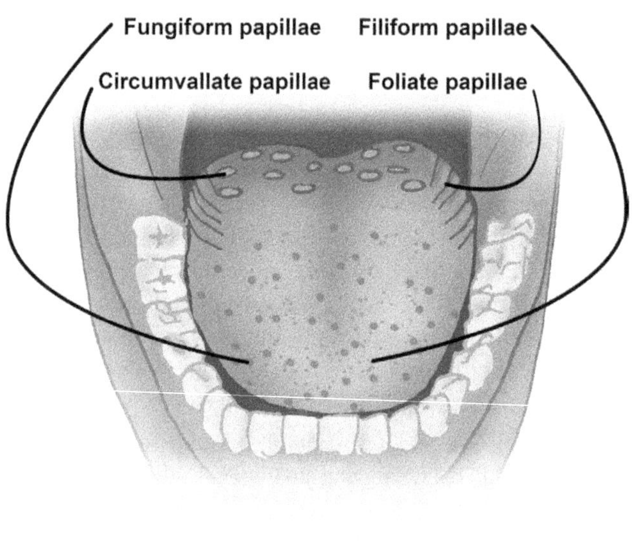

Figure 2.3 Representation of papillae in the human tongue.

Receptors involved in bitter, sweet, and umami taste

Discovery of the T2Rs was published first, of which there are known to be 25 receptors which can detect a wide range of bitter compounds (Behrens and Meyerhof 2006; Chandrashekar et al. 2000). Each T2R receptor can respond to more than one bitter tastant; however, some are 'generalists' that can detect a wide diversity of structurally different compounds (such as TAS2R14), whereas others are much more specific, for example TAS2R38 which only responds to compounds containing a thiouracil (N-C=S) group (Behrens and Meyerhof 2006). The T1R receptor family is responsible for detection of sweet and umami tastes. The finding that both T1R2 and T2R3 combined to functions as a sweet receptor was published in 2001, with the authors concluding that the T1R2-T1R3 heterodimer could detect sweetness from both natural sugars and high-intensity sweeteners (Nelson et al. 2001). Similarly, the T1R1-T1R3 heterodimer was found to detect most L-amino-acids (Nelson et al. 2002). Amino acids can lead to a variety of taste responses, whereas only glutamate and aspartate markedly cause the savoury umami taste, and there is a strong synergistic response to glutamate taste in the presence of 5-nucleotides (Yamaguchi et al. 1971). Nelson et al. (2002) found that the T1R1-T1R3 response to amino acids was greatly enhanced in the presence of the 5′-nucleotide inosine monophosphate (IMP), making it a candidate umami taste receptor. The two glutamate-selective GPCRs (mGluR4 and mGluR1) are predominantly activated by glutamate, yet they do not respond to 5-nucleotides, suggesting that both mGluRs and the T1R1-T1R3 heterodimer may be involved in the perception of umami taste (Chaudhari et al. 2009).

Receptors involved in salty taste

The most common stimulus for salty taste is sodium chloride, so common that it is named 'salt'. Although it is the cation, Na^+, that is responsible for the taste sensation, it is strongly influenced by the anion (Cl^-), and indeed other salts, such as potassium chloride, can provide salty taste. It has been proposed for many years that there are various ion channels that can be involved in salt taste transduction (Lindemann 2001); however, more recently it has been shown that the primary ion channel involved is the epithelial sodium channel (ENaC) (Bigiani 2020; Liman 2020; Staehler et al. 2008). Research in rodent models has used the amiloride sensitive nature of ENaC to demonstrate the importance of its role in salt perception. Where the amiloride drug is used to block the receptor, the rodents no longer respond to salt taste, although this response is contested in humans (Bigiani 2020; Staehler et al. 2008). Certainly, all four ENaC subunits (α, β, γ, δ) are found in human fungiform and circumvallate papillae (Staehler et al. 2008) and further evidence of the involvement of ENaC in human salt taste perception comes from directly relating sensory and cell studies. Salt taste resulting from sodium chloride has been shown to be enhanced by the addition of basic amino acids (especially L-arginine and L-lysine) as well as choline chloride, and it has been shown that such addition also results in increased Na^+ currents in ENaC expressing cells (Staehler et al. 2008). Genetic differences in the coding for the ENaC β-subunit, referred to as *single nucleotide polymorphisms* (SNPs), have been shown to influence suprathreshold taste sensitivity to salt (sodium chloride) solutions, again supporting a fundamental role of ENaC in human salt taste perception (Dias et al. 2013). However, in the same study, the authors additionally researched SNPs of the nonspecific cation channel TRPV1, likewise finding that SNPs of the genes coding for this channel also modified suprathreshold sensitivity to salt. TRPV1 is a transient receptor potential (TRP) cation subfamily V member (TRP vanilloid-1), it is known to respond to a variety of cations, not just sodium, and it is insensitive to amiloride (Lyall et al. 2004). TRPV1 responds to a variety of minerals which may partly explain why other minerals can contribute to salty taste. It is activated more efficiently by divalent cations, such as calcium and magnesium, than by monovalent cations such as sodium and lithium (Ahern et al. 2005). However, the response of TRPV1 to divalent salts, such as $CuSO_4$, $ZnSO_4$, and FeO_4, leads to not only salty taste perception, but to metallic taste sensations in addition (Riera et al. 2007).

An interesting point for a food perspective is that anion of the mineral salt has a clear impact on the taste perception, so sodium chloride is saltier than other sodium salts. As molecular weight increases, with increased anion size, the taste reaction time is slower (Delwiche et al. 1999). The mechanism proposed to support this phenomenon is that there is a diffusion potential across a paracellular shunt barrier in the tight junction between taste-bud cells. It is proposed that this barrier behaves as a weak cation exchange membrane, where cations pass through tight junctions leaving their larger, less permeable anions behind (Delwiche et al. 1999; Ye et al. 1993). More recently, the need for a trans-epithelial potential for this anion effect to occur has been tested in mouse circumvallate papillae. It was found that, in amiloride-insensitive type III taste cells that there was still an anion effect in the absence of a diffusion potential, suggesting more than one mechanism might be responsible (Lewandowski et al. 2016). Whatever the mechanism, when reducing the salt content of foods, it is not only the sodium of salt that needs to be considered from the perspective of taste. Although when considering cheese, we have to realise that salt has many functions other than taste as it can be used to stop growth of bacterial start cultures, to draw whey out from curd, and in the development of rind.

Receptors involved in sour taste

Sour taste results from the perception of acids, in the case of dairy foods most commonly lactic acid (Mullin and Emmons 1997). The definition of an acid is a compound that can donate a proton, and as protons are toxic to most cells, it is interesting to consider the role of sour taste and the receptors cells responsible for sour taste transduction. Sour taste acts as a warning signal of potentially harmful foods where acidity is a result of spoilage; however, acidity can of course differentiate products and be sensorially attractive as exemplified by fresh acidic cheeses and yoghurt. Two research groups identified TRP channels known as the polycystic-kidney disease-like ion channels (PKDs) as likely candidates for mammalian sour taste receptors (Huang et al. 2006; Ishimaru et al. 2006). In mice studies, they found PKD1L3 and PKD2L1 to be expressed in taste cells, with the two being co-expressed in circumvallate and foliate papillae. Mice engineered not to express PKD2L1 did not respond to sour taste but continued to respond to other taste modalities (Huang et al. 2006). When co-expressed, PKD1L3-PKD2L1 were found to respond to both organic acids (acetic, citric, malic, succinic and tartaric acid) and inorganic acids (hydrochloric, sulphuric and phosphoric acid) and not to other tastants (Inada et al. 2008; Ishii et al. 2009; Ishimaru et al. 2006). Activation has been shown to be pH dependent and to occur at where the pH was less than 3.1 (Inada et al. 2008; Ishii et al. 2009). More specifically, the response to pH has been described as an 'off-response' where PKD1L3-PKD2L1 channels are activated by exposure to acids but only open after the removal of the acid stimuli at a pH of less than approximately 3 (Inada et al. 2008). However, PKDs are not the only mechanism reported as responsible for sour taste. More recently, Otopetrin1 (Otop1) was identified as a sour taste receptor (Tu et al. 2018; Zhang et al. 2019). Through a series of experiments, Otop1 was found to be essential for sour tasting in mice. Where Otop1 was knocked out, sour taste receptor cells (TRCs) did not respond to acidic stimuli; additionally, where Otop1 was engineered into sweet cells mice responded to sour stimuli (Zhang et al. 2019). Finally, sour taste was found to have a dedicated neuronal line from TRCs to the brain (Zhang et al. 2019). These findings challenge the importance of PKD1L3-PKD2L1, yet further investigations have concluded that it is a non-selective cation channel that can have both Ca^{2+} and acid-induced activity (Su et al. 2021). Taken together, current research of sour taste receptors suggests that the organic acids found in dairy foods will be sensed through specialist cells where activation is pH dependent.

2.3.2 *Oleogustus, a taste response to fatty acids*

Sensory perception of fat is often appreciated as a mouthfeel response to the texture of fat; however, humans also have a taste response to fatty acids, now considered to be our sixth taste, 'oleogustus' (Running et al. 2015). Detection thresholds for emulsified oils were first established in 1992, for a longer chain soybean oil, a medium-chain triglyceride oil and a mineral oil, where aroma was minimised through wearing nose-clips (Schiffman et al. 1992). More recently, research has considered that it is FFAs that are the relevant stimuli, with thresholds varying between short-, medium-, and long-chain fatty acids (Richard D. Mattes 2009). This led to the consideration of the receptors involved, with Richard D. Mattes (2011) reviewing the potential involvement of a specific receptor CD36 (commonly considered a FFA transporter), the GPCRs known to bind FFAs as part or metabolic cell signalling pathways (e.g. GPR 40, 41, 43, 84 and 120) or the blocking of delayed rectifying potassium channels (DRKs). *In vitro* studies found high concentration of linoleic acid (C18:2) to induce signalling in human fungiform taste-bud cells via both GPCR120 and CD36; with only CD36 responding at lower concentrations (Ozdener et al. 2014). Studies in mice taste cells have confirmed that GPR84 responds to medium-chain saturated FFAs (caproic (C6:0), caprylic acid (C8:0), capric acid (C10:0), undecanoic acid (C11:0), lauric acid (C12:0), and arachidic acid (C20:0)) (Liu et al. 2021). More recently, the role of transient receptor potential canonical (TRPC) channels in fat taste perception has been investigated (Murtaza et al. 2021). In mice studies, it was shown that TRPC3 channels were essential for mice to respond to long-chain fatty acids (linoleic acid [C18:2], canola oil which predominantly comprises oleic acid [C18:1] and linoleic acid [C18:2]). Inactivation of TRPC3 reduced FA induced signalling.

While it is clear that FFA (i.e. non-esterified fatty acids) are taste stimuli, it is less clear what specific taste quality they evoke. A recent study explored the taste qualities of C2–C18 FFAs, finding as chain length increased the ratings for taste quality shifted from sour (C2–C4), through irritation ('spicy') (C8–C12), to bitter (C18) (Cheon and Mattes 2021). It was proposed that the lack of the term 'fatty' by tasters may be because it is a complex sensation (perhaps a combination of irritation and bitter), or that a 'fatty' sensation is only perceived as a multi-modal effect where viscosity or mouthfeel is also present. This questions how the perception of 'fatty taste' might influence our perception of real foods. Of specific relevance to dairy fats was study which investigated the perception of fat intensity of milk/cream mixtures, at seven different fat levels (0–20% fat) under four conditions; mouthfeel odour-masked, mouthfeel-masked, odour-masked and no masking (Zhou et al. 2016). Tasters were able to distinguish between fat levels on taste along (where mouthfeel and odour were masked), suggesting that a taste response to fat is relevant within real foods, and in dairy products. Mouthfeel and odour further enhanced the perception of fat. Reviewing the relationship between taste and macronutrient intake, Mattes (2021) discusses both the olfactory and taste cues of FFA. As aroma compounds, short chain fatty acids or volatile compounds derived from degradation of fatty acids are generally considered to be unpleasant as they signal for rancidity. Therefore, levels of FFA in foods and lipid degradation (oxidation and hydrolysis) are carefully controlled in food manufacture. There are of course exceptions to the rule, and indeed these particularly relate to dairy foods. Butyric acid (C4:0) and caproic acid (hexanoic acid, C6:0) are short chain fatty acids that while in isolation might be considered to have an unpleasant odour (sometimes described as 'sweaty' and 'goat/barnyard', respectively), they are often a desirable characteristic of mature cheese (Aparicio et al. 1997; Christensen and Reineccius 1995). However, Mattes suggests that low concentrations of long-chain FFA, known to evoke a taste response, may contribute positively to the overall flavour profile of a food, which might explain why some studies have found that where humans are insensitive to the taste of FFA, this can lead them to have a higher intake of dietary fat (Liu et al. 2016; Mattes 2021).

2.3.3 *Taste perception of peptides and diketopiperazines and their relevance to cheese*

The taste of peptides was first reported by Kirimura et al. (1969). A total of 60 peptides were classified into three different groups based on their taste attributes, namely sour (rich in acidic residues), bitter (rich in hydrophobic residues) and those having little or no taste which had a more balanced composition. Furthermore, Kirimura and co-workers also noticed that these peptides had weaker taste intensities than their individual amino acid constituents.

Since then, extensive research work has been carried out and peptides, similarly to amino acids, do have a taste covering the five taste modalities of salty, sour, umami, sweet, and bitter, although contrary to amino acids, charged terminals and/or charged side chains are responsible for the sweet and bitter taste of peptides rather than their electronic or conformational features (Temussi 2011).

In the context of cheese, one of the most important organoleptic attributes of cheese is its taste, which develops during ripening and is heavily dependent on both, the production method and the microorganisms responsible for the fermentation process (McSweeney 1997). These processes have been linked to the formation of bitter peptides (Guigoz and Solms 1976) which in cheese mainly originates from the enzymatic hydrolysis of the main protein present in cheese, casein. Low level of bitterness is a typical and desirable attribute in some cheeses, but at higher levels it is a flavour defect (Lemieux and Simard 1992). Bitter peptides in cheese are numerous and varied. Produced during proteolysis, bitter casein fragments ranging from 2 to 23 amino acids in length have been identified in water-soluble extracts from cheeses (Engel et al. 2000; Fallico et al. 2005; Toelstede and Hofmann 2008b; Karametsi et al. 2014). The bitter peptides found in cheese vary by cheese variety and are thought to be influenced by the cultures used in cheesemaking and the duration of cheese maturation. Longer maturation periods are associated with increased formation of short chain peptides, as cheese proteinases and peptidases act on cheese proteins to produce smaller peptides and amino acids.

Diketopiperazines (DKPs) are cyclic dipeptides with bitter, astringent or metallic taste (Borthwick and Da Costa 2017). They are found in fermented products such as: saki (Takahashi et al. 1974), beer (Gautschi et al. 1997), yeast extract (Da Costa et al. 2010) where they are formed by bacteria, yeasts and fungi, and cooked products such as: cocoa (Stark and Hofmann 2005), coffee (Ginz and Engelhardt 2000), chicken essence (Chen et al. 2004) where their formation is the thermally induced cyclisation of di-and-tripeptides (Borthwick and Da Costa 2017; Otsuka et al. 2019; Rizzi 1989).

The presence of DKPs in cheese was first reported by Roudot-Algaron and co-workers (Roudot-Algaron et al. 1993), who reported the presence of cyclo(Pro-Pro), cyclo(Pro-Val), cyclo(Pro-Phe), cyclo(Pro-Leu) and cyclo(Pro-Ala) in Comté. Proline-containing DKPs are the most abundant class of DKPs due to the cis conformation of the proline amino acid with respect to the tertiary amide bond in the precursor peptide. This confirmation makes 2,5 DKP formation more successful in proline-containing peptides (Borthwick and Da Costa 2017; Yaron and Naider 1993).

All DKPs detected by Roudot-Algaron were subthreshold in raw Comté. However, above-threshold concentrations of DKPs have recently been detected in cooked cheeses (Sullivan 2022). Furthermore, matured cheeses (parmesan, mature cheddar) developed higher concentrations of DKPs during cooking than younger cheeses (mild cheddar, mozzarella). This is likely due to formation of small chain peptides as products of proteolysis during maturation, which are the precursors to DKP formation when cooked.

2.3.4 *The perception of kokumi and its relevance to cheese*

Kokumi is a common and well accepted Japanese term derived from *koku* (rich) and *mi* (taste). It is a taste sensation described as rich, thick, long-lasting mouthfeel and delicious taste (Ueda et al. 1990). Kokumi compounds, although tasteless in isolation (Ueda et al. 1990, 1994), are also able to enhance the sensation of sweet, salty and umami tastes (Ohsu et al. 2010; Rhyu et al. 2020).

Kokumi was first identified by Ueda and colleagues (Ueda et al. 1990) in an aqueous extract of garlic, where sulphur-containing compounds and their γ-glutamyl peptides namely alliin, (+)-S-methyl-L-cysteine sulfoxide, γ-L-glutamyl-S-allyl-L-cysteine, and glutathione (γ-Glu-Cys-Gly, GSH) were able to enhance the mouthfeel of an umami solution composed of monosodium glutamate (MSG) and disodium inosinate, while minimally eliciting any taste on their own. Further work from Ueda and co-workers led to the identification of two other kokumi compounds, trans-(+)-S-propenyl-L-cysteine sulfoxide (PeCSO) and its γ-glutamyl peptide (γ-Glu-PeCSO) in aqueous extracts from onion with similar properties (Ueda et al. 1994).

Since then, several other kokumi compounds, both amino-acid-containing and non-amino-containing compounds, have been identified in a variety of foods, either naturally present or as a result of processes such as fermentation, heat treatment and ageing. Most commonly reported kokumi compounds are, however, amino-acid-containing compounds, namely sulphur-containing amino acids, peptides and their derivatives, examples of which have been reported in raw and cooked foodstuffs (Ueda et al. 1997), raw and processed scallop products (Kuroda et al. 2012a), fish sauces (Kuroda et al. 2012b), soy sauces (Kuroda et al. 2013), yeast extract (Liu et al. 2015), brewed alcoholic beverages (Miyamura et al. 2015), sourdough (Zhao and Ganzle 2016), soybean seeds (Shibata et al. 2017), durians (Pinsorn et al. 2018), mushroom (Feng et al. 2019), fermented cocoa beans (Salger et al. 2019), ripened Italian salami (Rocchetti et al. 2021), whole soybeans and the defatted soybeans fermented soy sauce (Zhu et al. 2021). Kokumi active peptides have also been identified in cheese thus contributing to the kokumi sensation in cheese (Table 2.1).

Roudot-Algaron et al. (1994) were the first to report the presence of γ-glutamyl peptides in cheese (Roudot-Algaron et al. 1994). They identified γ-Glu-Phe, γ-Glu-Tyr, and γ-Glu-Leu in the water-soluble fraction of Comté cheese although this was unexpected as no γ-glutamyl bonds occur in caseins and, therefore, peptides formed as a result of the proteolysis of caseins containing only α bonds. They explained that their formation could

Table 2.1 Peptides identified in cheese showing kokumi taste activity.

Compound	Threshold (mM)	References
γ-Glu-Glu	0.018	Beksan et al. (2003), Toelstede et al. (2009), and Toelstede and Hofmann (2009)
γ-Glu-Gly	0.018	Beksan et al. (2003), Toelstede et al. (2009), and Toelstede and Hofmann (2009)
γ-Glu-His	0.01	Beksan et al. (2003), Toelstede et al. (2009), and Toelstede and Hofmann (2009)
γ-Glu-Leu	0.005	Beksan et al. (2003), Toelstede et al. (2009), Toelstede and Hofmann (2009), and Sforza et al. (2006)
γ-Glu-Met	0.005	Beksan et al. (2003), Toelstede et al. (2009), and Toelstede and Hofmann (2009)
γ-Glu-Gln	0.008	Beksan et al. (2003), Toelstede et al. (2009), and Toelstede and Hofmann (2009)
γ-Glu-Val	0.003	Beksan et al. (2003), Toelstede et al. (2009), and Toelstede and Hofmann (2009)
γ-Glu-Tyr	2.5	Beksan et al. (2003) and Toelstede et al. (2009)

Source: Adapted from (Zhao et al. 2016).

be as a result of the ɣ-glutamyl transferase (GGT) present in bovine milk, catalysing the transfer of the ɣ-glutamyl moiety of glutathione (or other ɣ-Glu peptide) to a variety of amino acids as well as by the action of synthetases which intervene in glutathione synthesis. These ɣ-glutamyl peptides were described as sour and slightly salty (ɣ-Glu-Tyr) and umami, slightly sour, salty and metallic (ɣ-Glu-Phe), with taste thresholds between 200 and 500 ppm, in the same order of magnitude as MSG. However, it is worth mentioning that relative amounts of these peptides ranged between 9 and 70 mg/Kg suggesting that they contributed to the complex flavour of cheese rather than on their own.

In 2009, Toelstede and co-workers were the first ones to make a direct correlation of ɣ-glutamyl peptides as the key kokumi molecules responsible for the enhancing mouthfulness and complex taste continuity of cheese (Toelstede et al. 2009). Using a sensomic approach, water-soluble extracts of a 4- and 44-week ripened Gouda cheese allowed the identification of a series of α-L-glutamyl and γ-L-glutamyl dipeptides. Significantly higher levels (between 10-fold and >100-fold) of ɣ-Glu-Glu, ɣ-Glu-Gly, ɣ-Glu-Gln, ɣ-Glu-Met, ɣ-Glu-Leu and ɣ-Glu-His were detected in the 44-week mature cheese compared to the 4-week mature cheese contributing to the perceived enhanced mouthfulness and complexity taste of the more mature cheese, which highlighted ripening as having a positive effect in the formation of these ɣ-glutamyl peptides. However, they observed that none of the α-L-glutamyl peptides detected (where only onefold to eightfold increase was observed) were active, thus were unable to induce a kokumi sensation.

Sullivan and co-workers (Sullivan 2022) also reported an increase in ɣ-glutamyl peptides concentration during maturation from mild to mature cheddar. Comparison of low-, medium-, and high-fat cheddars demonstrated that fat content made no significant difference to the formation of ɣ-glutamyl peptides during maturation. Their work also demonstrated a decrease in concentration of ɣ-glutamyl peptides when cheddar is cooked. This change is attributed to thermally induced reactions of the peptides as part of the Maillard reaction.

The mechanism of formation of these ɣ-glutamyl peptides in cheese has been described in the work by Toelstede and Hofmann (2009). They confirmed GGT present in raw milk as the main catalyst for the production of ɣ-glutamyl peptides involving free amino acids released upon proteolysis, such as glutathione or L-glutamine, which then combine with other amino acids or peptides, in agreement with previous observations (Roudot-Algaron et al. 1994). Furthermore, these authors also identified *Penicillium roquefortii* as another source of GGT responsible for the formation of high levels of ɣ-glutamyl peptides in Blue Shropshire, a blue-veined cheese. In this case, pasteurised bovine milk rather than raw milk was used for the production of this cheese, thus destroying the naturally occurring GGT during pasteurisation. It was observed that ɣ-glutamyl peptides were produced by a strain of *P. roquefortii* in the presence of the amino acid L-glutamine (donor amino acid) and a mixture of other amino acids. Toelstede and Hofmann (2009) also observed that ripening had an effect on the ratio and distribution of the γ- and α-glutamyl dipeptides within the cheese wheel. ɣ-Glutamyl peptides were predominantly present in the outer parts of the cheese as the lack of water may have influenced the GGT-catalysed transpeptidation reaction, whereas the higher content of water in the centre of the cheese may be required for the proteolysis of the caseins thus favouring the formation and the higher yield of the α-glutamyl peptides.

Unpublished work carried out by our research group looking at Jersey milk inclusions in the elaboration of Cheddar-cheese, as an alternative to the traditional Holstein-Friesian milk, led to the identification of several γ-glutamyl peptides. These cheeses were made at pilot scale from blends of Holstein-Friesian and Jersey milk (0, 50, and 100% inclusion rate of Jersey milk) and ripened for 7, 11, and 16 months. Among the γ-glutamyl peptides identified, γ-Glu-Phe, γ-Glu-Leu, γ-Glu-Val and γ-Glu-Met were the most abundant and their concentration increased with ripening, as previously reported (Toelstede et al. 2009). However, this was only observed for the cheeses without Jersey milk inclusion whereas this pattern was not observed in samples with Jersey milk inclusions, suggesting for the first time, the impact of the animal breed milk on the γ-glutamyl dipeptides formation in cheese. Previous research has reported a lower plasmin activity in Jersey milk compared to Holstein-Friesian milk thus affecting proteolysis (Richardson 1983), which may explain our results.

Kokumi-active compounds, mainly glutathione and some γ-glutamyl peptides, have been linked to the activation of the extracellular calcium-sensing receptor (CaSR) in humans, a class C, G-protein receptor (Maruyama et al. 2012; Ohsu et al. 2010). This receptor is located in the taste buds on the human tongue, but also in parathyroid glands, the kidneys and in many other tissues, such as the liver, heart, lungs, alimentary canal and pancreas. Kokumi compounds, as previously mentioned, have little or no inherent taste; however, they may modify the five basic tastes, especially sweet, salty and umami, which are also mediated by GPCRs, namely T1Rs for umami and sweet tastes and T2Rs for bitter taste (Spaggiaria et al. 2020). The expression of CaSR is considered to be independent from T1R3 (umami and sweet tastes). It may instead be co-expressed with T2Rs and act as an umami taste receptor (Maruyama et al. 2012; Brennan et al. 2014; Zhang et al. 2019) although further investigation is needed.

2.4 Mouthfeel perception

While aroma and taste play a key part in our sensory perception of dairy products, mouthfeel has an important role to play in the overall sensory profile of all dairy products. Mouthfeel is the broad term used to refer to sensations arising from tactile and chemical stimulations perceived in the mouth. Within the context of sensory evaluation, the physical sensations are classified as texture (e.g. density or viscosity), while mouthfeel also encompasses sensations such as astringency. Texture refers to the perception of physical properties of stimuli, and can be broken down into mechanical, geometrical, surface and body attributes (ISO 2008).

Tactile stimulations can occur throughout many phases of eating: first bite; mastication; mixing with saliva; bolus formation; swallowing; and oral retention post-consumption. The sensations perceived change depending on the phase of eating and the way in which it is manipulated in the mouth, referred to as 'oral processing'.

2.4.1 *Physiology of mouthfeel perception*

Mouthfeel is perceived through the somatosensory system, which mainly comprises mechanoreceptors. These are nerve endings in cells throughout the body, including the mouth, which respond to different mechanical changes such as pressure, vibration, and friction and project to the brain via the trigeminal nerve. Other receptors in the system include thermoreceptors, which respond to temperature changes and chemoreceptors, which respond to chemical stimuli. Texture is perceived through kinaesthesis and auditory receptors, alongside the somatosensory system.

Mechanoreceptors have differing abilities in spatial perception, allowing different detail resolutions to be differentiated between. Different mechanoreceptors can be classed as slow adapting or fast adapting, relating to whether the receptor maintains the response to a continuous stimuli, or adapts after a brief response (Johnson 2001).

The oral cavity is a highly innervated region of the body both in terms of variety and density of receptors, which leads to high oral sensitivity (Mount-castle 1974). Low thresholds for stimulation of mechanoreceptors in the mouth lead to the ability to distinguish between subtly different tactile sensations (Engelen and Van Der Bilt 2008). Individual variation in lingual nerve endings, linked to papillae density, can influence the ability to perceive mouthfeel sensations.

The stimulation of mechanoreceptors arises from contact with foods and pressure from food on oral surfaces or teeth, but also can occur due to movement of food across these surfaces. The interactions of food with saliva, oral mucosal surfaces, and the teeth influence the types of stimulations, and consequently the mouthfeel of a product.

2.4.2 *The perception of key mouthfeel attributes relevant to dairy products*

The types of mouthfeel attributes in dairy products are largely affected by the state of the product, for example a solid product such as a hard cheese varies greatly in mouthfeel from a semi-solid, such as yoghurt, or a liquid, such as milk. There are a wealth of different mouthfeel attributes that can be associated with dairy products, from chewiness to creaminess, and from chalky to smooth.

Hardness is a textural attribute important to the assessment of solid products such as cheeses. This arises from the oral processes applied to the cheese, often by the compression of teeth. This is perceived through the force required by the jaw to compress the food in the mouth, alongside tactile sensations on the soft surfaces of the mouth and the pressure sensations on the teeth. In the context of sensory evaluation, this is often ascribed to the first bite of a food product, however for consumer consumption of a product, this will be a dynamic attribute, changing throughout oral processing. Perceived hardness, or firmness, of a cheese is directly related to the deformation of the sample between the teeth (Xiong et al. 2002).

When chewing a food, the pattern of acceleration and deceleration of mastication and the resultant breakage characteristics are what lead to the sensory perception of the texture of that food. The resistance caused by the food influences the force required to accelerate and decelerate, therefore, leading to the breakage patterns (van der Bilt et al. 2006). Mastication also results in the stimulation of saliva flow, which can alter the perception of mouthfeel through physical and chemical interactions with the bolus.

An attribute more applicable to liquid and semi-solid dairy products, such as yoghurt, is creaminess. This is a complex sensation, influenced by the lubrication of the oral mucosa, directly linked to the viscosity of samples. Creaminess is often linked to fat content but can also relate to the smoothness and particle size of the food.

While creaminess perception appears to be comparable for 6-*n*-propylthiouracil (PROP) supertasters and nontasters, supertasters tended to use a more varied vocabulary when describing the creaminess of milk (Kirkmeyer and Tepper 2003). PROP supertasters have more taste papillae innervated by the trigeminal nerve, and therefore may have a more intense or varied sensation.

The link between fat content and creaminess has also led to the discussion of creaminess as a complex sensation involving tactile, chemosensory, gustatory and olfactory elements.

2.4.3 *The influence of oral processing and saliva*

Oral processing can change how we perceive textural attributes, as we apply different forces using different muscles resulting in different physical sample behaviour and different kinaesthetic responses. As discussed earlier in this chapter, the perceived hardness of a cheese directly relates to the chewing behaviour of the consumer, and this can vary among individuals (Xiong et al. 2002). Likewise, tongue movements can lead to differing perceptions of mouthfeel, for example in whey protein beverages (Carter and Drake 2021).

The way in which individuals process their food has been shown to be due to preferences known as Mouth Behaviour (Jeltema et al. 2015). In this model, consumers are shown to fall into one of four categories: crunchers, chewers, smooshers and suckers. The way in which the different groups manipulate foods varies both in jaw and tongue movement, which leads to not only different food choice, but different perceived textures (Jeltema et al. 2020).

Interactions with the saliva and oral mucosa can also influence the perception of mouthfeel. The oral mucosa comprises the surfaces of the mouth covered in mucosal tissue, including the tongue, cheeks, gums, and palate. These surfaces are coated with a film of saliva which is bound to the tissue, referred to as the 'salivary-mucosal pellicle'.

Stimulated and non-stimulated saliva vary in composition, viscosity and flow rate, and these variations can alter the perception of mouthfeel attributes through the chemical and physical interactions of those different components. Differences in salivary protein content, mucin and amylase have been shown to correlate to mouthfeel attributes in custard: mucin content negatively correlated to creaminess; amylase activity negatively correlated to slippery lip-tooth feel, thick mouthfeel and creamy after feel; total protein concentration negatively corresponded to slippery lip-tooth feel and fatty after feel (Engelen et al. 2007).

2.4.4 *The perception of mouth-drying in high-protein dairy products*

A phenomenon observed in liquid dairy products, particularly those with high protein content, is mouth-drying. Mouth-drying is a mouthfeel attribute that is often interchangeably used with the term 'astringency'. Whereas astringency specifically refers to 'the complex of sensations due to shrinking, drawing or puckering of the epithelium as a result of exposure to substances such as alums or tannins' (ASTM 2004); mouth-drying is a less substance-specific term, and as such is suitable to use for the description of the sensation of lack of lubrication in the oral cavity. This attribute can be an unpleasant one in dairy products (Beecher et al. 2008), leading to a reduction in consumer liking.

Mouth-drying sensations have been reported in both casein and whey proteins, the insoluble and soluble protein fractions of milk (Harwalkar and Elliott 1971; Lemieux and Simard 1994). Work by the authors has investigated the binding of both casein and whey protein to the oral mucosa as a mechanism for mouth-drying (Bull et al. 2017; Withers et al. 2013).This corresponds to a theory regarding the disruption of salivary structure, through the lack of a continuous lubricative layer (Gibbins and Carpenter 2013). This disruption of lubrication is thought to arise from interactions between salivary and whey proteins at low pH, leading to large particle sizes, which correlated with increased astringency (Vardhanabhuti and Foegeding 2010; Ye et al. 2011).

Work by the authors has shown that a build-up of whey protein over repeated consumption leads to higher mouth-drying due to oral retention (Bull et al. 2017; Norton et al. 2021; Withers et al. 2014). Elsewhere, the observation of structural interactions between whey protein and salivary proteins implies that these are mostly hydrophilic in nature (Celebioglu et al. 2019) and the binding of certain whey proteins to human oral epithelial cells indicates an adherence to the oral cavity (Ye et al. 2012).

2.5 Chemesthesis

Chemesthesis describes our sensory response to irritants, predominantly in the nasal mucosa, oral cavity and eyes, and mostly stimulating the trigeminal nerve to produce sensations such as burning, stinging, tingling, and cooling (Green 2016). Such chemosensation occurs in the skin or mucosa due to the activation of TRP family of cation channels. For example, the vanilloid TRPV1 is activated by pungent compounds (such as capsaicin in chilli) and the melastatin TRPM8 is activated by cooling compounds (such as menthol) (Alpizar et al. 2016). Trigeminal sensations in cheese, such as pungency, have been found to increase with ripening (Alvarez et al. 2007; Fresno and Alvarez 2012). The compounds responsible for pungency in cheese are predominantly the volatile fatty acids, such as acetic, propionic and butyric acids (C2:0, C3:0, C4:0) through to caproic acid (hexanoic acid, C6:0). It has been demonstrated in rat studies that fatty acids can activate trigeminal neurons (Bryant and Moore 1995; Yu et al. 2012), more specifically weak acids (including fatty acids) have been shown to stimulate the ion channel TRPA1 (Wang et al. 2011). This further discussion of the perception of short chain fatty acids persists to question the relative importance of these key dairy stimuli to the taste, olfactory and trigeminal systems.

2.6 The influence of individual differences in phenotype and genotype and their relevance to the perception of dairy products

The discussions in this chapter have focused on how the human senses perceive taste, aroma, mouthfeel and chemesthesis, particularly in relation to dairy foods. However, we cannot assume that we all perceive stimuli in the same way or to the same extent. Individual differences in perception can relate to either differences in intensity of perception, or differences in the sensorial quality, what it is that is perceived. For both taste and oral tactile sensitivity, differences in perceived intensity are thought to relate to the density of papillae, and consequently the relative abundance of taste buds and TRCs and trigeminal neurons that are located around the papillae. Higher fungiform papillae density (FPD) has been shown to relate to increased oral tactile sensitivity and increased taste sensitivity to the FA oleic acid (Zhou et al. 2021). FPD and oral tactile sensitivity were subsequently related to mouthfeel perception of biscuits. Individual differences in taste sensitivity are often characterised by the phenotypic response to 6-*n*-propylthiouracil (PROP) (or phenylthiocarbamide, PTC), a variation first reported in 1932 (Fox 1932). To some people, PROP tastes intensely bitter, while to others, it is tasteless. A standard method is not commonly used to classify people into 'nontasters', 'medium tasters' and 'supertasters' according to their relative sensitivity to PROP and sodium chloride (Tepper et al. 2001). PROP sensitivity has been associated with sensitivity to other tastants in solutions (Dinnella et al. 2018; Prescott et al. 2001), to the taste within real foods (Dinehart et al. 2006), to differences in sensitivity to oral irritants (Prescott et al. 2004) and to texture (de Wijk and Prinz 2007). However, in a study of over 1000 individuals, PROP sensitivity was not found to correlate with FPD (Dinnella et al. 2018). This is perhaps not surprising, FPD is a surrogate marker for taste sensitivity; it indicates the likelihood that that there will be more taste cells available to detect stimuli but does not directly measure that. This does not account for genetic diversity in taste receptors. In the case of PROP, genetic variation in the bitter taste receptor gene *TAS2R38* has been found to explain up to 85% of individual variation in PTC/PROP taste sensitivity (Kim et al. 2003). The *TAS2R38* gene is located on chromosome 7q36, and has two common molecular forms: proline-alanine-valine (PAV) and alanine-valine-isoleucine (AVI). PAV is associated with tasting the bitter thiouracil grouping (found in PTC, PROP, and also in glucosinolates which occur in brassica vegetables) and is associated with nontasting; although less common haplotypes also exist. The most common genotypes are PAV homozygotes (PAV/PAV) which correspond generally to 'supertasters', PAV/AVI heterozygotes ('medium tasters'), and AVI homozygotes (AVI/AVI) ('nontasters') (Duffy et al. 2004). In addition to variation in *TAS2R38*, PROP sensitivity may be moderated by variation in taste-bud density, for which polymorphisms in the taste-bud trophic factor gustin gene have been explored. The presence of allele A in gustin polymorphism *rs2274333* has been shown to enhance detection of PROP at low concentrations (Calo et al. 2011). PROP nontasters were most frequently of *TAS2R38*-Gustin *rs2274333* genotype AVI/AVI–GG, and supertasters most commonly either PAV/AVI–AA or PAV/PAV–AA (Calo et al. 2011).

This could be considered the 'tip of the iceberg' in understanding individual differences in sensory perception of foods. Genetic variation (SNPs) can be found in genes for numerous taste receptors (see reference to SNPs for the ENaC salt receptor earlier in this chapter), for odour receptors, and for genes that encode for other vital parts of transduction pathways. Precone et al. (2019) reviewed articles up to 2018 that had researched genes related to taste, olfaction and texture perception. In taste perception, they refer not only to *T2R38,* but also variation in the bitter receptor gene *T2R16*, and to research investigating SNPs in the T1R GPCR genes (*T1R1, T1R2* and *T1R3*; influencing sweet and umami taste sensitivity), the *SNNP* genes that encode for the ENaC salt receptor, in

TRPV1 (an amiloride-insensitive salt), and in *CD36* involved in oral detection of fatty acids. Considering olfaction, the review puts into perspective that OR genes are from the largest gene family in vertebrates and distributed in 51 loci across 21 human chromosomes (Precone et al. 2019). As elucidated earlier, a pattern of different ORs can encode for a specific odorant, and hence genetic variation in a single receptor can substantially influence the perception of an odorant. Of direct relevance to dairy foods, variants of the *OR11H7P* gene have been individual variation in sensitivity to isovaleric acid (i.e. 3-methylbutyric acid) (Menashe et al. 2007), a pungent cheesy volatile compound commonly found in cheese (Thierry et al. 2004).

Perhaps the key point to note here is never to rely in single individuals and their perception of a food, with the plethora of individual differences in perception it is surely good practice the use sensory panels that are screened for sensitivity to a wide range of stimuli, and to carry out consumer tests with a very wide range participants as their sensitivities will not be known even if their familiarity can be accounted for.

References

Ahern, G.P., Brooks, I.M., Miyares, R.L., and Wang, X.B. (2005). Extracellular cations sensitize and gate capsaicin receptor TRPV1 modulating pain signaling. *Journal of Neuroscience* 25 (21): 5109–5116. https://doi.org/10.1523/jneurosci.0237-05.2005.

Alpizar, Y., Voets, T., and Talavera, K. (2016). Molecular mechanisms that underpin the tTole of TRP channels in chemesthesis. In: *Chemesthesis: Chemical Touch in Food and Eating* (ed. S.T. McDonald, D.A. Bolliet and J.E. Hayes). Wiley-Blackwell.

Alvarez, S., Fresno, M., Mendez, P. et al. (2007). Alternatives for improving physical, chemical, and sensory characteristics of goat cheeses: the use of arid-land forages in the diet. *Journal of Dairy Science* 90 (5): 2181–2188. https://doi.org/10.3168/jds.2006-506.

Amoore, J.E. (1946). Current status of the steric theory of odour. *Annals of the New York Academy of Sciences* 116: 457–476.

Aparicio, J., SulbaranDeFerrer, B., Ferrer, A. et al. (1997). Short-chain organic acids in aged Blanco cheese. *Journal of Food Science and Technology-Mysore* 34 (3): 204–207. <Go to ISI>://WOS:A1997XB65100004.

ASTM (American Society for Testing and Materials) (2004). Standard definitions of terms relating to sensory evaluation of materials and products. In: *Annual Book of ASTM Standards*. Philadelphia, PA: ASTM.

Bachmanov, A.A. and Beauchamp, G.K. (2007). Taste receptor genes. *Annual Review of Nutrition* 27: 389–414. https://doi.org/10.1146/annurev.nutr.26.061505.111329.

Beecher, J.W., Drake, M.A., Luck, P.J., and Foegeding, E.A. (2008). Factors regulating astringency of whey protein beverages. *Journal of Dairy Science* 91: 2553–2560.

Behrens, M. and Meyerhof, W. (2006). Bitter taste receptors and human bitter taste perception. *Cellular and Molecular Life Sciences* 63 (13): 1501–1509. https://doi.org/10.1007/s00018-006-6113-8.

Beksan, E., Schieberle, P., Robert, F. et al. (2003). Synthesis and sensory characterization of novel umami-tasting glutamate glycoconjugates. *Journal of Agricultural and Food Chemistry* 51: 5428–5436.

Belitz, H.D., Grosch, W., and Schieberle, P. (2009). *Food Chemistry*, 4e. Berlin: Springer.

Bigiani, A. (2020). Does ENaC work as sodium taste receptor in humans? *Nutrients* 12 (4): https://doi.org/10.3390/nu12041195.

van der Bilt, A., Engelen, L., Pereira, L.J. et al. (2006). Oral physiology and mastication. *Physiology & Behavior* 89 (1): 22–27. https://doi.org/10.1016/j.physbeh.2006.01.025.

Borthwick, A. and Da Costa, N. (2017). 2,5-diketopiperazines in food and beverages: taste and bioactivity. *Critical Reviews in Food Science and Nutrition* 57 (4): 718–742. https://doi.org/10.1080/10408398.2014.911142.

Brennan, S.C., Davies, T.S., Schepelmann, M., and Riccardi, D. (2014). Emerging roles of the extracellular calcium-sensing receptor in nutrient sensing: control of taste modulation and intestinal hormone secretion. *British Journal of Nutrition* 111: S16–S22. https://doi.org/10.1017/S0007114513002250.

Bryant, B.P. and Moore, P.A. (1995). Factors affecting the sensitivity of the lingual trigeminal nerve to acids. *American Journal of Physiology – Regulatory, Integrative and Comparative Physiology* 268 (1): R58–R65. <Go to ISI>://WOS:A1995QB49900009.

Buck, L. and Axel, R. (1991). A novel multigene family may encode odorant receptors – a molecular-basis for odor recognition. *Cell* 65 (1): 175–187. https://doi.org/10.1016/0092-8674(91)90418-x.

Bull, S.P., Hong, Y., Khutoryanskiy, V.V. et al. (2017). Whey protein mouth drying influenced by thermal denaturation. *Food Quality and Preference* 56: 233–240. https://doi.org/10.1016/j.foodqual.2016.03.008.

Calo, C., Padiglia, A., Zonza, A. et al. (2011). Polymorphisms in TAS2R38 and the taste bud trophic factor, gustin gene co-operate in modulating PROP taste phenotype. *Physiology & Behavior* 104 (5): 1065–1071. https://doi.org/10.1016/j.physbeh.2011.06.013.

Carter, B.G. and Drake, M. (2021). Influence of oral movement, particle size, and zeta potential on astringency of whey protein. *Journal of Sensory Studies* 36 (3): https://doi.org/10.1111/joss.12652.

Celebioglu, H.Y., Lee, S., and Chronakis, I.S. (2019). Interactions of salivary mucins and saliva with food proteins: a review. *Critical Reviews in Food Science and Nutrition* 60: 1–20. https://doi.org/10.1080/10408398.2018.1512950.

Chandrashekar, J., Mueller, K.L., Hoon, M.A. et al. (2000). T2Rs function as bitter taste receptors. *Cell* 100 (6): 703–711. https://doi.org/10.1016/s0092-8674(00)80706-0.

Chaudhari, N., Pereira, E., and Roper, S.D. (2009). Taste receptors for umami: the case for multiple receptors. *American Journal of Clinical Nutrition* 90 (3): 738S–742S. https://doi.org/10.3945/ajcn.2009.27462H.

Chen, Y., Liou, S., and Chen, C. (2004). Two-step mass spectrometric approach for the identification of diketopiperazines in chicken essence. *European Food Research and Technology* 218: 589–597.

Cheon, E. and Mattes, R.D. (2021). Perceptual quality of nonesterified fatty acids varies with fatty acid chain length. *Chemical Senses* 46: 1–7. https://doi.org/10.1093/chemse/bjab023.

Christensen, K.R. and Reineccius, G.A. (1995). Aroma extract dilution analysis of aged cheddar cheese. *Journal of Food Science* 60 (2): 218–220. https://doi.org/10.1111/j.1365-2621.1995.tb05641.x.

Connor, E.E., Zhou, Y., and Liu, G.E. (2018). The essence of appetite: does olfactory receptor variation play a role? *Journal of Animal Science* 96 (4): 1551–1558. https://doi.org/10.1093/jas/sky068.

Da Costa, N., Chen, M., Merritt, D. et al. (2010). Methionine-containing cyclic dipeptides: occurrence in natural products, synthesis, and sensory evaluation. *Controlling Maillard Pathways To Generate Flavors* 1042: 111–120.

Delwiche, J.F., Halpern, B.P., and Desimone, J.A. (1999). Anion size of sodium salts and simple taste reaction times. *Physiology & Behavior* 66 (1): 27–32. https://doi.org/10.1016/s0031-9384(98)00273-x.

Dias, A.G., Rousseau, D., Duizer, L. et al. (2013). Genetic variation in putative salt taste receptors and salt taste perception in humans. *Chemical Senses* 38 (2): 137–145. https://doi.org/10.1093/chemse/bjs090.

Dinehart, M.E., Hayes, J.E., Bartoshuk, L.M. et al. (2006). Bitter taste markers explain variability in vegetable sweetness, bitterness, and intake. *Physiology & Behavior* 87 (2): 304–313. https://doi.org/10.1016/j.physbeh.2005.10.018.

Dinnella, C., Monteleone, E., Piochi, M. et al. (2018). Individual variation in PROP status, fungiform papillae density, and responsiveness to taste stimuli in a large population sample. *Chemical Senses* 43 (9): 697–710. https://doi.org/10.1093/chemse/bjy058.

Duffy, V.B., Davidson, A.C., Kidd, J.R. et al. (2004). Bitter receptor gene (TAS2R38), 6-n-propylthiouracil (PROP) bitterness and alcohol intake. *Alcoholism, Clinical and Experimental Research* 28 (11): 1629–1637. https://doi.org/10.1097/01.Alc.0000145789.55183.D4.

Dyson, G.M. (1938). The scientific basis of odour. *Journal of the Society of Chemical Industry* 57: 647–651.

Engel, E., Nicklaus, S., Septier, C. et al. (2000). Taste active compounds in a goat cheese water-soluble extract. 2. Determination of the relative impact of water-soluble extract components on its taste using omission tests. *Journal of Agricultural and Food Chemistry* 48 (9): 4260–4267. https://doi.org/10.1021/jf991364h.

Engelen, L. and Van Der Bilt, A. (2008). Oral physiology and texture perception of semisolids. *Journal of Texture Studies* 39 (1): 83–113. https://doi.org/10.1111/j.1745-4603.2007.00132.x.

Engelen, L., van den Keybus, P.A.M., de Wijk, R.A. et al. (2007). The effect of saliva composition on texture perception of semi-solids. *Archives of Oral Biology* 52 (6): 518–525. https://doi.org/10.1016/j.archoralbio.2006.11.007.

Fallico, V., McSweeney, P., Horne, J. et al. (2005). Evaluation of bitterness in Ragusano cheese. *Journal of Dairy Science* 88 (4): 1288–1300. https://doi.org/10.3168/jds.S0022-0302(05)72795-8.

Feng, T., Wu, Y., Zhang, Z. et al. (2019). Purification, identification, and sensory evaluation of Kokumi peptides from *Agaricus bisporus* mushroom. *Food* 8 (2): https://doi.org/10.3390/foods8020043.

Fox, A.L. (1932). The relationship between chemical constitution and taste. *Proceedings of the National Academy of Sciences of the United States of America* 18: 115–120.

Fresno, M. and Alvarez, S. (2012). Chemical, textural and sensorial changes during the ripening of Majorero goat cheese. *International Journal of Dairy Technology* 65 (3): 393–400. https://doi.org/10.1111/j.1471-0307.2012.00842.x.

Gautschi, M., Schmid, J., Peppard, T. et al. (1997). Chemical characterization of diketopiperazines in beer. *Journal of Agricultural and Food Chemistry* 45 (8): 3183–3189. https://doi.org/10.1021/jf9700992.

Gibbins, H.L. and Carpenter, G.H. (2013). Alternative mechanisms of astringency – what is the role of saliva? *Journal of Texture Studies* 44 (5): 364–375. https://doi.org/10.1111/jtxs.12022.

Ginz, M. and Engelhardt, U. (2000). Identification of proline-based diketopiperazines in roasted coffee. *Journal of Agricultural and Food Chemistry* 48: 3528–3532. https://doi.org/10.1021/jf991256v.

Green, B.G. (2016). Introduction: What is chemesthesis? In: *Chemesthesis: Chemical Touch in Food and Eating* (ed. S.T. McDonald, D.A. Bolliet and J.E. Hayes). Wiley-Blackwell.

Guigoz, Y. and Solms, J. (1976). Bitter peptides, occurrence and structure. *Chemical Senses and Flavour* 2: 71–84.

Harwalkar, V.R. and Elliott, J.A. (1971). Isolation of bitter and astringent fractions from cheddar cheese. *Journal of Dairy Science* 54 (1): 8–11. https://doi.org/10.3168/jds.S0022-0302(71)85772-7.

Hasin, Y., Olender, T., Khen, M. et al. (2008). High-resolution copy-number variation map reflects human olfactory receptor diversity and evolution. *PLoS Genetics* 4 (11): https://doi.org/10.1371/journal.pgen.1000249.

Hasin-Brumshtein, Y., Lancet, D., and Olender, T. (2009). Human olfaction: from genomic variation to phenotypic diversity. *Trends in Genetics* 25 (4): 178–184. https://doi.org/10.1016/j.tig.2009.02.002.

Hayden, S. and Teeling, E.C. (2014). The molecular biology of vertebrate olfaction. *The Anatomical Record: Advances in Integrative Anatomy and Evolutionary Biology* 297 (11): 2216–2226. https://doi.org/10.1002/ar.23031.

Huang, A.L., Chen, X., Hoon, M.A. et al. (2006). The cells and logic for mammalian sour taste detection. *Nature* 442 (7105): 934–938. https://doi.org/10.1038/nature05084.

Inada, H., Kawabata, F., Ishimaru, Y. et al. (2008). Off-response property of an acid-activated cation channel complex PKD1L3-PKD2L1. *EMBO Reports* 9 (7): 690–697. https://doi.org/10.1038/embor.2008.89.

Ishii, S., Misaka, T., Kishi, M. et al. (2009). Acetic acid activates PKD1L3-PKD2L1 channel-A candidate sour taste receptor. *Biochemical and Biophysical Research Communications* 385 (3): 346–350. https://doi.org/10.1016/j.bbrc.2009.05.069.

Ishimaru, Y., Inada, H., Kubota, M. et al. (2006). Transient receptor potential family members PKD1L3 and PKD2L1 form a candidate sour taste receptor. *Proceedings of the National Academy of Sciences of the United States of America* 103 (33): 12569–12574. https://doi.org/10.1073/pnas.0602702103.

ISO (2008). International Organization for Standardization. *Sensory analysis – Vocabulary* ICS 67.240, 01.040.67.

Jaeger, S.R., McRae, J.F., Bava, C.M. et al. (2013). A Mendelian trait for olfactory sensitivity affects odor experience and food selection. *Current Biology* 23 (16): 1601–1605. https://doi.org/10.1016/j.cub.2013.07.030.

Jeltema, M., Beckley, J., and Vahalik, J. (2015). Model for understanding consumer textural food choice. *Food Science & Nutrition* 3 (3): 202–212. https://doi.org/10.1002/fsn3.205.

Jeltema, M., Beckley, J., Vahalik, J., and Garza, J. (2020). Consumer textural food perception over time based on mouth behavior. *Journal of Texture Studies* 51 (1): 185–194. https://doi.org/10.1111/jtxs.12479.

Johnson, K.O. (2001). The roles and functions of cutaneous mechanoreceptors. *Current Opinion in Neurobiology* 11 (4): 455–461. https://doi.org/10.1016/s0959-4388(00)00234-8.

Karametsi, K., Kokkinidou, S., Ronningen, I., and Peterson, D. (2014). Identification of bitter peptides in aged Cheddar cheese. *Journal of Agricultural and Food Chemistry* 62 (32): 8034–8041. https://doi.org/10.1021/jf5020654.

Keller, A., Zhuang, H.Y., Chi, Q.Y. et al. (2007). Genetic variation in a human odorant receptor alters odour perception. *Nature* 449 (7161): 468–U466. https://doi.org/10.1038/nature06162.

Kim, U.K., Jorgenson, E., Coon, H. et al. (2003). Positional cloning of the human quantitative trait locus underlying taste sensitivity to phenylthiocarbamide. *Science* 299 (5610): 1221–1225. https://doi.org/10.1126/science.1080190.

Kirimura, J., Shimizu, A., Kimizuka, A. et al. (1969). Contribution of peptides and amino acids to the taste of foodstuffs. *Journal of Agricultural and Food Chemistry* 17 (4): 689–695.

Kirkmeyer, S.V. and Tepper, B.J. (2003). Understanding creaminess perception of dairy products using free-choice profiling and genetic responsivity to 6-n-propylthiouracil. *Chemical Senses* 28 (6): 527–536. https://doi.org/10.1093/chemse/28.6.527.

Kuroda, M., Kato, Y., Yamazaki, J. et al. (2012a). Determination of gamma-glutamyl-valyl-glycine in raw scallop and processed scallop products using high pressure liquid chromatography-tandem mass spectrometry. *Food Chemistry* 134 (3): 1640–1644. https://doi.org/10.1016/j.foodchem.2012.03.048.

Kuroda, M., Kato, Y., Yamazaki, J. et al. (2012b). Determination and quantification of gamma-Glutamyl-valyl-glycine in commercial fish sauces. *Journal of Agricultural and Food Chemistry* 60 (29): 7291–7296. https://doi.org/10.1021/jf3012932.

Kuroda, M., Kato, Y., Yamazaki, J. et al. (2013). Determination and quantification of the kokumi peptide, gamma-glutamyl-valyl-glycine, in commercial soy sauces. *Food Chemistry* 141 (2): 823–828. https://doi.org/10.1016/j.foodchem.2013.03.070.

Laureati, M., Sandvik, P., Almli, V.L. et al. (2020). Individual differences in texture preferences among European children: development and validation of the Child Food Texture Preference Questionnaire (CFTPQ). *Food Quality and Preference* 80: https://doi.org/10.1016/j.foodqual.2019.103828.

Lemieux, L. and Simard, R.E. (1991). Bitter flavor in dairy-products. 1. A review of the factors likely to influence its development, mainly in cheese manufacture. *Le Lait* 71 (6): 599–636. <Go to ISI>://WOS:A1991GT66100001.

Lemieux, L. and Simard, R.E. (1992). Bitter flavor in dairy-products. 2. A review of bitter peptides from caseins – their formation, isolation and identification, structure masking and inhibition. *Le Lait* 72 (4): 335–382. <Go to ISI>://WOS:A1992JQ99600001.

Lemieux, L. and Simard, R. (1994). Astringency, a textural defect in dairy products. *Le Lait* 74: 217–240.

Lewandowski, B.C., Sukumaran, S.K., Margolskee, R.F., and Bachmanov, A.A. (2016). Amiloride-insensitive salt taste is mediated by two populations of type III taste cells with distinct transduction mechanisms. *Journal of Neuroscience* 36 (6): 1942–1953. https://doi.org/10.1523/jneurosci.2947-15.2016.

Liman, E.R. (2020). Salty taste: from transduction to transmitter release, hold the calcium. *Neuron* 106 (5): 709–711. https://doi.org/10.1016/j.neuron.2020.05.012.

Lindemann, B. (1999). Receptor seeks ligand: on the way to cloning the molecular receptors for sweet and bitter taste. *Nature Medicine* 5 (4): 381–382. https://doi.org/10.1038/7377.

Lindemann, B. (2001). Receptors and transduction in taste. *Nature* 413 (6852): 219–225. https://doi.org/10.1038/35093032.

Liu, J., Song, H., Liu, Y. et al. (2015). Discovery of kokumi peptide from yeast extract by LC-Q-TOF-MS/MS and sensomics approach. *Journal of the Science of Food and Agriculture* 95 (15): 3183–3194. https://doi.org/10.1002/jsfa.7058.

Liu, D.L., Archer, N., Duesing, K. et al. (2016). Mechanism of fat taste perception: association with diet and obesity. *Progress in Lipid Research* 63: 41–49. https://doi.org/10.1016/j.plipres.2016.03.002.

Liu, Y., Xu, H., Dahir, N. et al. (2021). GPR84 is essential for the taste of medium chain saturated fatty acids. *Journal of Neuroscience* 41 (24): 5219–5228. https://doi.org/10.1523/jneurosci.2530-20.2021.

Lunde, K., Egelandsdal, B., Skuterud, E. et al. (2012). Genetic variation of an odorant receptor OR7D4 and sensory perception of cooked meat containing Androstenone. *PLoS One* 7 (5): https://doi.org/10.1371/journal.pone.0035259.

Lyall, V., Heck, G.L., Vinnikova, A.K. et al. (2004). The mammalian amiloride-insensitive non-specific salt taste receptor is a vanilloid receptor-1 variant. *Journal of Physiology (London)* 558 (1): 147–159. https://doi.org/10.1113/jphysiol.2004.065656.

Malnic, B., Hirono, J., Sato, T., and Buck, L.B. (1999). Combinatorial receptor codes for odors. *Cell* 96 (5): 713–723. https://doi.org/10.1016/s0092-8674(00)80581-4.

Maruyama, Y., Yasuda, R., Kuroda, M., and Eto, Y. (2012). Kokumi substances, enhancers of basic tastes, induce responses in calcium-sensing receptor expressing taste cells. *PLoS One* 7 (4): https://doi.org/10.1371/journal.pone.0034489.

Mattes, R.D. (2009). Oral detection of short-, medium-, and long-chain free fatty acids in humans. *Chemical Senses* 34 (2): 145–150. https://doi.org/10.1093/chemse/bjn072.

Mattes, R.D. (2011). Accumulating evidence supports a taste component for free fatty acids in humans. *Physiology & Behavior* 104 (4): 624–631. https://doi.org/10.1016/j.physbeh.2011.05.002.

Mattes, R.D. (2021). Taste, teleology and macronutrient intake. *Current Opinion in Physiology* 19: 162–167. https://doi.org/10.1016/j.cophys.2020.11.003.

McRae, J.F., Mainland, J.D., Jaeger, S.R. et al. (2012). Genetic variation in the odorant receptor OR2J3 is associated with the ability to detect the 'Grassy' smelling odor, cis-3-hexen-1-ol. *Chemical Senses* 37 (7): 585–593. https://doi.org/10.1093/chemse/bjs049.

McSweeney, P. (1997). The flavour of milk and dairy products: III. Cheese: taste. *International Journal of Dairy Technology* 50 (4): 123–128. https://doi.org/10.1111/j.1471-0307.1997.tb01752.x.

Menashe, I., Abaffy, T., Hasin, Y. et al. (2007). Genetic elucidation of human hyperosmia to isovaleric acid. *PLoS Biology* 5 (11): 2462–2468. https://doi.org/10.1371/journal.pbio.0050284.

Miyamura, N., Iida, Y., Kuroda, M. et al. (2015). Determination and quantification of kokumi peptide, gamma-glutamyl-valyl-glycine, in brewed alcoholic beverages. *Journal of Bioscience and Bioengineering* 120 (3): 311–314. https://doi.org/10.1016/j.jbiosc.2015.01.018.

Mullin, W.J. and Emmons, D.B. (1997). Determination of organic acids and sugars in cheese, milk and whey by high performance liquid chromatography. *Food Research International* 30 (2): 147–151. https://doi.org/10.1016/s0963-9969(97)00026-4.

Murtaza, B., Hichami, A., Khan, A.S. et al. (2021). Implication of TRPC3 channel in gustatory perception of dietary lipids. *Acta Physiologica* 231 (2): https://doi.org/10.1111/apha.13554.

Negoias, S., Visschers, R., Boelrijk, A., and Hummel, T. (2008). New ways to understand aroma perception. *Food Chemistry* 108 (4): 1247–1254. https://doi.org/10.1016/j.foodchem.2007.08.030.

Nelson, G., Hoon, M.A., Chandrashekar, J. et al. (2001). Mammalian sweet taste receptors. *Cell* 106 (3): 381–390. https://doi.org/10.1016/s0092-8674(01)00451-2.

Nelson, G., Chandrashekar, J., Hoon, M.A. et al. (2002). An amino-acid taste receptor. *Nature* 416 (6877): 199–202. https://doi.org/10.1038/nature726.

Nor, N.D.M., Houston-Price, C., Harvey, K., and Methven, L. (2021). The effects of taste sensitivity and repeated taste exposure on children's intake and liking of turnip (Brassica rapa subsp. rapa); a bitter Brassica vegetable. *Appetite* 157: https://doi.org/10.1016/j.appet.2020.104991.

Norton, V., Lignou, S., and Methven, L. (2021). Whey protein derived mouthdrying found to relate directly to retention post consumption but not to induced differences in salivary flow rate. *Food* 10 (3): https://doi.org/10.3390/foods10030587.

Ohsu, T., Amino, Y., Nagasaki, H. et al. (2010). Involvement of the calcium-sensing receptor in human taste perception. *Journal of Biological Chemistry* 285 (2): 1016–1022. https://doi.org/10.1074/jbc.M109.029165.

Olender, T., Waszak, S.M., Viavant, M. et al. (2012). Personal receptor repertoires: olfaction as a model. *BMC Genomics* 13: https://doi.org/10.1186/1471-2164-13-414.

Otsuka, Y., Arita, H., Sakaji, M. et al. (2019). Investigation of the formation mechanism of proline-containing cyclic dipeptide from the linear peptide. *Bioscience, Biotechnology, and Biochemistry* 83: 2355–2363. https://doi.org/10.1080/09168451.2019.1659718.

Ozdener, M.H., Subramaniam, S., Sundaresan, S. et al. (2014). CD36-and GPR120-mediated Ca2+ signaling in human taste bud cells mediates differential responses to fatty acids and is altered in obese mice. *Gastroenterology* 146 (4): 995–U544. https://doi.org/10.1053/j.gastro.2014.01.006.

Pauling, I. (1946). Analogies between antibodies and simpler chemical substances. *Chemical and Engineering News Archive* 24: 1064–1065.

Pinsorn, P., Oikawa, A., Watanabe, M. et al. (2018). Metabolic variation in the pulps of two durian cultivars: unraveling the metabolites that contribute to the flavor. *Food Chemistry* 268: 118–125. https://doi.org/10.1016/j.foodchem.2018.06.066.

Precone, V., Beccari, T., Stuppia, L. et al. (2019). Taste, olfactory and texture related genes and food choices: implications on health status. *European Review for Medical and Pharmacological Sciences* 23 (3): 1305–1321. <Go to ISI>://WOS:000459178100049.

Prescott, J., Ripandelli, N., and Wakeling, I. (2001). Binary taste mixture interactions in PROP non-tasters, medium-tasters and super-tasters. *Chemical Senses* 26 (8): 993–1003. https://doi.org/10.1093/chemse/26.8.993.

Prescott, J., Soo, J., Campbell, H., and Robells, C. (2004). Responses of PROP taster groups to variations in sensory qualities within foods and beverages. *Physiology & Behavior* 82 (2–3): 459–469. https://doi.org/10.1016/j.physbeh.2004.04.009.

Rhyu, M., Song, A., Kim, E. et al. (2020). Kokumi Taste Active Peptides Modulate Salt and Umami Taste. *Nutrients* 12 (4): https://doi.org/10.3390/nu12041198.

Richardson, B.C. (1983). Variation of the concentration of plasmin and plasminogen in bovine milk with lactation. *New Zealand Journal of Dairy Science and Technology* 18: 247–252.

Riera, C.E., Vogel, H., Simon, S.A., and le Coutre, J. (2007). Artificial sweeteners and salts producing a metallic taste sensation activate TRPV1 receptors. *American Journal of Physiology – Regulatory, Integrative and Comparative Physiology* 293 (2): R626–R634. https://doi.org/10.1152/ajpregu.00286.2007.

Rizzi, G.P. (1989). Heat-induced flavor formation from peptides. In: *Thermal Generation of Aromas. ACS Symposium Series*. 409 Chapter 16: 172–181. https://doi.org/10.1021/bk-1989-0409.ch016.

Rocchetti, G., Rebecchi, A., Dallolio, M. et al. (2021). Changes in the chemical and sensory profile of ripened Italian salami following the addition of different microbial starters. *Meat Science* 180: 108584. https://doi.org/10.1016/j.meatsci.2021.108584.

Roudot-Algaron, F., Le Bars, D., Einhorn, J. et al. (1993). Flavor constituents of aqueous fraction extracted from Comte cheese by liquid carbon dioxide. *Journal of Food Science* 58: 1005–1009.

Roudot-Algaron, F., Kerhoas, L., Lebars, D. et al. (1994). Isolation of gamma-glutamyl peptides from Comté cheese. *Journal of Dairy Science* 77 (5): 1161–1166. https://doi.org/10.3168/jds.S0022-0302(94)77053-3.

Running, C.A., Craig, B.A., and Mattes, R.D. (2015). Oleogustus: the unique taste of fat. *Chemical Senses* 40 (7): 507–516. https://doi.org/10.1093/chemse/bjv036.

Salger, M., Stark, T.D., and Hofmann, T. (2019). Taste modulating peptides from overfermented cocoa beans. *Journal of Agricultural and Food Chemistry* 67 (15): 4311–4320. https://doi.org/10.1021/acs.jafc.9b00905.

Schiffman, S.S., Graham, B.G., Vance, A.R. et al. (1992). Detection thresholds for emulsified oils in young and elderly subjects. *Chemical Senses* 17 (5): 693. <Go to ISI>://BCI:BCI199344074794.

Sforza, S., Galaverna, G., Schivazappa, C. et al. (2006). Effect of extended aging of Parma dry-cured ham on the content of oligopeptides and free amino acids. *Journal of Agricultural and Food Chemistry* 54: 9422–9429.

Shibata, M., Hirotsuka, M., Mizutani, Y. et al. (2017). Isolation and characterization of key contributors to the 'kokumi' taste in soybean seeds. *Bioscience Biotechnology and Biochemistry* 81 (11): 2168–2177. https://doi.org/10.1080/09168451.2017.1372179.

Spaggiaria, G., Di Piziob, A., and Cozzinia, P. (2020). Sweet, umami and bitter taste receptors: state of the art of in silico molecular modeling approaches. *Trends in Food Science & Technology* 96: 21–29.

Staehler, F., Riedel, K., Demgensky, S. et al. (2008). A role of the epithelial sodium channel in human salt taste transduction? *Chemosensory Perception* 1 (1): 78–90. https://doi.org/10.1007/s12078-008-9006-4.

Stark, T. and Hofmann, T. (2005). Structures, sensory activity, and dose/response functions of 2,5-diketopiperazines in roasted cocoa nibs (*Theobroma cacao*). *Journal of Agricultural and Food Chemistry* 53 (18): 7222–7231. https://doi.org/10.1021/jf051313m.

Su, Q., Chen, M., Wang, Y. et al. (2021). Structural basis for Ca2+ activation of the heteromeric PKD1L3/PKD2L1 channel. *Nature Communications* 12 (1): https://doi.org/10.1038/s41467-021-25216-z.

Sullivan, R. (2022). The flavour of cooked cheese; Chapter 3: 'Characterisation of cooked cheese flavour'. Doctoral Thesis. University of Reading, UK.

Takahashi, K., Tadenuma, M., Kitamoto, K., and Sato, S. (1974). L-Prolyl-L-Leucine anhydride a bitter compound formed in aged sake. *Agricultural and Biological Chemistry* 38: 927–932.

Temussi, P.A. (2011). The good taste of peptides. *Journal of Peptide Science* 18 (2): 73–82. https://doi.org/10.1002/psc.1428.

Tepper, B.J., Christensen, C.M., and Cao, J. (2001). Development of brief methods to classify individuals by PROP taster status. *Physiology & Behavior* 73 (4): 571–577. https://doi.org/10.1016/s0031-9384(01)00500-5.

Thierry, A., Richoux, R., and Kerjean, J.R. (2004). Isovaleric acid is mainly produced by *Propionibacterium freudenreichii* in Swiss cheese. *International Dairy Journal* 14 (9): 801–807. https://doi.org/10.1016/j.idairyj.2004.02.002.

Toelstede, S. and Hofmann, T. (2008a). Quantitative studies and taste re-engineering experiments toward the decoding of the nonvolatile sensometabolome of Gouda cheese. *Journal of Agricultural and Food Chemistry* 56 (13): 5299–5307. https://doi.org/10.1021/jf800552n.

Toelstede, S. and Hofmann, T. (2008b). Sensomics mapping and identification of the key bitter metabolites in Gouda cheese. *Journal of Agricultural and Food Chemistry* 56 (8): 2795–2804. https://doi.org/10.1021/jf7036533.

Toelstede, S. and Hofmann, T. (2009). Kokumi-active Glutamyl peptides in cheeses and their biogeneration by *Penicillium roquefortii*. *Journal of Agricultural and Food Chemistry* 57 (9): 3738–3748. https://doi.org/10.1021/jf900280j.

Toelstede, S., Dunkel, A., and Hofmann, T. (2009). A series of Kokumi peptides impart the long-lasting mouthfulness of matured Gouda cheese. *Journal of Agricultural and Food Chemistry* 57 (4): 1440–1448. https://doi.org/10.1021/jf803376d.

Trivedi, B.P. (2012). The finer points of taste. *Nature, UK* 486 (7403): S2–S3. <Go to ISI>://FSTA:2013-04-Ag1953.

Tu, Y.-H., Cooper, A.J., Teng, B. et al. (2018). An evolutionarily conserved gene family encodes proton-selective ion channels. *Science* 359 (6379): 1047–1050. https://doi.org/10.1126/science.aao3264.

Turin, L. (1996). A spectroscopic mechanism for primary olfactory reception. *Chemical Senses* 21 (6): 773–791. https://doi.org/10.1093/chemse/21.6.773.

Ueda, Y., Sakaguchi, M., Hirayama, K. et al. (1990). Characteristic flavour constituents in water extract of garlic. *Agricultural and Biological Chemistry* 54: 163–169.

Ueda, Y., Tsubuku, T., and Miyajima, R. (1994). Composition of sulfur-containing components in onion and their flavor characters. *Bioscience, Biotechnology, and Biochemistry* 58 (1): 108–110.

Ueda, Y., Yonemitsu, M., Tsubuku, T. et al. (1997). Flavour characteristics of glutathione in raw and cooked foodstuffs. *Bioscience, Biotechnology, and Biochemistry* 61: 1977–1980.

Vardhanabhuti, B. and Foegeding, E.A. (2010). Evidence of interactions between whey proteins and mucin their implication on the astringency mechanism of whey proteins at low pH. *Gums and Stabilisers for the Food Industry* 15: 137–146. <Go to ISI>://WOS:000285024600014.

Wang, Y.Y.Y., Chang, R.B., Allgood, S.D. et al. (2011). A TRPA1-dependent mechanism for the pungent sensation of weak acids. *Journal of General Physiology* 137 (6): 493–505. https://doi.org/10.1085/jgp.201110615.

de Wijk, R.A. and Prinz, J.F. (2007). Fatty versus creamy sensations for custard desserts, white sauces, and mayonnaises. *Food Quality and Preference* 18 (4): 641–650. https://doi.org/10.1016/j.foodqual.2006.09.004.

Withers, C.A., Cook, M.T., Methven, L. et al. (2013). Investigation of milk proteins binding to the oral mucosa. *Food & Function* 4: 1668–1674.

Withers, C.A., Lewis, M.J., Gosney, M.A., and Methven, L. (2014). Potential sources of mouth drying in beverages fortified with dairy proteins: a comparison of casein- and whey-rich ingredients. *Journal of Dairy Science* 97: 1233–1247.

Xiong, R., Meullenet, J.F., Hankins, J.A., and Chung, W.K. (2002). Relationship between sensory and instrumental hardness of commercial cheeses. *Journal of Food Science* 67 (2): 877–883. https://doi.org/10.1111/j.1365-2621.2002.tb10693.x.

Yamaguchi, S., Yoshikawa, T., Ikeda, S., and Ninomiya, T. (1971). Measurement of relative taste intensity of some L-alpha-amino acids and 5′-nucleotides. *Journal of Food Science* 36 (6): 846–849. https://doi.org/10.1111/j.1365-2621.1971.tb15541.x.

Yaron, A. and Naider, F. (1993). Proline-dependent structural and biological properties of peptides and proteins. *CRC Critical Reviews in Biochemistry* 28: 31–81.

Ye, Q., Heck, G.L., and Desimone, J.A. (1993). Voltage-dependence of the rat chorda tympani response to Na+ salts – implications for the functional-organization of taste receptor-cells. *Journal of Neurophysiology* 70 (1): 167–178. <Go to ISI>://WOS:A1993LM76000014.

Ye, A., Streicher, C., and Singh, H. (2011). Interactions between whey proteins and salivary proteins as related to astringency of whey protein beverages at low pH. *Journal of Dairy Science* 94: 5842–5850.

Ye, A., Zheng, T., Ye, J.Z., and Singh, H. (2012). Potential role of the binding of whey proteins to human buccal cells on the perception of astringency in whey protein beverages. *Physiology and Behaviour* 106: 645–650.

Yu, T., Shah, B.P., Hansen, D.R. et al. (2012). Activation of oral trigeminal neurons by fatty acids is dependent upon intracellular calcium. *Pflügers Archiv/European Journal of Physiology* 464 (2): 227–237. https://doi.org/10.1007/s00424-012-1116-9.

Zarzo, M. (2007). The sense of smell: molecular basis of odorant recognition. *Biological Reviews* 82 (3): 455–479. https://doi.org/10.1111/j.1469-185X.2007.00019.x.

Zhang, J., Jin, H., Zhang, W.Y. et al. (2019). Sour sensing from the tongue to the brain. *Cell* 179 (2): 392–492. https://doi.org/10.1016/j.cell.2019.08.031.

Zhao, C. and Ganzle, M. (2016). Synthesis of taste-active gamma-Glutamyl dipeptides during sourdough fermentation by *Lactobacillus reuteri*. *Journal of Agricultural and Food Chemistry* 64 (40): 7561–7568. https://doi.org/10.1021/acs.jafc.6b02298.

Zhao, C., Schieber, A., and Ganzle, M. (2016). Formation of taste-active amino acids, amino acid derivatives and peptides in food fermentations – a review. *Food Research International* 89: 39–47. https://doi.org/10.1016/j.foodres.2016.08.042.

Zhou, X., Shen, Y., Parker, J.K. et al. (2016). Relative effects of sensory modalities and importance of fatty acid sensitivity on fat perception in a real food model. *Chemosensory Perception* 9 (3): 105–119. https://doi.org/10.1007/s12078-016-9211-5.

Zhou, X., Yeomans, M.R., Thomas, A. et al. (2021). Individual differences in oral tactile sensitivity and gustatory fatty acid sensitivity and their relationship with fungiform papillae density, mouth behaviour and texture perception of a food model varying in fat. *Food Quality and Preference* 90: https://doi.org/10.1016/j.foodqual.2020.104116.

Zhu, X., Sun-Waterhouse, D., Chen, J. et al. (2021). Comparative study on the novel umami-active peptides of the whole soybeans and the defatted soybeans fermented soy sauce. *Journal of the Science of Food and Agriculture* 101 (1): 158–166. https://doi.org/10.1002/jsfa.10626.

3 Sensory Data Analysis and Future Developments

Matthew McSweeney

3.1 Introduction

Sensory properties are an important characteristic of dairy products. From the raw materials to the finished products, the sensory properties need to be evaluated. In some cases, a measurement of product quality or consistency is all that is required. However, for the majority of products and particularly for research, more detailed information is needed. Sensory analysis allows food product developers to make educated decisions and evaluate the sensory properties of new food products. Additionally, food product developers need a clear understanding of which method or technique should be used concerning the objective of the testing, experimental design, validity and reliability (Tuorila and Monteleone 2009).

Sensory analysis and its application proliferated in the second half of the twentieth century, as did the production of processed food and consumer products (Lawless and Heymann 2013). Sensory analysis is a multidisciplinary science structured in the scientific principle related to different areas of knowledge, including food science, psychology, human physiology, statistics, sociology and knowledge about product preparation practices (Stone and Sidel 2004). It has been defined as a scientific method used to evoke, measure, analyse and interpret those responses to food products as perceived through the senses of sight, smell, touch, taste and hearing (Stone and Sidel 2004). Specific methods have been developed to accurately and reproducibly evaluate the sensory properties of food products. A trained panel or descriptive analysis can be used to generate results that are correlated to instrumental methods. Rapid sensory methods (projective mapping [PM], check-all-that-apply [CATA]) can gather information about food items quickly and efficiently, directly from their consumers (Varela and Ares 2012). Furthermore, sensory testing can investigate consumer perceptions, including acceptability, and investigate the effect of the storage environment and packaging, ingredient substitution and process variability. Each of these specific methods has its statistical analysis techniques as well. To properly utilise sensory analysis, an understanding of both the sensory methodology and statistical analysis techniques are needed.

Sensory Profiling of Dairy Products, First Edition. Edited by John J. Tuohy.
© 2023 John Wiley & Sons Ltd. Published 2023 by John Wiley & Sons Ltd.

3.2 Scoring methods

3.2.1 *Affective testing*

Affective testing, also called acceptance testing, asks consumers to identify if they like a product or not. The most common method used to assess the acceptability of food items is hedonic scales structured with 5, 7 or 9 points (Lévy and Köster 1999). An example of the 9-point hedonic scale is shown in Table 3.1. These tests' objective is to determine or infer the likes and dislikes of consumers (Cruz et al. 2010). Results from the hedonic scales can be used for many other purposes, including being used in preference mapping (Greenhoff and MacFie 1994). Preference mapping allows for the visualisation of the directions for product preferences in spatial models of a product set (Lawless and Heymann 2013). The products are represented by points in space and products that are perceived to be similar by the consumers are positioned close together (Lawless and Heymann 2013). Preferences of individual consumers can be projected as vectors on the spatial model to show directions of increased liking and present directions for product optimisation (Lawless and Heymann 2013). It is suggested that acceptance tests include at least 50 consumers, but most studies recruit over 100 consumers (Cardello and Jaeger 2010). In acceptability tests, there are two important aspects that need to be answered: how the participants perceive the sensory characteristics of the food item and how much the variation in the sensory characteristic perceived influenced the participants' responses (Villegas et al. 2008).

A good example of acceptance testing using 9-point hedonic scales is seen in the study by Souza et al. (2019) investigating two strawberry flavoured whey beverages. The consumers (n = 100) evaluated their liking of the appearance, aroma, flavour, texture and overall liking on a 9-point hedonic scale (1 = disliked extremely and 9 = liked extremely). The results were then evaluated using an analysis of variance (ANOVA) and Tukey's honestly significantly different (HSD) test. The ANOVA provides a global analysis of the overall variance and evaluates whether the variance of one or more data sets differs. The ANOVA does not assess which variables differ (Granato et al. 2014). To determine if the variables differ, a posthoc test must be conducted. In the study earlier in this chapter (Souza et al. 2019), the authors used a Tukey's HSD test to determine if the means differed statistically.

Other posthoc tests can be used, but it is important to determine which posthoc test will be used in advance so that bias is not attributed to any one set of data (Granato et al. 2014). Other posthoc tests include Duncan's multiple range test (MRT) and Fisher's least significant difference (LSD). The Tukey HSD is a single-step multiple comparison test that is generally used in conjunction with an ANOVA. It compares all possible pairs of means and is a very useful test for multiple comparisons (Granato

Table 3.1 Number representation of the 9-point hedonic scale.

el	2	3	4	5	6	7	8	9
Dislike extremely	Dislike very much	Dislike moderately	Dislike slightly	Neither like nor dislike	Like slightly	Like moderately	Like very much	Like extremely

et al. 2014). Unfortunately, the method is not statistically robust as it requires the means to follow a normal distribution (Nunes et al. 2015). However, many researchers choose to use it to avoid Type II errors (false negative).

Fisher's LSD test is used when the sample sizes are small, and the residues' distribution is normal (Granato et al. 2014). Fisher's LSD test is much more robust than Tukey's HSD but is the most sensitive to Type I errors (false positives) (Granato et al. 2014). Duncan's MRT does not need the data to conform strictly to a normal distribution, but similar to Fisher's LSD test; it is prone to Type I errors. Another test to consider is the Tukey–Kramer HSD, which is more conservative than the original Tukey HSD test and can be used with unequal group sizes. It is stricter than other tests, and it is not prone to Type I errors (Keppel 1991).

3.2.2 *Cluster analysis*

Another statistical analysis method that can be applied to affective testing is cluster analysis. Cluster analysis aims to group (people or products) into homogenous subsets. In terms of preference mapping, the goal is to understand consumer liking drivers and summarise the differences among consumers in their likes and dislikes (MacFie 2007). There are two common types of clustering methods: hierarchical and partitioning methods (Piggott 1986). The most common form is hierarchical clustering; the analysis begins by assigning each object its own cluster. Then, agglomeration starts, and the most similar objects are combined to form a new cluster and then followed by successively less similar objects. At each step in the hierarchy, the number of clusters decreases until all objects are included in one cluster. Hierarchical clustering produces a tree-like graph, referred to as a 'dendrogram'. The major limbs of the dendrogram identify the major object groups (Serrano-Megías and López-Nicolás 2006).

In the partitioning or criterion approach, the researcher or investigator specifics the number of clusters to be created (Dahl and Næs 2004). Initially, the objects are assigned to the clusters on a random basis or based on prior knowledge. Then the algorithm reassigns each object to clusters until the within-cluster homogeneity cannot be improved. This cluster analysis can be repeated based on the number of clusters of interest to the researcher/investigator.

Researchers can also combine cluster analysis and affective testing results with intensity ratings (attribute diagnostics) and descriptive analysis (both methods are discussed later in this chapter). A broad spectrum of chocolate milks (n = 28) was evaluated using descriptive analysis and consumer testing (Thompson et al. 2004). Thirteen of the chocolate milks evaluated in the descriptive analysis trial were also evaluated in consumer testing (n = 120). The consumers were asked to evaluate each milk sample for the overall liking, overall flavour liking, overall thickness/mouthfeel liking, and the perceived intensity and liking on a 9-point hedonic scale with 9 = 'high intensity/like extremely' and 1 = 'low intensity/dislike extremely'.

The results of the descriptive analysis and consumer testing were analysed individually and together. An ANOVA and principal component analysis (PCA) were used to analyse the descriptive analysis results (discussed in Section 3.2.4). Two techniques were then employed to classify the consumers and identify drivers of chocolate milks' acceptance. Firstly, a two-step cluster analysis (TCA) followed by generalised Procrustes analysis (GPA) was performed. The TCA/GPA was completed by segmenting consumers using the results of TCA's acceptance scores, and then the overall acceptance profiles of segments were identified and used to select the eight most significantly different samples. The eight milks' raw acceptance scores within each segment were subjected to GPA with the attribute means from descriptive analysis (Thompson et al. 2004). GPA is a technique based on a standardising profile concerning rotation, reflection, isotropic scaling and translation, to find a so-called 'consensus configuration' or 'better average' (Dahl and Næs 2004).

Thompson et al. (2004) also used landscape segmentation analysis (LSA), followed by TCA. LSA unfolds the acceptance space and estimates the individual ideal locations in the space. Vectors of the descriptive attributes are then overlaid on the acceptance space with significant linear correlations between the orders of samples projected on the vectors and the attribute-mean orders generated by the descriptive analysis to determine drivers of acceptance. The TCA then segments consumers on their ideal points estimated by LSA. The authors concluded that both LSA/TCA and TCA/PA revealed complementary conclusions, but LSA/TCA was able to identify drivers of acceptance with a simpler format (two dimensions) than TCA/GPA (four dimensions).

3.2.3 *Just about right*

The just about right (JAR) scale has been included in sensory test and market research questionnaires for decades. It is usually used in conjunction with liking or hedonic scales (Gacula et al. 2007). The JAR scale is bipolar, with both ends anchored by opposite sensory descriptions, and the middle category is anchored with the phrase 'just about right', denoting the most acceptable (Gacula et al. 2007). The JAR scale can have three, five, seven or nine categories; however, the middle category needs to denote the most acceptable or 'just about right' response. The end-anchors are usually 'too little' or 'too much' of a specific attribute. The JAR scales give information that can be diagnostic or explanatory if the food item is not acceptable (Lawless and Heymann 2013). However, there are limitations to the JAR scale, and a useful review of them can be found in van Trijp et al. (2007).

Zhi et al. (2016) evaluated flavoured liquid milk using hedonic scales and JAR scales. The participants (n = 656) evaluated the flavoured liquid milk for overall acceptability on a 5-point hedonic scale (1 = 'dislike very much', 5 = 'like very much') and the level of suitability of sourness, sweetness, fresh milk flavour, thickness and smoothness using 5-point JAR scales (1 = much too little, 3 = just about right, 5 = much too much). Penalty analysis (or mean drop analysis) combined the overall liking scores and the results from the JAR scale to estimate the impact of the attribute not being just-right and how this may detract from the overall acceptance of the product. The penalty analysis consisted of three steps: (i) Collapse the 5-point JAR scale into three categories (too little, JAR and too much) for each attribute; (ii) Calculate the mean overall liking score from the hedonic scale, as well as the percentage of the respondents represented in each of the three categories and (iii) Calculate the mean drops by subtracting the mean overall liking score for the JAR group from the mean overall liking score of the 'too little' or 'too much' groups (Rothman 2007).

Balthazar et al. (2015) also applied JAR scales, hedonic liking and penalty analysis to assess ovine milk yoghurt with inulin addition. The participants (n = 100) evaluated the acceptability of the appearance, aroma, taste, texture and overall impression on 9-point hedonic scales (1 = dislike extremely, 9 = like extremely). They also evaluated the consistency, acid taste, sweet taste, milk flavour and salty taste on 9-point JAR scales (1 = extremely less, 5 = just about right, 9 = extremely more). The acceptance data were grouped by PCA (discussed later in this chapter). In both studies, penalty analysis was able to determine whether a certain attribute was significantly related to different results in the acceptability scores.

3.2.4 *Intensity scales*

In conjunction with hedonic scales, researchers and investigators can use both category scales and line scales to identify the intensity of different sensory properties. The idea is to present an easily understandable term, for instance, sourness, and ask the participant to evaluate the attribute's perceived intensity (Lawless and Heymann 2013). The low and the high ends of the scale must be labelled as well. Various scales range from nine points on the category scale to a 15 cm line scale to an unlabelled check-box scale.

The line scales involved the participants making a mark or a slash on a line to indicate the attribute's intensity (Lawless and Heymann 2013). A line scale is also called a graphic-rating scale or a visual analogue scale. Once the participant has made their evaluation on the line scale, the response is recorded as the mark's distance from one end of the scale (usually, the lower end) (Lawless and Heymann 2013). Line marking differs from the category scale as the participants' choices are continuous and offer many more options than the category scales. When using line scales, usually only the endpoints are labelled, and the end-anchors are usually indented to avoid endpoint biases from the participants.

In a study on Minas Frescal cheese, a wide variety of different statistical analyses was utilised to evaluate the results from intensity scales and a hedonic scale (Rocha et al. 2020). The study asked 100 consumers to evaluate the cheese for different attributes on 9 cm unstructured scales (Rocha et al. 2020). The consumers also evaluated their overall liking of the cheese using a 9-point hedonic scale (1 = disliked very much, 9 = liked very much). These scales allowed several different data analyses to be completed. Firstly, the results were analysed using an ANOVA and a Fisher LSD test. A PCA was performed using a correlation matrix with ellipse confidence generated using the bootstrap technique.

PCA is the most widely used multivariate statistical tool in sensory analysis (Borgognone et al. 2001). It is a type of factor analysis that creates a simplified view of a multidimensional data set. The data is reduced to a smaller set of underlying factors based on the original variables' correlations (Lawless and Heymann 2013). PCA uses a linear transformation to generate a set of uncorrelated principal components, and each principal component is a linear combination of the original variables. The first principal component explains the largest variation in the data; the next greatest is assigned to the second principal component and continues like that. When the PCA is applied to the acceptability studies, the data consists of a sample (rows) by consumers (columns) matrix, and the result is called internal preference mapping (Greenhoff and MacFie 1994). PCA can also be applied to descriptive analysis, and the input data, in that case, would be a sample (rows) by descriptor (columns) matrix. It is usually constructed from the mean values over judges (Borgognone et al. 2001). The PCA's main objective is to explain the original data's variability with as few principal components as possible (Borgognone et al. 2001). PCA calculations can be performed using many different statistical software packages, and there is the option of using the covariance matrix or the correlation matrix (as seen in Rocha et al. (2020)). PCA is used to provide a visualisation of the relationship among samples and the sensory attributes. In the study by Rocha et al. (2020), the PCA explained 78.38% of the variance and visualised that the samples were well-separated and well-discriminated products.

3.3 Descriptive analysis

Descriptive analysis is one of the most sophisticated tests used in sensory evaluation. It can provide complete descriptions of a food item, identify similarities and differences between products, and be paired with acceptance testing to determine which sensory properties correlate to consumer liking. The panellists are asked to quantify the qualitative aspects (aroma, appearance, flavour, texture, aftertaste, sound properties) of the food item (Murray et al. 2001). Descriptive analysis panels provide an objective description of the products being tested and the results can be compared to instrumental measurements (Murray et al. 2001). Furthermore, descriptive analysis can be used for quality control, as well as product matching and can investigate if changes in ingredients, processing or packaging affect the sensory perception of the manufactured products.

Generally, descriptive analysis requires between 8 and 12 panellists to be trained using reference standards to evaluate a food product's specific sensory properties. These panellists need to understand the definitions of the sensory properties to be tested. They evaluate the products for these properties on a quantitative scale for intensity, and then these results are statistically analysed. The panellists are not asked questions about the acceptability of these products or their hedonic responses to these food items. The panellists need to be screened to evaluate their health status, allergies, availability, personality, motivation, dietary habits and previous experiences (Murray et al. 2001). Once the panellists have been selected, they must be trained, and a common language must be developed. The first step is attribute generation, and the panellists are exposed to a wide number of products in the category and asked to describe them (Lawless and Heymann 2013). The panellists then choose the attributes, and then a common 'frame of reference' is created. To create this 'frame of reference', many researchers have recommended using reference standards (Civille and Lawless 1986; Murray and Delahunty 2000; Nielsen et al. 1998). Reference standards can include any chemical, ingredient, spice or product (Rainey 1986).

The judgement consistency can be evaluated by serving the panellists samples in triplicate and then analysing the results. The sensory scientist or investigator should study the significance level of the interaction effects associated with the panellists (Lawless and Heymann 2013). In a well-trained panel, these effects would not be significantly different among judges (Lawless and Heymann 2013). Additionally, feedback calibration (graphical computerised feedback on performance) can be used, which has led to a reduction in training time and excellent panel performance. A three-way ANOVA with main effects (product, panellist and replication) and their interaction effects (panellist by product, panellist by replication and product by replication) can be used to determine if there are any issues with panellist performance (Lawless and Heymann 2013). Also, a PCA of each attribute will indicate the agreement among the panellists for that attribute. If there is a high level of agreement between the panellists, then the first dimension should explain most of the variance (50–70%) (Lawless and Heymann 2013).

Once results from the panellists are consistent, then they move on to evaluating the samples. Standard sensory practices should be applied (randomised serving sequences, individual booths, sample coding) (Lawless and Heymann 2013). The panellists should evaluate samples in triplicate. Samples are served monadically, and all attributes for a specific sample are evaluated before the panellists move on to the next sample (Lawless and Heymann 2013).

There are several different methods of descriptive analysis, including flavour profile method, spectrum method, texture profile method, free choice profiling and quantitative descriptive analysis. An explanation of these methods can be found in Lawless and Heymann (2013). Additionally, generic descriptive analysis (GDA) is frequently used in practical applications, which combines approaches from all of these methods. Usually, an ANOVA analyses the results, but multivariate techniques can be used as well. Multivariate techniques aim to present information from the product-attribute matrix (Lawless and Heymann 2013). These techniques can include PCA, partial least squares regression (PLSR), multivariate ANOVA, canonical variate analysis, GPA and preference mapping. PCA and GPA are discussed earlier in this chapter, but other examples of multivariate techniques are presented later in this chapter.

Trained panel results can allow for better product understanding and understanding of the consumer. Researchers can combine the affective test results (discussed earlier in this chapter) and the trained panel's results. The combination of these two sensory tests allows us to understand which attributes are driving consumer acceptability. The two sets of data are combined in a multivariate statistical technique, generally called preference mapping. To complete preference mapping, generally, six products are considered to be the minimum number of products. This method allows the researcher to identify what consumers like and dislike about the products. Preference mapping can be performed either by internal or external analysis. In internal preference mapping, the map is based on the acceptance results from the consumers. External preference mapping uses the results from an external source (descriptive analysis) to create the preference map (called the product space), and then the consumers' acceptance results are projected into this space through polynomial regression (Lawless and Heymann 2013). The internal preference map can produce good information for marketing and new product creation, while external preference mapping is important for technological developments (Van Kleef et al. 2006). Good examples of preference mapping can be found in Dooley et al. (2010), Lovely and Meullenet (2009) and Young et al. (2004).

The results from a trained panel can also be compared to instrumental measurements. To ensure a relationship between the trained panel and the instrument's results, the researcher must ensure that the attributes in the sensory panel are well defined.

Clear definitions and reference standards for the attributes facilitate a comparison between the trained panel's results and instrumental measurements. A study by Panouillé et al. (2011) aimed to better understand salt release and perception in model dairy products. Both sensory and instrumental methods were used to quantify and relate texture to salt mobility and perception. The eight model dairy products were evaluated by a quantitative descriptive analysis panel and instrumental measures, including a small-amplitude oscillation test, texture profile analysis, and a compression test. The trained panel results were related to the instrumental data using a PLSR. PLSR is a method that can relate one data matrix to another data matrix via linear low-rank regression modelling. In this study, Panouillé et al. (2011) identified that the salty perception was related to rheological data. The rheological data was assigned to be the X-variables, and the sensory data set was designated as the Y-variables in the PLSR analysis. The authors identified that the more gelled the sample, the less the perception of saltiness.

3.4 Rapid sensory analysis

Rapid sensory analysis, rapid sensory profiling or novel sensory methods result from sensory scientists needing methods that are not as expensive, time-consuming or resource-heavy as descriptive analysis. They are more rapid, more flexible and agile compared to sensory description in terms of training and time requirements (Varela and Ares 2012). Rapid sensory analysis methods are well suited to gather information about food items quickly and efficiently, directly from consumers. The methods include flash profiling (Dairou and Sieffermann 2002), napping® or PM (Risvik et al. 1994), sorting (Lawless et al. 1995) and CATA (Adams et al. 2007). This chapter will discuss two of the most common methods, CATA and PM.

3.4.1 *Check-all-that-apply (CATA)*

CATA can be used to determine which sensory attributes affect the acceptability of a product. CATA asks participants to select from a set of attributes perceived in the sample. This method can be used to ask participants about sensory attributes and product positioning, emotions and usage occasions. A CATA question is easy for participants to understand and requires minimal instruction (Lado et al. 2010). The data obtained from a CATA questionnaire is multivariate binary, meaning an individual observation is ambiguous (Meyners and Castura 2014). Each participant makes a subjective decision regarding each descriptor's applicability for each food item being evaluated (Meyners and Castura 2014). The participant perceives a checked descriptor and unchecked descriptor, indicating that it was or was not perceived by the participant, but it is considered inappropriate to describe the sample (Meyners and Castura 2014).

A CATA study should consider using approximately 100 consumers and a full cross-over design where each participant evaluates each sample in a sequential monadic presentation (Meyners and Castura 2014). It is essential to consider the number of attributes included in the questionnaire and the attribute position in the CATA list. The CATA questionnaire's attributes can be based on studies in the literature, the result of a focus group or the result of other sensory trials, including PM and descriptive analysis. It is recommended to randomise the attributes based on the participant to avoid bias (Meyners and Castura 2016). There are many different approaches to evaluating the data collected from a CATA questionnaire, but most of them involve contingency tables, bar charts, significance testing and correspondence analysis (Meyners and Castura 2014). Multifactor analysis, multidimensional analysis and multidimensional scaling can also be used (Meyners and Castura 2014). If the acceptability data is also collected, then penalty lift analysis can also be conducted. An example of the results from a penalty lift analysis can be seen in Figure 3.1. In the example, participants (n = 101) evaluated five different cheddar cheeses for their overall liking (9-point hedonic scale) and answered a CATA question consisting of sensory attributes.

Galli et al. (2019) used CATA, a hedonic scale and penalty analysis to evaluate Camembert-type cheese. The trial involved 80 participants, and the CATA was composed of terms from a flash profile task. Before evaluating the cheese, the participants were asked to complete a CATA questionnaire for their ideal Camembert-type cheese. The participants evaluated their overall liking of the cheese samples on a 9-point scale and then answered the CATA questionnaire. The results were then subjected to a penalty analysis. The ideal Camembert-type cheese was described as having a characteristic taste and aroma, creaminess, ammoniacal aroma, bitter taste, white colour, shiny and uniform texture. The penalty analysis identified the attributes, shiny appearance and ammoniacal taste as necessary for Camembert-type cheese.

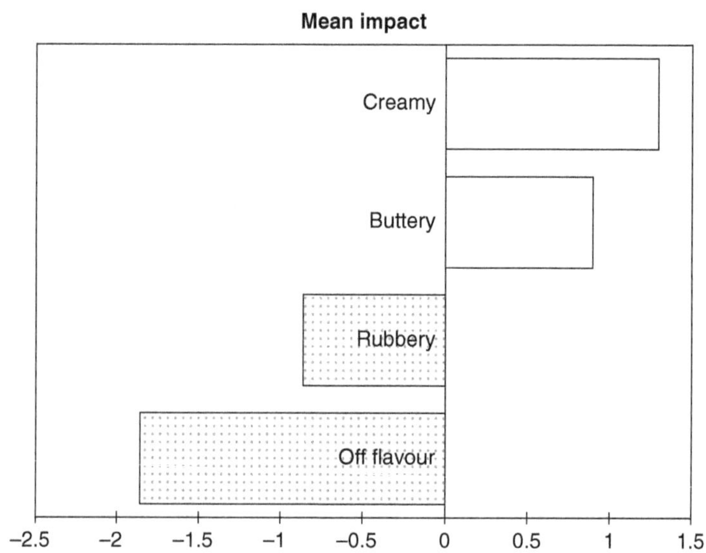

Figure 3.1 Penalty lift analysis of the sensory attributes (CATA question) and overall liking based on the participants' evaluation of five different commercially produced cheddar cheeses (data from unpublished study by author).

3.4.2 *Projective mapping*

PM is a method of assessing and describing differences among samples. The samples are presented to the study participants, who are then required to place samples on a two-dimensional space in a way that reflects their perception of the samples' differences. Thus, participants are to place similar samples closer together, and those that are different are placed further apart (Risvik et al. 1994). It is a method in which consumers can assess samples comprehensively and straightforwardly, resulting in a graphical representation of the product's relationships as perceived by the participants (Varela and Ares 2012). PM is often used with ultra-flash profiling (UFP) to provide descriptions of each sample. UFP asks the panellists to identify their own set of attributes. Participants list adjectives next to each sample after the PM task.

For data analysis, the coordinates of the products' location for each participant are measured, and the attributes given (UFP) are counted across the consumers (Varela and Ares 2012). The data is structured into three tables (example of a data table is shown in Table 3.2) and then, as proposed by Pagès (2005), it can be analysed using multiple-factor analysis (MFA). MFA is based on PCA, and it is used to analyse multiple sets of data about the same set of products. During the first step of an MFA, a PCA is performed on a group of variables, and then the groups are normalised by dividing all the variables within a group by the first eigenvalue for that group. After all of the variables are combined into a single data set, and global PCA is completed, MFA can combine a group of quantitative variables (the samples' placement on the biplot in the PM task) with qualitative variables (the attributes identified by the participants).

Esmerino et al. (2017) compared eight concepts of fermented dairy beverages. The participants (n = 50) were asked to use PM and UFP on an A4 white sheet of paper (210×297 mm). Data was then obtained from the Cartesian coordinates X and Y of each concept for each participant. The lower left was considered the origin of the coordinate system. MFA was then conducted to evaluate the results. The authors found that PM was an effective tool to evaluate dairy products, especially for dairy companies with limited financial and human resources.

Table 3.2 Data table based on the results from a projective mapping and ultra-flash profile task. After structuring the results in this table, the results can be evaluated using multiple-factor analysis (MFA).

Sample	X1	Y1	X2	Y2	...	Xn	Yn	Descriptor 1	Descriptor 2	...	Descriptor n
S1											
S2											
S3											
...											
Sn											
Coordinates from participants								Frequency of attributes listed by participants			

3.5 Conclusions

Sensory analysis allows sensory scientists and food product developers to make educated decisions and evaluate sensory properties of products. It includes various methodologies, from acceptance testing to descriptive analysis, to rapid sensory analysis methods. It is important to select the appropriate method and statistical treatment to create meaningful and reproducible results. Appropriate sensory analysis techniques allow for consumer understanding, the relationship to chemical and instrumental analysis, and knowledge of product variability and stability.

References

Adams, J., Williams, A., Lancaster, B., and Foley, M. (2007). Advantages and uses of check-all-that-apply response compared to traditional scaling of attributes for salty snacks. *7th Pangborn Sensory Science Symposium. Minneapolis*, USA (12–16 August 2007).

Balthazar, C.F., Gaze, L.V., da Silva, H.L. et al. (2015). Sensory evaluation of ovine milk yoghurt with inulin addition. *International Journal of Dairy Technology* 68 (2): 281–290. https://doi.org/10.1111/1471-0307.12189.

Borgognone, M.G., Bussi, J., and Hough, G. (2001). Principal component analysis in sensory analysis: covariance or correlation matrix? *Food Quality and Preference* 12 (5): 323–326. https://doi.org/10.1016/S0950-3293(01)00017-9.

Cardello, A.V. and Jaeger, S.R. (2010). 7 – Hedonic measurement for product development: new methods for direct and indirect scaling. In: *Consumer-Driven Innovation in Food and Personal Care Products* (ed. S.R. Jaeger and H. MacFie), 135–174. Woodhead Publishing https://doi.org/10.1533/9781845699970.2.135.

Civille, G.V. and Lawless, H.T. (1986). The importance of language in describing perceptions. *Journal of Sensory Studies* 1 (3–4): 203–215. https://doi.org/10.1111/j.1745-459X.1986.tb00174.x.

Cruz, A.G., Cadena, R.S., Walter, E.H.M. et al. (2010). Sensory analysis: relevance for prebiotic, probiotic, and synbiotic product development. *Comprehensive Reviews in Food Science and Food Safety* 9 (4): 358–373. https://doi.org/10.1111/j.1541-4337.2010.00115.x.

Dahl, T. and Næs, T. (2004). Outlier and group detection in sensory panels using hierarchical cluster analysis with the Procrustes distance. *Food Quality and Preference* 15 (3): 195–208. https://doi.org/10.1016/S0950-3293(03)00058-2.

Dairou, V. and Sieffermann, J.M. (2002). A comparison of 14 jams characterized by conventional profile and a quick original method, the flash profile. *Journal of Food Science* 67 (2): 826–834. https://doi.org/10.1111/j.1365-2621.2002.tb10685.x.

Dooley, L., Lee, Y., and Meullenet, J.F. (2010). The application of check-all-that-apply (CATA) consumer profiling to preference mapping of vanilla ice cream and its comparison to classical external preference mapping. *Food Quality and Preference* 21 (4): 394–401. https://doi.org/10.1016/j.foodqual.2009.10.002.

Esmerino, E.A., Ferraz, J.P., Filho, E.R.T. et al. (2017). Consumers' perceptions toward 3 different fermented dairy products: insights from focus groups, word association, and projective mapping. *Journal of Dairy Science* 100 (11): 8849–8860. https://doi.org/10.3168/jds.2016-12533.

Gacula, M., Rutenbeck, S., Pollack, L. et al. (2007). The just-about-right intensity scale: functional analyses and relation to hedonics. *Journal of Sensory Studies* 22 (2): 194–211. https://doi.org/10.1111/j.1745-459X.2007.00102.x.

Galli, B.D., Baptista, D.P., Cavalheiro, F.G., and Gigante, M.L. (2019). *Lactobacillus rhamnosus* GG improves the sensorial profile of Camembert-type cheese: an approach through flash-profile and CATA. *LWT* 107: 72–78. https://doi.org/10.1016/j.lwt.2019.02.077.

Granato, D., de Araújo Calado, V.M., and Jarvis, B. (2014). Observations on the use of statistical methods in Food Science and Technology. *Food Research International* 55: 137–149. https://doi.org/10.1016/j.foodres.2013.10.024.

Greenhoff, K. and MacFie, H.J.H. (1994). Preference mapping in practice. In: *Measurement of Food Preferences* (ed. H.J.H. MacFie and D.M.H. Thomson), 137–166. Springer US https://doi.org/10.1007/978-1-4615-2171-6_6.

Keppel, G. (1991). *Design and Analysis: A researcher's Handbook*, 3e (pp. xiii, 594. Prentice-Hall, Inc.

Lado, J., Vicente, E., Manzzioni, A., and Ares, G. (2010). Application of a check-all-that-apply question for the evaluation of strawberry cultivars from a breeding program. *Journal of the Science of Food and Agriculture* 90 (13): 2268–2275. https://doi.org/10.1002/jsfa.4081.

Lawless, H.T. and Heymann, H. (2013). *Sensory Evaluation of Food: Principles and Practices.* Springer Science & Business Media.

Lawless, H.T., Sheng, N., and Knoops, S.S.C.P. (1995). Multidimensional scaling of sorting data applied to cheese perception. *Food Quality and Preference* 6 (2): 91–98. https://doi.org/10.1016/0950-3293(95)98553-U.

Lévy, C.M. and Köster, E.P. (1999). The relevance of initial hedonic judgements in the prediction of subtle food choices. *Food Quality and Preference* 10 (3): 185–200. https://doi.org/10.1016/S0950-3293(99)00016-6.

Lovely, C. and Meullenet, J.-F. (2009). Comparison of preference mapping techniques for the optimization of strawberry yogurt. *Journal of Sensory Studies* 24 (4): 457–478. https://doi.org/10.1111/j.1745-459X.2009.00221.x.

MacFie, H. (2007). 23—Preference mapping and food product development. In: *Consumer-Led Food Product Development* (ed. H. MacFie), 551–592. Woodhead Publishing https://doi.org/10.1533/9781845693381.3.551.

Meyners, M. and Castura, J. (2014). Check-all-that-apply questions. In: *Novel Techniques in Sensory Characterization and Consumer Profiling* (ed. G. Ares and P. Varela), 271–305. CRC Press https://doi.org/10.1201/b16853-12.

Meyners, M. and Castura, J.C. (2016). Randomization of CATA attributes: should attribute lists be allocated to assessors or to samples? *Food Quality and Preference* 48: 210–215. https://doi.org/10.1016/j.foodqual.2015.09.014.

Murray, J.M. and Delahunty, C.M. (2000). Selection of standards to reference terms in a Cheddar-type cheese flavor language. *Journal of Sensory Studies* 15 (2): 179–199. https://doi.org/10.1111/j.1745-459X.2000.tb00264.x.

Murray, J.M., Delahunty, C.M., and Baxter, I.A. (2001). Descriptive sensory analysis: past, present and future. *Food Research International* 34 (6): 461–471. https://doi.org/10.1016/S0963-9969(01)00070-9.

Nielsen, R.G., Zannoni, M., Lavanchy, P. et al. (1998). Progress in developing an international protocol for sensory profiling of hard cheese. *International Journal of Dairy Technology* 51 (2): 57–64. https://doi.org/10.1111/j.1471-0307.1998.tb02510.x.

Nunes, C.A., Alvarenga, V.O., de Souza Sant'Ana, A. et al. (2015). The use of statistical software in food science and technology: advantages, limitations and misuses. *Food Research International* 75: 270–280. https://doi.org/10.1016/j.foodres.2015.06.011.

Pagès, J. (2005). Collection and analysis of perceived product inter-distances using multiple factor analysis: application to the study of 10 white wines from the Loire Valley. *Food Quality and Preference* 16 (7): 642–649. https://doi.org/10.1016/j.foodqual.2005.01.006.

Panouillé, M., Saint-Eve, A., de Loubens, C. et al. (2011). Understanding of the influence of composition, structure and texture on salty perception in model dairy products. *Food Hydrocolloids* 25 (4): 716–723. https://doi.org/10.1016/j.foodhyd.2010.08.021.

Piggott, J.R. (1986). *Statistical Procedures in Food Research.* Elsevier Applied Science https://agris.fao.org/agris-search/search.do?recordID=US8908186.

Rainey, B.A. (1986). Importance of reference standards in training panelists. *Journal of Sensory Studies* 1 (2): 149–154. https://doi.org/10.1111/j.1745-459X.1986.tb00167.x.

Risvik, E., McEwan, J.A., Colwill, J.S. et al. (1994). Projective mapping: a tool for sensory analysis and consumer research. *Food Quality and Preference* 5 (4): 263–269. https://doi.org/10.1016/0950-3293(94)90051-5.

Rocha, R.S., Calvalcanti, R.N., Silva, R. et al. (2020). Consumer acceptance and sensory drivers of liking of Minas Frescal Minas cheese manufactured using milk subjected to ohmic heating: performance of machine learning methods. *LWT* 126: 109342. https://doi.org/10.1016/j.lwt.2020.109342.

Rothman, L. (2007). 17—The use of just-about-right (JAR) scales in food product development and reformulation. In: *Consumer-Led Food Product Development* (ed. H. MacFie), 407–433. Woodhead Publishing https://doi.org/10.1533/9781845693381.3.407.

Serrano-Megías, M. and López-Nicolás, J.M. (2006). Application of agglomerative hierarchical clustering to identify consumer tomato preferences: influence of physicochemical and sensory characteristics on consumer response. *Journal of the Science of Food and Agriculture* 86 (4): 493–499. https://doi.org/10.1002/jsfa.2392.

Souza, F.P., Balthazar, C.F., Guimarães, J.T. et al. (2019). The addition of xyloligoosaccharide in strawberry-flavored whey beverage. *LWT* 109: 118–122. https://doi.org/10.1016/j.lwt.2019.03.093.

Stone, H. and Sidel, J.L. (2004). 1—Introduction to sensory evaluation. In: *Sensory Evaluation Practices*, 3e (ed. H. Stone and J.L. Sidel), 1–19. Academic Press https://doi.org/10.1016/B978-012672690-9/50005-6.

Thompson, J.L., Drake, M.A., Lopetcharat, K., and Yates, M.D. (2004). Preference mapping of commercial chocolate milks. *Journal of Food Science* 69 (9): S406–S413. https://doi.org/10.1111/j.1365-2621.2004.tb09958.x.

van Trijp, H.C.M., Punter, P.H., Mickartz, F., and Kruithof, L. (2007). The quest for the ideal product: comparing different methods and approaches. *Food Quality and Preference* 18 (5): 729–740. https://doi.org/10.1016/j.foodqual.2007.01.005.

Tuorila, H. and Monteleone, E. (2009). Sensory food science in the changing society: opportunities, needs, and challenges. *Trends in Food Science & Technology* 20 (2): 54–62. https://doi.org/10.1016/j.tifs.2008.10.007.

Van Kleef, E., Trijp, H., and Luning, P. (2006). Internal versus external preference analysis: an exploratory study on end-user evaluation. *Food Quality and Preference* 17: 387–399. https://doi.org/10.1016/j.foodqual.2005.05.001.

Varela, P. and Ares, G. (2012). Sensory profiling, the blurred line between sensory and consumer science. A review of novel methods for product characterization. *Food Research International* 48 (2): 893–908. https://doi.org/10.1016/j.foodres.2012.06.037.

Villegas, B., Carbonell, I., and Costell, E. (2008). Effects of product information and consumer attitudes on responses to milk and soybean vanilla beverages. *Journal of the Science of Food and Agriculture* 88 (14): 2426–2434. https://doi.org/10.1002/jsfa.3347.

Young, N.D., Drake, M., Lopetcharat, K., and McDaniel, M.R. (2004). Preference mapping of Cheddar cheese with varying maturity levels. *Journal of Dairy Science* 87 (1): 11–19. https://doi.org/10.3168/jds.S0022-0302(04)73136-7.

Zhi, R., Zhao, L., and Shi, J. (2016). Improving the sensory quality of flavored liquid milk by engaging sensory analysis and consumer preference. *Journal of Dairy Science* 99 (7): 5305–5317. https://doi.org/10.3168/jds.2015-10612.

4 Application of Multivariate Statistical Analysis and Machine Learning to Sensory Data Analysis

Ana C.M. Pinheiro, Elenilson G. Alves Filho, Michele N. Ribeiro, Jéssica F. Rodrigues, Tatiana C. Pimentel, Lorena M.A. Silva, Sueli Rodrigues, Erick A. Esmerino and Adriano G. da Cruz

4.1 Introduction

Dairy foods are important sources of nutrients, are beneficial for gastrointestinal health, the immune system and for sensory satisfaction. Sensory analysis uses the human as a measuring instrument to assess the quality of foods and materials. Therefore, it is an essential tool for the industry in the development and quality control of products. Consumers and/or recruited assessors are selected depending on the methods/tests used. However, data analysis techniques are paramount for data validation and decision-making based on sensory investigation data. Sensory data analysis can be performed using univariate and multivariate statistical techniques and algorithms that can be automatically improved based on previous experience.

In recent years, machine learning (ML, data analysis area inserted in artificial intelligence [AI]) has been applied in sensory data analysis. ML has enabled the development of algorithms and models that can be widely used to predict sensory responses and product classification as a function of sensory response patterns. This chapter deals with the use of multivariate techniques and ML for analysing data originating from the sensory analysis of dairy products.

4.2 Multivariate analysis applied to data from sensory assessment of dairy products

Raw datasets obtained from experiments often have little significance until helpful information and relationships between factors are obtained. In some cases, univariate analysis may be insufficient to deal with sensory data due to the matrix complexity and the many compounds and attributes present in foods (Zielinski et al. 2014). Perrot et al. (2006) highlighted that a single food property (such as texture or flavour) can be related

Sensory Profiling of Dairy Products, First Edition. Edited by John J. Tuohy.
© 2023 John Wiley & Sons Ltd. Published 2023 by John Wiley & Sons Ltd.

to several sensory attributes perceived by the human brain. Moreover, it is noteworthy that depending on the product and its characteristics, there is a synergistic effect that can mask the presence of some attributes, especially between flavour and aroma characteristics, thus creating a nonlinear relationship between these factors (Noble and Ebeler 2002). Therefore, all the factors combine and form a complex relationship, which univariate methods cannot analyse (Yu et al. 2018). Multivariate analysis is, therefore, an essential tool when dealing with sensory data.

Studies using different approaches to multivariate statistical analysis are essential to understand the trends and patterns of some sensory and compositional attributes of dairy foods and to identify the parameters most relevant to consumers' purchase intent (Balthazar et al. 2021). Multivariate statistical methods study the behaviour of three or more variables simultaneously. The use of multivariate statistical methods can provide trends and patterns within multiple datasets. It is a tool that provides the means to manage and evaluate this large data volume efficiently. These methods are mainly used to find the least representative variable and eliminate it, thereby simplifying the statistical models. In this way, it is used when the number of variables becomes a problem in understanding the relationship between the various variables (Izenman 2008; Yu et al. 2018).

Multivariate statistical analyses of Big Data (BD) are considered the basis for the development of systems in AI and ML (Fuentes et al. 2020). Given the wide use of multivariate techniques in the sensory field, a list of some of the most recently published works where these techniques were applied to sensory data from dairy products is presented in Table 4.1. The studies show a range of objectives, such as product characterisation, quality control and studying consumer perception.

Table 4.1 Recent studies where multivariate techniques were used for the analysis of data from the sensory analysis of dairy products.

Objective	Sensory test	Statistical analysis	References
To determine the drivers of liking of dairy products	CATA[a] Sorting Task	• MFA[b] • MDA[c] • DISTATIS[d] • PCA[e]	Rodrigues et al. (2021a) Rodrigues et al. (2020)
To characterise the sensory profile of dairy products by sensory temporal analysis	TI[f], TDS[g]	PCA[e] PARAFAC[h]	Bemfeito et al. (2016)
To access sensory acceptance	Acceptance test by Preference mapping	• MFA[b] • DISTATIS[d]	Barbosa et al. (2017) Pereira et al. (2020)
To optimise a stevia mix in a high-protein natural yoghurt	Napping	• Mixture design and regression models	Ribeiro et al. (2020)
Study of cross-modal interactions between aroma and flavour	TI[f], TDS[g] and acceptance test	• PCA[e]	Oliveira et al. (2021)
Determine the sensory acceptance and emotional profile associated with sheep milk kefir sweetened with different sugars	Emotion and acceptance test	• PCA[e]	Larosa et al. (2021)

Table 4.1 (Continued)

Objective	Sensory test	Statistical analysis	References
Analyse children's emotional responses associated with the consumption of probiotic fermented milks	Emotional test with emoji list	• CA[i]	Cruz et al. (2021)
Segment consumers based on their overall liking of yoghurt	Consumer acceptance and purchase intent	• HCA[j]	Cruz et al. (2011)
Segment Dulce de leche based on the manufacturing process	PAE[k]	GPA[l] HCA[j]	Silva et al. (2020)
Segment consumers of Coalho cheese based on intra-cultural differences in Brazil	PAE[k]	GPA[l] MFA[b]	Soares et al. (2020)
Compare descriptive analysis and PAE[k] in the case of probiotic yoghurt sweetened with xylitol and with added prebiotics	PAE[k] and descriptive analysis	GPA[e] MFA[b]	Costa et al. (2020)
Compare descriptive analysis and CATA[a] in the case of spreadable processed cheese	CATA[a] and descriptive analysis	CA[i] MFA[b]	Torres et al. (2017)

[a] Check-all-that-apply.
[b] Multiple-factor analysis.
[c] Multidimensional alignment.
[d] Derived technique from STATIS to evaluate the similarity between the distance matrices.
[e] Principal component analysis.
[f] Time-intensity.
[g] Temporal dominance of sensations.
[h] PARAllel FACtor analysis.
[i] Correspondence analysis.
[j] Hierarchical cluster analysis.
[k] Preferred attribute elicitation.
[l] Generalised Procrustes analysis.

4.2.1 *Principal component analysis*

Principal component analysis (PCA) is widely used for sensory data analysis. Consumer preference can be assessed using PCA by applying preference mapping (Fernández-Vázquez et al. 2018). For example, Barbosa et al. (2017), Pereira et al. (2020) and Bemfeito et al. (2016) used the preference mapping technique to assess the preferences of consumers for nutritionally enriched strawberry yoghurt, low-sodium mozzarella coated with zein and essential oils, and Serra da Canastra artisanal cheeses (texture), respectively. At the same time, Oliveira et al. (2021) used PCA in sensory data of time-intensity (TI) and temporal dominance of sensations (TDSs) analyses to verify the relationship between flavours (vanilla and strawberry) and sweet taste with the aim of sugar reduction in yoghurt. Likewise, Larosa et al. (2021) used PCA to analyse the emotional profiling and sensory acceptance of sheep milk kefir sweetened with different sugars. Thus, PCA may be used for multiple objectives in sensory analysis. PARAFAC (PARAllel FACtor analysis) is a multi-variate statistical technique that can also be associated with TDS (Bemfeito et al. 2016).

4.2.2 *Correspondence analysis*

Correspondence analysis (CA) is commonly used in studies with dairy products and CATA (check-all-that-apply) methodology. For example, Cruz et al. (2021) measured children's emotional responses using a product-specific emoji list for probiotic fermented milks. The consumers were asked to select all emojis that could describe their emotional experience with the product using the CATA method. The CA map allowed the visualisation of the relationship between the different probiotic fermented milk formulations and the emojis used by consumers. Oliveira et al. (2017) used the CATA questionnaire to reformulate Minas Frescal cheese. The CA map allowed the discrimination between full-fat and low-fat cheeses. Torres et al. (2017) used rapid sensory studies with consumers, including CATA, to characterise spreadable processed cheeses. The CA map grouped the formulations according to the presence of starch and/or vegetable fat.

4.2.3 *Hierarchical cluster analysis*

Hierarchical clustering analysis (HCA) is commonly used in studies with dairy products. Cruz et al. (2011) used HCA to segment yoghurt consumers based on their overall liking. Silva et al. (2020) was able to group different Dulce de leche formulations based on the thermal processing technology used; conventional pasteurisation (72–75 °C/15 seconds) versus ohmic heating and processing parameter variation (electric field strength applied during ohmic heating). Vieira et al. (2020) were able to group different goat whey orange juice beverages based on the processing (microfiltration or pasteurisation) and processing parameters (microfiltration temperature).

4.2.4 *Generalised Procrustes analysis*

Generalised Procrustes analysis (GPA) is another multivariate technique used in sensory studies. Several studies used GPA to analyse the preferred attribute elicitation (PAE) methodology data, a descriptive sensory method that uses consumers as assessors. Silva et al. (2020) evaluated Dulce de leche submitted to ohmic heating treatment and observed that the GPA map separated the formulations based on the processing type (ohmic heating or conventional pasteurisation). Soares et al. (2020) investigated intra-cultural differences in Brazil using Coalho cheese and observed that GPA separated the samples based on the fat content. Costa et al. (2020) evaluated probiotic yoghurt sweetened with xylitol and added prebiotic components and observed that GPA could separate the formulations.

4.2.5 *Multiple-factor analysis*

Rodrigues et al. (2020) applied multiple-factor analysis (MFA) to determine the differential sensory patterns of artisanal cheeses produced in nearby regions but with different geographical indications (Canastra and Araxá). For this, consumers evaluated cheeses from both regions through a Sorting test. The data obtained were evaluated by MFA to define the characteristic sensory attributes of each region. It involved establishing a matrix where the rows were the artisanal cheeses. The columns were the consumers, allocating each product to the particular chosen group and the attributes used by the consumers to describe the artisanal cheeses. Moreover, multidimensional alignment (MDA) was also applied to investigate the association of the sensory descriptors with each sample. The authors noted that the Sorting task and consumer description were practical to characterise and define the sensory profiles of artisanal cheeses from the Canastra and Araxá regions.

MFA may also be used for CATA questionnaire responses, as performed by Rodrigues et al. (2021a). The authors observed that MFA enabled the characterisation of the sensory attributes of Prato cheeses. Soares et al. (2020) used MFA with PAE data of Coalho cheeses and observed that their fat content could separate the cheeses. Costa et al. (2020) used MFA to evaluate the similarity between data from descriptive analysis and PAE of probiotic yoghurts. The authors observed that the products were characterised similarly using both methodologies. Finally, Torres et al. (2017) used MFA to evaluate the similarity between data from descriptive analysis and CATA for spreadable processed cheese. The authors observed that the products were characterised similarly using both methodologies.

4.2.6 *DISTATIS*

Another application of multivariate analysis to sensory data is to determine the attributes that drive the sensory acceptance of the product. Rodrigues et al. (2021b) evaluated 10 samples of Dulce de leche using Preference Sorting. For data evaluation, a three-way multidimensional analysis called DISTATIS (derived from STATIS to evaluate the similarity between the DIstance matrices) was conducted with grouping or sorting of the data based on the preferences and (dis)similarities. The data was arranged in a co-occurrence matrix for each assessor in each approach. In a co-occurrence matrix, the rows and columns represent the samples, and a value of 1 between a row and a column indicates that the assessor grouped the samples. In contrast, a value of 0 indicates that the samples were not grouped. The co-occurrence matrices were transformed into distance matrices by subtracting the co-occurrence matrices from a conformable matrix filled with 1's. The distance matrices were then transformed into cross-product matrices and aggregated to create a compromise cross-product matrix, representing the assessors. MFA evaluated data obtained from bidimensional maps for each sorting approach. MFA showed a greater separation among the samples of the Sorting task, allowing the evaluation of both the attributes that drive product acceptance and comparison between the different methodological approaches proposed in the study (Rodrigues et al. 2020).

Therefore, as noted, there is a wide variety of multivariate analysis options that can be applied to dairy sensory data. Consequently, it is essential to observe the peculiarities and limitations of each one of them, the characteristics of the data set and the objectives of the study to choose the appropriate analysis to be applied.

4.3 Machine learning

Machine learning (ML) is a research area that uses statistical methods to build intelligent algorithms that can be automatically improved through experience and data usage (Sindhu et al. 2020). ML algorithms take one or more inputs (measured variables – parameters that define the problem) and generate one or more outputs representing the problem solution (Gambella et al. 2021). The primary purpose of ML is to teach algorithms from the data, and, in this way, they can make decisions with minimal human intervention. In addition, ML can 'learn by example' by analysing existing datasets (Liu et al. 2021).

We can divide ML algorithms based on three main approaches: supervised learning, unsupervised learning and reinforcement learning (Simeone 2018). Supervised and unsupervised learning differ mainly because supervised learning involves mapping input to output, while unsupervised learning does not produce an output response to the input. Reinforcement learning is from examples and needs human interaction or the environment to learn (Gouse and Helini 2019).

The unsupervised approach is often used with sensory data and applied in preprocessing, exploratory analysis, HCA or dimensionality reduction algorithms. Several researchers have used PCA, CA, MFA and HCA (also classified as multivariate statistical analyses) to assess sensory data, as demonstrated in the previous section of this chapter. Supervised learning has also been frequently used to predict an output based on input values. Furthermore, this approach is capable of predicting discrete or continuous values (classes). Figure 4.1 illustrates the classification techniques used in ML.

ML has been frequently used for calibration in chemometrics and tasks involving pattern recognition, classification and regression. Several studies demonstrate that

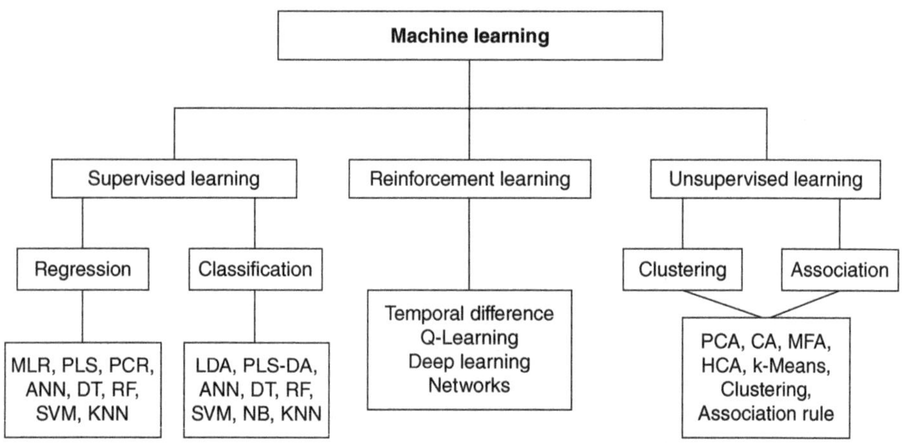

Figure 4.1 Machine learning approaches and their algorithms (**MLR:** multiple linear regression; **PLS:** partial least squares regression; **PCR:** principal component regression; **ANN:** artificial neural network; **DT:** decision tree; **RF:** random forest; **SVM:** support vector machine; **KNN:** k-nearest neighbours; **LDA:** linear discriminant analysis; **PLS-DA:** partial least squares – discriminant analysis; **NB:** naive bayes; **PCA:** principal component analysis; **CA:** correspondence analysis; **MFA:** multiple-factor analysis; **HCA:** hierarchical cluster analysis).

chemical, physical, physicochemical, biometric and neurological measurements are correlated with sensory measurements. Most ML applications have focused on developing regression and classification models from product composition data. Furthermore, ML has been used to predict sensory responses in several products, such as green tea, wine, meat, bread, coffee, fruit and beer (Yu et al. 2018). Examples of dairy product studies using ML are summarised in Table 4.2.

Some reports in the literature on dairy products (Table 4.2) propose chemical measurements to complement or replace sensory measurements based on a predictive relationship. For example, Yang et al. (2015) used the data obtained from the electronic nose sensor array to predict the intensity of goat milk flavour through the PLS algorithm. Tian et al. (2020) observed that the combination of data obtained by the electronic nose and the application of the back propagation of neural networks (BPNN) and RF (random forest) algorithms was effective in predicting yoghurt acceptability.

Table 4.2 ML applications in sensory studies of dairy products.

Objective	Sensory test	Algorithm	Reference
Predicting changes in sensory profiles during UHT milk storage	• Trained panellists • Intensity analysis of sensory attributes	ANN[a]	Singh et al. (2009)
Correlate descriptive attributes and hedonic judgements and investigate the effects of ice cream composition on the sensory profile	• Trained panellists and consumers • Quantitative descriptive analysis (QDA) and overall liking	PLS[b]	Soukoulis et al. (2010)
Classification of cheeses based on maturation times	• Electronic nose aroma analysis	ANN[a]	Cevoli et al. (2011)
Prediction of consumer acceptance and purchase intent for yoghurt	• Consumers • Overall liking using a 9-point hedonic scale and purchase intent	ANN[a]	Oliveira et al. (2021)
Prediction of consumer acceptance of ice cream	• Trained panellists • Sensory evaluations of appearance, flavour, body and texture, and global acceptance were performed using the 9-point scale	ANN[a]	Bahramparvar et al. (2014)
Prediction of yoghurt flavour acceptance from electronic nose data	• Electronic nose flavour analysis	Nonlinear chemometric BPNN[c] and RF[d] models	Tian et al. (2020)
Predicting consumer acceptance and sensory drivers of liking for Minas Frescal cheese	Consumers Overall liking using a 9-point hedonic scale and evaluation of the intensity of cheese attributes using a 9 cm unstructured scale	RF[d], GBT[e] and ELM[f]	Rocha et al. (2020)

(*Continued*)

Table 4.2 (Continued)

Objective	Sensory test	Algorithm	Reference
Prediction of cheese sensory attributes using near-infrared spectroscopy	Trained panellists ADQ	ANN[a]	Curto et al. (2020)
Butter classification regarding quality and authentication issues	Trained panellists Butter classification in good quality or not	PLS-DA[g]	Van ruth et al. (2008)
Characterise and determine the intensity of goat milk flavour	Trained panellists Intensity scale	PLS[b]	Yang et al. (2015)

[a] Artificial Neural Networks.
[b] Partial least squares.
[c] Back-propagation neural network.
[d] Random forest.
[e] Gradient boosted trees.
[f] Extreme learning machine.
[g] Partial least square – discriminant analysis.

Curto et al. (2020) used the artificial neural network (ANN) algorithm to predict the cheese sensory attributes (taste, texture, appearance, smell and other relevant sensations) through data obtained by near-infrared spectroscopy (NIR). The authors noted that ANN made reliable predictions of sensory responses from instrumental measurements such as NIR spectra.

Singh et al. (2009) developed kinetic models, based on the Arrhenius relationship for physicochemical parameters, as indicators of quality deterioration in UHT milk during storage to predict its sensory quality. In addition, ANN models were developed with physicochemical parameters to predict milk taste. The authors concluded that the predictive performance of ANN models was better than that of kinetic models.

The prediction of consumer acceptability was attempted based on the scores of sensory attributes that characterised samples, evaluated by Cruz et al. (2011). Soukoulis et al. (2010) also attributes to build PLS models to quantify and predict consumer acceptability. They noted that probiotic ice cream acceptability was effectively related to specific flavour and texture attributes. Bahramparvar et al. (2014) and Rocha et al. (2020) subjected data from descriptive and hedonic sensory judgements to ML analysis to understand how sensory characteristics can be used to interpret consumer acceptance. The authors used sensory attributes (aroma, flavour, texture, among others) to predict the acceptability of ice cream and Minas Frescal cheese, respectively.

Unlike regression, where the output is continuous, the classification of samples in known groups considers problems where the output is discrete, corresponding to distinct category labels. Van Ruth et al. (2008) evaluated the potential of proton transfer reaction mass spectrometry (PTR-MS) to classify milk fat samples according to their quality. The criterion for good quality samples was that they should be of 'good quality and specific to the product in question'. If the samples did not comply, the assessors were asked to describe the defects found. The authors concluded that PTR-MS spectral data successfully classified milk fat samples according to their sensory quality. Cevoli et al. (2011) used data obtained by the electronic nose to classify cheese into three categories: hard cheeses, semi-hard cheeses and Pecorino cheese, using the ANN algorithm. The authors concluded that the use of ANN was very effective for the evaluation of Pecorino cheese.

4.4 Conclusions

This chapter discussed the application of multivariate techniques and ML for analysing data from the sensory testing of dairy products. The main multivariate techniques used with the sensory data of dairy products were PCA, CA, HCA, GPA and MFA, allowing the determination of the drivers of liking, characterisation of the sensory profile, identification of significant sensory acceptability parameters and highlighting the relationship between the emotional profile of consumers and sensory acceptability. These multivariate techniques could be used for segmentation of consumers or products based on overall liking, processing treatments or intra-cultural differences. Finally, a comparison of different sensory methodologies could be performed. ML application to predict sensory responses can be an extremely suitable technique, given the importance of establishing quality standards for dairy products to ensure consumer acceptance. This review of the literature indicates that it is possible to predict sensory acceptability from the interaction of physical and chemical characteristics. It can also be concluded that it is possible to predict consumer responses from descriptive tests. Thus, ML can be a valuable and efficient tool for predicting sensory responses to dairy products.

References

Bahramparvar, M., Salehi, F., and Razavi, S.M. (2014). Predicting total acceptance of ice cream using artificial neural network. *Journal of Food Processing and Preservation* 38: 1080–1088.

Balthazar, C.F., Guimarães, J.T., Rocha, R.S. et al. (2021). Nuclear magnetic resonance as an analytical tool for monitoring the quality and authenticity of dairy foods. *Trends in Food Science & Technology* 108: 84–91.

Barbosa, B.T., Rodrigues, J.F., and Bastos, S.C. (2017). Sensory optimization of nutritionally enriched strawberry yogurt. *British Food Journal* 119: 301–310.

Bemfeito, R.M., Rodrigues, J.F., e Silva, J.G., and Abreu, L.R. (2016). Temporal dominance of sensations sensory profile and drivers of liking of artisanal Minas cheese produced in the region of Serra da Canastra, Brazil. *Journal of Dairy Science* 99: 7886–7897.

Cevoli, C., Cerretani, L., Gori, A. et al. (2011). Classification of pecorino cheeses using electronic nose combined with artificial neural network and comparison with GC–MS analysis of volatile compounds. *Food Chemistry* 129: 1315–1319.

Costa, G.M., de Paula, M.M., Costa, G.N. et al. (2020). Preferred attribute elicitation methodology compared to conventional descriptive analysis: a study using probiotic yogurt sweetened with xylitol and added with prebiotic components. *Journal of Sensory Studies* 35: e12602.

Cruz, A.G., Cadena, R.S., Faria, J.A. et al. (2011). Consumer acceptability and purchase intent of probiotic yogurt with added glucose oxidase using sensometrics, artificial neural networks and logistic regression. *International Journal of Dairy Technology* 64: 549–556.

Cruz, M.F., Rocha, R.S., Silva, R. et al. (2021). Probiotic fermented milks: children's emotional responses using a product-specific emoji list. *Food Research International* 143: 110269.

Curto, B., Moreno, V., García-Esteban, J.A. et al. (2020). Accurate prediction of sensory attributes of cheese using near-infrared spectroscopy based on artificial neural network. *Sensors* 20: 3566.

Fernández-Vázquez, R., Stinco, C.M., Hernanz Vila, D. et al. (2018). Internal preference mapping of milk–fruit beverages: influence of color and appearance on its acceptability. *Food Science & Nutrition* 6: 27–35.

Fuentes, S., Gonzalez Viejo, C., Cullen, B. et al. (2020). Artificial intelligence applied to a robotic dairy farm to model Milk productivity and quality based on cow data and daily environmental parameters. *Sensors* 20: 2975.

Gambella, C., Ghaddar, B., and Naoum-Sawaya, J. (2021). Optimization problems for machine learning: a survey. *European Journal of Operational Research* 290: 807–828.

Gouse, S. and Helini, K. (2019). A comparative performance analysis of different unsupervised and reinforcement learning algorithms of machine learning using python. *Algorithms 15* (16): 17.

Izenman, A.J. (2008). Modern multivariate statistical techniques. *Regression, Classification and Manifold Learning* 10: 978–980.

Larosa, C.P., Balthazar, C.F., Guimarães, J.T. et al. (2021). Sheep milk kefir sweetened with different sugars: sensory acceptance and consumer emotion profiling. *Journal of Dairy Science* 104: 295–300.

Liu, H., Fu, Z., Yang, K. et al. (2021). Machine learning for glass science and engineering: a review. *Journal of Non-Crystalline Solids* 557: 119419.

Noble, A.C. and Ebeler, S.E. (2002). Use of multivariate statistics in understanding wine flavor. *Food Reviews International* 18: 1–20.

Oliveira, E.W., Esmerino, E.A., Carr, B.T. et al. (2017). Reformulating Minas Frescal cheese using consumers' perceptions: insights from intensity scales and check-all-that-apply questionnaires. *Journal of Dairy Science* 100: 6111–6124.

Oliveira, A.A.A., Andrade, A.C., Bastos, S.C. et al. (2021). Use of strawberry and vanilla natural flavors for sugar reduction: a dynamic sensory study with yogurt. *Food Research International* 139: 109972.

Pereira, L.A.S., Bemfeito, R.M., Bemfeito, C.M. et al. (2020). Acceptability of low-sodium mozzarella coated with zein and essential oils. *British Food Journal* 122: 2939–2952.

Perrot, N., Ioannou, I., Allais, I. et al. (2006). Fuzzy concepts applied to food product quality control: a review. *Fuzzy Sets and Systems* 157: 1145–1154.

Ribeiro, M.N., Rodrigues, D.M., Rocha, R.A.R. et al. (2020). Optimising a stevia mix by mixture design and napping: a case study with high protein plain yoghurt. *International Dairy Journal* 110: 104802.

Rocha, R.S., Calvalcanti, R.N., Silva, R. et al. (2020). Consumer acceptance and sensory drivers of liking of Minas Frescal Minas cheese manufactured using milk subjected to ohmic heating: performance of machine learning methods. *LWT* 126: 109342.

Rodrigues, J.F., Mangia, B.A., Silva, J.G. et al. (2020). Sorting task as a tool to elucidate the sensory patterns of artisanal cheeses. *Journal of Sensory Studies* 35: e12562.

Rodrigues, J.F., Andrade, R.D.S., Souza, V.R.D. et al. (2021a). Drivers of linking of Prato cheeses: an evaluation using the check all that apply (CATA) and temporal dominance of sensations (TDS) tools. *Food Science and Technology International* https://doi.org/10.1177/10820132211018037.

Rodrigues, J.F., de Souza, V.R., de Sousa Amorim, I. et al. (2021b). Preference sorting as a tool for Dulce de Leches' drivers of liking determination. *Journal of Sensory Studies* 36: e12634.

Silva, R., Rocha, R.S., Guimarães, J.T. et al. (2020). Dulce de leche submitted to ohmic heating treatment: consumer sensory profile using preferred attribute elicitation (PAE) and temporal check-all-that-apply (TCATA). *Food Research International* 134: 109217.

Simeone, O. (2018). A very brief introduction to machine learning with applications to communication systems. *IEEE Transactions on Cognitive Communications and Networking* 4 (4): 648–664.

Sindhu, V., Nivedha, S., and Prakash, M. (2020). An empirical science research on bioinformatics in machine learning. *Journal of Mechanics of Continua and Mathematical Sciences* 7: 86–94.

Singh, R.R.B., Ruhil, A.P., Jain, D.K. et al. (2009). Prediction of sensory quality of UHT milk – a comparison of kinetic and neural network approaches. *Journal of Food Engineering* 92: 146–151.

Soares, E.K., Silva, R., da Silva, W.P. et al. (2020). An intra-cultural investigation in Brazil using Coalho cheese and preferred attribute elicitation. *Journal of Sensory Studies* 35: e12543.

Soukoulis, C., Lyroni, E., and Tzia, C. (2010). Sensory profiling and hedonic judgement of probiotic ice cream as a function of hydrocolloids, yogurt and milk fat content. *LWT - Food Science and Technology* 43 (9): 1351–1358.

Tian, H., Liu, H., He, Y. et al. (2020). Combined application of electronic nose analysis and back-propagation neural network and random forest models for assessing yogurt flavor acceptability. *Journal of Food Measurement and Characterization* 14: 573–583.

Torres, F.R., Esmerino, E.A., Carr, B.T. et al. (2017). Rapid consumer-based sensory characterization of requeijão cremoso, a spreadable processed cheese: performance of new statistical approaches to evaluate check-all-that-apply data. *Journal of Dairy Science* 100 (8): 6100–6110.

Van Ruth, S.M., Koot, A., Akkermans, W. et al. (2008). Butter and butter oil classification by PTR-MS. *European Food Research and Technology* 227: 307–317.

Vieira, A.H., Balthazar, C.F., Rocha, R.S. et al. (2020). The free listing task for describing the sensory profiling of dairy foods: a case study with microfiltered goat whey orange juice beverage. *Journal of Sensory Studies* 35: e12594.

Yang, C.J., Ding, W., Ma, L.J., and Jia, R. (2015). Discrimination and characterization of different intensities of goaty flavor in goat milk by means of an electronic nose. *Journal of Dairy Science* 98: 55–67.

Yu, P., Low, M.Y., and Zhou, W. (2018). Design of experiments and regression modelling in food flavour and sensory analysis: a review. *Trends in Food Science & Technology* 71: 202–215.

Zielinski, A.A., Haminiuk, C.W., Nunes, C.A. et al. (2014). Chemical composition, sensory properties, provenance, and bioactivity of fruit juices as assessed by chemometrics: a critical review and guideline. *Comprehensive Reviews in Food Science and Food Safety* 13: 300–316.

5 Projective Sensory Evaluation Methods for Dairy Products

Adriana Gambaro, Marcelo Miraballes, Adriano G. da Cruz and Erick A. Esmerino

5.1 Introduction

Understanding consumer perception of a product is extremely important for the dairy industry. As the choice and consumption of foods is a highly complex phenomenon, it is essential for the commercial success of a product to understand consumer expectations regarding sensory attributes, but also about extrinsic features such as: packaging, brand, price, etc. and even socio-economic and ethical aspects (Andrade et al. 2016; Fonseca et al. 2016). Sensory characteristics by themselves are not sufficient to satisfy consumer requirements in today's highly competitive and rapidly evolving markets (Enneking et al. 2007).

Among the affective methodologies found and used in sensory analysis are the qualitative methods which mainly aim to measure consumer emotions, attitudes and perceptions (Jervis and Drake 2014). These techniques mostly come from psychology and have been adapted to food applications. Qualitative research works with people, situations, observations, stories and behaviours using, basically, words, and not numbers, to communicate findings (Miles and Huberman 1994). The objective of qualitative research is to obtain information about attitudes and opinions of a group of individuals with similar habits, necessities and interests (Barrios and Costell 2004).

Qualitative research allows a deeper exploration of consumers' perceptions in a natural environment which requires little or no previous training (Jervis and Drake 2014) and that helps to identify market opportunities, generate ideas, hypotheses, explore and develop new concepts, as well as to understand consumer perspectives (Guerrero et al. 2010; Lawless and Heymann 2013).

These methodologies involve unstructured stimuli with the objective to obtain, deeply felt responses which are not easily revealed and that make consumers express their opinions, points of view and motivations more simply (Banović et al. 2016; Vidal et al. 2013). Due to the changing character of consumer tastes and preferences, qualitative research has increased remarkably. The most commonly used methodologies are: deep interviews, focus groups and projective techniques.

Sensory Profiling of Dairy Products, First Edition. Edited by John J. Tuohy.
© 2023 John Wiley & Sons Ltd. Published 2023 by John Wiley & Sons Ltd.

Projective techniques are qualitative tools of indirect inquiry that make the subject 'project' in response to the presentation of a stimulus, what he/she thinks or feels. They are a disguised means of formulating questions that allows the respondent to project beliefs and feelings to either a third party or an inanimate object or a situation (Masson et al. 2016; Pontual et al. 2017).

By directly asking people about their own opinions, feelings and motivations, inaccurate results can easily be obtained. Imagine this situation: I want to develop a dairy dessert with a high content of sucrose. Through an in-depth interview or a focus group, I inquire about the opinions of consumers concerning that kind of product. Many times, although the real behaviour of the consumer before the purchase decision will be different, when an interviewer or a moderator asks a direct question, the consumer responds with a 'socially correct' answer, which in this case would be, for example, that he/she would not like the product because it has a lot of sugar, which is bad for health. Misleading results can occur due to the inability of participants to explain and express certain aspects such as their own behaviour or motivations due to social barriers, or even their fear of being considered irrational or not smart enough (Donoghue 2000).

Therefore, projective techniques are used to encourage people to express their private emotions and feelings, that if asked directly, would not guarantee that the responses coincided with the real feelings of respondents. Through providing a verbal or visual stimulus, respondents are encouraged to express themselves beyond the rational and stimulated to have more creative, liberal and metaphorical thinking (Malhotra and Birks 2006). These techniques also allow to access deeply held or subjacent attitudes or emotions of the individual, thereby revealing any unconscious or non-openly accepted motivations under which a purchase decision is made. They help consumers overcome resistance to providing the information that is sought (Steinmann 2009; Vidal et al. 2013).

In summary, the projective techniques allow researchers to explore deeply, but in a relaxed and uninhibited way, the feelings, thoughts and ideas of the consumer, while the participants' experience of the activity is like playing a game. These techniques can be extremely advantageous for the dairy sector, which is a highly competitive market, by helping companies to understand the necessities and expectations of different groups of consumers in order to direct the development of their products (Busse and Siebert 2018; Hovart et al. 2019).

The methods can be subdivided in five categories according to the type of response that was generated: association, construction, completion, choice ordering and expressive (Donoghue 2000; Hofstede et al. 2007), as illustrated in Figure 5.1.

5.2 Categories of projective methods

5.2.1 *Word association*

Within this category, the so-called 'word association' (WA) technique is one of the most commonly used projective techniques due to its easy application and effectiveness in exploring consumer perception (Mesías and Escribano 2018). In this technique, a stimulus is presented to participants and then they are asked to express the first words, images

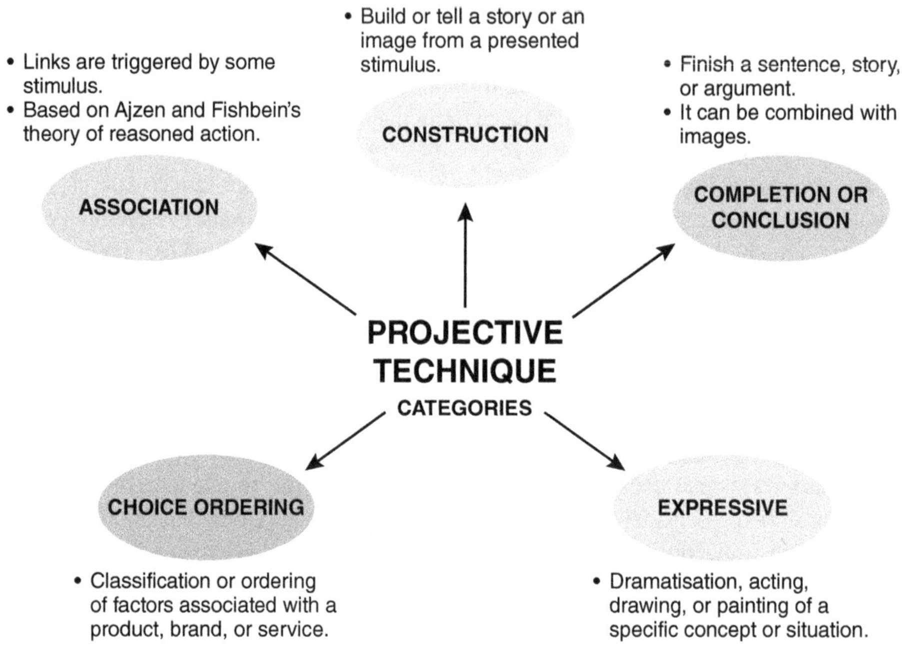

Figure 5.1 The projective techniques categories.

or thoughts that come in to their minds (Banović et al. 2016; Guerrero et al. 2010; Latorres et al. 2016).

This methodology considers that the purchase decisions made by consumers are directly related to the first associations they have had with the different stimuli presented (Guerrero et al. 2010). Also, the spontaneity with which participants express their ideas means that the methodology has less bias and constraints than other approaches such as personal interviews or pre-defined surveys (Andrade et al. 2016; Ávila et al. 2020; Pontual et al. 2017).

WA was used by Ares et al. (2010) to study consumers' expectations regarding chocolate dairy desserts enriched with antioxidants. Hence a stimulus (in text format) was presented to participants: 'chocolate milk dessert enriched with antioxidants'. Then, they had to express the first four images, associations, thoughts or feelings that came into their minds. Researchers found that the terms which were most frequently mentioned by participants were related to health benefits associated with antioxidant ingestion, expectations related to sensory characteristics of the product as well as hedonic aspects. On the other hand, a cluster analysis performed on the data enabled the identification of three groups of consumers with different expectations associated with the stimulus. One of the clusters mainly associated the product with sensory and hedonic expectations. The second cluster mentioned to a large degree, aspects related to health, also to the terms 'interesting' and 'off-flavour', illustrating the perception that enrichment with antioxidant could possibly affect product flavour. Consumers in the third cluster mentioned sensory attributes related to health, showing their expectations as regards a beneficial product for health and according to their likes. Moreover, these results showed how enrichment of dairy desserts with antioxidants has a positive impact on how healthy the product is perceived.

Da Silva et al. (2014) studied through WA consumer perception in relation to different types of ice cream: traditional, light, zero sugar, zero fat, fibre enriched, enriched with bioactive proteins and enriched with omega 3. In this case, each stimulus was presented under the image of an ice cream and a text in which the product was described. The only difference in each card was the text. The results allowed the identification of different associations toward the concepts studied. Traditional ice cream was characterised by seasonable consumption and high caloric content due to its association with terms such as: refreshing, summer, calories, good appeal, sugar sweet. The concepts of light ice cream, zero sugar, zero fat were mainly associated with a loss of sensory quality, diet, good fitness and light. The latter is in accordance with previous studies in which sensory aspects have been seen as extremely important when a consumer chooses functional foods (Lyly et al. 2007; Tuorila and Cardello 2002). The concept of an ice cream enriched with fibre made consumers think about its beneficial health effects, mainly due to improving digestive function. Finally, ice cream enriched with omega 3 had perceptions associated with health, fish and price.

Using WA, Pinto et al. (2018) studied the perceptions and beliefs regarding different fermented milk products (traditional, with probiotic, with prebiotic, light, lactose-free and organic). In this study, the stimuli were presented to consumers in text form and along with the formulations of each product. The products containing probiotics and prebiotic fibre were associated to a larger extent with categories such as healthy and functional, possibly due to their perceived beneficial health effects such as: improved intestinal function, intestinal flora regulation and combating pathogenic bacteria, showing that consumers perceive that the effect of these ingredients is linked to the benefits for the gastrointestinal system (Ares et al. 2008; Bruhn et al. 2002). The results also showed that fermented milk is perceived as a product with good sensory characteristics and due to its positive reputation for health effects, fermented milks associate well with public interest in health and well-being.

Esmerino et al. (2017) assessed and compared the use of different exploratory qualitative techniques, WA among them, to collect consumer information about different fermented dairy products: full-fat yoghurt, low-fat yoghurt, probiotic yoghurt, full-fat fermented milk, low-fat fermented milk, probiotic fermented milk, fermented milk beverage containing whey and a probiotic fermented milk beverage. In this study, eight cards were used as the stimulus. Each card was designed with an identical image of a milky-looking liquid that contained different specifications, including technical nomenclature and legal lettering requirements from Brazil. The information that the text contained was the only difference among the cards, thus any effect in the data obtained would happen due to the messages given. For each of the cards presented, each consumer who participated in the WA task was asked to write spontaneously the first four words, terms, phrases, thoughts, feelings, images and associations they saw when reading them. In the results obtained, the category most frequently mentioned was sensory characteristics, showing that this is the main dimension that differentiates this type of products. At the same time, emotions and feelings were important categories in both methodologies which differentiated the different stimuli. The stimuli corresponding to samples with probiotics were also perceived in relation to the perception of the health benefits associated with the product. In addition, probiotic yoghurt was associated with more positive feelings than the functional versions of the other products, possibly due to greater familiarity with the product. The probiotic fermented milk beverage was mainly associated with negative feelings, although the benefits of the product have been mentioned. It is believed that participants were unaware of the significance of composition, and the association with the other products investigated, creating a disconnection with their expectations. Stimuli corresponding to low-fat products were mostly associated with categories of a technological process, positive feelings, related to diet, consumption and marketing. Finally, the stimuli described as 'full fat' were mainly characterised to be associated with the composition and sensory attributes.

According to several authors, creaminess is an important attribute that has a great effect on how dairy products are perceived by consumers (Kirkmeyer and Tepper 2003; Tournier et al. 2007). Antmann et al. (2011) studied the concept of creaminess with consumers from two different countries using the WA technique. The stimulus in this case was triggered by a card with a text format: 'creaminess'. Interestingly, in both of the countries studied, the comments with the highest frequency were related to dairy products (dairy desserts, ice cream and yoghurt). These authors point out that the high frequency of comments obtained could indicate that consumers might reject dairy products such as desserts, ice cream or yoghurt if they are not perceived as creamy products.

Mazon et al. (2020) studied perceptions of an ice cream that was made with the incorporation of ora-pro-nóbis (a south American native plant) as a functional ingredient and substitute for emulsifiers and stabilisers. A sample identified as ora-pro-nóbis ice cream was presented to the participants and the WA technique was applied. The authors found that WA was useful in identifying the potential of the experimental ingredient used in the product. This was because the product was associated in general with positive terms within the frequently mentioned categories: tasty, creamy, innovation, nostalgia and healthy. A large number of consumers associated the product within the innovation category, which was favourable to the development of new products in the future.

5.2.2 *Construction task*

In the construction category, subjects are asked to make a story or an image from a presented stimulus. One of the approaches in this type of technique is the Haire's shopping list (Haire's SL). This was the first projective technique used in a published study and was developed by Haire (1950). During the performance of the task, respondents are asked to project from a situation in which they depict the consumer who bought foods from a list, providing a brief description of the consumer's personality and character.

Pinto et al. (2018) studied the perceptions and beliefs about different types of fermented milk by means of the application of Haire's SL with the objective to identify and characterise the product's profile as well as that of different types of fermented milk. Four hundred and fifty people assessed six shopping lists (75 for each list). The six lists contained items which are commonly present in the population's diet, differing only on the type of fermented milk: traditional, with probiotic, with prebiotic, light, lactose-free and organic. Participants were invited to read the list and then, in a blank space, provide a brief description about the profile of the buyer of the list in their own words (Haire 1950; Vidal et al. 2013). They were told that there were no correct or incorrect answers; thus, they were free to describe whatever came to their minds.

The buyer of fermented milk with prebiotic fibre obtained a lower-than-expected value of frequency of mentions compared to the 'basic buyer' category. Moreover, the buyer of the list which contained prebiotic fermented milk received high mentions compared to a profile that showed concern with the nutritional and functional benefits of the product. The perception of a profile that showed 'concern with nutrition and health' was likewise correlated with probiotic and organic formulations. Answers such as: People who (i) seek well-being through nutrition, (ii) are concerned about health and quality of life, (iii) seek a healthy and nutritive nutrition and (iv) are concerned about the quality of nutrition were linked to these products.

5.2.3 *Completion task*

In completion of task techniques, the participant is given an unfinished sentence, story, argument or conversation and is asked to complete it (Donoghue 2000). These techniques can be divided into three types: (i) completion of sentences, (ii) completion of stories and (iii) completion of dialogues. In the completion of sentences, the respondents are provided with unfinished sentences and asked to complete them using the first word or phrase that comes up to the mind. In the completion of stories, the respondents are provided with part of a story which is enough to draw the attention to a particular topic and to suggest the end of it. They are asked to give a conclusion with their own words. Generally, in the completion of dialogues, stimuli are presented as speech bubble designs in which the respondents are asked to fill out the characters´ thoughts (Donoghue 2000; Steinmann 2009).

When planning a study that involves figures or images, it is important to consider the quality of the images and the words used, as they play an important role in the results. It is necessary to find open Internet images with good resolutions, good colours which represent consumers that are relatively common among the population. These characters should be as neutral as possible and their faces must look unconcerned, not so smiley and without showing anxiety. For the construction of phrases, complicated terms or uncommon words should be avoided as these affect the participation of people with a lower education level. Neither should regional terms be used as they could be more common in specific areas of the country. Kind words should be used in order to make a friendly questionnaire and an authoritative tone of voice should be avoided (Viana et al. 2016).

Eldesouky et al. (2015) used the completion task with three different stimuli to study consumer's expectations regarding cheese packaging. Images with unfinished phrases and dialogues were presented along with the statement: 'Please complete the dialogues as you deem appropriate'. An online survey, completed by 200 people, was used to collect data.

For the first stimulus, a consumer in a supermarket was depicted. The dialogue guided the participants to complete it mentioning the potential characteristics of the cheese package that could promote its consumption: 'I would buy this cheese if the packaging. . . .' For the second stimulus, participants were suggested a dialogue between two people in order to mention the negative characteristics of the product that discouraged its purchase: 'I like this cheese'. 'But I don't like it because the packaging is. . . .' Finally, for the third stimulus, the dialogue guided respondents to indicate positive characteristics of the cheese packaging that encouraged its purchase: 'Usually, I buy this cheese because the packaging is. . . .'

The categories identified in the completion task for the first stimulus (potential characteristics of the packaging to promote purchase of the cheese) were easy to open/resealable package and transparent/not hiding the product, despite 10% of participants mentioning negative characteristics, such as unconcerned about package. The categories identified in the completion task for the second stimulus (negative characteristics that constrained the purchase of packaged cheese) were difficult to open and close, opaque/hiding the product, impractical/inconvenient and plastic package. Finally, the most important aspects mentioned by the respondents which encouraged the purchase of the product were easy to open and close with resealability, correct preservation and transparent/not hiding the product (Eldesouky et al. 2015).

This study concluded that consumers have an interest in food safety and also in some of the packaging characteristics such as: easy to open and close, resealable, suitable size and transparency of package material. The authors concluded that this methodology was suitable for determining consumer perceptions about cheese packaging and its influence on their purchase decisions, thus enabling food companies to use it as a consumer research tool.

In an online study, with 263 participants, Torres et al. (2020) used the completion task with three different stimuli to investigate consumer knowledge about Petit-Suisse cheese and the perceptions involved in the consumption and purchase processes, in order to identify the requirements needed to enable the dairy industry to expand the market share of this cheese variety. The first stimulus was related to the identification of the purchase decision motivations through the following dialogue between two people: Person 1: 'I like Petite Suisse'. Person 2: 'I like it too, I usually buy it because. . ..' The second stimulus sought to identify the reason why people do not purchase the product, even though they like it. Again, through a dialogue between two people: Person 1: 'I like Petite Suisse'. Person 2: 'I like it too, but I don't buy it because. . ..' Finally, the respondents were stimulated to reveal what condition would make them purchase the product: Person 2: 'I would buy these Petit-Suisse if' (unfinished phrase expressed by Person 2).

Through the first stimulus of the completion task, it was possible to identify that the main reason why consumers buy a product is 'acceptability' (I like it, it is tasty, it is good). Fifty-three percent of participants bought Petit-Suisse cheese because they liked it. One group of respondents (10.3%) considered Petit-Suisse cheese to be a healthy food and that was the reason why they purchased the cheese. The other reasons for purchasing were: 'Have children' (my children like it; my granddaughter likes it; the kids like it; my family likes it) and 'affective memory' (it reminds me of my childhood; it tastes like childhood). Both purchase motivations strengthen the idea that Petit-Suisse cheese is for children; therefore, its consumption is only during childhood. These results suggested that a marketing plan to promote the consumption of Petit-Suisse cheese by adults should be considered.

The second stimulus provided a bigger variety of categories, for quite different reasons. The main category with a significant difference compared to the others was 'High price' (it is expensive; it is expensive to eat daily; it is expensive; and the portion size is small) (22.8%). The authors presumed that the perception of being costly is related to the small portion size of the product that is traded, commonly around 40 g. Although consumers perceive the product as healthy, the price turns out to be a barrier, especially for low-income consumers (Steenhuis et al. 2011).

Other important reasons for deciding not to buy Petit-Suisse were the perception of 'unhealthy' (it is not healthy; it is bad for my health, there is no nutritional value, I prefer healthier food), 'sugar content' (high sugar content, it is not a 'diet' food), 'intended for children' (it is for children; I don't have children; it is baby food; I'm an adult), 'don't like the product' (I don't like it; my children don't like it anymore), 'presence of additives and preservatives' (it has a lot of additives; it contains additives/preservatives, it contains food colouring), 'preference for other products' (I prefer to buy another product that I like more; I prefer to buy a product with better nutritional value; I prefer other dairy products), 'fattening' (it contains fat, it contains too much fat; I'm on a diet; it contains too many calories) and 'industrialised product' (it is artificial; it is ultra-processed; it is an industrialised product; I prefer natural food). The industry should consider these consumer perceptions when identifying issues to work on such as how to deconstruct this image of the product, or create variations of the product that address these perceptions and widen the public appeal (Torres et al. 2020).

The third and last stimulus provided information about possible future improvements: what would awaken consumer interest in Petit-Suisse cheese to make them purchase the product. As expected, many of the points obtained were the opposite of the ones obtained by the second stimulus. Among the important points, the standouts were 'accessible price', 'more natural product', 'less sugar' and 'healthy'. Nevertheless, some of the new categories provided deeper information and suggested different flavours and versions such as: dietary, light and organic or lactose-free. Another interesting and important factor was the 'suitability of the packaging'. According to the respondents, the product should be traded in larger portions, or the packaging information should be appeal to the adult public.

5.2.4 *Choice ordering task*

In the choice ordering task, respondents are asked to classify or place in order factors associated with a product, brand or service, explaining why there are certain things that are more or less important. One of the approaches of this type of technique is projective mapping (PM). PM is a holistic methodology based on the assessment of global similarities and differences among products. In this technique, the participants receive a paper sheet, a group of samples, and then they are asked to assess the samples and locate them on the sheet according to their degree of similarity and difference. Therefore, the samples which are perceived as similar should be located close to one another, whereas the samples perceived as different should be located separately (Mielby et al. 2014). After locating the stimuli, participants are asked to write on the paper sheet, next to the stimuli, three to five words that come into their minds and help to differentiate each product or group of products (Varela et al. 2014).

Data are obtained from X and Y Cartesian coordinates from each stimulus and for each participant. The lower left corner is considered as the origin of the system of coordinates. Multiple-factor analysis (MFA) is used to analyse the data considering the values of coordinates of the samples in each individual map generated by the participants (Valentin et al. 2012).

Esmerino et al. (2017) used this technique to collect information about three different fermented dairy products: yoghurt; a fermented milk beverage containing whey; and fermented milk. Fifty consumers received eight cards as stimuli, with an identical image of a liquid with a milky appearance, containing the following information: full-fat yoghurt, low-fat yoghurt, probiotic yoghurt, full-fat fermented milk, low-fat fermented milk, probiotic fermented milk, fermented milk beverage that is not a yoghurt and contains whey, and probiotic fermented milk beverage that is not a yoghurt and contains whey. The participants could determine 10 different categories associated with fermented dairy products. The most frequently mentioned categories were: sensory attributes, positive feelings, sensory defects and related to diet. The full-fat samples were characterised according to the categories of composition, consumption, sensory attributes and positive feelings. Probiotic samples were characterised by marketing and benefits. Low-fat samples were characterised according to the categories related to diet and sensory defects. Regarding the fermented dairy beverage, the categories most frequently assigned were negative feelings and detriments. According to these authors, PM would be an effective tool for dairy companies with constrained financial and human resources, to enable them to define the most effective and economic strategy to position their products on the market with a minimal risk of failure.

5.2.5 *Expressive task*

When expressive techniques are applied, participants have to dramatise, act, draw or paint a specific concept or situation. One of the approaches of this type of technique is to ask consumers to draw an 'ideal' product. Unfortunately, there are no published studies with the application of expressive task in dairy products.

5.3 Comparison of projective techniques in dairy products case studies

Using the WA technique and the Haire's SL, Pinto et al. (2018) studied the perceptions and beliefs regarding different types of fermented milk (traditional, with probiotics, with prebiotics, light, lactose-free and organic). Comparing both methods (Haire's SL and WA), these authors concluded that both methodologies were effective for the identification of participant perceptions in relation to the products and their buyers. Haire's SL proved more suitable for obtaining the profile of fermented milk buyers but not for information of the product itself, whereas WA was more suitable for obtaining descriptions of the product.

Esmerino et al. (2017) used WA and PM techniques to collect information on three different fermented dairy products: yoghurt; a fermented milk beverage containing whey; and fermented milk. The most mentioned categories were very similar with both techniques although MP mentioned less categories. Possibly due to its experimental design and holistic basis, the participants tended to focus on the most relevant stimuli when performing MP. In this sense, MP presents high-quality results because it exposes the most objective dimensions that are considered relevant for product characterisation and differentiation, especially when the product is not submitted to sensory evaluation (Fiszman et al. 2015; Varela et al. 2014).

Eldesouky et al. (2015) used WA and completion task techniques to study consumer's expectations about cheese packaging. WA enabled the identification of a larger number of items compared to the other method and the respondents could easily associate the food package with any other things they considered. This lack of constraints could have encouraged the respondents to mention a wider spectrum of terms which would suggest that the WA methodology provides more information. However, with the WA task, only positive characteristics of cheese packaging were mentioned, whereas with the completion task the dialogues dictate that consumers mention positive or negative characteristics. This suggests that the completion task methodology, combined with images, could generate a higher projection of the innermost feelings and thoughts of consumers because the stimuli guided participants to concentrate on different aspects in order to allow them to externalise their perceptions.

5.4 Analysis of projective technique data

As with any qualitative technique, one of the key steps for data analysis is the examination of the results for meanings, intentions, consequences, context and the transformation of unstructured data into a format for analysis and the application of statistical tools (Elo and Kyngäs 2008; Soares et al. 2017). The interpretation of the consumer responses is the main objective of the analysis of contents, and it is generally performed by grouping the responses into categories (Krippendorff 2004).

To analyse data, it is necessary to perform a search for recurring terms for each stimulus removing all connectors, auxiliary terms and adverbs from each comment. Terms with similar meanings are independently grouped into categories by three researchers, which is called 'triangulation' (Esmerino et al. 2017; Guerrero et al. 2010). The final categories are obtained by a consensus of the researchers. The frequency of each category is defined by the number of participants that used relative terms to in response to the stimuli that were presented. To avoid a great loss of information, the categories mentioned by more than 5% of participants are considered significant (Guerrero et al. 2010; Vidal et al. 2013). The Chi-square test is commonly used to prove the independence between rows and columns of the contingency table in order to clearly visualise the relationship between the categories and the stimuli used. Correspondence analysis (CA) is applied to the table of frequencies (Fonseca et al. 2016). Differences between different consumer segments can also be highlighted from the use of the partial least squares – discriminant analysis (PLS-DA) (Soares et al. 2017).

5.5 Online versus paper-based surveys

With the increasing popularity of the Internet, the use of online surveys has significantly increased in recent years. Using online questionnaires has proven useful due to the possibility of collecting information from respondents who are geographically dispersed within countries (especially beneficial for large countries) and also at low cost compared with face-to-face surveys. Other important benefits of online surveys are the quick collection of responses and data registration and convenient access to the survey data at any time (Viana et al. 2016). Nevertheless, the use of online methodology for data collection also has some limitations, especially regarding sample representability (Koutsimanis et al. 2012; Wright 2005).

5.6 Conclusions

Projective techniques have shown to be very valuable, versatile and beneficial for consumer researchers investigating dairy products. They provide a greater understanding of what people really think and feel about the products. They allow the identification of relevant attributes for the consumer acceptability of a product and could be used by food companies to undertake a wide range of consumer research at a relatively low cost.

More research is needed to validate how well they reflect consumer perceptions and to improve reliability (repeatability of findings using different consumer groups). There are also many projective techniques that have not yet been applied to consumer research on dairy products, and hence great opportunities exist in this field of research.

References

Andrade, J.C., Sobral, L.A., Ares, G., and Deliza, R. (2016). Understanding consumers' perception of lamb meat using free word association. *Meat Science* 117: 68–74.

Antmann, G., Ares, G., Salvador, A. et al. (2011). Exploring and explaining creaminess perception: Consumers' underlying concepts. *Journal of Sensory Studies* 26: 40–47.

Ares, G., Gimenez, A., and Gambaro, A. (2008). Uruguayan consumers' perception of functional foods. *Journal of Sensory Studies* 23: 614–630.

Ares, G., Barreiro, C., Deliza, R. et al. (2010). Consumer expectations and perception of chocolate milk desserts enriched with antioxidants. *Journal of Sensory Studies* 25: 243–260.

Ávila, B.P., da Rosa, P.P., Fernandes, T.A. et al. (2020). Analysis of the perception and behaviour of consumers regarding probiotic dairy products. *International Dairy Journal* 106: 104703.

Banović, M., Krystallis, A., Guerrero, L., and Reinders, M.J. (2016). Consumers as co-creators of new product ideas: an application of projective and creative research techniques. *Food Research International* 87: 211–223.

Barrios, E. and Costell, E. (2004). Review: use of methods of research into consumers opinions and attitudes in food research. *Food Science and Technology International* 10 (6): 359–371.

Bruhn, C.M., Bruhn, J.C., Cotter, A. et al. (2002). Consumer attitudes toward use of probiotic cultures. *Journal of Food Science* 67: 1969–1972.

Busse, M. and Siebert, R. (2018). The role of consumers in food innovation processes. *European Journal of Innovation Management* 21: 20–43.

Da Silva, V.M., Minim, V.P.R., Ferreira, M.A.M. et al. (2014). Study of the perception of consumers in relation to different ice cream concepts. *Food Quality and Preference* 36: 161–168.

Donoghue, S. (2000). Projective techniques in consumer research. *Journal for Family Ecology and Consumer Sciences* 28: 47–53.

Eldesouky, A., Pulido, A.F., and Mesias, F.J. (2015). The role of packaging and presentation format in consumer's preference for food: an application of projective techniques. *Journal of Sensory Studies* 30: 360–369.

Elo, S. and Kyngäs, H. (2008). The qualitative content analysis process. *Journal of Advanced Nursing* 62: 107–115.

Enneking, U., Neumann, C., and Henneberg, S. (2007). How important intrinsic and extrinsic product attributes affect purchase decision. *Food Quality and Preference* 18: 133–138.

Esmerino, E.A., Ferraz, J.P., Tavares Filho, E.R. et al. (2017). Consumers' perceptions toward 3 different fermented dairy products: insights from focus groups, word association, and projective mapping. *Journal of Dairy Science* 100: 8849–8812.

Fiszman, S., Carrillo, E., and Varela, P. (2015). Consumer perception of carriers of a satiating compound. Influence of front-of-package images and weight loss-related information. *Food Research International* 78: 88–95.

Fonseca, F.G., Esmerino, E.A., Tavares Filho, E.R. et al. (2016). Novel and successful free comments method for sensory characterization of chocolate ice cream: a comparative study between pivot profile and comment analysis. *Journal of Dairy Science* 99: 3408–3420.

Guerrero, L., Claret, A., Verbeke, W. et al. (2010). Perception of traditional food products in six European regions using free word association. *Food Quality and Preference* 21: 225–233.

Haire, M. (1950). Projective techniques in marketing research. *Journal of Marketing* 14: 649–656.

Hofstede, A., Van Hoof, J., Walenberg, N., and De Jong, M. (2007). Projective techniques for brand image research: two personification-based methods explored. *Qualitative Marketing Research* 10: 300–309.

Hovart, A., Granato, G., Fogliano, V., and Luning, P.A. (2019). Understanding consumer data use in new product development and the product life cycle in Europeans food firms – an empirical study. *Food Quality and Preference* 76: 20–32.

Jervis, M.G. and Drake, M.A. (2014). The use of qualitative research methods in quantitative science: a review. *Journal of Sensory Studies* 29: 234–247.

Kirkmeyer, S.V. and Tepper, B.J. (2003). Understanding creaminess perception of dairy products using free-choice profiling and genetic responsivity to 6-n-propylthiouracil. *Chemical Senses* 28: 527–536.

Koutsimanis, G., Getter, K., Behe, B. et al. (2012). Influences of packaging attributes on consumer purchase decisions for fresh produce. *Appetite* 59: 270–280.

Krippendorff, K. (2004). *Content Analysis: An Introduction to Its Methodology*. Sage Publications.

Latorres, J.M., Mitterer-Daltoé, M.L., and Queiroz, M.I. (2016). Hedonic and word association techniques confirm a successful way of introducing fish into public school meals. *Journal of Sensory Studies* 31 (3): 206–212.

Lawless, H.T. and Heymann, H. (2013). *Sensory Evaluation of Food: Principles and Practices*. Springer Science & Business Media.

Lyly, M., Roininen, K., Honkapaa, K. et al. (2007). Factors influencing consumers' willingness to use beverages and ready-to-eat frozen soups containing oat b-glucan in Finland, France and Sweden. *Food Quality and Preference* 18: 242–255.

Malhotra, N.K. and Birks, D.F. (2006). *Marketing Research. An Applied Approach*, 1e. Harlow, England: Pearson Education.

Masson, M.J., Delarue, S., Bouillot, J.M. et al. (2016). Beyond sensory characteristics, how can we identify subjective dimensions? A comparison of six qualitative methods relative to a case study on coffee cups. *Food Quality and Preference* 47: 156–165.

Mazon, S., Menin, D., Cella, B.M. et al. (2020). Exploring consumers' knowledge and perceptions of unconventional food plants: case study of addition of *Pereskia aculeata* Miller to ice cream. *Food Science and Technology* 40 (1): 215–221.

Mesías, F.J. and Escribano, M. (2018). Projective techniques. In: *Methods in Consumer Research*, vol. 1 (ed. G. Ares and P. Varela), 79–102. Woodhead Publishing.

Mielby, L.H., Hopfer, H., Jensen, S. et al. (2014). Comparison of descriptive analysis, projective mapping and sorting performed on pictures of fruit and vegetable mixes. *Food Quality and Preference* 35: 86–94.

Miles, M. and Huberman, M. (1994). *Qualitative Data Analysis*, 2e. England: Sage Publications.

Pinto, L.D.P.F., Silva, H.L., Kuriya, S.P. et al. (2018). Understanding perceptions and beliefs about different types of fermented milks through the application of projective techniques: a case study using Haire's shopping list and free word association. *Journal of Sensory Studies* 33: e12326.

Pontual, I., Amaral, G.V., Esmerino, E.A. et al. (2017). Assessing consumer expectations about pizza: a study on celiac and non-celiac individuals using the word association technique. *Food Research International* 94: 1–5.

Soares, E.K.B., Esmerino, E.A., Ferreira, M.V.S. et al. (2017). What are the cultural effects on consumers' perceptions? A case study covering coalho cheese in the Brazilian northeast and southeast area using word association. *Food Research International* 102: 553–558.

Steenhuis, I.H.M., Waterlander, W.E., and de Mul, A. (2011). Consumer food choice: the role of price and pricing strategies. *Public Health Nutrition* 14: 2220–2226.

Steinmann, R.B. (2009). Projective techniques in consumer research. *International Bulletin of Business Administration* 5: 37–45.

Torres, F.R., Da Silva, H.L., Cutrim, C.S., and Cortez, M.A.S. (2020). Consumer perception of Petit-Suisse cheese: identifying market opportunities for the Brazilian dairy industry. *Food Science and Technology* 40: 653–660.

Tournier, C., Martin, C., Guichard, E. et al. (2007). Contribution to the understanding of consumers' creaminess concept. A sensory and a verbal approach. *International Dairy Journal* 17: 555–564.

Tuorila, H. and Cardello, A.V. (2002). Consumer response to an off-flavor in juice in the presence of specific health claims. *Food Quality and Preference* 13: 561–569.

Valentin, D., Chollet, S., Lelièvre, M., and Abdi, H. (2012). Quick and dirty but still pretty good: a review of new descriptive methods in food science. *International Journal of Food Science & Technology* 47: 1563–1578.

Varela, P., Tárrega, A., Salvador, A. et al. (2014). Diabetic and non-diabetic consumers' perception of an apple juice beverage supplemented with a Fraxinus excelsior L. seed extract having potential glucose homeostasis benefits. *LWT – Food Science and Technology* 57: 648–655.

Viana, M.M., Silva, V.L.S., Deliza, R., and Trindade, M.A. (2016). The use of an online completion test to reveal important attributes in consumer choice: an empirical study on frozen burgers. *Food Quality and Preference* 52: 255–261.

Vidal, L., Ares, G., and Giménez, A. (2013). Projective techniques to uncover consumer perception: application of three methodologies to ready-to-eat salads. *Food Quality and Preference* 28: 1–7.

Wright, K.B. (2005). Researching Internet-based populations. Advantages and disadvantages of online survey research, online questionnaire authoring software packages, and web survey services. *Journal of Computer-Mediated Communication* 10: 1–2.

6 Sensory Attributes of Liquid Milk Products

Elson R. Filho, Amanda A. Prestes, Maria H. Canella,
Elane S. Prudencio, Mônica Q. Freitas, Tatiana C. Pimentel,
Erick A. Esmerino and Adriano G. da Cruz

6.1 Introduction

High-quality milk is a balanced, delicate and pleasantly sweet food. There is also a tendency toward neutrality and the absence of aftertaste. Such characteristics are reflections of a balanced nutritional composition, rich in proteins, fats, minerals and sugar (lactose) (Clark et al. 2009). However, the inherent blandness means that milk does not have highly pronounced attributes, which has led to the preoccupation of sensory investigations with possible deviations from this neutrality.

At the beginning of the twentieth century, the sensory analysis of milk and other dairy products focused on quality measurement using standardised methods, such as the Dairy Product Judging and the American Dairy Science Association (ADSA) scorecard system (Schiano et al. 2017). Such methods focused their efforts on identifying the most likely defects and off-flavours in the milk by expert judges, who analysed the samples in search of signs of contamination or deterioration.

The perishability of milk is caused by its high water activity and nutritional richness. Therefore, sensory deviations resulting from contamination are quickly manifested in the 'flavour' of the product, deviating it from the expected neutrality. Although in common sense, flavour and taste are interchangeable words, they have different meanings in sensory analysis. In this chapter, we present a basic overview of the definitions of taste and flavour due to the growing trend in the use of sensory methods based on consumer perception and the need to increasingly reduce the training hours of specialists (trained tasters/panellists). In both cases, confusion about these definitions can impair the performance of sensory methods.

The concept of flavour can be better understood when broken down into five related terms (taste, flavour, somatosensation, odour and aroma) and two human senses: taste and smell. Taste can be described as the phenomenon resulting from the stimulation of taste buds distributed in the mouth, especially those on the tongue. Although not restricted to these, in sensory analysis, taste is primarily related to the manifestation of its five primary forms: sweet, sour, salty, bitter and umami. Other somatosensations such as metallic, greasy, spicy, refreshing and astringent are also perceived in the oral cavity when stimulated by some foods. Somatosensations can be understood as

Sensory Profiling of Dairy Products, First Edition. Edited by John J. Tuohy.
© 2023 John Wiley & Sons Ltd. Published 2023 by John Wiley & Sons Ltd.

complementary, not because they are less important than basic tastes in terms of consumer acceptance, but because they are perceived temporally after primary tastes.

The sense of smell results from the stimulation, by volatile chemicals, of the olfactory receptor neurons located mainly in the anterior region of the nose. There are two main ways of stimulating olfaction: the orthonasal route and the retronasal route. The first happened when the molecules aspirated through the nose came into contact with the olfactory mucosa. The second, retronasal route, occurs during the act of eating when odorant molecules present in food travel from the mouth to the nose through the nasopharynx. The phenomenon arising from orthonasal stimulation in the sensory analysis is called 'odour'. While the phenomenon originating from the retronasal route is called 'aroma'. Because they are closely related, taste and retronasal aroma are perceived together. So, the phenomenon arising from these stimuli, secondarily associated with somatosensations, is called 'flavour' (Simon et al. 2006) (Figure 6.1).

Sensory analysis was born as the science responsible for investigating the interaction of human senses with food and beverages, especially concerning the sensory attributes involved with appearance, odour (orthonasal), flavour (taste and retronasal aroma) and texture. As the core of sensory analysis, analytical techniques were developed initially. They were intended to calibrate human perception, making it robust and with a high degree of precision for the sensory attributes present in the food product. The idea of working with experts and trained judges was consolidated in the first phase of sensory science. Then, given the expansion of the food industry and consumers' demand, affective techniques were developed, aimed at using the perception of consumers as a complement to that of expert judges. The affective techniques were intended to complement the information produced in the analytical approaches.

However, as a result of increased competition in the food market, it has become necessary to understand more and more deeply the food behaviour of consumers, their motivations, ideas and perceptions not restricted to sensory attributes (according to the focus of analytical and affective techniques). Thus, in the last decades, sensory analysis began to encompass more and more consumer study techniques, which contributed to the advancement of the area. A possible organisation of sensory approaches can be seen in Figure 6.2. Moskowitz et al. (2012) published 'Sensory and consumer research in food product design and development', which stood out as the first book willing to

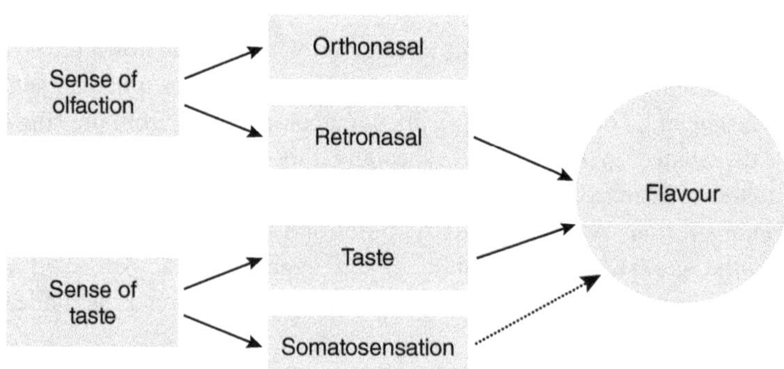

Figure 6.1 Stimuli related to flavour perception.

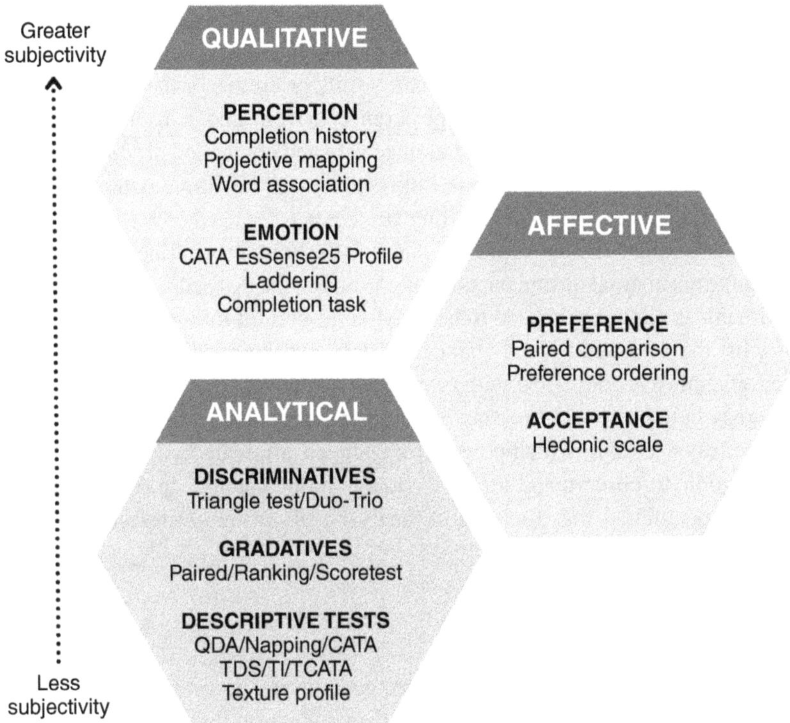

Figure 6.2 Organisation of sensory analysis and consumer study techniques.

formalise the definitive link between sensory analysis and market research, already well explored in industrial practice and scientific publications (Bi 2005; Dijksterhuis 1995; Guinard 2000). From the book's publication to the present day, the link between sensory analysis and consumer studies has strengthened even more, including in research involving fluid milk (de Graaf et al. 2016; McCarthy et al. 2017a; Schiano and Drake 2021).

It is essential to think of sensory analysis as a strategic part of the production and marketing of fluid milk. In this way, it is essential to know the different issues associated with milk to plan a route of sensory tests capable of raising the necessary answers for the improvement of the product. Because just monitoring the emergence and identification of milk defects, as done at the beginning, is no longer a guarantee of commercial success. In recent years, it can be seen that major advances in good manufacturing practices and sanitary legislation have resulted in a general improvement in marketing of fluid milk. However, fluid milk consumption has declined in recent decades notwithstanding the elimination of classic sensory defects, while consumption of non-dairy alternatives has grown (McCarthy et al. 2017b). In this way, it is necessary to increase the understanding of the motivations of consumers for buying or not buying fluid milk and the reasons for the growing preference for non-dairy alternatives. Modern sensory analysis should not be restricted to quality investigation but also to track new trends.

Two sensory approaches can be strategically applied for fluid milk at three important points in the production chain. With a focus on quality, the first should be used to monitor defects and sensory deviations resulting from handling problems on the farm. The second concerns changes resulting from processing, always comparing the product present in the production line to a high-quality milk reference or the market leader. The third approach may identify product preference drivers and consumer attitudes and motivations. The first phase seeks to guarantee safe and controlled production, while the second will allow inferring points of improvement to be implemented, both in the product and in the marketing and labelling strategies.

Regardless of the sensory approach chosen, the sensory routine begins with constructing an experimental design capable of producing the desired responses. The design of experiments is a technique used to help define important information for the project, such as what the data will be, the quantity of data and the conditions of how it is to be collected. Planning is concerned with two major objectives: the greatest possible statistical accuracy in the response and the lowest operational cost. A designed experiment is a test, or series of tests, in which purposeful changes are made to the input variables so that it is possible to observe and identify corresponding changes in the response. Stone et al. (2020) postulated that for sensory analysis, the primary elements in industrial planning involve the following:

- Identify the target population.
- Calculate the sample size.
- Establish the region and location for testing, acquiring and storing products.
- Define sample preparation protocols and data collection instrument.
- Design data collection and analysis methods.
- Define documentation (reports).

A robust and well-documented sensory strategy will allow sensory information to reach the chain of command, increasing product quality and, consequently, commercial success. It is important to emphasise the importance of the strategic choice of tests and the importance of rigorous execution and documentation of the analysis to create a traceable and verifiable basis for the inferences produced by the sensory team. In the milk production chain, the sensory team must be attentive to every production stage, given the high perishability and sensitivity of the raw material. However, the entire process can be divided into three general phases: collecting raw material on the farms, processing at the factory and storage in commercial establishments.

Processing and storage conditions have a significant effect on the quality of milk. Because of the perishability of milk and the nature of milk production/handling procedures, the development of off-flavours/odours is not uncommon. Once the quality of raw milk is impaired, it cannot be improved during processing, and defects often become more pronounced (Mounchili et al. 2005). Applying specific nutritional programmes or management practices on the farm can sometimes cause off-flavour problems in raw milk. These off-flavours are a significant control problem for the industry, as consumer acceptance can be significantly affected. To prevent flavour/odour defects in milk, proper milk-handling procedures from the farm to the consumer are essential (Ishler and Roberts 1991).

Milk is a pleasant and satisfying food after correct production and processing. For this reason, milk is evaluated by sensory analysis after:

1. On-farm milk production – control of the raw milk (effect of the forages used in cow's feed, health and environment of the herd)
2. Processing – effects of processing methods on the finished product (temperature of heating and time), the impact of packaging materials and inter-batch consistency
3. Marketing – consumer preferences, understanding how the product performs against the competition, determining whether or not consumers can detect differences between products and developing innovative value-added milk products.

This chapter will deal solely with the sensory properties of commercially traded heat-treated fluid milk products of bovine origin. Thus, the sensory properties of raw milk and milk of other species will not be discussed.

6.2 Sensory evaluation of heat-treated fluid milk

The ADSA scorecard became the standard for evaluating commercial fluid milk in the early 1930s. The instrument was standardised to identify the most likely milk defects and off-flavours. The ADSA scorecard looked at 16 undesirable attributes related to milk flavour (Table 6.1). Some of these attributes, e.g. cowy, barny, unclean, musty/stale and disinfectant, have practically fallen into disuse with advances in good manufacturing

Table 6.1 Undesirable flavours investigated in the first decades of milk analysis.

Attribute	Description
Barny, cowy	Conveys the suggestion of an unclean, poorly ventilated cow stable
Bitter	Associated with milk from cows far advanced in lactation
Cardboard	Resembles freshly dampened cardboard
Cooked	Suggestive of boiled milk; results from improper pasteurisation
Disinfectant	Differs by disinfectant
Feedy, silage	Characteristic cleanliness after expectoration distinguishes feed from barny and cowy flavours
Flat, watery	A reference may be made by adding water to a sample of milk
Garlic, leek, onion	Characterised by intensity and offensiveness
Lacking richness	Sufficient flavour but lacking the creamy smoothness of whole milk. Typically, the removal of butterfat results in slightly greater sweetness
Malty	Malty, walnut or maple flavour, which may be attributable to the action of microorganisms
Metallic	Puckery feeling obtained when a piece of tinfoil or new metal is rotated within the mouth
Musty, stale	Suggests a damp, mouldy, poorly ventilated cellar
Rancid, strong	Undesirable; often sour, soapy or bitter
Salty	Associated with milk from cows in advanced stage of lactation or cows with mastitis
Sour	Detectable by odour sooner than taste due to fermentation
Unclean	Characterised by an unclean, unpleasant and unwholesome aftertaste

Source: Data extracted from Schiano et al. (2017) and Nelson and Trout (1934).

practices. However, a good part was kept in the current analyses, while ADSA added other attributes to update the scorecard. Table 6.2 presents the main attributes used in milk analysis today.

The first advances in the sensory analysis of milk were the development of questionnaires focused on identifying defects and off-flavours to be used in sensory profiling. This was followed by the increased use of descriptive tests to quantify positive attributes. The change in the focus of the analysis was primarily motivated by advances in good manufacturing practices and conservation technologies, which reduced the prevalence of sensory defects and consequently promoted improvements in the general quality of marketed fluid milk.

Among the methodologies for sensory profiling, descriptive analysis, especially quantitative descriptive analysis (QDA), has stood out in recent decades due to its strong analytical character, good reproducibility, robustness and reliability (Stone et al. 2020). For example, several studies have successfully used descriptive analysis to assess the sensory properties of fluid milk (Chapman and Boor 2001; Chung et al. 2008; Claassen and Lawless 1992; Francis et al. 2005; Lawless and Claassen 1993; Lee et al. 2017; McCarthy et al. 2017b; Phillips et al. 1995; Watson and McEwan 1995). Furthermore, descriptive analysis data can be paired with consumer panel data in a preference mapping technique to understand better drivers of liking a product (MacFie and Thomson 1988; Thompson et al. 2004).

Table 6.2 Main undesirable flavours and off-flavour adopted in the sensory analysis of milk today.

Attribute	Description
Acid	Classical definition of the taste promoted by substances with acidic pH. In milk, acidity usually results from the action of fermenting bacteria fermenting lactose to lactic acid.
Astringent	Sensation popularly known as 'dry mouth', is caused by the shrinkage of the mucous membranes of the palate and tongue.
Bitter	Refers to the taste imparted by bitter substances such as quinine, caffeine and certain peptides derived from milk proteins. In milk, it is often associated with the growth of proteolytic microorganisms in milk.
Cooked	A sulphur-like odour detectable immediately after processing. In milk, the cooked flavour usually appears immediately after heating to 73.8 °C (164.8 °F) or higher. Suggestive of boiled milk; results from improper pasteurisation.
Feeding	Flavour related to compounds that come from cattle feed. In milk, the United States Department of Agriculture (USDA) traces back to the flavour of kibble, alfalfa, sweet clover, silage or similar kibble.
Flat	Identified by the absence of the usual intensity of the characteristic flavour of milk. Common when water is added to milk.
Foreign	Any off-flavours that are not expected in high-quality milk and that are none of the other known defects. In the past, it was more common due to contamination with disinfectants.
Metallic	Chemical sensation on the tongue, reminiscent of steel, iron, copper and/or silver spoons.
Oxidised	It resembles wet cardboard, oily substances or aged beef tallow.
Rancid	Rancidity in dry dairy products often exhibits a strong, pungent odour that may be accompanied by a soapy aftertaste. These sensory properties are mainly due to the generation of short-chain and small fatty acids resulting from the hydrolysis of milk triglycerides.

Source: Authors.

The QDA is a method segmented into indispensable and interconnected steps: pre-selection of panellists with sensory acuity tests; definition of descriptor terms with screening methods (e.g. Kelly's repertory grid); consensus meetings; exhaustive training; and sample profiling. Due to the panellists' greater training and calibration load, QDA can produce robust data with fewer panellists, ranging from 8 to 12 (Stone and Sidel 2004). The extent of training tends to vary according to the complexity of the sensory attributes to be profiled.

Lawless and Claassen (1993) conducted a critical study on milk quality, in which they decided to compare the performance of descriptive analysis with traditional dairy judgement terminology. Some differences were noted between the methodologies. Such as the absence of acceptability inferences by the descriptive analysis; while there is acceptability prediction in the traditional methods. The initial questionnaires infer the high acceptability of samples with few defects. Another important divergence point between descriptive analysis and classical terminology concerns the terminology adopted in the analysis. In the descriptive analysis, the descriptors (attributes) used to start from simple concepts thought in the survey stage, represented in training by standard everyday life products.

In the traditional analysis, complex and subjective associations sometimes occur, often related to processing, and there is no attribute in itself, such as 'oxidised', which in traditional terminology can be connected to several terms such as *metallic*, *oily*, *cappy*, *cardboardy*, *stale*, *tallowy*, *painty* and *fishy*. Unlike conventional dairy terminology, descriptive analysis does not group several flavours into one profiling descriptor. Instead, descriptive analysis would provide the panellist with separate rating scales for each individual perceived characteristics/flavour. The descriptive analysis also makes no effort to quantify attributes that are not recognisable in the product, inputting in the analysis only what was raised and prioritised in the screening stages.

In the study by Lawless and Claassen (1993), experts using defect-oriented sensory terminology system used the terms *light oxidised*, *metallic*, *oxidised* and *rancid* to assess the quality of untreated, artificially oxidised samples artificially rancid milk. However, in the descriptive analysis used to evaluate the same samples, the attributes *sweet*, *cream*, *plastic*, *cardboard*, *metallic*, *old oil*, *butter*, *cheese*, *mouthcoating* and *mouthdrying* were selected. This study exemplifies the difference in approach between the methods, mainly because descriptive analysis fragments the descriptions across all the attributes perceived by the panellists, including the positive ones.

Technological advances incorporated in the milk production chain have raised the quality standard of the final product sold. Factors such as the implementation of good manufacturing practices, improvements in aseptic filling and the diversification of conservative methods allowed the milk marketed to have more subtle defects. The mitigation of the classic defects (described in the 1930s) made sensory profiling also quantify desirable attributes. Another important point for the sensory profiling of milk marketed today is the greater attention to descriptors related to the heat treatment of the product. This was due to the range of preservation processes (e.g. low-temperature long-time [LTLT], high-temperature short-time [HTST] and ultra-high-temperature [UHT]) that expose milk to different temperature values during processing.

Table 6.3 Main descriptors used in the descriptive analysis of milk.

Visual	Aroma	Taste/Flavour	Aftertaste	Texture
Whiteness	Cheese	Cooked	Cardboard	Viscosity
Visual thickness	Cooked	Butter	Sweet	Drying
	Cream	Sweet	Metallic	
	Hay	Baby formula	Sour/acid	
	Grain			
	Fruity	Flat	Bitter	
	Old oil	Nutty		
	Sulphur	Rancid		
	Rancid	Sweet		
	Plastic	Caramelised		
	Sour/fermented	Sour		
	putrid			

Source: Extracted and compiled from studies by Lee et al. (2017), Fromm and Boor (2004), Martin et al. (2016) and Chapman et al. (2001).

The ADSA Committee on Flavour Nomenclature and Reference Standards recommended in its technical report the occurrence of four types of flavours caused by heating milk: boiled/sulphur, heated, caramelised and burned (Clark et al. 2009). The notes from heating, especially cooked and caramelised, do not result in unpleasant changes when present in a moderate way. On the contrary, consumers often appreciate them and, therefore, should not be considered a defect. Furthermore, Calvo and De La Hoz (1992) noted that when a heated flavour occurs in milk or cream products, an oxidised flavour will rarely be present. The main attributes related to the descriptive analysis of milk can be seen in Table 6.3.

Most of the attributes in descriptive analysis are inherited from the defect-oriented sensory terminology system (Tables 6.1 and 6.2). The panellists generally recognise and define the others in the survey phase and consensus of the sensory attributes. Descriptors not listed in traditional terminology are briefly explained below.

Whiteness: the definition of colours in the descriptive analysis is usually based on colour definition and standardisation tools, such as the Munsell Books of Colour.

Visual thickness can be defined as the product's readiness to move to form a flow when its container is moved.

Cheese aroma and flavour are associated with cured cheeses. During the descriptive analysis, the panellists choose a cheese to be defined as a reference by consensus.

The cream aroma is associated with cream and fresh unfermented full-fat dairy products.

Hay aroma can be described as the characteristic aroma of alfalfa.

Grain aroma is associated with grains in animal feed such as corn, oats or barley. In the case of grains, panellists must decide by consensus to which grain the identified note corresponds.

Fruit aroma can be defined as aromatics associated with different fruits. Fresh pineapple is usually used for training.

Old oil aroma is associated with the smell of heated vegetable oil, corn or soy.

Sulphur aroma is the smell of sulphur-containing amino acids. Characteristic of overcooked egg yolks.

Plastic aroma can be defined as the characteristic smell of polypropylene, especially when heated.

Putrid aroma is associated with the strong, sometimes pungent smell present in some types of cheese, such as Limburger.

Butter flavour is the characteristic buttery flavour found in butter.

Sweet aroma and flavour can be defined as the characteristic flavour or aroma of products sweetened with sugars such as glucose, sucrose and fructose.

Baby formula was found in only one study (Fromm and Boor 2004). Therefore, it was defined only as a characteristic of infant formulas.

The nutty flavour is the non-specific flavours of oilseeds such as walnuts, almonds and peanuts.

Caramelised is defined as the flavour derived from whole milk or butter cooking when added with sugar.

Viscosity (texture) can be understood as the cohesive internal forces manifest in its tendency to resist the flow. Viscosity can be inversely related to flat and directly associated with mouth-filling sensation.

Drying (texture) is generally associated with an astringent flavour, consisting of the experience of drying, puckering and shrinking in the oral cavity.

In addition to sensory techniques involving expert judges and trained panellists, a consumer-oriented approach can also be used for the sensory evaluation of fluid milk. The quality of information from consumer tests is related to the smaller distance between the answers found and attitudes related to purchasing and consumption. Classically in sensory analysis, tests with consumers were mainly structured to measure responses of an affective nature. For example, Martin et al. (2016) investigated the impact of exposing fluid milk to LED light. Following the classic route, the authors performed descriptive techniques with trained panellists for sample profiling (analytical response) and consumer test to obtain affective responses. To investigate the affective response, the researchers opted for a self-report questionnaire with acceptance questions assessed on 9-point hedonic scales. Just-about-right (JAR) preference scales and purchase intent questions were also used with consumers.

Acceptance testing is an indispensable resource for every sensory programme. In general terms, acceptance analysis indirectly measures consumer ‘like’ or preference for a product compared to market rivals or variations in the formula. In this way, consumers can provide feedback that is more focused on the intrinsic characteristics of the product, without the biases inherent to market tests that the marketing department usually carries out on a large scale. It is the role of the sensory to provide an estimate of product acceptance (Stone et al. 2020). Acceptance tests have been increasingly consolidated with the hedonic scale due to their intuitiveness for the consumer. Overall, the acceptance assessment is carried out with five questions that ask the consumers how much (intensity) they like or dislike the appearance, aroma, flavour, texture and overall impression of the evaluated product. Consumers oriented their response on a structured or unstructured (preferably) scale anchored on the left with ‘dislike extremely’ and on the right with ‘liked extremely’.

There is also the possibility of correlating acceptance data with data from descriptive analysis through the external preference map (EPM) statistical technique.

Preference maps make it possible to identify and quantify the sensory attributes that influence the preference for a product. For this, statistical methods such as principal components analysis and multiple regression are used. The EPM correlates the measures resulting from the descriptive analysis, relating them to acceptance data. In addition, the EPM indicates which attributes evaluated in the descriptive analysis are most linked to consumer preference.

The JAR scale is also widely used with consumers. In these tests, consumers evaluate how ideal the intensity of strategic attributes of the product is. For example, Potts et al. (2017) use the JAR scale to investigate how consumers perceived the intensity of flavour and mouth sensations (thickness and viscosity) of milk samples exposed to light. JAR scales usually work with five possibilities of intensity: 'much too little'; 'too little'; 'just about'; 'right'; 'too much' and 'much too much'.

Consumer studies may also focus on qualitative aspects of the consumer experience. Qualitative consumer data are often collected in the form of free-response comments, check-all-that-apply questions or, in some cases, organised interview methods such as focus groups. One advantageous consumer testing question type about product development is the JAR scale. This method is an excellent tool for understanding the influence of individual sensory attributes or product qualities on overall liking. This methodology has been used extensively in studies investigating consumer acceptance of functional or flavoured milk beverages, such as probiotic milk (Villegas et al. 2010), regionally flavoured milk (Zhi et al. 2016) and coffee-flavoured milk (Li et al. 2014, 2021), as well as different pasteurisation methods (Chapman and Boor 2001; Lee et al. 2017).

Although they have already been used in other dairy products (e.g. yoghurt, ice cream and cheese, functional or flavoured milk), more exploratory sensory techniques such as check-all-that-apply, temporal dominance of sensations, napping and flash profile have not yet been used in sensory analysis of fluid milk. Nevertheless, these methods are essential and represent advances in sensory analysis that can provide accurate information about milk perception. Therefore, the use of exploratory methods means a possibility for advancing the sensorial study of fluid milk.

The sensory team needs to be familiar with the different analysis methods in order to know how to rationally decide which ones can be used to produce the answers sought by the company. Knowing the different nature of the problems to be solved will facilitate the choice of sensory approach. In general, discriminative tests can verify the existence of global or directional, generally qualitative, differences between samples and assess the limits of perception of stimuli. In the dairy production chain, discriminative tests can verify whether specific changes such as changes in animal management, machinery or process parameters have manifested in significant changes in milk as a final product.

Discriminative tests are useful in the initial phases of the sensory routine, especially when you want a quick response about possible changes in the product. Discriminative tests produce reliable results with teams of 20–30 tasters. If the difference in milk is identified after any change in the production chain, and if there is a desire to know more deeply the nature of the change, descriptive tests can be carried out to investigate the attributes that describe the samples thoroughly. As a rule, descriptive tests are conducted with teams of trained tasters (ranging from 8 to 12) (Stone 2020). There is, however, a growing tendency to conduct descriptive tests (such as CATA [check-all-that-apply]) with groups of at least 50 consumers (the ideal being around 120) (Stone 2020). Affective tests, in turn, seek to directly access the consumer's opinion regarding the preference or acceptability of the product. In the fluid milk industry, this type of information can be used to identify whether sensory changes resulting from changes in the production chain tend to improve or worsen the commercial performance of milk. Sensory changes in the product are not always associated with worsening acceptability. For example, the presence of soft caramel notes in heat-treated milk is considered positive for some consumers. For affective tests, groups of at least 50 consumers can be used, with 120 being recommended (Stone 2020).

Finally, when the focus is not to evaluate a sample of milk itself but the attitudes and motivations of its consumer public, qualitative methodologies such as projective maps, word association, hard laddering and completion tasks can be used. Such methods seek the free expression of the consumer, bringing with it a greater subjective load than other sensory approaches. For qualitative techniques, it is recommended to use at least 100 consumers.

6.3 Influence of heat treatment on sensory characteristics of milk

Liquid milk products include pasteurised milk, sterilised milk, UHT milk, ESL (extended-shelf-life) milk and flavoured milk (Table 6.4). According to Sarkar (2020), milk is considered the primary food for every class of human beings for nutrition and

Table 6.4 Categories of heat-treated fluid milk products.

Liquid milk product	Method	Treatment conditions (°F/°C)	Bacterial kill	Product storage	Shelf life
Pasteurised	Batch (LTLT, low-temperature long-time)	145 °F (62.8 °C) for 30 min	Some survival	Refrigerated	Days
	Continuous (HTST, high-temperature short-time)	161.6 °F (72 °C) for 15 seconds	Some survival	Refrigerated	Days
Sterilised		239–248 °F (105–120 °C) for 10–40 min	Nearly 'sterile'	Room temperature	Month
UHT (ultra-high temperature)	Direct Indirect	280–293 °F (138–145 °C) for 2–10 s	Nearly 'sterile'	Room temperature	Month
ESL (extended shelf life)	Thermal processing	Less severe than UHT and more severe than pasteurisation processes	Some survival	Refrigerated	Days/month
	Non-thermal treatments	With or without a pasteurisation step	Some survival	Refrigerated	Days/month

Source: Authors.

health. Milk is also considered ideal for the growth and multiplication of diverse microorganisms. Therefore, thermal treatments extend the shelf life and ensure high-quality liquid milk products.

Many different conditions can be applied during thermal processing to achieve liquid milk storage stability and microbiological safety. Liquid milk may be heat-treated by conventional pasteurisation processes, such as batch LTLT or continuous HTST processes. Pasteurised milk is efficiently heat-treated at a sufficiently high temperature for an appropriate time to ensure the destruction of all pathogenic microorganisms and to enable the product to be transported, distributed and consumed as liquid milk. Sterilised milk is heated in a sealed container to generate long-life milk by destroying microorganisms of public health significance and potential spoilage organisms. Besides being heat-treated to UHT temperatures (138–145 °C [280–293 °F] for 2–10 seconds), UHT or long-life milk is homogenised, filled and sealed aseptically into sterile retail containers to achieve commercial sterility. ESL milk can be produced by a variety of technologies with heat treatment. Hence, ESL milk is commonly loosely defined as milk with an extended shelf life beyond pasteurised milk. Flavoured milk usually contains added sugar, colouring and permitted flavouring agents and additives and can be heat-treated by various processes.

The heat treatment of milk brings another important aspect, the action of heat both as an enzymatic accelerator and as an inactivator, depending on the temperature reached and the time maintained. Enzyme inactivation is one of the treatment goals, seeking to increase the product's shelf life. However, the role of enzymes in the formation of aromatic and flavourful compounds must also be considered. In milk, enzymes can be derived from mammary tissue, somatic (and blood) cells or colonising microorganisms. Most of the endogenous enzymes in milk are inactivated at 70°C. However, exogenous enzymes produced by contaminating bacteria can be more resistant, even at higher temperatures. In this way, the sensory analysis must always be based on the idea of high-quality milk obtained using good manufacturing methods (Deeth 2021). Poorly handled milk will carry bacterial enzymes that can promote unpredictable deleterious sensory changes in the product that has left the factory with the approval of the sensory team.

The 'time and temperature' binomial produces more stable milk when the temperatures and heating time are higher. However, in addition to higher energy consumption, excessive milk heating will cause intense sensory changes in the product. For example, volatile sulphur compounds formed on the heating manifest as cooked or sulphurous off-flavour aromas that can severely impair acceptance (Al-Attabi et al. 2014) of milk subjected to more intense heat treatment, for example, UHT.

Claassen and Lawless (1992) compare defect-oriented analysis and descriptive analysis in heat-treated milk samples. The authors observed that the terms raised in the descriptive analysis were more sensitive and effective to describe the sample than the complex terms of the defect-based method. By breaking down the samples into all perceived attributes, descriptive analysis can be useful to monitor the effect of thermal processing on the sensory attributes of milk, especially if the sample set is assembled comparing a raw milk sample with others treated in different thermal parameters. The association of descriptive data with affective data (acceptance testing) can be useful to identify the influence of heat treatment on product acceptance (Greenhoff and MacFie 1994).

6.3.1 *Sensory profile of pasteurised milk*

The first pasteurised milk ordinance was published in 1924 (Rankin et al. 2017) and defined milk pasteurisation as a thermal process of not less than 61.1 °C (142 °F) for 30 minutes in approved equipment to inactivate *Mycobacterium tuberculosis*. Although the 'batch' method (heating milk to 63.5 °C [146.3 °F] for 30 minutes) was the most widely used, plate heat exchangers were coming on the market and being applied for the HTST continuous pasteurisation method. Therefore, the standard of 71.7 °C (161 °F) for 15 seconds was agreed upon after considering the effect of HTST treatment on the creaming ability of milk, practical experience and numerous other investigations. Recently, Deeth (2022) reaffirmed that the pasteurisation process used in the dairy industry today inactivates the most heat-stable pathogenic non-spore-forming bacterium found in raw milk (*Coxiella burnetii*) while at the same time causing little detrimental change to the nutrient components of milk. Cole et al. (2020) showed an in-depth definition of pasteurised milk. According to these authors, milk pasteurisation is realised to ensure the safety of fluid milk by killing pathogens known to occur in milk and prolonging the shelf life by destroying undesirable enzymes and reducing the number of viable spoilage microorganisms. The guidance was also given that an additional 3 °C (5 °F) be added when products containing more fat than found in whole fluid milk or containing added sugar need to be pasteurised (FDA 2019). Cooling is prescribed as an integral part of the pasteurisation process. Pasteurised milk is usually considered pathogen-free except for *Bacillus cereus* spores if present in large numbers (Holsinger et al. 1997). It is noteworthy that the cold milk chain is important to maintain its quality, keeping the milk at a relatively low temperature during storage, transportation and distribution (<5 °C [41 °F]).

The dairy industry has been applying a pasteurisation process, HTST, which uses a binomial time temperature of 72 °C (161.6 °F) for 15 seconds. Under the most favourable processing and storage conditions, this HTST pasteurisation process can extend the shelf life of milk for days, depending on the initial microbiological quality of the raw milk. Exposing pasteurised milk to temperatures above approx. 7 °C (45 °F) will shorten the shelf life dramatically; the colder the milk, the longer it will last, the warmer the milk, the quicker it will spoil. Ideal storage temperatures for milk and dairy products are approx. 1–4 °C (34–39 °F). Under ideal refrigeration, most pasteurised milk will remain fresh for two to five days after its sell-by date. Once opened, pasteurised milk should be used as soon as possible for the best quality and taste (Cornell University 2007).

If carried out with technical rigour, pasteurisation does not cause significant sensory changes in the milk (Coolbear et al. 2022). However, if the equipment used is not accurate in promoting homogeneous heating and ends up excessively heating the milk, there will be the formation of aroma and cooked flavour in the milk, which, if prominent, will impair the acceptance of the product. Cooked aroma and flavour should not be seen directly as a sensory defect because some consumers tend to value the subtle presence of this aroma in fluid milk (Gandy et al. 2008). In the study by Gandy et al. (2008), it was noted that in the range of temperatures investigated for pasteurisation (77–85 °C for 15 seconds of heating and 7 °C for storage), 79 °C was the most accepted by consumers immediately after pasteurisation. However, on the sixth day of storage, the preferences of consumers were slightly different. When heated at 82 and 85 °C, the cooked flavour and aroma were intense and impaired acceptance after six days storage but for milk heated at 77 °C or 79°, the flavour was mild and desirable.

In the case of some more conservative consumers, pasteurisation is considered to impair the natural flavour of milk. Such concepts are closely related to the emotional and attitudinal burden of the consumers, since little or no sensory changes are seen in blind tests comparing raw and pasteurised milk samples (Coolbear et al. 2022). In this scenario, studies of nutritional appeals and labelling may be welcome if there is an intention to understand this more conservative portion of consumers. In general, consumers trust the safety of pasteurisation, paying more attention to the fat content and the size of the package, compared to claims on the label (McCarthy et al. 2017b). However, in regions where the marketing of raw milk is still allowed, further investigations around consumers' perceptions of thermal processing would be interesting.

6.3.2 *Sensory profile of ESL (extended-shelf-life) milk*

ESL milk, often called ultra-pasteurised milk, is produced by thermal processing using conditions between traditional HTST pasteurisation and those used for the UHT process. Therefore, ESL milk is known as one that has been treated to reduce the microbial count beyond standard pasteurisation, packaged under highly hygienic conditions and which have a defined prolonged shelf life under refrigeration conditions (Sarkar 2020). ESL milk must be stored in the fridge regardless of whether it is opened or unopened for up to one month when sealed (unopened). In contrast, the shelf life of pasteurised milk is some days in the fridge, and the shelf life of UHT milk is three to six months at ambient temperature.

Although there are no laws or regulations which prescribe a particular production method for ESL milk, Sarkar (2020) reported that the current methods used commercially are based on two principal methodologies: (i) thermal processing but less and more severe than UHT and pasteurisation processes, respectively; and (ii) non-thermal treatments such as microfiltration and centrifugation/bactofugation.

No specific temperature–time combination for ESL milk has been specified in the literature. Sarkar (2020) stated that in USA for ESL milk is recommended a thermal treatment ≥138 °C (280.4 °F) for ≥2 seconds; however, commercially, ESL milk processing is realised at 123–127 °C (253.4–260.6 °F) for 1–5 seconds. These authors also affirmed that ESL milk processed at 120–125 °C (248–257 °F) for 1 second, resulted in a shelf life greater than 37 days at a temperature of 7 °C (4.6 °F). Billington et al. (2020) reported that the most frequent temperature and time combinations for commercially producing ESL milk by heat alone are in the range of 123–127 °C (253.4–260.6 °F) for 1–3 seconds.

For ESL milk, the sensory analysis team must look beyond the basic routine. Also focusing on shelf-life tests, since this product intends to remain commercial for longer than basic pasteurised milk. ESL is a more emerging and, therefore, less consolidated technological approach. Thus, different processing parameters can be used by different factories and therefore, this type of production may require greater customisation of the sensory routine.

With a more modest approach, Johnson et al. (2015) used a series of triangular tests to evaluate, among others, the sensory quality of 2% fat fluid milk poured into high-density polyethylene (HDPE) bottles, containing different concentrations of TiO_2 and within a period of up to 36 days of cold storage (3 °C), under fluorescent lighting, simulating retail storage. Hedegaard et al. (2006) investigated the development of off-flavour in milk, correlating data from descriptive analysis with instrumental chemical analyses.

6.3.3 *Sensory profile of UHT (ultra-high-temperature) milk*

The UHT process exposes milk to an intense heat treatment under which the relevant microorganisms and most of the enzymes are inactivated, enabling UHT milk to be stored for months under ambient temperature conditions. UHT milk processing consists of treating the milk at 138–145 °C (280–293 °F) for 2–10 seconds, and then it is immediately cooled to <32 °C and filled aseptically in sterile sealed packaging. Unlike sterilised milk from the batch autoclaves, rotating autoclaves or continuously operating hydrostatic sterilisers, the UHT process heats the milk by continuous flow, which causes fewer chemical changes. Fewer chemical changes are due to shorter holding times (seconds) because the milk is immediately cooled and packed after the thermal treatment, compared to the use of autoclaves or hydrostatic steriliser that sterilises milk already filled in cans or bottles.

There are two main systems to obtain UHT milk, direct and indirect methods. Direct heating can take place using steam injection into the milk. In the indirect system, the milk and heating medium do not have contact; for example, the plates (stainless steel) of a plate heat exchanger serve as a barrier. Although there is no consensus among the community, some authors claim that electric tube heating systems darken milk samples less, promoting more subtle flavour changes than other indirect heating systems (Deeth 1999; Coolbear et al. 2022). There is also the possibility of working with mixed heating systems.

Rosenberg (2020) affirmed that the accumulation of different volatile and non-volatile flavour and aroma compounds might originate from the effects of heat treatment and storage conditions on the proteins, carbohydrates and lipid constituents of milk reflecting in the sensory quality of UHT milk. On the other hand, most of the enzymatic activity in raw milk is reduced to a hardly or not detectable level under sterilisation conditions. Li et al. (2021), studying the association of seasonal variations with physicochemical changes and age gelation in stored UHT milk, observed that UHT milk produced in the early season had the lowest protein and calcium contents, the smallest mean particle size when fresh and the lowest propensity for age gelation. These authors also concluded that the heat-induced aggregation among partially κ-casein-depleted casein micelles might play a part in the physicochemical gelation of UHT milk during storage. In addition to the seasonal factors, Datta and Deeth (2001) also reported that other significant factors in the UHT milk could influence the age gelation, such as the proteolysis during storage, the initial milk composition, quality and storage temperature.

For full-fat and semi-skimmed UHT milk, another possible detrimental change during storage is the agglomeration of the fat globules with the formation of a floating fat layer or a fat-rich protein aggregate that is dense enough to sediment. Fat separation during the shelf life of UHT milk has been commonly recognised to be caused by insufficient homogenisation because a small number of fat globules will remain large and represent a significant volume of the total fat. The relatively short shelf life of pasteurised milk led to the development of UHT milk for ambient distribution, which has gained widespread acceptance and popularity in many countries. At the beginning of the commercialisation of UHT milk, consumers perceived a 'cooked' taste in this product, but this is not highlighted today by current milk consumers. ESL milk has been available in Europe and North America since the early 1960s.

Several changes can result from the high-temperature heating of milk. However, greater attention from the sensory team should be given to the formation of flavours and odours related to the presence of sulphur compounds, such as cooked, cabbage and sulphur (Coolbear et al. 2022). The attention of the sensory team should be divided into two important points of the UHT milk production line, the sensory quality of the product immediately after processing and the sensory quality after a few days of storage.

Many of the sensory attributes related to sulphur compounds, that are apparent after heating, disappear after a few days (Manning and Nursten 1985). It is also necessary to be concerned with the browning promoted by the compounds from the Maillard reaction. Due to the uncertainties about the consumers' relationship with the development of cooked flavour in milk, it is suggested to combine descriptive analyses with affective consumer analyses. In particular, the generation of 'liking' or 'penalty' helps to identify the impact on the sensory quality of the product of the compounds generated in the UHT process (Desai et al. 2013; Pagès et al. 2014).

6.3.4 *Sensory profile of sterilised milk*

Manners and Craven (2003) highlighted that the term 'sterilised' is used to describe milk that has been packed in airtight containers and subjected to temperatures above 100 °C (212 °F). In this case, the product is packed before heat treatment. These authors confirm that this type of sterilisation heat treatment of the milk can result in a caramelised, burnt flavour and a brown colour. Typical heating parameters for packaged sterilised milk are between 115 and 120 °C (239–248 °F) for 10–20 minutes. However, the temperature and time conditions applied to obtain sterilised milk depend on the composition of the product, the type and dimensions of the package and the desired degree of caramelisation (Manners and Craven 2003). Hinrichs et al. (2022) reported that for in-container sterilisation of milk temperatures range from 105 to 120 °C (239 to 248 °F) with an adequate treatment time of 10–40 minutes, and the higher temperatures being associated with shorter times. However, these authors also reported that the term 'sterilisation' does not ensure that the product is sterile in the strict microbiological sense because absolute sterility cannot be obtained by standard heat treatment if statistics are taken into consideration.

Manners and Craven (2003) also affirmed that the consumption of sterilised milk products has declined since the advent of UHT processing and improved chilled storage and distribution supply chains for pasteurised and ESL milk.

6.4 Sensory profile of flavoured milks

Flavoured milks are milk-based products with added food colouring, flavouring and sweetened with sugar or other sweeteners. Bisig and Kelly (2017) defined flavoured milks as ready-to-drink products made from unfermented milk of different fat contents, mixed with sugar (or other sweeteners), other ingredients and additives. Flavoured milk beverages have increased in popularity and are packaged in single-service, closable plastic containers. The most popular flavours include vanilla, strawberry, chocolate, banana, coffee, cocoa and other fruity flavours that appeal to children. Generally, flavoured milk is prepared by standardising dairy and non-dairy ingredients in appropriate concentrations.

Initially, flavouring was intended to increase the palatability of the product, aiming to increase milk consumption, especially among children (Li and Drake 2015). Despite commercial success, the principal nutritional disadvantage of flavoured milk is the content of added sucrose, often as much as 5–9% w/v (Bisig and Kelly 2017). Therefore, the possibility of reducing the concentration of sugar in these products is of sensory interest. Flavoured milk can also be prepared by adding plant materials or components, resulting in functional milk products. Functional foods are defined as foods containing bioactive compounds that show health benefits and exhibit no toxic effects (Arab et al. 2019). Currently, the production of flavoured milk is not exclusively aimed at increasing palatability but also the possibility of producing nutritionally enriched milk. However, the addition of functional compounds can negatively impact the sensory characteristics of conventional fluid milk. The flavouring process can lead to overlapping of undesirable flavours often inherent with substances of nutritional value such as vitamins, probiotics, prebiotics and minerals. It is the role of the sensory team to monitor possible sources of sensory deviation, which in flavoured milk are: the base fluid milk, the set of substances added during formulation (and in relation to their specific functions, e.g. thickeners on texture, sweeteners on taste, colouring on appearance) and nutritional additives when present.

To guide the development of sensory-enriched flavoured milk, Zhi et al. (2016) used the hedonic scale and JAR methods directly on consumers. Thus, they conducted penalty analysis and partial least squares (PLS) regression on dummy variables to obtain valuable directional information for product reformulation. Penalty analysis makes it possible to identify the attributes that cause an increase or a decrease in consumer acceptance (Pagès et al. 2014). While, with PLS, a better prediction of general preference can be made compared to the original variables (Mohammadi Moghaddam et al. 2016).

In CATA tests conducted with consumers to produce the sensory profile of chocolate-flavoured milk, the following descriptors were used: dark brown colour, light brown colour, brightness (capacity to reflect light on a surface), chocolate aroma, milk aroma, sweet aroma, caramel aroma, sweet taste, milk flavour, vanilla flavour, chocolate flavour, caramel flavour, chocolate flavour, little flavour, undissolved chocolate, bitter, watery, thick, viscosity (on pouring from the glass into the mouth), liquid and rough (Della Lucia et al. 2016; Oliveira and Deliza 2021). The previous example illustrates the model of attributes that can be used for chocolate-flavoured products. Other flavours will follow the same structure, replacing unnecessary attributes with those raised by sensory panellist teams about general characteristics (liquid and sweet dairy base) and especially those related to the specific sensory attributes of the flavouring in question. Sweet taste is a vital acceptance criterion for flavoured milk, so it is essential that the sensory team carefully monitor it. In flavoured milk, sweet taste relates to more than the concentration of sweetener added. Sweetness perception can be increased by vanilla, caramel or fruity aromas, at concentrations above and below their own detection threshold (Boakes and Hemberger 2012; Labbe and Martin 2009).

The significant concerns of a segment of consumers with the healthiness of products has motivated the production of fortified foods with functional characteristics. In flavoured milk, the association between the addition of functional substances and reducing the sugar load has not been uncommon. When this type of association is made, attention must be paid to the two primary sources of sensory variation: the impact of the added functional substance and the effect of replacing sugar with low-energy sweeteners.

In products with sugar substitution, aftertaste is often found, especially a sweet aftertaste. Dynamic (or temporal) sensory methods are used to detect changes in stimuli throughout the tasting process. Oliveira et al. (2015) conducted temporal check-all-that-apply (TCATA) with trained consumers and tasters to verify the effect of sugar reduction in chocolate-flavoured milk. In TCATA, evaluators receive a list of terms and are asked to select all the terms they recognise in the sample at each time point in the evaluation. Evaluators can check the terms when the sensory attribute is applicable and uncheck them when it is not applicable (Castura et al. 2016). Marsanasco et al. (2015) studied the addition of encapsulated bioactive compounds (omega-3, omega-6, vitamin E and encapsulating folic acid) in flavoured milk. The authors observed with acceptance tests that the addition of the compounds did not significantly affect consumer acceptance. No differences were found by panellists between the samples using discriminative tests. In informed tests, a positive attitude change among potential consumers can be observed when they know that they are drinking functional milk, which means an opportunity for the sale of such products.

Significant advancements in product quality have been made, many of these arising from improved knowledge of the functional properties of ingredients and their impact on structure and texture (Goff and Griffiths 2006). Flavoured milk may be sold as a refrigerated pasteurised product or as a UHT-treated product that does not require refrigeration. In this scenario, adding flavours or aromas to milk can increase the diversity of dairy product tastes offerings. It makes milk consumption more attractive for some demographic groups, particularly children, youths and even some adult groups.

References

Al-Attabi, Z., D'Arcy, B.R., and Deeth, H.C. (2014). Volatile sulfur compounds in pasteurised and UHT milk during storage. *Dairy Science & Technology* 94 (3): 241–253.

Arab, M., Razavi, S.H., Hosseini, S.M. et al. (2019). Production and characterisation Mortazavi an of functional flavored milk and flavored fermented milk using microencapsulated canthaxanthin. *LWT- Food Science and Technology* 114: 108373.

Bi, J. (2005). Similarity testing in sensory and consumer research. *Food Quality and Preference* 16 (2): 139–149.

Billington, C., Powell, J., and Soboleva, T. (2020). *Safety Profile of Extended Shelf Life Refrigerated Milk*, 53. New Zealand: Crown Copyright.

Bisig, W. and Kelly, A. (2017). Liquid milk products: flavored milks. In: *Reference Module in Food Science*. Elsevier ISBN: 9780081005965. https://doi.org/10.1016/B978-0-08-100596-5.21868-4.

Boakes, R.A. and Hemberger, H. (2012). Odour-modulation of taste ratings by chefs. *Food Quality and Preference* 25 (2): 81–86.

Calvo, M.M. and de la Hoz, L. (1992). Flavour of heated milks. A review. *International Dairy Journal* 2 (2): 69–81.

Castura, J.C., Antúnez, L., Giménez, A., and Ares, G. (2016). Temporal Check-All-That-Apply (TCATA): a novel dynamic method for characterizing products. *Food Quality and Preference* 47: 79–90.

Chapman, K.W. and Boor, K.J. (2001). Acceptance of 2% ultra pasteurised milk by consumers, 6 to 11 years old. *Journal of Dairy Science* 84: 951–954.

Chapman, K.W., Lawless, H.T., and Boor, K.J. (2001). Quantitative descriptive analysis and principal component analysis for sensory characterization of ultrapasteurized milk. *Journal of Dairy Science* 84 (1): 12–20.

Chung, S.J., Lim, C.R., and Noh, B.S. (2008). Understanding the sensory characteristics of various types of milk using descriptive analysis and electronic nose. *Korean Journal of Food Science and Technology* 40: 47–55.

Claassen, M. and Lawless, H.T. (1992). Comparison of descriptive terminology systems for sensory evaluation of milk. *Journal of Food Science* 57: 596–621.

Clark, S., Costello, M., Drake, M., and Bodyfelt, F. (ed.) (2009). *The Sensory Evaluation of Dairy Products*. Springer Science & Business Media.

Cole, S., Goetze, A., and Meunier-Goddik, L. (2020). Pasteurized milk. In: *Encyclopedia of Dairy Sciences*, 3e (ed. P.L.H. McSweeney and J.P. McNamara), 444–450. Oxford: Academic Press https://doi.org/10.1016/B978-0-12-818766-1.00142-2.

Coolbear, T., Janin, N., Traill, R., and Shingleton, R. (2022). Heat-induced changes in the sensory properties of milk. *International Dairy Journal* 126: 11. ISSN: 0958-6946. https://doi.org/10.1016/j.idairyj.2021.105199.

Cornell University (2007). Dairy Foods Science Notes. Department of Food Science. Version 10-10-07. Available in: https://foodsafety.foodscience.cornell.edu/sites/foodsafety.foodscience.cornell.edu/files/shared/documents/CU-DFScience-Notes-Milk-Pasteurisation-UltraP-10-10.pdf. Date accessed: 2 April 2022.

Datta, N. and Deeth, H.C. (2001). Age gelation of UHT milk-a review. *Food and Bioproducts Processing* 79: 197–210.

Deeth, H.C. (1999). Electrical heating using 'current passage tube' technology. *Australian Journal of Dairy Technology* 54 (1): 66.

Deeth, H.C. (2021). Heat-induced inactivation of enzymes in milk and dairy products: a review. *International Dairy Journal* 121: 105104. ISSN: 0958-6946. https://doi.org/10.1016/j.idairyj.2021.105104.

Deeth, H.C. (2022). Heat treatment of milk: pasteurization (HTST) and thermization (LTLT). In: *Encyclopedia of Dairy Sciences*, 3e (ed. P.L.H. McSweeney and J.P. McNamara), 645–654. Academic Press ISBN 9780128187678, https://doi.org/10.1016/B978-0-12-818766-1.00133-1.

Della Lucia, F., Do Carmo, J.R., Morais, C.S.N. et al. (2016). Physicochemical and sensory quality of several commercial Brazilian chocolate milk beverages. *International Journal of Dairy Technology* 69 (3): 364–371.

Desai, N.T., Shepard, L., and Drake, M.A. (2013). Sensory properties and drivers of liking for Greek yogurts. *Journal of Dairy Science* 96 (12): 7454–7466.

Dijksterhuis, G. (1995). Multivariate data analysis in sensory and consumer science: an overview of developments. *Trends in Food Science & Technology* 6 (6): 206–211.

FDA (2019). *Grade "A" Pasteurized Milk Ordinance (2019 Revision)*. U.S. Department of Health and Human Services Public Health Service Food and Drug Administration https://www.fda.gov/media/140394/download.

Francis, L.L., Chambers, D.H., Kong, S.H. et al. (2005). Serving temperature effects on milk flavor, milk aftertaste, and volatile compound quantification in non-fat and whole milk. *Journal of Food Science* 70: S413–S418.

Fromm, H.I. and Boor, K.J. (2004). Characterization of pasteurized fluid milk shelf-life attributes. *Journal of Food Science* 69 (8): M207–M214.

Gandy, A.L., Schilling, M., Coggins, P. et al. (2008). The effect of pasteurisation temperature on consumer acceptability, sensory characteristics, volatile compound composition, and shelf-life of fluid milk. *Journal of Dairy Science* 91: 1769–1777.

Goff, H.D. and Griffiths, M.W. (2006). Major advances in fresh milk and milk products: fluid milk products and frozen desserts. *Journal of Dairy Science* 89: 1163–1173.

de Graaf, S., Van Loo, E.J., Bijttebier, J. et al. (2016). Determinants of consumer intention to purchase animal-friendly milk. *Journal of Dairy Science* 99 (10): 8304–8313.

Greenhoff, K. and MacFie, H.J.H. (1994). Preference mapping in practice. In: *Measurement of Food Preferences* (ed. H.J.H. MacFie and D.M.H. Thomson). Boston, MA: Springer https://doi.org/10.1007/978-1-4615-2171-6_6.

Guinard, J.X. (2000). Sensory and consumer testing with children. *Trends in Food Science & Technology* 11 (8): 273–283.

Hedegaard, R., Kristensen, D., Nielsen, J. et al. (2006). Comparison of descriptive sensory analysis and chemical analysis for oxidative changes in milk. *Journal of Dairy Science* 89: 495–504.

Hinrichs, J., Wedel, C., and Atamer, Z. (2022). Heat treatment of milk: Sterilization. In: *Encyclopedia of Dairy Sciences*, 3e, vol. 4 (ed. P.L.H. McSweeney and J.P. McNamara), 659–670. Academic Press. https://doi.org/10.1016/B. ISBN: 978-0-12-818766-1.00232-4.

Holsinger, V.H., Rajkowski, K.T., and Stabel, J.R. (1997). Milk pasteurisation and safety: a brief history and update. *Revue scientifique et technique* 16 (2): 441–451.

Ishler, V. and Roberts, B. (1991). *Guidelines for Preventing Off-flavors in Milk*. Pennsylvania, PA: The Dairy Practices Council.

Johnson, D.S., Duncan, S.E., Bianchi, L.M. et al. (2015). Packaging modifications for protecting flavor of extended-shelf-life milk from light. *Journal of Dairy Science* 98 (4): 2205–2214.

Labbe, D. and Martin, N. (2009). Impact of novel olfactory stimuli at supra and subthreshold concentrations on the perceived sweetness of sucrose after associative learning. *Chemical Senses* 34 (8): 645–651.

Lawless, H.T. and Claassen, M.R. (1993). Validity of descriptive and defect-oriented terminology systems for sensory analysis of fluid milk. *Journal of Food Science* 58 (1): 108–112.

Lee, A.P., Barbano, D.M., and Drake, M.A. (2017). The influence of ultra-pasteurisation by indirect heating versus direct steam injection on skim and 2% fat milks. *Journal of Dairy Science* 100: 1688–1701.

Li, X.E. and Drake, M. (2015). Sensory perception, nutritional role, and challenges of flavored milk for children and adults. *Journal of Food Science* 80 (4): R665–R670.

Li, B., Hayes, J.E., and Ziegler, G.R. (2014). Just-about-right and ideal scaling provide similar insights into the influence of sensory attributes on liking. *Food Quality and Preference* 37: 71–78.

Li, S., Ye, A., and Sing, H. (2021). Physicochemical changes and age gelation in stored UHT milk: seasonal variations. *International Dairy Journal* 118: 105028.

MacFie, H.J.H. and Thomson, D.M.H. (1988). Preference mapping and multidimensional scaling. In: *Sensory Analysis of Foods* (ed. J.R. Piggott), 381–409. New York, NY: Elsevier Applied Science.

Manners, J. and Craven, H. (2003). Milk. Processing of liquid milk. In: *Encyclopedia of Food Sciences and Nutrition*, 2e (ed. B. Caballero, L.C. Trugo and P.M. Finglas), 3951–3957. London: Academic Press.

Manning, D.J. and Nursten, H.E. (1985). Flavour of milk and milk products. In: *Developments in Dairy Chemistry—3*, 217–238. Dordrecht: Springer.

Marsanasco, M., Márquez, A.L., Wagner, J.R. et al. (2015). Bioactive compounds as functional food ingredients: characterization in model system and sensory evaluation in chocolate milk. *Journal of Food Engineering* 166: 55–63.

Martin, N., Carey, N., Murphy, S. et al. (2016). Exposure of fluid milk to LED light negatively affects consumer perception and alters underlying sensory properties. *Journal of Dairy Science* 99 (6): 4309–4324.

McCarthy, K.S., Parker, M., Ameerally, A. et al. (2017a). Drivers of choice for fluid milk versus plant-based alternatives: what are consumer perceptions of fluid milk? *Journal of Dairy Science* 100 (8): 6125–6138.

McCarthy, K.S., Lopetcharat, K., and Drake, M.A. (2017b). Milk fat threshold determination and the effect of milk fat content on consumer preference for fluid milk. *Journal of Dairy Science* 100: 1702–1711.

Mohammadi Moghaddam, T., Razavi, S., Taghizadeh, M., and Sazgarnia, A. (2016). Sensory and instrumental texture assessment of roasted pistachio nut/kernel by partial least square (PLS) regression analysis: effect of roasting conditions. *Journal of Food Science and Technology* 53 (1): 370–380.

Moskowitz, H.R., Beckley, J.H., and Resurreccion, A.V. (2012). *Sensory and Consumer Research in Food Product Design and Development*, 448. Iowa, USA: Wiley, Blackwell Publishing Ltd., and The Institute of Food Technologists ISBN-13: 978-0813813660.

Mounchili, A., Wichtel, J.J., Bosset, J.O. et al. (2005). HS-SPME gas chromatographic characterisation of volatile compounds in milk tainted with off-flavour. *International Dairy Journal* 15: 1203–1215.

Nelson, J.A. and Trout, G.M. (1934). Judging market milk. In: *Judging Dairy Products*, 31–49. Milwaukee, WI: Olsen.

Oliveira, D. and Deliza, R. (2021). Comparison of consumer-based methodologies for optimizing the development of new products: a case study with probiotic chocolate flavored milk. *Food Science and Technology International* 27 (6): 539–553.

Oliveira, D., Antúnez, L., Giménez, A. et al. (2015). Sugar reduction in probiotic chocolate-flavored milk: impact on dynamic sensory profile and liking. *Food Research International* 75: 148–156.

Pagès, J., Berthelo, S., Brossier, M., and Gourret, D. (2014). Statistical penalty analysis. *Food Quality and Preference* 32: 16–23.

Phillips, L.G., McGiff, M.L., Barbano, D.M., and Lawless, H.T. (1995). The influence of fat on the sensory properties, viscosity, and color of low-fat milk. *Journal of Dairy Science* 78: 1258–1266.

Potts, H.L., Amin, K.N., and Duncan, S.E. (2017). Retail lighting and packaging influence consumer acceptance of fluid milk. *Journal of Dairy Science* 100 (1): 146–156.

Rankin, S.A., Bradley, R.L., Miller, G., and Mildenhall, K.B. (2017). A 100-year review: a century of dairy processing advancements—pasteurization, cleaning and sanitation, and sanitary equipment design. *Journal of Dairy Science* 100 (12): 9903–9915.

Rosenberg, M. (2020). UHT sterilized milks. In: *Encyclopedia of Dairy Sciences*, 3e (ed. P.L.H. McSweeney and J.P. McNamara), 477–488. Academic Press ISBN: 9780128187678.

Sarkar, S. (2020). UHT processing – best technology for shelf-life extension of milk. *International Journal of Food Sciences and Nutrition Dietetics* 9 (1e): 1–2.

Schiano, A.N. and Drake, M.A. (2021). Consumer understanding of fluid milk and cheese processing and composition. *Journal of Dairy Science* 104 (8): 8644–8660.

Schiano, A.N., Harwood, W.S., and Drake, M.A. (2017). A 100-year review: sensory analysis of milk. *Journal of Dairy Science* 100 (12): 9966–9986.

Simon, S.A., de Araujo, I.E., Gutierrez, R., and Nicolelis, M.A.L. (2006). The neural mechanisms of gustation: a distributed processing code. *Nature Reviews. Neuroscience* 7: 890–901. https://doi.org/10.1038/nrn2006.

Stone, H. and Sidel, J.L. (2004). *Introduction to Sensory Evaluation. Sensory Evaluation Practices*, 3e, 1–19. San Diego: Academic Press.

Stone, H., Bleibaum, R.N., and Thomas, H.A. (2020). *Sensory Evaluation Practices*, 5e, 465. London: Academic Press.

Thompson, J.L., Drake, M.A., Lopetcharat, K., and Yates, M.D. (2004). Preference mapping of commercial chocolate milks. *Journal of Food Science* 69: S406–S413.

Villegas, B., Tárrega, A., and Costell, E. (2010). Optimising acceptability of new prebiotic low-fat milk beverages. *Food Quality and Preference* 21: 234–242.

Watson, M.P. and McEwan, J.A. (1995). Sensory changes in liquid milk during storage and the effect on consumer acceptance. *Journal of the Society of Dairy Technology* 48: 1–8.

Zhi, R., Zhao, L., and Shi, J. (2016). Improving the sensory quality of flavored liquid milk by engaging sensory analysis and consumer preference. *Journal of Dairy Science* 99: 5305–5317.

7 Sensory Profile of Yoghurt and Related Products

Amanda A. Prestes, Elane S. Prudencio, Maria H. Canella, Mônica Q. Freitas, Elson R. Filho, Tatiana C. Pimentel, Erick A. Esmerino and Adriano G. da Cruz

7.1 Introduction

Fermented milks have been manufactured and enjoyed from antiquity. Fermentation has been used for milk preservation since it was discovered that fermented milk was a stable and nutritious food that could be safely stored for a period of time. Fermented milks production expanded and improved over the centuries with an evolution from domestic production to industrial manufacturing with defined strain starter cultures, production protocols and large-scale hygienically designed equipment.

Fermented milk products are produced by the fermentation of lactose (milk sugar) by lactic acid bacteria (LAB), via various metabolic pathways without the involvement of oxygen or its agents (Tamime 2006). LAB strains are able to hydrolyse the lactose molecules, into its constituent glucose and galactose monosaccharides, from which lactic acid is produced. Lactic acid development reduces the pH of the milk until the isoelectric point of casein around pH 4.6 is reached and gelation occurs, due to casein precipitation and binding with calcium and phosphorus. Gelation is the basis of texture development while flavour and aroma are primarily attributable LAB. Importantly, acidification by LAB is responsible for inhibiting the growth of spoilage and pathogenic bacteria and contributing to the shelf life of fermented milk products.

Ferment milk products can be classified into two broad categories, based on fermentation temperature: cultured milks, such as cultured buttermilk, kefir, koumiss and a range of indigenous Scandinavian cultured milk products, are produced using starter cultures comprising mesophilic LAB and typical incubation temperatures in the range 20–25 °C and yoghurts, produced using LAB capable of fast growth at typical incubation temperatures in the range 37–45 °C. This chapter deals exclusively with the sensory properties of yoghurt and yoghurt products and explores the application of sensory analysis methodology in relation to product acceptability, quality improvement and innovative new product development.

Sensory Profiling of Dairy Products, First Edition. Edited by John J. Tuohy.
© 2023 John Wiley & Sons Ltd. Published 2023 by John Wiley & Sons Ltd.

7.2 Yoghurt and related products

Yoghurt has the highest per capita consumption of all fermented milks in Europe and North America. Yoghurt market revenue has been growing globally since 2015, reflecting the product's consumption growth. Even during the Covid-19 pandemic and the Russia–Ukraine war, there was an increase in consumption, and there are estimates of continued growth until 2027 (STATISTA 2022). This increase in consumption may be related to the functional and healthy properties which are attributed to fermented milks, since consumers in these countries have been focusing on diet as part of a healthy lifestyle. Consumer perception of yoghurt has also been improving due to more awareness in relation to digestibility and the activity of LAB in hydrolysing milk proteins into peptides that are better assimilated in the gastrointestinal tract and some of which are known to have various bioactive properties (Nguyen et al. 2020).

Originating in the Middle East, the discovery of yoghurt is reported among the first Neolithic civilisations who observed a change in raw milk appearance when it was stored for a number of hours at high-ambient temperatures in that region. At the beginning, this change occurred accidentally due to the introduction of LAB into the raw milk from contact surfaces during milking from the surrounding environment. This type of sour milk was tasty and could be stored for several days. The word 'yoghurt' is derived from a Turkish verb '*jugurt*' around the eighth century, which means 'coagulated' or 'curdled' and its consumption was related to stories about a long and prosperous life (Tamime and Robinson 1999, Gawai et al. 2017).

7.2.1 *Natural yoghurt*

According to the Codex Alimentarius (FAO 2003), yoghurt is derived from the fermentation of milk, with or without composition modifications, by the symbiotic action of two specific thermophilic strains: *Streptococcus salivarius* subsp. *thermophilus* and *Lactobacillus delbrueckii* subsp. *bulgaricus* which must be abundant (minimum of 10^7 CFU/g – Colony-Forming Units), active and viable in the product. If the product is heat-treated after the fermentation, the requirement for viable cells cannot be met. In some countries such as: Portugal, Spain, USA, France, Belgium or Sweden, there is a particular legislation that allows the choice and the use of *Lactobacillus bulgaricus* or *Streptococcus thermophilus*.

From a sensory perspective, the two fundamental types of natural yoghurt (without added flavouring or sweeteners) are set yoghurt and stirred yoghurt. The first is incubated directly in the pot in which it will be sold, conserving the structure of the gel formed by the milk fermentation process. Stirred yoghurt is incubated in large fermentation tanks and the natural gel is broken by mechanical stirring until a smooth and viscous consistency is achieved suitable for different types of packaging. Syneresis and gel surface homogeneity are important sensory attributes in set yoghurt. Syneresis, leading to the appearance of whey on the top surface, is a classical defect in stirred yoghurts. To prevent syneresis, the milk is given a high-heat treatment before addition of the starter cultures to increase the water binding capacity of the milk proteins.

7.2.2 *Concentrated yoghurts*

The labels 'Greek' or 'Greek-style' yoghurt came to prominence in 1990s in the United States, creating a new yoghurt market segment there and internationally. The origin of concentrated yoghurt products lies in the Balkan peninsula (from Bulgaria to Turkey) and concentrated fermented milk products have long been important products in the Middle East with such diverse names as *labneh, labaneh* or *lebneh* (in some Arab regions), *mast/mastou* in Iraq, *laben zeer* and *tan/than* in Egypt and Armenia regions, respectively (Tamime 2003). In the traditional process, after fermentation, the yoghurt was filled into muslin bags which were hung up to allow whey to drained slowly from the gel resulting a concentrated product. In America and Western Europe, the 'Greek' word on the packing appears to have enhanced sales because consumers associated it with a Mediterranean diet and healthy eating habits (Litopoulou-Tzanetaki and Tzanetakis 2014; Pimentel et al. 2017).

Codex Alimentarius recognises fermented milks as concentrated when the protein content has been increased to a minimum of 5.6% (w/v) after the fermentation process – about 2–2.5 times that of traditional yoghurt (FAO 2003). Nowadays, the concentration is achieved using centrifugal separators or by membrane filtration, typically ultrafiltration. Alternatively, powdered milk also can be added to enrich the solids level in the milk prior to fermentation.

The concentration step causes an increase in the total solids content from 14% to 21–23% and containing approximately 10% fat. Concentration results in sensory changes as a result of alterations to the physicochemical structure of the gel to create a creamy, smooth and viscous texture. Although traditional concentrated yoghurt has a high-fat content, several reduced-fat versions have been introduced in response to consumer trends (Robinson 2002; Chandan and O'Rell 2006). In addition to the concentration step, other process adaptations are necessary to obtain a product with the ideal sensory characteristics (Pimentel et al. 2017).

Skyr is another popular variant obtained by the concentration process and can be classified as a yoghurt in its strained form or a fresh acid-curd cheese from skim milk. The Skyr process is very similar to the yoghurt procedure, with the use of thermophilic cultures containing: *L. bulgaricus*, *S. thermophilus* and *Lacticaseibacillus casei* (basonym *Lactobacillus casei*) in the starter culture. In Skyr, acid development by the starter cultures is used to curdle the milk but rennet may also be added to the milk to promote gel formation and avoid wheying-off.

7.2.3 *Sweetened and flavoured yoghurts*

While Natural yoghurt usually contains only milk, added milk solids and starter cultures (Prestes et al. 2021; Tamime 2006), a diverse range of sweetened and flavoured yoghurt products are now marketed by international dairy companies. In accordance with regulatory controls in different countries, as well as milk solids, the addition of fruit pulp, fruit juices, other symbiotic LAB, potable water, thickeners, stabilisers, sweeteners, emulsifiers, colourants and flavourings are permitted.

7.2.4 *Drinking yoghurt*

Although classical yoghurt products are consumed with a spoon, used as topping or in savoury sauces, drinking-type yoghurts have grown in popularity following the trends for food on the go and lifestyle. Drinking yoghurts typically combine new fruits, juices, sweeteners and flavours with functional ingredients such as probiotics, fibres, vitamins and minerals to appeal to young and health-conscious consumers. Drinkable yoghurts are popular due to their ease of transport (practical packaging) and because they can be consumed out of the home.

The drinking yoghurt manufacturing process is similar to that for stirred yoghurt but the breaking of the coagulum after the fermentation includes a homogenisation step. Consequently, the yoghurt viscosity is reduced to a smoother, more fluid state, which makes it enjoyable to drink. The separation and release of the whey during storage is, however, a challenge and stabilisers such as carboxymethylcellulose or gelatine are often incorporated during manufacture (Prestes et al. 2021).

The composition of drinking yoghurt typically comprises 9% non-fat milk solids (proteins, carbohydrates, minerals), 1.5% fat, up to 8% sugar, 0.5% stabilisers and 5–15% fruit pulp, juice or syrup (Gawai et al. 2017).

7.2.5 *Frozen yoghurt*

In the Codex classification, fermented milks with the term 'frozen' is applied when the fermented dairy product is submitted to freezing operation and the specific starter cultures can be reactivated in considerable counts by thawing (FAO 2003).

Frozen yoghurt is consumed in the frozen state, similar to ice cream in texture but with a characteristic yoghurt flavour (Goff 2011; Chandan et al. 2017).

Frozen yoghurt composition varies according to the manufacturer or between countries. The typical composition is 1.8–6.0% fat, 3.5–3.8% protein, 28.8–34.0% total solids and pH 6.4–7.1 (Tamime 2003). Depending on the formulation, frozen yoghurt contains less calories compared to common ice creams and approximately 70% less fat.

7.3 Sensory profile of yoghurt and related products

The sensory analysis of yoghurt and its related products has to take into account the wide range of products within the category, e.g. natural yoghurt, set yoghurt, stirred yoghurt, concentrated yoghurt, flavoured yoghurt, high-protein yoghurt, frozen yoghurt, among others. Thus, the analyst must keep in mind that the choice of sensory techniques will depend on the objective to be investigated and the type of product to be evaluated. The combination of sensory and consumer study techniques seems to assume greater importance in the yoghurt market than for other dairy product categories. It appears that product segmentation in the yoghurt market is matched by a corresponding segmentation among yoghurt consumers, creating marketing associations:

- Natural yoghurts without additives for individuals concerned about food purity;
- Organic yoghurts aimed at individuals with ecological concerns;
- Sweetened and flavoured yoghurt presented in playful packaging for children;
- Yoghurts enriched with whey protein aimed at the athletic and fitness segment.

Milk, as the fundamental ingredient in yoghurt, can vary widely in composition (species, whole milk, semi-skimmed, skim milk, lactose-free, organic) and quality. Regardless of the type of milk, flavour attributes such as softness, sweetness, acidity and the absence of extraneous flavours are fundamental to the sensory profile of yoghurt. These central sensory attributes are not adequate to address the particular sensory characteristics of the wide range of yoghurt categories and products. The sensory team must also be aware of consumer trends and changes in presentation or form of consumption of yoghurt products (portioned pots for individual consumption with a spoon, multi-serving pots, beverage bottles, etc.). The primary sensory characteristics for the main yoghurt categories will be discussed according to the specifics and market considerations of each category.

7.3.1 *Sensory profile of natural yoghurt (set and stirred type)*

Natural yoghurt is generally marketed as 'set yoghurt' or 'stirred yoghurt'. The first type is traditionally fermented in the pot it is sold in, resulting in a firmer texture than stirred yoghurt but also the potential for texture variation from pot to pot. In stirred type, there is batch standardisation, so inter-pot variation can be minimised.

Natural yoghurt is characterised by its soft and viscous structure and slightly acidic, delicate flavour with nutty notes (Kosikowski 1982). Such characteristics are attributable to the symbiotic relationship developed by colonies of *S. salivarius ssp. thermophilus* and *L. delbrueckii ssp. bulgaricus*, and also by the acid concentration in the matrix. In his study, Crawford (1962) postulated that titratable acidity values between 0.85% and 0.90% (expressed as lactic acid) make the final ready-to-eat product most desirable. Acetic acid, diacetyl and acetaldehyde contribute to yoghurt's mild, slightly sour taste, and the balance between them is critical (Cheng 2010). Any increase in acidity in the supply chain post-production, due to secondary fermentation, is considered objectionable and is recognised as a shelf-life limiting factor.

Sensory evaluation of the yoghurt and other milk fermentation products began from the perspective of identifying sensory defects and off-flavours. In this way, it was already known since the beginning of the industrialisation of milk processing that the use of low-quality ingredients and technical errors made during processing would be reflected in several sensory defects in the final product. Regulatory agencies such as the United States Department of Agriculture (USDA) developed standardised tools for judging the quality of milk and milk products based on the detection of the important defects found in the final product and correlation with deviations in the quality of the raw milk and/or processing procedures (Schiano et al. 2017). Trout et al. (1939) published the document *Official Flavor Criticisms of Dairy Products Judged in the National Contest*, which became one of the landmarks for the standardised evaluation of milk, butter, cheese and ice cream and was also used as a basis for the evaluation of yoghurt.

Although the dairy products score card was also generally applicable to yoghurts, it was not included in the first version of the guidelines. Instead, the USDA published in 2001 a specification for yoghurt, non-fat yoghurt and low-fat yoghurt, which presented critical sensory attributes for these products. The USDA (2001) specified that yoghurt should have pleasant clean acid flavour. In addition, yoghurt must not have a bitter, rancid, oxidised, stale, yeasty or unclean taste. In relation to texture, the USDA (2001) recommended that yoghurt should be full-bodied and firm and, like cream, should be smooth and homogeneous (i.e. without grains). Yoghurt should have a uniform colour with a smooth and velvety appearance. The default colour of natural yoghurt should range from a glossy white to a duller off-white. A visual homogeneity is also desired, with only minor separation of the serum acceptable. Mould growth or surface discoloration/fading were ground for rejection.

In recent decades, descriptive and quantitative descriptive analysis (QDA)® (Stone 1974) has been widely used for yoghurts and related products. So, it is possible to observe the refinement and sophistication of the evaluated attributes, especially when seeking to meet the increasing demands of consumers. The following are examples of studies that have applied descriptive analysis to yoghurts over the last five decades.

Modler et al. (1983) used descriptive analysis to evaluate the effect of adding milk proteins to yoghurt. In the study, the authors analysed only the attributes 'softness', 'firmness', 'acidity' and appearance. Harper et al. (1991) used a team of 10 trained panellists to evaluate the intensity of 17 commercial natural yoghurts and 153 consumers to evaluate the acceptability of the same yoghurts. The authors correlated descriptive data with consumer responses and analytical measurements. For this, they chose the attributes 'sour', 'astringent', 'sally', 'sweet', 'cooked milk', 'buttery', 'bitter', 'yeasty', 'fruity', 'acetaldehyde', 'cheesy', 'milky' and 'caramel'. Isleten and Karagul-Yuceer (2006) compared the physical and sensory properties fat-free yoghurts containing added powdered milk. In the study, the authors profiled yoghurts through the attributes: free whey, lumpiness, thickness, chalkiness, aftertaste, cooked, whey, creamy, cereal, animal-like, cardboard, fermented, sour, salty, sweet, astringent.

Grygorczyk et al. (2013) studied the texture preferences of yoghurt consumers by comparing preferred attribute elicitation methods and conventional descriptive analysis. The attributes used in the descriptive analysis were 'stringiness', 'sour aroma', 'sweet', 'sour', 'dairy', 'thickness', 'chalkiness', 'aftertaste intensity' and 'colour'. Kycia et al. (2020) used descriptive analysis to investigate the effect of adding pullulan (a natural extracellular polysaccharide) on the sensory properties of low-fat yoghurt. The attributes used for sensory profiling were fermented milk aroma, heated milk aroma, sweet aroma, irritating aroma, different aroma, colour, thickness, smoothness, viscosity and for the flavour, fermented milk taste, sterilisation taste, acid taste, sweet taste, salty taste, bitter taste, chalky taste, impure taste and unnatural/different taste.

Thus, it can be seen that some attributes are central and common to the sensory evaluation of yoghurt and related products (Table 7.1) and related to the basic fundamentals of the yoghurt fermentation process

Table 7.1 Key attributes for yoghurt profiling and unflavoured or modified counterparts.

Attributes	Definition
Appearance	
Whiteness	White colour intensity.
Brightness	Ability of the surface to reflect light.
Homogenous	Product uniformity.
Thick	The ability of the product to flow from the spoon.
Aroma	
Sweet aroma	Specific aroma of products sweetened with sucrose or glucose.
Acid aroma	Aroma that refers to products with low pH.
Milky	Characteristic aroma of high-quality fresh milk.
Cooked	Suggestive of overheated milk, slightly sulphureous, and caramelised.
Flavour	
Acid taste	Classical definition of the taste promoted by substances with acidic pH.
Sweet taste	Specific taste of sucrose or glucose.
Bitter taste	Taste related to quinine or caffeine.
Creamy	Sour cream flavour, slightly greasy.
Cooked	Suggestive of boiled milk. Refers to sulphur compounds.
Astringent	Sensation by the shrinkage of the mucous membranes of the palate and tongue.
Milky	Features of high-quality fresh milk.
Yoghurt	Characteristics of unflavoured full-fat yoghurt.
Mouthfeel	
Thick	Ability to spread in the mouth with consistency.
Graininess	Presence of granular solids.
Creamy	Ability to fill the mouth pleasantly.
Firm	Ability to resist mouth movements.
Smoothness	Ability to demonstrate lightness when in the mouth.
Aftertaste	
Milky	Persistent milky flavour.
Sour milk	Persistent fermented taste after swallowing.
Astringent	Mouthfeeling tied after swallowing.
Bitter	Persistent bitter taste after swallowing.

Source: From the author.

There is also the possibility of generating sensory profiling of yoghurts with more agile methodologies that use untrained consumers to the detriment of highly trained panellists, e.g. check-all-that-apply (CATA) (Cruz et al. 2013) and rate-all-that-apply (RATA) (Tan et al. 2020).

The CATA technique consists of a structured questionnaire that asks consumers to select, from a closed list, the attributes that best describe the sample. In this test, the consumer describes the samples based only on perceived attributes. Unlike the classic descriptive analysis, the consumer does not need to select, and, therefore, evaluate, attributes that they do not recognise in the product. Therefore, CATA has been considered more intuitive and consumer-friendly, minimising cognitive processing (NG et al. 2013). RATA follows the same logic as CATA, but differs by asking consumers to indicate the intensity of the attribute recognised in the sample (Ares et al. 2014).

Consumer-based methods are also valuable for investigating the commercial perception of yoghurts. Cruz et al. (2013) evaluated the perception of yoghurt consumers with the techniques of projective mapping, CATA questionnaire and intensity scale. In the 1990s, Risvik et al. (1994) proposed the projective map, a technique that asks participants to position products in a two-dimensional space, bringing together those of more significant similarity. Esmerino et al. (2017a) tested pivot profile, projective mapping and CATA methods on Greek yoghurt samples. The idea of pivot is to obtain the relative meaning of the description of the products, collecting the free description of the consumers when placed to compare a sample tested against another considered pivot. The study highlighted the ability of all the three methods to describe Greek yoghurt samples in detail.

Acceptance tests using the hedonic scales are highly recommended as an essential part of any sensory analysis strategy for the analysis of yoghurts and related products. The hedonic classification represents a natural characteristic of the human response to food and its stimuli, involving the likes and dislikes of a person and relating different attitudes, such as preference between two or more products, acceptability of products, frequency of consumption and purchase intention (Moskowitz 1983).

The term ‘all-natural’ on food labels, including yoghurt, is an essential factor to be explored in consumer tests, where market development is the objective. It is necessary to consider, a priori, the absence of precise regulation involving the terminology ‘natural’ on food labels in general. The Food and Drug Administration (FDA) in the United States considers as natural a product that does not contain any added artificial substance: including colouring, texturisers and flavourings and does not associate ‘natural’ with any nutritional or other health benefits (FDA 2018).

The appeal of ‘natural’ on the label of food products can refer to a positive feeling related to the idea of health and well-being (Amos et al. 2019; Dominick et al. 2018). Consumers can sometimes associate products with ‘all-natural’ labels with characteristics of higher sensory and nutritional quality, which can be reflected in a greater willingness to pay a higher price for these products (Umberger et al. 2009). However, generalisations like this can be misleading and it is necessary to verify the importance of the natural appeal for specific products and markets. In yoghurt, for example, when studied in a population of primarily young women, the ‘all-natural’ designation received significantly lower acceptance scores compared to ‘high-protein’, ‘low-fat’ or ‘made with stevia’ designations on the label (Li and Dando 2019). Thus, it is essential to include techniques in the sensory methodology that address consumer motivations.

The Food Choice Questionnaire (FCQ) is an instrument designed to assess the level of importance given by individuals to nine factors related to food choices: health, mood, convenience, sensory appeal, natural content, price, weight control, familiarity and ethical concerns. The FCQ has been successfully used a few times to understand the motivations of yoghurt consumers (Pohjanheimo and Sandell 2009; Ares and Gámbaro 2007; Vecchio et al. 2016).

7.3.2 *Sensory profile of concentrated yoghurts*

Greek-style yoghurt is the best-known example of concentrated yoghurt on the market. In common with set-type yoghurts, this category of yoghurt is specifically designed for consumption with a spoon. Therefore, the sensory analysis of concentrated

Table 7.2 Texture attributes of concentrated yoghurts.

Attribute	Definition
Adhesiveness	Ability of the sample to adhere to the spoon and the side of the pot.
Apparent viscous	Ability of the sample to flow when subjected to tilting.
Consistent	Ability of the sample to maintain its shape and not deform while in the spoon.
Creamy	Creamy sensation when spreading when stirring the sample with the spoon. Refers to aerated and fat-rich products.
Density	Ability of the sample to resist spooning.
Grainy	Presence of visual granules in the sample.
Homogeneous	Uniformity of appearance on the spoon.
Jiggle	Gelatine-like movement of a product on the spoon while holding its shape.
Meltaway	Rate at which the sample dissolves or melts.
Spoon ropy	Visible threading off the spoon as the product is spooned.
Velvety	Velvety appearance during spooning. Refers to the combination of creamy texture and smooth appearance.

Source: Adapted from studies of Desai et al. (2013); Esmerino et al. (2017b); Megalemou et al. (2017).

Table 7.3 The main flavour attributes of concentrated yoghurts.

Attribute	Definition
Dairy sour	Specific dairy acidity.
Free fatty acid	Associated with short-chain fatty acids (butyric acid).
Milkfat	Associated with sugar-free heavy cream.
Soapy	Refers to the flavour of unflavoured bar soap.
Sour	Associated with low pH values.
Bitter	Basic taste associated with caffeine and alkaloids.
Greasy	Greasy sensation in the mouth after swallowing.
Rancid	Aroma associated with oxidised, old oil.

Source: Adapted from studies of Desai et al. (2013); Esmerino et al. (2017b); Megalemou et al. (2017).

yoghurts should involve using a spoon (Table 7.2) but should also include texture attributes detected by sight, hand movement and mouthfeel. Of course, the sensory team must be aware of traditional yoghurt sensory properties (see Section 7.3.1) because of the possibility of yoghurt concentration promoting an increase in the intensity of some sensory attributes of conventional yoghurt in a negative way. Table 7.3 lists some attributes related to flavour and odour to be monitored more selectively in concentrated yoghurts.

7.3.3 *Sensory profile of sweetened and flavoured yoghurt products*

The USDA general recommendations are still relevant for the quality assurance of commercial yoghurt products. However, the intensification of competition in the dairy market has driven the search for more diversified products to different consumer niches.

At the beginning of large-scale industrial production, yoghurt was generally sold in its natural form, i.e. without the addition of sweeteners, flavourings or colouring agents. Over time, yoghurts flavoured with fruits were successfully introduced to market and now represent more than two-thirds of yoghurt sales. In 2007, the top 10 flavours utilised in the production of flavoured yoghurt products were: vanilla, strawberry, mixed berries, blueberry, peach, raspberry, strawberry/banana, cherry, lemon and lime (Tribby 2009). Glanbia Nutritionals have recently reported that among recently launched flavoured yoghurt products, 32% were berry flavoured, 6% were citrus and 32% were other fruits such as: pineapple, mango and banana (Glanbia Nutritionals 2021).

The addition of jams and fruits to yoghurt led to the need to expand the sensory attributes analysed to encompass all the attributes of flavoured yoghurt in the early 1980's. Bodyfelt (1981) published a study questioning whether the American Dairy Science Association (ADSA) dairy scorecard was consistent with contemporary sensory assessment principles and proposed a revised version of the ADSA scorecard (Table 7.4) suitable for the assessment of flavoured yoghurts.

At the end of the decade, the same author (Bodyfelt et al. 1988) published an updated and expanded version of the list of attributes to be adopted in the scoring of flavoured yoghurts (Table 7.5), keeping the focus of the sensory analysis on the presence or absence of specifically defined defects. Later, the USDA specified that in the case of flavoured and coloured yoghurts the distribution of colouring or flavouring must be homogeneous to achieve a natural appearance (USDA 2001).

Table 7.4 Revised ADSA scorecard for flavoured yoghurts.

Appearance	Defects
Foamy (nonhomogeneous)	Shrunken
Free whey	Surface growth
Colour or hue	
Atypical colour	Too light
Colour "bleeding"	Too dark
Fruit frequency	
Excess fruit	Lacks fruit
Flavour defects	
Unnatural	Lacks flavouring
Lacks fine flavour	
Acidity level and microbiological characteristics	
High acid	Low acid
Green	Fruity/fermented
Sweetened	
Lacks sweetness	Too sweet
Dairy ingredients	
Cooked	Oxidised
Lacks freshness	Rancid
Old ingredient	Unclean
Other	
Bitter	Stabiliser

Source: Adapted from Bodyfelt (1981).

Table 7.5 American Dairy Science Association–approved attributes for strawberry yoghurt evaluation.

Appearance	Body/texture	Flavour
Atypical colour	Gel-like	High acid
Colour leaching	Too firm	Low acid
Excess fruit	Weak	Acetaldehyde
Lacks fruit	Grainy	Cooked
Lumpy	Ropy	Lacks fine flavour
Free whey		Too-high flavour
Shrunken		Unnatural flavour
Surface growth		Lacks sweetness
		Too sweet
		Stabiliser flavour
		Lacks freshness
		Unclean
		Oxidised
		Rancid
		Storage

Source: Adapted from Bodyfelt et al. (1988).

There are a range of aspects to be considered when formulating sweetened and flavoured yoghurt products, in particular:

- The choice of natural or synthetic flavourings
- The choice of sweetener (natural or synthetic)
- The choice of a light, diet or conventional formulations
- Whether to add fibre or other probiotic ingredients

Yoghurt flavouring began with the addition of fruit jams or jellies. However, advances in the flavours industry made the flavouring process more reproducible, and most yoghurt producers soon adopted the use of concentrated flavour extracts or synthetic flavours. From a sensory point of view, attention must be paid to the different flavouring designations used on yoghurt labelling. For example, Janiaski et al. (2016) analysed artificially flavoured strawberry yoghurts, so the list of evaluated attributes contained the terms 'Artificial strawberry odour', 'Particles (related to the colouring agent)' and 'Artificial strawberry flavour'.

Sweetness is a critical attribute in flavoured yoghurts. De Souza et al. (2021) established, via an acceptance test with a hedonic scale, the levels of sucrose that could be added in the production of strawberry yoghurt without compromising product acceptability (compromised acceptance threshold) or risk of sensory rejection (hedonic rejection threshold).

On the other hand, increasing consumers concerns about the intake of sugars and the need to serve the diabetic public motivated the expansion of versions of flavoured yoghurt products sweetened with sugar alternatives. To match the intensity of the sweet taste between the standard sample (usually, sucrose) and the sugar-free sample can be challenging, due to the difference in potency between the sweeteners used. Even minor variations in the concentration of high-potency sweeteners can impact consumer acceptability. The relative potency of the primary sweeteners used in the yoghurt industry is summarised in Table 7.6

Table 7.6 Main sweeteners used in the food industry and their potency[a] relative to sucrose.

Sweetener	Potency
Acesulfame-K	100–200
Aspartame	100–200
Cyclamate	~30
Neotame	7000–13 000
Saccharin	300–500
Sucralose	~600
Stevia	300[b]
Xylitol	1
Sorbitol	0.6
Sucrose	1

[a] Potency is expressed as times sweeter than sucrose (m/m).
[b] As it is of natural source, it may vary according to the proportion of rebaudioside.
Source: Adapted from Beltrami et al. (2018); Genus (2003); Sheet et al. (2014).

Magnitude estimation and time-intensity techniques can be used to determine the equivalence of sweeteners in yoghurts (Reis et al. 2011; Esmerino et al. 2013). Another critical point to be monitored in yoghurts with added sweeteners is the persistence of a residual sweet aftertaste in the mouth. For this type of analysis, dynamic sensory methods can be of great help. Dynamic techniques, also known as temporal, do not collect information at a specific point in the consumption stage, as in classic descriptive analysis. Temporal methods collect data throughout the consumption process, even after swallowing, enabling the monitoring of aftertastes. Aiming to determine the temporal sensory profile with consumers, rather than trained tasters, Esmerino et al. (2017b) determined the dynamic sensory profile of different ready-to-drink fermented milk products, including yoghurt, using the temporal check-all-that-apply (TCATA), temporal dominance of sensations (TDSs) and progressive profile (PP) techniques.

7.3.4 *Drinking yoghurt*

Drinking yoghurt is a type of stirred yoghurt usually flavoured and sold in ready-to-drink bottles. There is a strong appeal for practicality in this type of product, which is often marketed in the form of single-serve portions suitable for transport and consumption outside the home. The drinking yoghurt market is dominated by flavoured and sweetened products. The focus of the sensory team for this type of product should not only be directed to the quality of the yoghurt used as the base (see Section 7.3.3) but also to the descriptors related to flavouring, sweetness and particularly texture, which must be appropriately adjusted for drinking. The primary texture descriptors of drinking yoghurt can be seen in Table 7.7.

Product diversification has been a feature of yoghurt market growth in general, but is particularly so in the yoghurt beverage segment where specific product lines for different consumer niches and packaging innovations have been explored.

Table 7.7 Main attributes related to the texture of drinking yoghurt.

Attribute (mouthfeel/ texture)	Definition
Airy	The sensation of lightness and low density in the mouth, similar to Chantilly.
Astringent	Sensation popularly known as "dry mouth", is caused by the shrinkage of the mucous membranes of the palate and tongue.
Chalky	Sensation of the presence of fine powders, powdery sensation that adheres to the mouth.
Mouth coating	Ability of the product to coat the mouth (tongue and palate), sensation of fullness in the oral cavity.
Smoothness	Smooth mouthfeel, opposite to sandy.
Thickness	Thick sensation of the liquid in the mouth, opposite of the fine texture of the water.
Viscous	Cohesive internal forces manifest in its tendency to resist the flow. Viscosity can be inversely related to flat and directly associated with mouth-filling sensation.

Source: Adapted from studies of Cruz et al. (2013); Gonzalez et al. (2011); Allgeyer et al. (2010); Gad and Mohamad (2014); Pohjanheimo and Sandell (2009).

Yoghurt beverages include traditional skimmed milk-based and sugar-free versions but also more specific versions such as protein-rich, lactose-free and products containing prebiotics, probiotics or synbiotics. In skimmed milk-based formulations, it is common to use gums and thickeners capable of mimicking the texture naturally have been imparted by the milkfat that has been removed from the product, so sensory analysis must include detailed texture comparisons. Products containing probiotics and prebiotics can undergo more significant changes in taste and odour. Although this may not be the case for all probiotics (Farag et al. 2021; Mohammadi and Mortazavian 2011), the sensory team must be aware of the potential impact of the formation of aromatic compounds by co-fermenting organisms.

Kárnyáczki and Csanádi (2017) reported that lactose-free milk in yoghurt production promotes a more intense sweet taste than regular milk. The other attributes do not seem to change, as long as the technology for obtaining lactose-free milk does not promote changes in colour and aroma. The quality of the lactose-free milk used in the manufacture of drinking yoghurt is a critical consideration and is a topic that merits further research.

Over time, the greater diversification of products and flavours in the drinking yoghurt and related dairy beverage markets has promoted significant diversification in manufacturing technologies, along with greater diversification in the application of sensory developments. Gonzalez (2011) investigated the effect of a prebiotic ingredient (fructooligosaccharide) and symbiotic ingredients (fructooligosaccharide and *Lactobacillus acidophilus*) on the sensory properties and consumer acceptability of peach-flavoured drinkable yoghurts. The authors observed that higher intensities of sour and fermented aroma and lower sweetness differentiated the symbiotic product from the control and the product containing only prebiotic. Consumer results indicated that whole milk-based yoghurt drinks were preferred over skim milk-based yoghurt drinks,

Table 7.8 Texture attributes relevant to frozen yoghurt.

Attribute	Definition
Creaminess	Creamy sensation when spreading or stirring the sample with ta spoon. Relevant to aerated and fat-rich products.
Crumbliness	Tendency to fragment and break into pieces as served.
Fluffiness	Airy sensation when tasting the sample in the mouth.
Greasy	Greasy sensation in the mouth after swallowing.
Hardness	Resistance of the sample to the entry of a spoon.
Melting rate	Rate at which the sample melts and dissolves.
Presence of crystals	Presence of ice crystals in the sample.
Sandiness	Presence of granularity in the sample.
Smoothness	Smooth mouthfeel, opposite to sandy.

Source: Adapted from studies of Isik et al. (2011); Skryplonek et al. (2017); Soukoulis and Tzia (2008).

with the full-fat-milk-based yoghurt containing the prebiotic being the most acceptable. Results such as these highlight the difficulty of implementing standardised models for the sensory investigation of yoghurt products and the importance of selecting the appropriate methodology for specific applications.

7.3.5 *Sensory profile of frozen yoghurt*

Frozen yoghurts (also called 'yoghurt ice cream') have a traditional yoghurt base but are adapted for consumption at low temperatures, similar to ice cream. Therefore, the sensory analysis team must be aware of attributes relevant to the evaluation of ice cream which are not commonly encountered with other categories of yoghurts. It is necessary to supplement the attributes and general sensory techniques for evaluating yoghurt (see Sections 7.3.1, 7.3.2 and 7.3.3) with ice cream texture evaluation (Table 7.8). Other sensory properties of frozen yoghurt are covered elsewhere in this book (Chapter 16).

Another consideration is that frozen yoghurts are often promoted as a healthier alternative to ice cream, so it is common for frozen yoghurts to have sugar and fat reduction in their formulations and the addition of fibre and other probiotics. Therefore, the sensory team must be aware of consumer behaviour and attitudes as well as the impact of nutrient substitution on the product attributes.

References

Allgeyer, L.C., Miller, M.J., and Lee, S.Y. (2010). Drivers of liking for yogurt drinks with prebiotics and probiotics. *Journal of Food Science* 75 (4): S212–S219.

Amos, C., Hansen, J.C., and King, S. (2019). All-natural versus organic: are the labels equivalent in consumers' minds? *The Journal of Consumer Marketing* 36 (4): 11.

Ares, G. and Gámbaro, A. (2007). Influence of gender, age and motives underlying food choice on perceived healthiness and willingness to try functional foods. *Appetite* 49 (1): 148–158.

Ares, G., Bruzzone, F., Vidal, L. et al. (2014). Evaluation of a rating-based variant of check-all-that-apply questions: rate-all-that-apply (RATA). *Food Quality and Preference* 36: 87–95.

Bodyfelt, F.W. (1981). Dairy product score cards: are they consistent with principles of sensory evaluation? *Journal of Dairy Science* 64 (11): 2303–2308.

Bodyfelt, F.W., Tobias, J., and Trout, G.M. (1988). *The Sensory Evaluation of Dairy Products*, 598. New York, USA: Van Nostrand Reinhold.

Carniel Beltrami, M., Döring, T., and De Dea Lindner, J. (2018). Sweeteners and sweet taste enhancers in the food industry. *Food Science and Technology* 38: 181–187.

Chandan, R.C. and O'Rell, K.R. (2006). Manufacture of various types of yogurt. In: *Manufacturing Yogurt and Fermented Milk* (ed. R.C. Chandan), 211–236. Blackwell Publishing.

Chandan, R.C., Gandhi, A., and Shah, N. (2017). Yogurt: historical background, health benefits, and global trade. In: *Yogurt in Health and Disease Prevention* (ed. N.P. Shah), 331–351. Elsevier.

Cheng, H. (2010). Volatile flavor compounds in yogurt: a review. *Critical Reviews in Food Science and Nutrition* 50 (10): 938–950.

Crawford, R.J.M. (1962). How to succeed with yogurt. *Dairy Engineering* 79: 4–10.

Cruz, A.G., Cadena, R.S., Castro, W.F. et al. (2013). Consumer perception of probiotic yogurt: performance of check all that apply (CATA), projective mapping, sorting and intensity scale. *Food Research International* 54 (1): 601–610.

Desai, N.T., Shepard, L., and Drake, M.A. (2013). Sensory properties and drivers of liking for Greek yogurts. *Journal of Dairy Science* 96 (12): 7454–7466.

Dominick, S.R., Fullerton, C., Widmar, N.J.O., and Wang, H. (2018). Consumer associations with the "all natural" food label. *Journal of Food Products Marketing* 24 (3): 249–262.

Esmerino, E.A., Cruz, A.G., Pereira, E.P.R. et al. (2013). The influence of sweeteners in probiotic petit Suisse cheese in concentrations equivalent to that of sucrose. *Journal of Dairy Science* 96 (9): 5512–5521.

Esmerino, E.A., Tavares Filho, E.R., Carr, B.T. et al. (2017a). Consumer-based product characterization using pivot profile, projective mapping and check-all-that-apply (CATA): a comparative case with Greek yogurt samples. *Food Research International* 99: 375–384.

Esmerino, E.A., Castura, J.C., Ferraz, J.P. et al. (2017b). Dynamic profiling of different ready-to-drink fermented dairy products: a comparative study using temporal check-all-that-apply (TCATA), temporal dominance of sensations (TDS) and progressive profile (PP). *Food Research International* 101: 249–258.

FAO-Food and Agriculture Organization (2003). *Codex Standard for fermented milks-* CODEX STAN 243-2003.

Farag, M.A., Saleh, H.A., El Ahmady, S., and Elmassry, M.M. (2021). Dissecting yogurt: the impact of milk types, probiotics, and selected additives on yogurt quality. *Food Reviews International* 1–17.

FDA "'Natural' on Food Labeling" (2018). Available online: https://www.fda.gov/food/food-labeling-nutrition/use-term-natural-food-labeling (Accessed on 11 February 2019).

Gad, A.S. and Mohamad, S.H. (2014). Effect of hydrocolloid type on physiochemical properties of nonfat drinkable yogurt fermented with ropy and non-ropy yogurt cultures. *Comunicata Scientiae* 5 (3): 318–325.

Gawai, K.M., Mudgal, S.P., and Prajapati, J.B. (2017). Stabilizers, colorants, and exopolysaccharides in yogurt. In: *Yogurt in Health and Disease Prevention* (ed. N.P. Shah), 49–68. Elsevier.

Geuns, J.M.C. (2003). Stevioside - molecules of interest. *Phytochemistry* 64: 913–921.

Glanbia Nutritionals (2021). Since flavor is a major driver in yogurt choice, understanding today's most popular and emerging yogurt flavors is key to grabbing consumers' attention. [S.l.]. The Most Popular Yogurt Flavors, 2021. https://www.glanbianutritionals.com/pt-br/node/5426#:~:text=For%20spoonable%20dairy%20yogurt%2C%20strawberry,and%20raspberry%20(both%204%25). Accessed: 1 Feb. 2022.

Goff, H.D. (2011). Ice cream and frozen desserts: product types. In: *Encyclopedia of Dairy Science* (ed. J.W. Fuquay, P.F. Fox and P.L.H. McSweeney), 893–898. Elsevier.

Gonzalez, N.J., Adhikari, K., and Sancho-Madriz, M.F. (2011). Sensory characteristics of peach-flavored yogurt drinks containing prebiotics and synbiotics. *LWT - Food Science and Technology* 44 (1): 158–163.

Grygorczyk, A., Lesschaeve, I., Corredig, M., and Duizer, L. (2013). Extraction of consumer texture preferences for yogurt: comparison of the preferred attribute elicitation method to conventional profiling. *Food Quality and Preference* 27 (2): 215–222.

Harper, S.J., Barnes, D.L., Bodyfelt, F.W., and McDaniel, M.R. (1991). Sensory ratings of commercial plain yogurts by consumer and descriptive panels. *Journal of Dairy Science* 74 (9): 2927–2935.

Isik, U., Boyacioglu, D., Capanoglu, E., and Erdil, D.N. (2011). Frozen yogurt with added inulin and isomalt. *Journal of Dairy Science* 94 (4): 1647–1656.

Isleten, M. and Karagul-Yuceer, Y. (2006). Effects of dried dairy ingredients on physical and sensory properties of nonfat yogurt. *Journal of Dairy Science* 89 (8): 2865–2872.

Janiaski, D.R., Pimentel, T.C., Cruz, A.G., and Prudencio, S.H. (2016). Strawberry-flavored yogurts and whey beverages: what is the sensory profile of the ideal product? *Journal of Dairy Science* 99 (7): 5273–5283.

Kárnyáczki, Z. and Csanádi, J. (2017). Texture profile properties, sensory evaluation, and susceptibility to syneresis of yogurt prepared from lactose-free milk. *Acta Alimentaria* 46 (4): 403–410.

Kosikowski, F. (1982). *Cheese and Fermented Milk Foods*, 2e. Brooktondale, New York: F. V. Kosikowski & Associates LLC 711p.

Kycia, K., Chlebowska-Śmigiel, A., Szydłowska, A. et al. (2020). Pullulan as a potential enhancer of Lactobacillus and Bifidobacterium viability in synbiotic low fat yogurt and its sensory quality. *LWT* 128: 109414.

Li, T. and Dando, R. (2019). Impact of common food labels on consumer liking in vanilla yogurt. *Food* 8 (11): 584.

Litopoulou-Tzanetaki, E. and Tzanetakis, N. (2014). Fermented milks - range of products. In: *Encyclopedia of Food Microbiology*, 2e (ed. C. Batt and P. Patel), 884–894. Oxford: Academic Press.

Megalemou, K., Sioriki, E., Lordan, R. et al. (2017). Evaluation of sensory and in vitro antithrombotic properties of traditional Greek yogurts derived from different types of milk. *Heliyon* 3 (1): e00227.

Modler, H.W., Larmond, M.E., Lin, C.S. et al. (1983). Physical and sensory properties of yogurt stabilized with milk proteins. *Journal of Dairy Science* 66 (3): 422–429.

Mohammadi, R. and Mortazavian, A.M. (2011). Technological aspects of prebiotics in probiotic fermented milks. *Food Reviews International* 27 (2): 192–212.

Moskowitz, H.R. (1983). *Product Testing and Sensory Evaluation of Foods: Marketing and R&D Approaches (No. MOS 641 (BV 905.7))*. Food & Nutrition Press, Inc.

Ng, M., Chaya, C., and Hort, J. (2013). Beyond liking: comparing the measurement of emotional response using EsSense profile and consumer defined check-all-that-apply methodologies. *Food Quality and Preference* 28 (1): 193–205.

Nguyen, H.T.H., Gathercole, J.L., Day, L., and Dalziel, J.E. (2020). Differences in peptide generation following in vitro gastrointestinal digestion of yogurt and milk from cow, sheep and goat. *Food Chemistry* 317, 126419.

Pimentel, T.C., Antunes, A.E.C., Zacarchenco, P.B. et al. (2017). Brazilian yogurt-like products. In: *Yogurt in Health and Disease Prevention* (ed. N.P. Shah), 331–351. Elsevier.

Pohjanheimo, T. and Sandell, M. (2009). Explaining the liking for drinking yogurt: the role of sensory quality, food choice motives, health concern and product information. *International Dairy Journal* 19 (8): 459–466.

Prestes, A.A., Vargas, M.O., Helm, C.V. et al. (2021). How to improve the functionality, nutritional value and health properties of fermented milks added of fruits bioactive compounds: a review. *Food Science and Technology (Brazil)* 42 (e17721): 1–12.

Reis, R.C., Minim, V.P., Bolini, H.M. et al. (2011). Sweetness equivalence of different sweeteners in strawberry-flavored yogurt. *Journal of Food Quality* 34 (3): 163–170.

Risvik, E., McEwan, J.A., Colwill, J.S. et al. (1994). Projective mapping: a tool for sensory analysis and consumer research. *Food Quality and Preference* 5 (4): 263–269.

Robinson, R.K. (2002). Yogurt: types and manufacture. In: *Encyclopedia of Dairy Science* (ed. J.W. Fuquay, P.F. Fox and P.L.H. McSweeney), 525–528. Elsevier.

Schiano, A.N., Harwood, W.S., and Drake, M.A. (2017). A 100-year review: sensory analysis of milk. *Journal of Dairy Science* 100 (12): 9966–9986.

Sheet, B.S., Artik, N., Mahmoud, A.A., and Abdulaziz, O.F. (2014). Some alternative sweeteners xylitol, sorbitol, sucralose and stevia. *Karaelmas Fen ve Mühendislik Dergisi* 4 (1): 63–70.

Skryplonek, K., Gomes, D., Viegas, J. et al. (2017). Lactose-free frozen yogurt: production and characteristics. *Acta Scientiarum Polonorum. Technologia Alimentaria* 16 (2): 171–179.

Soukoulis, C. and Tzia, C. (2008). Impact of the acidification process, hydrocolloids and protein fortifiers on the physical and sensory properties of frozen yogurt. *International Journal of Dairy Technology* 61 (2): 170–177.

de Souza, L.B.A., Pinto, V.R.A., Nascimento, L.G.L. et al. (2021). Low-sugar strawberry yogurt: hedonic thresholds and expectations. *Journal of Sensory Studies* 36 (3): e12643.

STATISTA (2022). Yogurt – Worldwide. Consumer Markets, Food, Dairy Products & Eggs. https://www.statista.com/outlook/cmo/food/dairy-products-eggs/yogurt/worldwide

Stone, H. (1974). Sensory evaluation by quantitative descriptive analysis. *Food Technology* 28 (11): 24.

Tamime, A.Y. (2003). Yogurt: yogurt-based products. In: *Encyclopedia of Food Sciences and Nutrition* (ed. B. Caballero), 6259–6264. Elsevier.

Tamime, A.Y. (2006). *Fermented Milks. Society of Dairy Technology Technical Series* (ed. A.K. Tamime). Blackwell.

Tamime, A.Y. and Robinson, R.K. (1999). Historical background. In: *Yogurt Science and Technology* (ed. A.K. Tamime and R.K. Robinson), 1–10. Woodhead.

Tan, V.W.K., Wee, M.S.M., Tomic, O., and Forde, C.G. (2020). Rate-all-that-apply (RATA) comparison of taste profiles for different sweeteners in black tea, chocolate milk, and natural yogurt. *Journal of Food Science* 85 (2): 486–492.

Tribby, D. (2009). Yogurt. In: *The Sensory Evaluation of Dairy Products*, 2e (ed. S. Clark, M. Costello, M. Drake and F. Bodyfelt), 191–224. Springer Science + Business Media, LLC.

Trout, G.M., White, W., Downs, P.A. et al. (1939). Official flavor criticisms of dairy products judged in the national contest. *Journal of Dairy Science* 23 (4): 325–330.

Umberger, W.J., Thilmany McFadden, D.D., and Smith, A.R. (2009). Does altruism play a role in determining US consumer preferences and willingness to pay for natural and regionally produced beef? *Agribusiness: An International Journal* 25 (2): 268–285.

USDA - United States Department of Agriculture (2001). USDA Specifications for Yogurt, Non-fat Yogurt and Low-fat Yogurt: Dairy Programs. United States of America: Agricultural Marketing Service AMS USDA GOV. 4 p. Disponível em: https://www.ams.usda.gov/sites/default/files/media/yogurtlowfatnonfat.pdf. Acesso em: 20 fev. 2022.

Vecchio, R., Van Loo, E.J., and Annunziata, A. (2016). Consumers' willingness to pay for conventional, organic and functional yogurt: evidence from experimental auctions. *International Journal of Consumer Studies* 40 (3): 368–378.

8 Sensory Profiles of Middle Eastern and Related Cheeses

Barbaros Özer, Şebnem Ö. Budak and Hamid Ghoddusi

8.1 Introduction

There are hundreds of cheese varieties available worldwide with varying physical and sensory characteristics. The sensory features of the cheeses are largely determined by the milk type, flora and fauna of the geographical area where milking animals are grazed and production practices. The vast majority of the recorded cheese varieties are produced locally and have limited commercial importance. It is a fact that artisanal cheeses are generally characterised by stronger flavour intensities than the industrialised cheese varieties. The diversity of aroma and flavour in industrialised cheeses is somewhat limited since a small number of starter strains are employed in large-scale cheese productions. Understanding the distribution and metabolic activities of the natural flora of artisanal cheeses is of critical importance in the industrialisation of artisanal cheeses. Sensory analysis is essential for quality control to verify conformity of raw materials, products during the production process or final products. It is intended to determine if products are consistent with sensory specifications or not. On the other hand, sensory evaluation plays a great role in research and development as well. Formulating new recipes to appeal to consumers makes it impossible to avoid a sensory analysis stage. This is to ensure that the innovative products meet consumer expectations or even to validate the stability and consistency of flavour and aroma of the product during the shelf life. There are many ways of analysing the organoleptic properties of cheese and the main classical techniques for sensory analysis of cheese are summarised in Figure 8.1.

8.2 Sensory evaluation of Middle Eastern and related cheeses

The safety and quality of cheese has always been of concern to consumers and the sensory properties of cheese are probably the most fundamental characteristics of a quality cheese. However, when it comes to sensory evaluation, attributes vary considerably. Therefore, it is rather hard, if not impossible, to standardise the sensory evaluation of cheese. Having said that the principal features can be, to a certain extent, consistent. There are a number of internationally recognised sensory evaluation protocols for milk products, including cheese. ISO 22935-1/IDF 99-1, for example, was published jointly

Sensory Profiling of Dairy Products, First Edition. Edited by John J. Tuohy.
© 2023 John Wiley & Sons Ltd. Published 2023 by John Wiley & Sons Ltd.

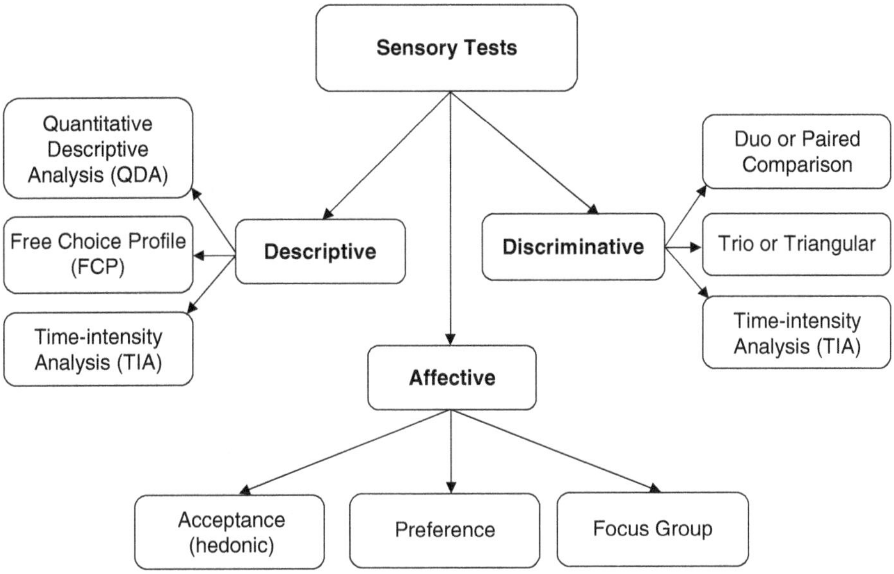

Figure 8.1 Main classical techniques for the sensory analysis of cheese. Based on Karimi et al. (2013).

by IDF and ISO and describes (i) the general guidance for the recruitment, selection, training and monitoring of assessors, (ii) recommended methods for sensory evaluation and (iii) guidance on a method for evaluation of compliance with product specifications for sensory properties by scoring. The American Dairy Science Association (ADSA) also established a guideline for the sensory judging of milk and milk products. Most of the Middle Eastern and North African countries do not have a specific regulatory standard for the sensory evaluation of dairy products and follow the guidelines mentioned earlier in this chapter. As an exception, the Iranian National Standards Organisation have introduced a guide for assessing certain important characteristics of some cheeses, including white brined cheese. For example, according to the Iranian National Standard (# 2324) on 'cheese-general specifications' (ISIRI 2002b) the sensory analysis of cheese should examine appearance (including colour), flavour and aroma, and body and texture. In another national standard on Brined Cheese-Specifications and Test Methods (# 2344-1) revised in 2016, more details are provided (ISIRI 2016). Sensory characteristics, which are scored between 0 and 5, include appearance (before and after cutting), texture, aroma (smell) and flavour. A total score of 15 or above is considered acceptable, subject to a minimum score of 4 for aroma and taste (flavour). This is perhaps the main reason why most of the studies on sensory evaluations of Iranian cheese varieties are limited to these sensory attributes. A closer look at the literature demonstrates that some of the studies include more attributes than the ones specified in ISIRI standard for Iranian cheese. It is also worth mentioning that in the published research the cut-off score of 4 for flavour and aroma recommended in the instructions by ISIRI is very rarely used. Table 8.1 summarises a selection of the attributes used in the sensory evaluation of Iranian cheese varieties.

Table 8.1 Some attributes considered in sensory analysis of Iranian cheese.

Attribute	References
Flavour	Pirouzian et al. (2012), Soltani et al. (2016)
Bitterness	Jalili (2016), Soltani et al. (2016)
Acidity (sourness)	Jalili (2016), Soltani et al. (2016), Hashemi et al. (2009)
Saltiness	Soltani et al. (2016), Hashemi et al. (2009)
Astringency	Jalili (2016)
Metallic	Jalili (2016)
Maturity	Hashemi et al. (2009)
Body & texture	Rashidi et al. (2015), Pirouzian et al. (2012), Barouei et al. (2011), Alizadeh and Lavasani (2013)
Firmness	Jalili (2016), Soltani et al. (2016)
Appearance	
Colour	Jalili (2016), Alizadeh and Lavasani (2013)
Odour (aroma)	Pirouzian et al. (2012), Barouei et al. (2011)
Creamy	Soltani et al. (2016), Hashemi et al. (2009)
Sweaty	Soltani et al. (2016)
Taste	Rashidi et al. (2015), Barouei et al. (2011)
Rancidity	Hashemi et al. (2009)
Overall acceptance/preference	Jalili (2016), Rashidi et al. (2015), Barouei et al. (2011), Alizadeh and Lavasani (2013)

Attributes in bold are recommended by the ISIRI cheese standard.

Table 8.2 Selected publications on techniques for the sensory analysis of Iranian cheese.

Cheese type	Sensory test	Details	References
UF-white cheese	Hedonic scale test	5-point	Zomorodi et al. (2011)
White brined cheese	Hedonic scale test	9-point	Barouei et al. (2011)
UF-white cheese	Commercial grader	20-point scale; colour & appearance (5), body & texture (5), odour & flavour (10)	Soltani et al. (2015, 2016)
UF Feta cheese	Hedonic scale test	5-point	Rashidi et al. (2015)
UF white cheese	Hedonic scale test	5-point	Pirouzian et al. (2012)
Processed cheese	Hedonic scale test	5-point texture, consistency, spreadability and flavour	Torkashvand (2010)

Reviewing the literature on the methods used for analysis of the sensory characteristics of Iranian cheeses reveals that in most of the studies, acceptance methods using hedonic 5- or 9-point scales were preferred (Table 8.2). An Overall Hedonic Score can be calculated for a cheese using the following formula based on authors' experience and dairy expert advice (Barouei et al. 2011).

$$\mathbf{OHS} = \frac{(\mathrm{TaS} \times 20) + (\mathrm{ArS} \times 10) + (\mathrm{TxS} \times 15)}{45}$$

Where OHS = Overall Hedonic Score; TaS = taste score; ArS = aroma score and TxS = texture score.

The sensory analysis of cheese, like many other foods, can be carried out at different levels. Sensory analysis by expert assessors, by selected panels and by consumer testing are common types of evaluation (Khosrowshahi et al. 2006). However, as Table 8.2 illustrates, most of the sensory analysis reported for Iranian cheese varieties has been limited to hedonic scale tests and the use of descriptive or discriminative analyses has not been reported. This is mainly due to the ease of recruiting panel members in an academic setting and the minimum amount of training needed for acceptance testing.

Most recently, Nottagh et al. (2020) demonstrated that the use of an edible coating for Iranian White brined UF cheese with chitosan (1.6%, w/v) and natamycin (18.5 mg/kg) did not affect the sensory properties of the cheese as evaluated using a 5-point hedonic test.

Over the past decades, a lot of advances have been made in instrumental sensory appraisal. While human sensory assessment of food is very valued, it is also recognised that it can be subjective and that sensory quality can also be measured by means of laboratory instruments, which eliminate or reduce the potential human errors. Examples of such instrumental techniques include colorimetry (Milani et al. 2017), texture profile analysis, (Milani et al. 2017) and electronic nose MS. Moreover, chemical analysis of lipolysis (e.g. acid degree value, volatile compounds) and proteolysis (e.g. free amino acids) have been used as an alternative to sensory analysis (Soltani et al. 2016; Hashemi et al. 2009). Nevertheless, instrumental and chemical analyses, at the very best, are generally considered as accompanying methods in assessing cheese sensory qualities rather than as alternative methods.

Numerous studies have been carried out to identify the volatile compound profiles of artisanal cheeses and to understand the factors determining the characteristic sensory properties of such cheeses. The sensory profiles of some selected Middle Eastern, North African, Greek-Cypriot, Turkish and Iranian cheese varieties are presented hereafter. Characterisation of the volatile aroma/flavour compounds found in the major cheese varieties of these regions and the factors affecting such compounds will also be discussed in detail. Since different regions of each country or countries sharing common cultural and geographical grounds have similar cheese varieties under different names, Middle Eastern and related cheeses have been grouped based on their manufacturing or ripening practices, i.e. cheeses ripened in brine, in goat skin, pasta-filata-type cheeses and so on. The volatile profiles of cheese varieties popular in the Middle East, the Balkans and North Africa are summarised in Table 8.3.

8.3 Cheeses ripened in brine

Brined-type white cheeses are particularly popular in the Balkans, the Middle East, the Mediterranean region and North Africa. The manufacturing procedures for these cheeses vary from one region to another, but they basically suit the hot climatic conditions of the regions where they are produced. The most widely known brined-type cheese varieties of the Balkans, Middle East and North Africa are Feta (Greece), Ezine, Edirne Beyaz and White cheese (Turkey, Iran), Halloumi (Cyprus and Greece), Akkawi (Lebanon, Syria and most Arab countries) and Domiati (Egypt). Images of some popular brine ripened cheese varieties are shown in Figure 8.2.

Table 8.3 Volatile compounds identified in some Middle Eastern and related cheeses[f].

	Acids	Aldehydes	Alcohols	Esters	Ketones	Miscellaneous	Remarks
Brined-type cheeses							
Ezine (Salum et al. 2018)[a]	**3.6** (butanoic and hexanoic acids)	**17** (benzaldehyde)	**13** (2-heptanol)	**78** (ethyl hexanoate)	**174** (2-octanone and 4-nonanone)	**13** (phenol)	Commercial cheeses with unknown ageing time
Feta (in wooden barrels) (Kondyli et al. 2012)[b]	**941** (butanoic, hexanoic and octanoic acids)	**4** (heptanal)	**111** (ethanol, 3-methyl-1-butanol and 2-butanol)	**77** (ethyl butanoate)	**18** (acetone)	**3** (d-nonalactone)	180-day-old experimental cheeses
Feta (in tin barrels) (Kondyli et al. 2012)[b]	**885** (octanoic and hexanoic acids)	**8** (hexanal and nonanal)	**69** (ethanol, 2-butanol)	**15** (hexyl formate)	**7** (acetone)	**3** (d-nonalactone)	180-day-old experimental cheeses
Van Otlu (Ocak et al. 2015)[a]	**70** (acetic and hexanoic acids)	**5** (tridecanal)	**8** (2-ethyl-1-hexanol)	**7** (ethyl hexanoate)	**3** (2-heptanone)	**6** (piperitone)	180-day old experimental cheeses
Urfa (Kirmaci et al. 2015)[a]	**182** (butanoic and hexanoic acids)	**3** (hexanal)	**117** (3-methyl-1-butanol)	**36** (ethyl acetate)	**123** (2-heptanone, 2-pentanone, 2-nonanone)	**48** (α-pinene)	180-day-old experimental cheeses
Civil (Gurkan et al. 2018)[a]	**399** (hexanoic and 2-methyl-1-butanoic acids)	**1** (3-methylbutanal)	**48** (3-methyl-1-butanol)	**22** (ethyl acetate)	**20** (3-hydroxy-2-butanone)	**23** (2,2-dimethyldecane)	90-day-old experimental cheeses
Scalded/Pasta-filata-type cheeses							
Kashkaval (Pappa et al. 2019)[c]	nd	**1** (hexanal)	**10** (ethanol and 3-methyl-butanol)	**60** (methyl hexanoate)	**25** (acetone)	nd	180-day-old experimental cheeses
Graviera Kritis (Bozoudi et al. 2016)[c]	**9** (butanoic and hexanoic acids)	—[d]	**1**	**1**	**1**	–	90-day-old experimental cheeses

GravieraNaxou[c]	**14** (hexanoic acid)	–	**1**	**2**	–	–	90-day-old experimental cheeses
Cheeses ripened in animal skin or pots							
Divle Tulum (Hayaloglu and Karabulut 2013)[a]	**91** (butanoic and hexanoic acids)	**5** (3-methylbutanal)	**390** (2-butanol)	**91** (methylbutanoate)	**1** (2-pentanone, 2-nonanone and 2-butanone)	**222** (hexane)	Commercial cheeses with unknown ageing time
Fresh cheese							
Kopanisti (Karali et al. 2013)[e]	nd	**71** (pentanal and butanal)	**2** (ethanol)	**854** (ethyl butanoate and ethyl propionate)	**86** (2-butanone and 2-pentanone)	nd	Commercial cheeses with unknown ageing time

[a] Expressed as μg/100 g.
[b] Expressed as peak area in TIC × 10^3.
[c] Expressed as peak area × 10^{-6}.
[d] Percentage of cyclohexanone peak is <1.
[e] Expressed as relative abundance as percentage of cyclohexanone peak.
[f] Figures in bold indicate the total concentration of each group of volatile compounds and chemicals in brackets indicate the most abundant volatile compounds of each group.

Figure 8.2 Middle Eastern brined cheese variety examples. (a) Ezine: full-fat, white brined cheese produced from a mixture of ewe's (40–50 %), goat's (at least 40%), and cow's (at most 15%) milk and native to north-western Turkey. *Source:* Moving Moment/Adobe Stock. (b) Edirne Beyaz: full-fat, white brined cheese produced from only one type of milk source (cow's, ewe's or goat's milk) and Native to Trakia region of Turkey. *Source:* nevigurme.com. (c) Feta PDO: full- or half-fat white cheese, produced from ovine milk or mixture of sheep's and goat's milk (up to 30%). Native to Greece. *Source:* MURRAY'S CHEESE. (d) Akkawi: Native to Syria, Palestine and most Arab countries. *Source:* Wall Street Foreign Trade Center.

8.3.1 *Ezine*

Ezine cheese is a full-fat, white brined cheese produced from a mixture of ewe's (approximately 40–50%), goat's (at least 40%) and cow's (at most 15%) milk. They are ripened in tinplate containers in large volumes (e.g. 17–20 kg), and then they are transferred to smaller tin plates or plastic vacuum packages towards the end of ripening. The total ripening period of this variety is about eight months. The main descriptive sensory features of this cheese variety are fatty acid, cooked, creamy, whey-like, animal-like, salty and acidic (Hayaloglu and Karagul-Yuceer 2011). Ezine cheese is characterised by a white colour and semi-soft to semi-hard texture with no rind (Figure 8.2a). Karagul-Yuceer et al. (2009) identified 35 aroma-active compounds from 12-month-old Ezine cheeses belonging to chemical groups of aldehydes, esters, acids, ketones, sulphur compounds and alcohols. Young Ezine cheeses had high-odour intensities of dimethyl sulphide (sulphur/corn-like note), diacetyl (buttery note) and ethyl butyrate (bubble gum odour). Intensities of methanethiol and methional, which are sulphur compounds, increased during ripening period. Acids are the main chemical group of volatile compounds in Ezine cheese accounting for 86.6 to 92% 1% of total volatile compounds (Salum et al. 2018) (Table 8.3). Especially, butanoic and hexanoic acids were reported to be the most dominant volatile acids in commercial Ezine cheeses (Salum et al. 2018). Karagul-Yuceer et al. (2007) applied a descriptive sensory analysis to investigate the sensory characteristics of commercial Ezine cheese samples (n = 22). A vocabulary containing 14 flavour and 10 texture terms were developed by a trained panel (n = 9). Regarding sensory properties, 20 out of 22 cheeses had similar characteristics to each other. Young cheeses were characterised as having cooked, creamy and whey-like flavours. Regarding the texture properties, the Ezine cheeses analysed were classified as soft or semi-hard.

8.3.2 *Edirne Beyaz*

Edirne Beyaz is one of the most famous traditional cheese varieties produced in Turkey (Anonymous. 2014). Edirne Beyaz cheese is made from only one type of milk source (cow's, ewe's or goat's milk). The milk is mildly heat-treated (similar to thermisation) and no starter culture is used. The ripening stage takes place in tinplate containers with brine solution containing 16–18% NaCl for 3–6 months. Edirne Beyaz is white in colour and has a soft to semi-hard texture (see Figure 8.2b). Salty acidic taste is the main characteristic of this variety often accompanied by a slight piquant flavour. Geographic conditions, local forage and production technologies applied determine the characteristic taste, aroma and texture of this variety. The major aroma compounds of mature Edirne cheese are carboxylic acids (Kamanlı-Can 2019).

8.3.3 *Feta*

Feta PDO cheese is the most typical product of the Greek cheese industry (Bozoudi et al. 2018). Its manufacturing practices are very much similar to Turkish White cheese (Edirne Beyaz). Preferably raw or mildly heat-treated milk is used in the manufacture of traditional Feta PDO cheese, and no starter culture is added. Traditional Feta PDO cheese is usually made from sheep's milk or a mixture of sheep's and goat's milk (up to 30%). The colour of Feta PDO cheese is snow-white and it has a semi-soft texture with small almond-like mechanical openings inside (see Figure 8.2c). Feta PDO cheese is slightly acidic and salty, and has a mild rancid taste. Pitso and Bester (2000) demonstrated that experimental Feta cheeses made from cow's milk or goat's milk or mixtures of both (65% cow's milk + 35% goat's milk or 35% cow's milk and 65% goat's milk) had indifferent sensory characteristics. The animal feeding system is the main factor determining the aroma and flavour characteristics of traditional Feta cheese. Natural pasture-based feeding system result in increases in the quantity of volatile compounds, mainly terpenes, in the end product (Morand-Fehr et al. 2007). The most abundant terpenoid compounds present in traditional Feta cheese were α-pinene, camphene, 2-β-pinene and α-fenchone, as reported by Bozoudi et al. (2018). The major volatile fractions of Feta cheeses are acids, esters, alcohols, hydrocarbons, aldehydes and ketones, with fatty acids being the most abundant (Bozoudi et al. 2018). Hexanoic acid is the most abundant fatty acid present in traditional Feta cheese made from raw milk. High levels of hexanoic acid are probably the result of the enhanced hexanoic-specific lipase activity of the wild lactic acid bacteria strains commonly found in Feta cheese. The characteristic aroma and flavour of Feta PDO cheese made from sheep's or goat's milk stems largely from branched-chain fatty acids (Curioni and Bosset 2002). Apart from their contribution to the characteristic Feta PDO cheese aroma and flavour, the fatty acids also act as precursors of alcohols, esters, methyl ketones and lactones (Curioni and Bosset 2002). Esters are the second most abundant group of volatile compounds in Feta cheese. Ethyl esters of hexanoic and octanoic acids were reported to be the most abundant ethyl esters in Feta PDO cheese produced in southwest Greece (Bozoudi et al. 2018). Feta PDO cheese contains high numbers of non-starter lactic acid bacteria (NSLAB) and lipases derived from these bacteria likely produce high levels of free fatty acids and esters (Kondyli et al. 2012). The degree of lipid hydrolysis occurring during the ripening period is among the major factors contributing the flavour characteristics of Feta cheese. Long-chain free fatty acids released during ripening as a result of metabolic activities of starter and/or NSLAB are particularly important (Zlatanos et al. 2002). Laskaridis et al. (2013) showed that the concentrations of saturated fatty acids increased during the ripening period of Feta cheese made from pasteurised milk with starter culture, whereas the levels of monounsaturated fatty acids (MUFA) and polyunsaturated fatty acids (PUFA) showed an opposite tendency. Generally, the levels of butyric, caproic, caprylic and capric acids are lower in Feta cheese made from pasteurised milk since the natural lipase of milk is destroyed by heat treatment (Kondyli et al. 2002). Also, full-fat Feta cheese was reported to have higher level of total free fatty acids (C_4-$C_{18:2}$) than half-fat Feta cheese (Kondyli et al. 2002).

The packaging material in which Feta cheese is ripened is another factor determining the type and concentration of volatile compounds in the end product. Feta cheese ripened in wooden barrels had higher levels of acetone and ethyl esters than those ripened in tin vessels (Kondyli et al. 2012) (Table 8.3). In case of industrial production, the choice of starter and adjunct cultures directly affects the volatile composition of Feta cheese. Bintsis and Robinson (2004) found that the incorporation of *Lactobacillus paracasei* subsp. *paracasei* and *Debaryomyces hansenii* as adjunct culture into Feta cheese contributed to the formation of a richer pattern of aroma compounds including alcohols, esters and aldehydes. In contrast, *Yersinia lipolytica* caused undesirable aroma and flavour in the final product. The use of commercial adjunct strains of *Lactococcus lactis* subsp. *lactis* and *L. lactis* subsp. *cremoris* in the manufacture of low-fat Feta cheese had a positive effect on the formation of organic acids (especially butyric acid and propionic acid), acetoin and on the overall sensory properties of the resulting cheese (Katsiari et al. 2002; Manolaki et al. 2006). Sarantinopoulos et al. (2002) investigated the possibility of using *Enterococcus faecium* FAIR-E 198 and *E. faecium* FAIR-E 243 either as a sole or as an adjunct culture in the production of Feta cheese. *E. faecium* strains contributed to the aroma/flavour, taste and colour of the experimental cheeses. The main volatile compounds produced by these adjunct cultures were ethanol, acetate, acetone, acetaldehyde, acetoin and diacetyl. Mesophilic starter cultures are usually preferred in the production of industrial Feta cheese. Pappa and Anifantakis (2001) showed that although the cheeses made with mesophilic starter cultures had higher flavour scores, grainy texture and acidic taste were the major drawbacks for the mesophilic cultures used.

8.3.4 *Lighvan and Iranian White cheeses*

Iranian White cheese has a closed texture and is ripened in brine solution similar to Edirne Beyaz cheese and Feta (Khosrowshahi et al. 2006). The end product is characterised by a desirable sour and salty taste, white to cream colour and a crumby semi-soft to semi-hard texture (Barouei et al. 2011). At an industrial level, the ripening period is 40–90 days (Azarnia et al. 1997). Lighvan, a traditional Iranian White cheese, is made with raw sheep's milk without the addition of starter cultures and ripened in the presence of around 18–24% NaCL brine for a minimum duration of six to eight months, to ensure the killing of potential pathogenic bacteria from the *Enterobacteriaceae* family, including *Escherichia coli*. This variety of cheese is characterised by a unique hardness (semi-hard), saltiness and spiciness (Pirouzian et al. 2012).

In recent decades, UF white cheese has become the dominant cheese variety in Iran. It is also known as Danish white cheese, as it was first imported from Denmark and gradually production lines were installed to produce the cheese in Iran. UF white cheese often does not go through a ripening stage and can be consumed 72 hours after production (Ghods-Rohani et al. 2009; ISIRI 2002a). The consumption of packaged cheese is becoming widespread and unpackaged products are becoming less popular, even in rural areas of Iran (Beldman et al. 2017). It is expected that the white cheese varieties remain the consumers' choice for breakfast in the years to come and recent western cheese varieties will only find their way into lunch and dinner consumption.

The picture of cheese production and consumption in Iran has dramatically changed over the past three decades. More varieties of white brined cheese are now commercially produced on a large scale and the manufacture of cheese from unpasteurised milk has declined, due to safety implications. Although the popularity of UF white cheese has been increasing among the Iranian consumers, traditional white brined cheese is still the preferred choice of many Iranians and consumed across the country. It is also fair to say that cheese consumption has been steadily on the rise among urban residents for the past decades (Anonymous 1991).

The varieties of cheese produced in Iran have expanded in recent decades. European varieties of cheese are now popular, at least among the urban affluent population. The range of cheeses produced in Iran now includes cream cheese, processed cheese, cheese with added flavours such as walnut, fresh cheese, pizza/mozzarella and Halloumi. European-type cheeses such as blue cheeses, butterkäse, Gouda, Parmesan, Prato, Camembert and Esrom are less popular and produced on a much smaller scale (Beldman et al. 2017).

8.3.5 *Akkawi*

Akkawi cheese is a fresh and semi-hard cheese variety (see Figure 8.2d) characterised by a chalky colour, firm texture and salty flavour (Kamleh et al. 2015). Salting and ageing critically affect the sensory properties of Akkawi cheese. Part replacement of NaCl by KCl (i.e. 70% NaCl and 30% KCl) was reported to result in an end product without any remarkable sensory problems (Kamleh et al. 2015). A time-dependent increase in the development of a fermented flavour was also reported by the same research group. To the best of our knowledge, no research on the volatile compounds profile of Akkawi cheese is available in the literature.

8.3.6 *Domiati*

Domiati is a soft and brined cheese variety native to Egypt but also popular in the Arab world. It is made from cow's milk, buffalo milk or an appropriate mixture of both. It can be consumed fresh or after pickling in salted whey for several months (Abou-Donia 2008). Although, consumption figures for this variety are high (hence its high commercial value), food safety issues associated with Domiati cheese on the Egyptian market have been a major concern and attempts have been made to develop strategies to improve the food safety standards Ibrahim and Awad (2018), for example, evaluated the anti-bacterial capacities of 40 lactic acid bacteria isolates against *Staphylococcus aureus* 6538 ATCC. Out of 40 isolates tested, 19 were found to inhibit *S. aureus* 6538 ATCC and these isolates were used in the manufacture of cheese. It has been reported that cheeses produced with *Lactobacillus rhamnosus* strains 130RZFAAU and 131RZFAAU had sensory properties similar to the control cheese. Mahgoub et al. (2013) studied the potential use of cold pressed black cumin seed oil (*Nigella sativa* sp.) in the manufacture of Domiati cheese to inhibit selected food-borne pathogens (*S. aureus*, *Escherichia coli*, *Listeria monocytogenes* and *Salmonella enteritidis*). In addition to the potent inhibitory effect on the pathogenic microorganisms, black cumin seed oil improved the sensory characteristics of the fresh and 30-day-old experimental cheeses, being most pronounced in the cheese supplemented with black cumin at a level of 0.2%. The enterococci, *E. faecium* and *Enterococcus faecalis*, are the most widely found strains in traditional Domiati and Ras cheeses. Malek et al. (2012) screened 35 indigenous strains of *E. faecium* isolated from two Egyptian cheeses for their technological suitability for industrial production of such cheeses. Some strains complying with the food safety standards were considered suitable for inclusion in cheese starter combinations. The major volatile compounds produced by these strains were acetoin, acetone, butanoic acid and hexanoic acid. Ayad (2009) developed a defined strain starter culture made up of 11 strains of *L. lactis* subsp. *lactis* and *L. lactis* subsp. *cremoris* for the production of Domiati cheese. The author demonstrated that the cheeses made by using defined strain starter culture resulted in a cheese with higher flavour scores than the control cheese. Domiati cheese was originally a high-fat cheese but in recent years due to consumer health concerns, reduced fat Domiati cheese has become popular. The reduction in milk fat eventually causes some textural and sensory defects in the end product. In order to overcome these problems, the addition of hydrocolloids to cheese milk was investigated by Abd El-Hamid (2013). Incorporation of carboxymethylcellulose (CMC) at levels of 0.4, 0.6 or 0.8% negatively affected the appearance, flavour and odour of the cheeses as assessed by trained sensory panellists. However, cheese made with 1.0% CMC was given sensory scores similar to the control cheese. As with many other brined-type cheeses, the overall sensory quality of Domiati cheese improves during the ripening period (El-Sheikh et al. 2001; Kebary et al. 2006). The impact of the addition of bacterial protease (derived from *Lactobacillus plantarum*) on the sensory characteristics of Domiati-type soft white cheese was studied by Abd El-Salam et al. (2019). Sensory evaluations were carried out by untrained panellists using a score sheet with maximum scores of 50 for flavour, 35 for body and texture, and 15 for appearance respectively. Although the number of panellists was insufficient to reach a fair judgement, the authors correlated the sensory evaluation scores with chemical analysis, and recommended the use of a bacterial protease with a specific activity of 2 U/mg protein at a level of 2% (w/v) to accelerate cheese ripening without impairing the sensory characteristics of the end product. Soryal et al. (2004) investigated the effect of supplementary feeding of grazing Alpine goats with different levels of alfalfa hay (0, 0.33 or 0.66 kg on a daily basis) on the sensory properties of Domiati cheese made from the goat's milk. Regarding the total and long-chain free fatty acids and overall sensory scores, no remarkable differences were noted between the feeding treatments. In all cheese groups, during the ripening period, the sensory characteristics of the cheeses improved along with development of a smother and creamier body and texture. Aly (1997) showed that conventional Domiati cheese had better aroma and flavour characteristics than UF-Domiati cheese and to overcome this shortcoming the addition of malt extract to UF-Domiati cheese milk at a level of 0.075% was proposed. On the contrary, Mehaia (2002) failed to detect any remarkable difference between fresh traditional Domiati cheese and UF-Domiati cheese regarding flavour and overall acceptability scores. Microfiltration may be a technological option to avoid the heat treatment of milk used in Domiati cheese production. Awad et al. (2010) demonstrated that Domiati cheese made from pasteurised milk had higher sensory scores than microfiltered (MF)-Domiati cheese at the ageing time of 60 days. However, when MF milk was inoculated with adjunct culture containing (*Lactobacillus delbrueckii* subsp. *lactis*, *L. paracasei* subsp. *paracasei*, *Lactobacillus casei*, *L. plantarum* and *E. faecium*), sensory scores of the MF-Domiati cheese increased. In order to exclude the whey separation stage from Domiati cheese production, a combination of whole cow's milk and whole milk powder was used to produce cheese by Gomah et al. (2019). The resulting product received higher sensory scores for taste, cheese cut, texture and overall acceptability compared to the control cheese. The volatile compound profiles of fresh, ripened and deteriorated Domiati cheeses were studied in detail by Collin et al. (1993). Acetaldehyde, acetone, ethanol, 1-propanol, 2-butanol, 2-methyl-1-propanol, ethyl formate and pentane were the most abundant volatile compounds in ripened Domiati cheeses. In the fully deteriorated cheese, the dominant volatile compounds, which were not present in the good quality cheeses, were acrolein, acetone and a range of sulphur compounds (dimethyl sulphide, dimethyl disulphide, methyl mercaptan, methyl thioacetate, methyl thiopropionate, methyl thiobutyrate and dimethyl trisulphide). Abd El-Mageed (1997) investigated the potential use of buffalo's milk with various fat levels (1.0, 3.5 or 7.0%) in the manufacture of Domiati cheese. The authors reported that the cheese made from milk with 7.0% milk became unacceptable after one month of ripening. The major volatile compounds in the buffalo Domiati cheese were acrolein (propenal), heptanal, acetone, 2-butanone, ethanol, 2-butanol, 2-methyl-1-propanol, 3-methyl-1-butanol, ethyl propionate, propyl propionate, pentane and octane. In the full-fat cheese, the flavour of which had deteriorated within a short time, the most abundant volatile compounds were recorded as methyl mercaptan, methyl thiopropionate and dimethyl trisulphide together with propyl butyrate. Although it is a far less common practice, the utilisation of camel milk in the production of Domiati-type fresh brined cheese is possible. In such case, the fat level of the camel milk should be reduced to 1.5% and the salt level of the brine should be set to 3% for acceptable organoleptic properties in the final product, as reported by Mehaia (1993).

8.3.7 *Mish*

Mish cheese, a soft pickled cheese popular in Egypt, Jordan, Sudan and Palestine, is characterised by a sharp, salty and pungent taste (Ahmad et al. 2016).

8.3.8 *Nabulshi*

Nabulshi, another cheese variety which is native to Middle Eastern countries, is a semi-hard unfermented cheese made from raw cow's milk often supplemented with mahlab (*Prunus mahaleb*) and mastic (*Pistacia lentiscus*) (Al-Dabbas et al. 2014). Both plants largely determine the sensory properties of Nabulshi cheese.

8.3.9 *Erzurum Civil*

Erzurum Civil cheese can be categorised both under acid and rennet curd cheeses. It is mostly produced in eastern part of Turkey. In Caucasus and Armenia, this cheese is famous under the name 'Cecil cheese'. Erzurum Civil cheese is usually made from skimmed raw cow's milk, and curd formation is achieved by heating the acidified milk (~pH 5.4) to 65–70 °C. The cheese blocks are hung from a platform with the aim of stretching under its own weight until acquiring its plastic and fibrous structure which is the characteristic textural feature of this cheese type. Finally, cheese is kept in a brine solution for ripening (Hayaloglu et al. 2008; Sengul et al. 2006). Hexanoic octanoic and butanoic acids are the dominant volatile compounds in Civil cheese (Gurkan et al. 2018). This variety is characterised by a slightly sweet taste, and mouldy, rancid flavour. The colour of Civil cheese is yellowish and the texture is elastic-like with a fibrous structure.

8.3.10 *Izmir Tulum*

İzmir Tulum cheese is traditionally manufactured from raw ewe's milk or mixtures of ewe's and goat's or cow's milk without a starter culture in the Aegean region of Turkey (western part of Turkey). In the industrial production of İzmir Tulum cheese, cow's milk is widely preferred. Izmir Tulum cheese is distinguished from other Tulum-type cheeses (discussed later in this chapter) by ripening under brine, having a more compact structure and higher salt levels (Akpinar et al. 2017; Hayaloglu and Karagul-Yuceer 2011). The principal sensory descriptors of İzmir Tulum cheese are understood to be salty, acidic, fermented, whey-like, animal-like, creamy and umami.

8.3.11 *Van Otlu*

Van Otlu (herby) cheese is a brined-type cheese which is traditionally produced in the eastern part of Turkey. This variety of cheese is manufactured from raw ewe's milk. In case of insufficient ewe's milk due to seasonal reasons, a mixture of ewe's, cow's or goat's milk can also be used. In traditional Van Otlu cheese production, no starter culture is used. However, in industrial production pasteurised milk is inoculated with mesophilic cheese cultures. Van Otlu cheese is a semi-hard variety, whitish in colour with a salty and slightly bitter taste. The characteristic taste, aroma and appearance of Van Otlu cheese is derived from herbs which are added to the curd before whey drainage. These herbs mainly include the species of *Allium*, *Thymus*, *Mentha*, *Ferula* and *Silene* which are naturally present in the mountains and plateaus of the Van province in Turkey. These herbs also provide an antimicrobial property to the cheese. Terpene compounds from the herbs comprise a high ratio of the volatile compounds found in this cheese variety (Hayaloglu and Fox 2008; Hayaloglu and Karagul-Yuceer 2011). Organic acids (e.g. butanoic, hexanoic, acetic and octanoic acids) are the most abundant group of volatile compounds, followed by the alcohols (e.g. 3-methyl-1-butanol).

8.3.12 *Urfa*

Urfa cheese is a brined-type semi-hard cheese variety. It is traditionally produced from raw milk of the Awassi sheep breed. This cheese variety is ripened under brine (16–18% NaCl, w/v) for about 6–8 mo. The dominant natural microflora of Urfa cheese was reported to contain salt-tolerant lactococci (*La. lactis* spp. and *Lactococcus garviae*) and enterococci (*Enterococcus durans*, *E. faecium* and *E. faecalis*) (Kirmaci 2010). A comprehensive study on the volatile compounds of Urfa cheese was carried out by Kirmaci et al. (2015). In total, 70 volatile compounds were identified from the cheese and acids were the most abundant chemical group of volatiles, with butanoic and hexanoic acids being the dominant. While 3-methyl-1-butanol was the most abundant alcohol in the 180-day-old cheese, 2-heptanone, 2-pentanone and 2-hexanone were identified as the dominant ketones. The most abundant terpene compound was α-pinene which gives Urfa cheese a pine and turpentine aroma. Urfa cheese is characterised by a closed texture, whitish colour and salty taste.

8.4 Scalded and pasta-filata-type cheeses

The major pasta-filata-type cheeses of the Middle East and Balkans are Kashkaval (Greece, Turkey, Bulgaria), Halloumi (Cyprus, Greece), Kars Kashar and Orgu (Turkey), and Graviera (Greece) cheeses. The manufacturing procedures for these varieties are close to each other but with slight variations in ripening conditions. *Kars Kashar cheese* is produced by scalding the curd in hot (72–78 °C) and salty (8–10% NaCl, w/v) water or cheese whey, and gains the required shape by kneading. Kars Kashar cheese is manufactured in the eastern region of Turkey, particularly in the province of Kars. Temperature and relative humidity of the ripening rooms are of critical importance for formation of the characteristic aroma and flavour of this variety. After a pre-ripening stage at 12–18 °C for about a week, the full maturation is achieved at 75% relative humidity and 6–8 °C for three to six months. Kars Kashar cheese is characterised by slightly pungent taste, rancid aroma, closed texture with a thick outer layer. Its own distinct characteristics arise from the treatments during production and ripening, in addition to the impact of the natural flora of the region's fauna (Cetinkaya 2020; Kamanlı-Can 2019).

8.4.1 *Kashkaval*

Kashkaval cheese is commonly consumed in Greece, Bulgaria and the north-western part of Turkey (Trakia region). Kashkaval cheese is made either from raw milk or pasteurised milk (cow's or sheep's milk or a mixture of both) and, therefore, the sensory characteristics of the final products vary depending on the production practices. Kashkaval cheese has a bright yellow colour, closed texture with elastic body and salty, piquant taste. Recently, Pappa et al. (2019) studied the volatile compounds profile of Kashkaval cheeses made from raw or pasteurised milk and found that butyric, myristic, palmitic, stearic and oleic acids were the most abundant free fatty acids in both cheeses. While ketones (acetone, 2-pentanone, 2-heptanone, acetoin and 2-nonanone) were the most abundant volatile compounds in Kashkaval cheese from pasteurised milk, esters (butanoic acid ethyl ester, hexanoic acid methyl ester, hexanoic acid ethyl ester, octanoic acid methyl ester and octanoic acid ethyl ester) was the largest chemical group of volatile compounds in the mature cheeses made from raw milk. Additionally, raw milk cheese had higher levels of total aldehydes than pasteurised milk cheese. Sulejmani and Hayaloglu (2016) investigated the effect of scalding temperature on the volatile compounds of Kashkaval cheeses made from pasteurised cow's milk. The cheese mass was scalded in hot brine (18% NaCl, w/v) at 60, 70 or 90 °C for 5 minutes and the cheeses were ripened for 120 days. SPME/GC-mass spectroscopy analysis revealed that esters and acids were the most abundant volatiles in the cheeses. Although there was no clear correlation between scalding temperature and the profile of volatile compounds, the cheese scalded at 90 °C for 5 minutes had higher levels of esters (ethyl acetate, ethyl butanoate), alcohols (ethanol) and hydrocarbons (3,3,5-Trimethyl heptane, 2,3,3-Trimethyl pentane).

8.4.2 *Halloumi*

Halloumi cheese is of Cypriot origin and characterised by a semi-hard to hard body with no holes and a skin/rind (Papademas and Robinson 1998, 2001). It can be produced from sheep's, goat's or cow's milk and the colour of the cheese is yellowish when the cow's milk is used as the main ingredient (Kaminarides et al. 2000). While fresh Halloumi cheese has a compact elastic texture, the mature cheese develops a hard and slightly crumbly body. Halloumi cheese is a cooked variety (at about 90 °C for 30–60 minutes) (Anifantakis and Kaminarides 1983; Galanakis et al. 2014) and, in most cases, crushed dry mint leaves are sprinkled on the cheese before it is folded in half. The moisture content of Halloumi cheese is about 42–44% and fat-in-dry matter content is close to 50%, except for caprine milk cheeses. Unlike other brined-type Greek or Cypriot cheeses, Halloumi cheese contains lactones at high concentrations. According to Partial Least Square regression analysis of volatile compounds, the minty flavour is predicted to arise from the terpenes pulegone and carvone, the former described as 'mint terpene' (Papademas and Robinson 1998). Regression analysis also strongly linked the creamy and milky flavours with nonanone and phenols. The organoleptic characteristics of traditional Halloumi cheese are derived from both the manufacturing procedure (raw milk is used and no starter is added) and the type of milk used in the manufacture (sheep's or goat's). However, as a result of increasing consumer demand for Halloumi cheese and the seasonal availability of sheep's or goat's milks, cow's milk is now being more widely used in the manufacture of Halloumi cheese. This has led to dramatic changes in the volatile composition of the cheese. Although traditional Halloumi cheese is made from raw milk, Robinson (1991) recommended the use of mesophilic (*L. lactis* subsp. *lactis* and *L. lactis* subsp. *cremoris*) or thermophilic (*Streptococcus thermophilus*) starter culture to enhance the flavour profile of the cheese. More recently, Kaminarides et al. (2007) investigated the volatile compounds and sensory properties of Halloumi cheese kept in brine solution. This group identified 25 volatile compounds present in the headspace fractions of 45-day-old cheese samples using GC/MS-Q. The volatile compounds identified belonged to the following chemical groups (the major compounds of each chemical group are given in brackets): alcohols (ethanol), aldehydes (acetaldehyde), ketones (acetone, diacetyl), volatile acids (acetic acid, butanoic acid, hexanoic acid), esters (butanoic acid methyl ester), hydrocarbons (1-methoxy-4[1-propenyl]-ben zene), sulphur compounds (1-pentanethiol) and other substances (acetonitrile). High level of ethanol in 45-day-old cheese were associated with the presence of thermotolerant bacteria and/or yeast contamination after cheese production (Kaminarides et al. 2007). Alcohols were also reported as the most abundant volatile chemical group in the cooked cheese (at 93–95 °C for 30 or 60 minutes in hot cheese whey), representing ca. 30% of total volatile compounds, and ethanol was the most abundant volatile alcohol of this chemical group (ethanol accounted for 71% of total alcohols) (Kaminarides et al. 2015). Interestingly, acetaldehyde which is commonly present in fermented dairy products and responsible for the development of sweet pungent smell, was the second most abundant volatile compound in Halloumi cheese, as reported by Kaminarides et al. (2007). However, Kaminarides et al. (2015) failed to detect acetaldehyde in Halloumi cheese cooked at 93–95 °C for 30 or 60 minutes. Diacetyl, which is characterised by a nut-like buttery flavour, was found in Halloumi cheese at all stages of ripening. Acetic ($C_{2:0}$), butyric ($C_{4:0}$), caproic ($C_{6:0}$), myristic ($C_{14:0}$), palmitic ($C_{18:0}$) and oleic ($C_{18:1}$) acids were the dominant free fatty acids in 45-day old Halloumi cheese (Kaminarides et al. 2007). Cooking the curd at high temperatures (i.e. >90 °C) for a long time (i.e. 30–60 minutes) resulted in marginal reductions in the volatile compounds of Halloumi cheese except for the furans (Kaminarides et al. 2015). The effects of milk type (sheep's or cow's milk) and maturation period (fresh or mature but time for maturation was not provided) on the free volatile fatty acid profiles of Halloumi cheese were investigated by Papademas and Robinson (2000). The milk type did not affect the ethanoic acid levels of the cheeses, but mature cheeses had much higher ethanoic acid concentrations. The reverse, decreases in propanoic acid and 3-methylbutanoic acid concentrations, was noted. The pentanoic acid levels in the fresh and mature sheep's milk cheeses were 47 and 89 mg/kg of cheese, respectively, but most interestingly while the fresh cow's milk did not contain pentanoic acid, its concentration jumped to 25 mg/kg of cheese in mature cheeses. A small number of published articles are available regarding the detailed organoleptic evaluation of Halloumi cheese. Papademas and Robinson (2000) ran a preliminary training session for 10 people, all from Cyprus, before the sensory evaluation of Halloumi cheese made from sheep's or cow's milk. A common vocabulary was generated by considering each panellist's descriptions of cheese samples provided. The vocabulary compromised 12 terms including six terms for taste and flavour (salty, bitter, acid, creamy, milky and minty), four terms for texture (crumbly, springy, moist, chewy) and the rest for appearance (white/yellow) and body (layered). The authors did not provide details of preliminary training session given to the tasters. Overall, the panel group described sheep's milk cheese as springier and chewier than its cow's milk counterpart. The authors also noted that the clear differences in the free volatile fatty acids profiles of the two cheeses were not perceived by the sensory panel. Mature cheeses were described as less springy and less moist, and the intensity of flavour notes were remarkably higher in the mature cheeses. Theophilou and Wilbey (2007) investigated how the fat-in-dry matter level of cheese affected the sensory characteristics of Halloumi cheese made from cow's milk. The sensory panel, made up of 15 people with an unknown training background, scored samples for texture (hardness, springiness, chewiness, moistness), taste (saltiness, creaminess, bitterness, aftertaste) and overall liking using a 100 mm scale for each parameter. The panel liked the cheeses with high fat in dry matter levels. Also, cheeses containing microparticulated whey proteins received higher preference scores than the control cheese. Mehyar et al. (2018) investigated the effects of a chitosan coating containing lysozyme or natamycin on the sensory properties of Halloumi cheese ripened in 5 or 10% brine solution. The sensory panel consisted of 20 trained panellists aged between 24 and 50 years and evaluated the cheese samples for colour, aroma, taste and overall acceptability based on a 9-point hedonic scale. The coatings did not affect the colour and aroma of the experimental cheeses. Chitosan with lysozyme positively affected the taste and overall acceptability scores of the cheeses kept in a 10% brine solution. Kamleh et al. (2012) applied both descriptive sensory analysis and hedonic evaluations to Halloumi cheeses ripened in brine solution in which sodium chloride was substituted by potassium chloride. Descriptive analysis was conducted with a panel of nine people (six females and three males) who were trained over three one-hour sessions. The panel developed a vocabulary containing 13 attributes (colour, fermented odour, milky odour, saltiness, bitterness, fermented flavour, milky flavour, hardness, crumbliness, squeakiness, moistness, bitter aftertaste and salty aftertaste). The authors found no major inconsistencies between the panellists' ratings. Hedonic evaluations were carried out by 72 panellists (mean age: 24.3 years, range: 18–57 years). The authors concluded that replacement of potassium chloride with sodium chloride did not cause any major sensory problems in the end product.

8.4.3 *Graviera*

Graviera cheese is a Greek hard cheese which is marketed under various names including Graviera Kritis, Naxou, Mytilinis and currently holds PDO status (Bozoudi et al. 2016). The source of the milk used in the manufacture of Graviera cheese varies depending on the region where it is produced. For example, Graviera Naxou PDO cheese is made from raw cow's milk and Graviera Kritis is produced from raw sheep's milk. Graviera cheese is characterised by high-fat and relatively low-salt levels. The mature cheese has a smear rind and irregular openings. Since Graviera cheese is made from raw milk, a large variety of microbial groups are usually present in the fresh or mature cheeses, including lactic acid bacteria, enterobacteriaceae, staphylococcus, coliforms and yeasts. Bozoudi et al. (2016) investigated the microbial profiles and volatile compositions of two traditional Graviera cheeses, namely Graviera Kritis and Graviera Naxou. They showed that both cheeses were rich in butanoic acid, hexanoic acid, acetic acid and octanoic acid as well as hexanoic acid ethyl esters. Similar to Feta cheese, the most abundant group of volatile compounds in both Graviera cheeses was acids followed by esters and alcohols. A total of 36 terpenoids were detected both in the pasture where the milking animals were fed and in the cheeses. Ethylbenzene, *p*-xylene, α-pinene, dl-limonene and isobornyl acetate were the most abundant terpenoid compounds present in both the forage and in mature Graviera Kritis cheese. In the mature Graviera Naxou cheese, α-pinene, eucalyptol and isobornyl acetate were the most abundant terpenoids. There is not much information available in the literature about Graviera cheese sensory evaluation methodology. Arvanitoyannis et al. (2011) evaluated the sensory characteristics of Graviera cheeses ripened under various modified atmosphere packaging (MAP) conditions. A trained panel evaluated the cheeses for external appearance (colour, homogeneity, mouldy and oily appearances), odour (normal, mouldy, oxidised), taste (saltiness, oily taste), after-chewing and texture properties. No detailed information was provided about the sensory evaluation test employed. Based on the sensory data, while four clusters were evident at day 1, six sensory clusters were recorded after 90 days of ripening. The authors concluded that the deterioration in sensory quality with time induced a better grouping of the attributes into clusters. Packaging of Graviera cheese under 60% CO_2/40% N_2 or 50% CO_2/50% N_2 was recommended for a better sensory quality Graviera cheese. In another study, grated Graviera cheeses ripened at 4 or 12 °C using active and MA packaging were subjected to an acceptability test by a panel consisted of 51 untrained panellists (20 females and 31 males) (Mexis et al. 2011). Based on the sensory evaluations which were supported by microbiological and chemical analysis, the authors concluded that the shelf life of grated Graviera cheese could be extended up to 8–8.5 weeks when a combination of an ethanol emitter (EE) and an oxygen absorber (OA) was included in high-barrier PET-SiOx//LDPE pouches.

8.4.4 *Diyarbakır Orgu*

Diyarbakır Orgu cheese is native to South-eastern Turkey. The word 'Orgu' comes from the 'braided' hair shape which describes the appearance of this variety. Although Orgu cheese is a pasta-filata-type cheese, unlike many other pasta-filata cheeses, it is ripened in brine solution (12–14%, w/v) at 4–6 °C for three to four months. The salty taste, closed texture and slight rancid flavour are the main features of this variety. The stretchability of Orgu cheese is reduced as the ripening progresses and a lipolytic aroma becomes more prominent.

8.5 Cheeses ripened in animal skins or pots

Divle Tulum cheese, Savak Tulum cheese and Yozgat Canak cheese are among the popular traditional cheese varieties of Turkey ripened either in animal skins or in pots. Industrial productions of these varieties are limited but small-scale dairies in Central and Eastern parts of Turkey produce such cheeses in fairly high quantities.

8.5.1 *Divle Tulum*

Divle Tulum cheese is a semi-hard variety manufactured from semi-skimmed raw ewe's milk. It is characterised by an intense flavour and a crumbly texture. Divle Tulum cheese is traditionally produced in the Central Anatolia region of Turkey in April–June when the ewe's milk is abundant. Divle Tulum cheese is ripened in goatskin bags, called *Tulum*, for four to six months in Divle cave which has a depth of 70 m and length of 230 m. The average temperature in the cave is 5–10 °C with a relative humidity of 85–90%. Divle cave is the only cave used for ripening of this cheese variety. During the ripening period, a wide variety of fungal strains grown on the surface of goatskin bags and contribute to the unique taste and aroma of this cheese variety by virtue of the enzymes and other metabolites that they synthesise. The colour of surface of the goatskin bags changes to red towards the end of ripening, which is the distinctive feature of Divle Tulum cheese. The majority of the aroma compounds found in the cheese during ripening mainly arise from the metabolic activities of the diverse bacterial and fungal community as well as from the natural forage crops of the region (Ozturkoglu-Budak et al. 2016a, 2016b). Alcohols (2-butanol), ketones (2-butanone, 2-pentanone), acids (butanoic, hexanoic, acetic acids) and esters (propionate, ethyl butanoate) were reported to be the dominant volatile compounds in Divle Tulum cheese (Hayaloglu and Karabulut 2013). Divle Tulum cheese is characterised by a slightly rancid aroma and mouldy taste.

8.5.2 *Savak Tulum*

Savak Tulum cheese is a semi-hard cheese with a crumbly texture and a buttery, sharp flavour (Cakir et al. 2016). Savak Tulum cheese is produced from raw ewe's milk without using starter culture in the eastern part of Turkey. In the traditional production method, home-made rennet is used and drained, and salted curd is filled into goatskin bags and left to ripen in natural caves for approximately three to six months. The goatskin bags let the whey drain out and oxygen enter owing to their permeable structure. In industrial production, plastic packages are preferred for ripening of this cheese (Hayaloglu et al. 2007; Hayaloglu and Karagul-Yuceer 2011).

8.5.3 *Yozgat Canak*

Yozgat Canak cheese is traditionally produced in the Central Anatolia region of Turkey and takes its name from the province of Yozgat. The milk is coagulated at room temperature with rennet of animal origin and the curd is hung in cloth bags to drain for a few hours. Then, in order to completely remove the whey, the cloth bags are pressed between flat stones of a certain weight. Afterwards, hand crumbled cheeses are dry-salted and filled into earthen pots in an airtight manner. The pots are turned upside down and left to ripening under the soil. After this process which lasts four to six months, the cheeses are removed from the pots, vacuum packaged and stored at 0–4 °C for two to three months. This cheese has a crumbly texture, acidic and salty taste and a strong flavour which is liked by consumers (Kamanlı-Can 2019).

8.6 Kopanisti cheese

Kopanisti cheese is a Greek PDO cheese made from cow's, sheep's or goat's milk or mixture of these milk types. This variety is characterised by a salty, peppery taste, Roquefort-like flavour and soft body. This variety is made from raw milk and no starter culture is added. Therefore, the aroma and flavour characteristics of Kopanisti cheese are largely determined by the milk source. A comprehensive study on characterisation of the aroma and flavour of Kopanisti was carried out by Karali et al. (2013). The authors identified a total of 81 volatile compounds from traditional Kopanisti cheeses. Ethanol was the most abundant alcohol, representing 83% of the total alcohols, followed by 3-Methyl-2-nonanol (3%) and 3-Methyl-3-nonanol (3%). Pentanal and butanal were the main aldehydes representing 45% and 43% of total aldehydes, respectively. Among the ketone fraction the major ones were 2-butanone (42%) and 2-pentanone (25%). Palmitic acid was the most abundant free fatty acid followed by acetic acid.

The characteristic aroma of each cheese variety mentioned earlier in this chapter is determined by a number of factors including milk type, climate, geography of the region, the natural fauna that animal feed on, the indigenous microflora of the raw milk, production methods and ripening practices.

References

Abd El-Hamid, A.M. (2013). Influence of carboxymethylcellulose on the physicochemical, rheological and sensory attributes of a low-fat Domiati cheese. *International Journal of Dairy Technology* 66 (4): 505–511.

Abd El-Mageed, M.A. (1997). Volatile compounds of Domiati cheese made from buffaloe's milk with different fat content. *Grasas y Aceitas* 48 (6): 391–396.

Abd El-Salam, Y.B., Ahmed, M.S., Yasser, M.M., and El-Din Taha, S.S. (2019). Soft white cheese ripening using bacterial protease enzyme. *Acta Scientiarum Polonorum. Technologia Alimentaria* 18 (4): 385–397.

Abou-Donia, S.A. (2008). Origin, history and manufacturing process of Egyptian dairy products: an overview. *Alexandria Journal of Food Science and Technology* 5: 51–62.

Ahmad, L.M., Al-Ismail, K.M., and Yamani, M.I. (2016). Influence of processing and storage on chemical and biochemical characteristics of Mish cheese traditionally produced in Jordan. *Quality Assurance and Safety of Crops & Foods* 8 (2): 243–248.

Akpinar, A., Yerlikaya, O., Kinik, O. et al. (2017). Some physicochemical characteristics and aroma compounds of Izmir Tulum cheese produced with different milk types. *Ege Universitesi Ziraat Fakultesi Dergisi* 54 (1): 27–35.

Al-Dabbas, M.M., Saleh, M., Habu-Ghoush, M. et al. (2014). Influence of storage, brine concentration and in-container heat treatment on the stability of white brined Nabulsi cheese. *International Journal of Dairy Technology* 67 (3): 427–436.

Alizadeh, M. and Lavasani, A.R.S. (2013). Effect of different types of milk on some physicochemical and sensory characteristics of Iranian white-brined cheese. *Annals of Biological Research* 4 (10): 67–70.

Aly, M.F. (1997). Malt extract for enhancing flavour development of ultrafiltered Domiati cheese. *Food Chemistry* 59 (1): 127–133.

Anifantakis, E.M. and Kaminarides, S.E. (1983). Contribution to the study of Halloumi cheese made from sheep's milk. *Australian Journal of Dairy Technology* 58: 29–31.

Anonymous (1991). *Cheese, Encyclopaedia Iranica*, Vol. V, Fasc. 4, 391–393. http://www.iranicaonline.org/articles/cheese-pers (accessed 20 May 2020).

Anonymous (2014). *Dairy Sector Statistics: World and Turkey*. Ankara, Turkey: National Dairy Council of Turkey.

Arvanitoyannis, I.S., Kargaki, G.K., and Hadjichristodoulou, C. (2011). Effect of several MAP compositions on the microbiological and sensory properties of Graviera cheese. *Anaerobe* 17: 310–314.

Awad, S., Ahmed, N., and El-Soda, M. (2010). Influence of microfiltration and adjunct culture of quality of Domiati cheese. *Journal of Dairy Science* 93: 1807–1814.

Ayad, E.H.E. (2009). Starter culture development for improving safety and quality of Domiati cheese. *Food Microbiology* 26 (5): 533–541.

Azarnia, S., Ehsani, M.R., and Mirhadi, S.S. (1997). Evaluation of the physico-chemical characteristics of the curd during the ripening of Iranian brine cheese. *International Dairy Journal* 7: 473–478.

Barouei, J., Karbassi, A., Ghoddusi, H. et al. (2011). Impact of native *Lactobacillus paracasei* subsp. *paracasei* and *Pediococcus* spp. as adjunct cultures on sensory quality of Iranian white brined cheese. *International Journal of Dairy Technology* 64: 526–535.

Beldman, A., Berkum, S., Kortstee, H., Zijlstra, J. (2017). Dairy farming and dairy industry in Iran. *Wageningen Economic Research Report No: 2017-010*, 52

Bintsis, T. and Robinson, R.K. (2004). A study of the effects of adjunct cultures on the aroma compounds of Feta-type cheese. *Food Chemistry* 88 (3): 435–441.

Bozoudi, D., Pavlidou, S., Kotzamanidis, C. et al. (2016). Graviera Naxou and Graviera Kritis Greek PDO cheeses: discrimination based on microbiological and physicochemical criteria and volatile organic compounds profile. *Small Ruminant Research* 136: 161–172.

Bozoudi, D., Kondyli, E., Claps, S. et al. (2018). Compositional characteristics and volatile organic compounds of traditional PDO Feta cheese made in two different mountainous areas of Greece. *International Journal of Dairy Technology* 71 (3): 673–682.

Cakir, Y., Cakmakci, S., and Hayaloglu, A.A. (2016). The effect of addition of black cumin (*Nigella sativa* L.) and ripening period on proteolysis, sensory properties and volatile profiles of Erzincan Tulum (Savak) cheese made from raw Akkaraman sheep's milk. *Small Ruminant Research* 134: 65–73.

Cetinkaya, A. (2020). Effect of different preservation and salting methods on some volatile compounds and sensory properties of Kashar cheese. *Kafkas Universitesi Veteriner Fakultesi Dergisi* 26 (3): 435–444.

Collin, S., Osman, M., Delcambre, S. et al. (1993). Investigation of volatile flavor compounds in fresh and ripened Domiati cheese. *Journal of Agricultural and Food Chemistry* 41: 1659–1663.

Curioni, P.M.C. and Bosset, J.O. (2002). Key odorants in various cheese types as determined by gas chromatography-olfactometry. *International Dairy Journal* 12: 959–984.

El-Sheikh, M.M., Farrag, A.F., Shahein, N.M., and El-Shibiny, S. (2001). Low fat Domiati cheese with particulated whey protein concentrate (PWPC). *Egyptian Journal of Dairy Science* 29: 331–342.

Galanakis, C.M., Chasiotis, S., Botsaris, G., and Gekas, V. (2014). Separation and recovery of proteins and sugars from Halloumi cheese whey. *Food Research International* 65: 477–483.

Ghods-Rohani, M., Mortazavi, S.A., Mazaheri Tehrani, M., and Razavi, S.M.A. (2009). Effect of processing conditions on physical, chemical and sensory properties of Ultra-filtrated Feta cheese made from cow's milk and soymilk blend. PhD. Thesis. Ferdowsi University of Mashhad. Iran.

Gomah, N.H., Mohamed, T.H., Ahmed, H.A., and Ragab, W.S. (2019). Production of soft cheese without whey separation. *Journal of Biochemical Technology* 10 (3): 105–109.

Gurkan, N.H., Yilmaztekin, M., Cakmakci, S., and Hayaloglu, A. (2018). Volatile compounds and biogenic amines during the ripening of mold-ripened civil cheese manufactured using three different strains of *Penicillium roqueforti*. *Journal of Food Safety* 38: e12568.

Hashemi, M., Azar, M., and Mazlumi, M.T. (2009). Effect of commercial adjunct lactobacilli on biochemical and sensory characteristics of Iranian white-brined cheese. *International Journal of Dairy Technology* 62: 48–55.

Hayaloglu, A.A. and Fox, P.F. (2008). Cheeses of Turkey: 3. Varieties containing herb or spices. *Dairy Science and Technology* 88: 245–256.

Hayaloglu, A.A. and Karabulut, I. (2013). SPME/GC-MS characterization and comparison of volatiles of eleven varieties of Turkish cheeses. *International Journal of Food Properties* 16 (7): 1630–1653.

Hayaloglu, A.A. and Karagul-Yuceer, Y. (2011). Utilization and characterization of small ruminants' milk and milk products in Turkey: current status and new perspectives. *Small Ruminant Research* 101: 73–83.

Hayaloglu, A.A., Cakmakci, S., Brechany, E.Y. et al. (2007). Microbiology, biochemistry, and volatile composition of Tulum cheese ripened in goat's skin or plastic bags. *Journal of Dairy Science* 90: 1102–1121.

Hayaloglu, A.A., Özer, H.B., and Fox, P.F. (2008). Cheeses of Turkey: 2. Varieties ripened under brine. *Dairy Science and Technology* 88: 225–244.

Ibrahim, A. and Awad, S. (2018). Selection and identification of protective culture for controlling *Staphylococcus aureus* in fresh Domiati like cheese. *Journal of Food Safety* 38: e12418.

ISIRI (2002a). *Fresh Cheese, Specification and Test Methods: No 6629*. Institute of Standards and Industrial Research of Iran.

ISIRI (2002b). *Cheese- General Specifications: No 2344*. Institute of Standards and Industrial Research of Iran.

ISIRI (2016). *Brined Cheese- Specifications and Test Methods; No 2344–1*. Institute of Standards and Industrial Research of Iran.

Jalili, M. (2016). Chemical composition and sensory characteristics of Feta cheese fortified with iron and ascorbic acid. *Dairy Science & Technology* 96: 579–589.

Kamanlı-Can, N. (2019). *Turkey's Richness of Cheese and Foreign Trade*. Ankara: Turkish Meat and Milk Administration Publishing.

Kaminarides, S., Rogoti, E., and Mallatou, H. (2000). Comparison of the characteristics of Halloumi cheese made from ovine milk, caprine milk or mixture of these milks. *International Journal of Dairy Technology* 53: 69–74.

Kaminarides, S., Stamou, P., and Massouras, T. (2007). Changes of organic acids, volatile aroma compounds and sensory characteristics of Halloumi cheese kept in brine. *Food Chemistry* 100: 219–225.

Kaminarides, S., Litos, I., Massouras, T., and Georgala, A. (2015). The effect of cooking time of curd composition and textural properties of sheep Halloumi cheese. *Small Ruminant Research* 125: 106–114.

Kamleh, R., Olabi, A., Toufeili, I. et al. (2012). The effect of substitution of calcium chloride with potassium chloride on the physicochemical, microbiological, and sensory properties of Halloumi cheese. *Journal of Dairy Science* 95: 1140–1151.

Kamleh, R., Olabi, A., Toufeili, I. et al. (2015). The effect of partial substitution of NaCl with KCl on the physicochemical, microbiological and sensory properties of Akkawi cheese. *Journal of the Science of Food and Agriculture* 95 (9): 1940–1948.

Karagul-Yuceer, Y., Isleten, M., and Uysal-Pala, C. (2007). Sensory characteristics of Ezine cheese. *Journal of Sensory Studies* 22: 49–65.

Karagul-Yuceer, Y., Tuncel, B., Guneser, O. et al. (2009). Characterization of aroma-active compounds, sensory properties, and proteolysis in Ezine cheese. *Journal of Dairy Science* 92: 4146–4157.

Karali, F., Georgala, A., Massouras, T., and Kaminarides, S. (2013). Volatile compounds and lipolysis levels of Kopanisti, a traditional Greek raw milk cheese. *Journal of the Science of Food and Agriculture* 93 (8): 1845–1851.

Karimi, R., Sohrabvandi, S., and Mortazavian, A.M. (2013). Sensory characteristics of probiotic cheese. *Comprehensive Reviews in Food Science and Food Safety* 12: 437–452.

Katsiari, M.C., Voutsinas, L.P., Kondyli, E., and Alichanidis, E. (2002). Flavour enhancement of low-fat feta-type cheese using a commercial adjunct culture. *Food Chemistry* 79 (2): 193–198.

Kebary, K.M.K., Hamid, A.I., Zedan, A.N., and El-Beheary, A.A.F. (2006). Manufacture of low fat Domiati cheese using Novagel. *Egyptian Journal of Dairy Science* 34: 175–184.

Khosrowshahi, A., Madadlou, A., Mousavi, M.E.Z., and Emam-Diomeh, Z. (2006). Monitoring the chemical and textural changes during ripening of Iranian white cheese made with different concentrations of starter. *Journal of Dairy Science* 89: 3318–3325.

Kirmaci, H.A. (2010). Isolation, molecular characterization of lactic acid bacteria in traditional Urfa cheese, and possible use of as starter culture. PhD Thesis. Institute of Applied and Natural Sciences, Harran University. Ankara, Turkey.

Kirmaci, H.A., Hayaloglu, A.A., Özer, H.B. et al. (2015). Effects of wild-type starter culture (artisanal strains) on volatile profile of Urfa cheese made from ewe milk. *International Journal of Food Properties* 18 (9): 1915–1929.

Kondyli, E., Katsiari, M.C., Massouras, T., and Voutsinas, L.P. (2002). Free fatty acids and volatile compounds of low-fat Feta cheese made with commercial adjunct culture. *Food Chemistry* 79: 199–205.

Kondyli, E., Pappa, E.C., and Vlachou, A.M. (2012). Effect of package type on the composition and volatile compounds of Feta cheese. *Small Ruminant Research* 108 (1–3): 95–101.

Laskaridis, K., Serafeimidou, A., Zlatanos, S. et al. (2013). Changes in fatty acid profile of Feta cheese including conjugated linoleic acid. *Journal of the Science of Food and Agriculture* 93: 2130–2136.

Mahgoub, S.A., Ramadan, F.M., and El-Zahar, K.M. (2013). Cold pressed *Nigella* sativa oil inhibits the growth of foodborne pathogens and improves the quality of Domiati cheese. *Journal of Food Safety* 33 (4): 470–480.

Malek, R., El-Attar, A., Mohamad, M. et al. (2012). Technological and safety properties display biodiversity among enterococci isolated from two Egyptian cheeses, "Ras" and "Domiati". *International Journal of Food Microbiology* 153 (3): 314–322.

Manolaki, P., Katsiari, M.C., and Alichanidis, E. (2006). Effect of a commercial adjunct culture on organic acid contents of low fat Feta-type cheese. *Food Chemistry* 98 (4): 658–663.

Mehaia, M.A. (1993). Fresh soft white cheese (Domiati-type) from camel milk: composition, yield, and sensory evaluation. *Journal of Dairy Science* 76 (10): 2845–2855.

Mehaia, M.A. (2002). Manufacture of fresh soft white cheese (Domiati-type) from ultrafiltered goat's milk. *Food Chemistry* 79 (4): 445–452.

Mehyar, G.F., Al Nabulsi, A.A., Saleh, M. et al. (2018). Effects of chitosan coating containing lysozyme or natamycin on shelf-life, microbial quality, and sensory properties of Halloumi cheese brined in normal or reduced salt solutions. *Journal of Food Processing and Preservation* 42: e13324.

Mexis, S.F., Chouliara, E., and Kontominas, M.G. (2011). Quality evaluation of Grated Graviera cheese stored at 4 and 12°C using active and modified atmosphere packaging. *Packaging Technology and Science* 24: 15–29.

Milani, J.M., Khedmati, S., Ghorbani, H.A., and Golkar, A. (2017). The effect of Tragacanth gum in physicochemical and textural properties of Lighvan cheese during ripening. *Journal of Research and Innovation in Food Science and Technology* 6: 103–114.

Morand-Fehr, P., Fedele, V., Decandia, M., and Le Frileux, Y. (2007). Influence of farming and feeding systems on composition and quality of goat and sheep milk. *Small Ruminant Research* 68: 20–34.

Nottagh, S., Hesari, J., Peighambardoust, S.H. et al. (2020). Effectiveness of edible coating based on chitosan and Natamycin on biological, physico-chemical and organoleptic attributes of Iranian ultra-filtrated cheese. *Biologia* 75 (4): 605–611.

Ocak, E., Javidipour, I., and Tuncturk, Y. (2015). Volatile compounds of Van Herby cheese produced with raw and pasteurized milks from different species. *Journal of Food Science and Technology* 52 (7): 4315–4323.

Ozturkoglu-Budak, S., Figge, M.J., Houbraken, J., and de Vries, R.P. (2016a). Microbial diversity and dynamics of Turkish Divle Cave cheese during production and ripening. *International Dairy Journal* 58: 50–54.

Ozturkoglu-Budak, S., Gursoy, A., Aykas, D.P. et al. (2016b). Volatile compound profiling of Turkish Divle Cave cheese during production and ripening. *Journal of Dairy Science* 99: 5120–5131.

Papademas, P. and Robinson, R.K. (1998). Halloumi cheese: the product and its characteristics. *International Journal of Dairy Technology* 51 (3): 98–103.

Papademas, P. and Robinson, R.K. (2000). A comparison of the chemical, microbiological and sensory characteristics of bovine and ovine Halloumi cheese. *International Dairy Journal* 10: 761–768.

Papademas, P. and Robinson, R.K. (2001). The sensory characteristics of different type of Halloumi cheese as perceived by tasters of different ages. *International Dairy Journal* 54: 94–98.

Pappa, E.C. and Anifantakis, E.M. (2001). Effect of different concentrated cultures on the proteolysis and organoleptic characteristics of Feta cheese. *Milchwissenschaft* 56 (7): 384–387.

Pappa, E.C., Kondyli, E., and Samelis, J. (2019). Microbiological and biochemical characteristics of Kashkaval cheese produced using pasteurised or raw milk. *International Dairy Journal* 89: 60–67.

Pirouzian, H.R., Hesari, J., Farajnia, S. et al. (2012). Effects of *Enterococcus faecalis* and *Enterococcus faecium*, isolated from traditional Lighvan cheese, on physicochemical and sensory characteristics of Iranian UF white cheese. *Journal of Agricultural Science and Technology* 14: 1023–1034.

Pitso, S. and Bester, B.H. (2000). Quality aspects of Feta cheese manufactured from mixtures of cow's and goat's milk. *Milchwissenschaft* 55 (8): 454–458.

Rashidi, H., Mazaheri-Tehrani, M., Razavi, S.M.A., and Ghods-Rohany, M. (2015). Improving textural characteristics of low-fat UF Feta cheese made with fat replacers. *Journal of Agricultural Science and Technology* 17: 121–132.

Robinson, R.K. (1991). Halloumi cheese. The product and its manufacture. In: *Feta and Related Cheeses* (ed. R.K. Robinson and A.Y. Tamime), 144–158. London: Ellis Horwood.

Salum, P., Govce, G., Kendirci, P. et al. (2018). Composition, proteolysis, lipolysis, volatile compound profile and sensory characteristics of ripened white cheeses manufactured in different geographical regions of Turkey. *International Dairy Journal* 87: 26–36.

Sarantinopoulos, P., Kalantzapoulos, G., and Tsakalidou, E. (2002). Effect of *Enterococcus faecium* on microbiological, physicochemical and sensory characteristics of Greek Feta cheese. *International Journal of Food Microbiology* 76 (1–2): 93–105.

Sengul, M., Gurses, M., Dervisoglu, M., and Yazici, F. (2006). A survey on the some chemical and biochemical properties of civil cheese, a traditional Turkish cheese. *International Journal of Food Properties* 9: 791–801.

Soltani, M., Guzeler, N., and Hayaloglu, A. (2015). The influence of salt concentration on the chemical, ripening and sensory characteristics of Iranian white cheese manufactured by UF-Treated milk. *Journal of Dairy Research* 82: 365–374.

Soltani, M., Sahingil, D., Gokce, Y., and Hayaloglu, A. (2016). Changes in volatile composition and sensory properties of Iranian ultrafiltered white cheese as affected by blends of *Rhizomucor miehei* protease or camel chymosin. *Journal of Dairy Science* 99: 7744–7754.

Soryal, K.A., Zeng, S.S., Min, B.R., and Hart, S.P. (2004). Effect of feeding treatments and lactation stages on composition and organoleptic quality of goat milk Domiati cheese. *Small Ruminant Research* 52 (1–2): 109–116.

Sulejmani, E. and Hayaloglu, A.A. (2016). Influence of curd heating on proteolysis and volatiles of Kashkaval cheese. *Food Chemistry* 211: 160–170.

Theophilou, P. and Wilbey, A. (2007). Effect of fat on the properties of Halloumi cheese. *International Journal of Dairy Technology* 60 (1): 1–4.

Torkashvand, Y. (2010). Processed cheese production based on milk protein co-precipitate and ultrafiltered cheese. *Veterinary Journal* 86 2–14.

Zlatanos, S., Laskaridis, K., Feist, C., and Sagredos, A. (2002). CLA content and fatty acid composition of Greek Feta and hard cheeses. *Food Chemistry* 78: 471–478.

Zomorodi, S., Asl, A.K., and Rohani, S.M.R. (2011). Survival of Lactobacillus casei, Lactobacillus plantarum and Bifidobacterium bifidum in free and microencapsulated forms on Iranian white cheese produced by ultrafiltration. *International Journal of Dairy Technology* 64: 84–91.

9 Sensory Profiles of Pan-American Fresh, Soft and Other Cheese Varieties

Callebe Camelo-Silva, Monique Juna Lopes Leite, Giordana Demaman Arend, Tatiana C. Pimentel, Marco Di Luccio, Silvani Verruck, Erick A. Esmerino and Adriano G. da Cruz

9.1 Introduction

According to the FAO (2017), by 2026 dairy product consumption will have grown by 16%. Cheese is one of the most consumed dairy products worldwide. According to the Global Trade Portal (GTP 2020), the United States (USA) is the leading cheese producer and consumer. However, the origin of cheese making is not from the Americas, but from Europe. During the colonial period, Europeans brought the first herds of cattle, sheep and goats to the Americas. From the milk of these animals, the production of a wide variety of cheeses began (González-Córdova et al. 2016). Subsequently, some of the production methods were modified by regional groups, for example; the type of milk used, ripening time (when applied), size, shape, texture, colour, flavour, moisture content, the use or not of starter cultures, among others.

Due to modifications and adaptations in the cheese production processes, several regions of the world have local or regional cheeses with peculiar sensory characteristics. There is a wide range of Pan-American cheeses on the American continents, and some of them are even known worldwide, such as Minas Frescal cheese, Coalho cheese and Queso Fresco. Some Pan-American-style cheeses that reflect their European origins include Cottage, Quark, Manchego and Mozzarella. On the other hand, some Pan-American cheeses use descriptive names, such as the most familiar Queso Fresco (fresh cheese) and Queso Blanco (white cheese). Still others refer to the region where they were processed, such as Queso Chihuahua (northern Mexican state), Oaxaca (southern Mexican state), Cotija (a town in the Mexican state of Michoacan) and Minas Frescal (south-eastern state of Brazil).

Several Pan-American cheeses are produced in an artisanal way, typically characterised by on-farm production on a small scale, and following traditional cheesemaking methods unique to the region (Kamimura et al. 2019). These products have different sensory and physicochemical characteristics that can be related to the producing area. As a result of maintaining popular traditions, they have high-cultural and economic value worldwide, demonstrating the importance of these products (Bemfeito et al. 2016;

Sensory Profiling of Dairy Products, First Edition. Edited by John J. Tuohy.
© 2023 John Wiley & Sons Ltd. Published 2023 by John Wiley & Sons Ltd.

Margalho et al. 2021; Souza et al. 2021). Artisanal cheeses can be produced using pasteurised or raw milk. However, there has been concern among regulators, ministries of agriculture and researchers regarding the use of raw milk, due to food safety issues. However, the replacement of raw milk by pasteurised milk has been reflected in a final product with different texture and flavour properties (Van Hekken et al. 2006), mischaracterising the sensory identity of cheese produced with raw milk. A possible alternative to this problem would be the use of native lactic acid bacteria isolated from raw milk cheese (Nájera-Domínguez et al. 2014).

Several similar cheeses may have different names since many Pan-American native cheeses evolved in rugged mountain and valley regions. Furthermore, cheeses of the same name often do not have the same flavour, texture or cooking properties. Therefore, this chapter will focus on the most popular and common Pan-American cheeses. The sensory profiles of Cottage cheese, Cotija cheese, Quark cheese, Manchego cheese, Queso Blanco, Queso Fresco, Mozzarella cheese, Oaxaca cheese, Queso Chihuahua, Minas Frescal cheese, Ultrafiltered Fresh cheese and Coalho cheese will be described and discussed with reference to the cheese production processes and their physicochemical composition.

9.2 Oaxaca cheese

Oaxaca cheese, also known as Threaded, Hebra, Bola or Quesillo cheese (Figure 9.1), is of Mexican origin, and is considered one of the most consumed dairy products in Mexico (González-Córdova et al. 2016; Carvajal-Moreno et al. 2019). The consumption of cheese per inhabitant in Mexico is 3.8 kg per year (Carvajal-Moreno et al. 2019) but when only cheese consumers are considered, the average consumption of Oaxaca cheese per cheese consumer is 17.5 kg per year (Estela Hernández et al. 2016). In 2010 it was the second most consumed cheese by Mexicans after Queso Fresco. Its name is believed to be related to the state of Oaxaca, where it probably originated (González-Córdova et al. 2016).

Oaxaca cheese is a *pasta filata* cheese that can be produced artisanally and also industrially. Generally, raw cow's milk is used to produce the artisanal type. The product is naturally acidified by the microflora present in the milk. At the same time, the

Figure 9.1 Oaxaca cheese. *Source:* Marcos/Adobe Stock.

industrial product is made with pasteurised cow's milk, acidified with commercial lactic acid bacteria. Usually, a mixture of *Lactococcus* spp. is used. *Streptococcus salivarius* ssp. *thermophilus* is preferred due to its more significant acidifying activity. However, direct acidification with acids can also be used (Sandoval-Copado et al. 2016). The strain of *Lactococcus lactis* subsp. *lactis* is the predominant species in Oaxaca cheese made with raw milk. This bacterium can produce 2-methylpropanal and 3-methylbutanal in larger quantities, contributing to improving the final product's flavour (Caro et al. 2020). The process of obtaining Oaxaca cheese comprises several steps, which can vary from one cheese factory to another. De Oca-Flores et al. (2009) visited 12 cheese factories in Mexico to verify the steps involved in the production of Oaxaca cheese and presented a standardised process flow diagram.

The physicochemical composition of Oaxaca cheese reported in the literature is quite variable, mainly due to the non-uniformity in the processing. According to De Oca-Flores et al. (2009), Oaxaca cheese may contain the following physicochemical composition: proteins (150.3–241.5 g/kg), fats (170–253.3 g/kg), moisture (124.1–610 g/kg), ash (18–41.8 g/kg), pH (4.9–5.8) and salt (8.4–36.2 g/kg).

The Oaxaca cheese sensory characteristics result from several factors, such as the milk physicochemical characteristics, processing conditions, acidification type, microflora present in the milk and storage conditions (Sandoval-Copado et al. 2016). Combining these factors results in peculiar sensory attributes, which are of paramount importance for the consumer at the time of purchase and consumption of the product. Unfortunately, the sensory characterisation of Oaxaca cheese has not been widely reported in the literature. More information about the sensory characteristics of this cheese would help consumers in their choices and the cheesemakers during the production of Oaxaca cheese.

Sandoval-Copado et al. (2016) trained 50 panellists following ISO 22935-1 to form an Oaxaca cheese evaluation sensory panel. Among the 50 people recruited, only 11 reached the level of competence required to join the sensory panel. Three types of cheese were characterised by the panellists: Type 1 cheese (MC) was manufactured using pasteurised milk and mesophilic LAB culture (*L. lactis* ssp. *lactis* and *L. lactis* ssp. *cremoris*; R-704, Chr Hansen, Horsholm, Denmark); type 2 cheese (TC), using pasteurised milk and thermophilic LAB culture (*S. salivarius* ssp. *thermophilus*; ST1.14, Chr Hansen) and type 3 cheese (RC), using naturally acidified raw milk. The MC and TC manufacturing processes represent industrially produced Oaxaca cheese, whereas the RC process represents artisanal Oaxaca cheese. The sensory panel defined 43 attributes common to the three Oaxaca cheeses. A total of 13 attributes contributed significantly to the discrimination of the three cheeses evaluated. The discriminant attributes perceived with greater intensity in RC cheese were moist (tactile), cow (orthonasal), salty, chewable, gummy, fibrous, moist (mouthfeel) and squeaky (auditory). In contrast, in MC cheese, the most intense were waxy, oily, empyreumatic, adherent and fatty. In TC, none of the attributes was perceived to be of high intensity; on the contrary, this cheese was characterised by the lowest intensity in the attributes waxy, salty, gummy and fibrous.

Fuentes et al. (2015) evaluated the sensory characteristics of Oaxaca cheeses from three small plants (Valle de Tucilango de Bravo, Hidalgo, Mexico) produced by the artisanal method. The cheeses were vacuum-packed and stored under refrigeration

(8 °C) for 24 days. A sensory panel consisting of nine trained judges performed a descriptive sensory analysis of taste, aroma and texture, as well as an acceptability analysis. Oaxaca cheeses were characterised as having a sour or buttery odour and a sour or bitter taste. These attributes showed no significant difference during the study period. Regarding texture, the cheeses had a creamy texture which tended to increase between days 8 and 16. However, the acceptability of the cheeses decreased during the 24 days of analysis. The results showed that the proportion of the samples that the panellists found to be unacceptable were 0% (d1), 6% (d8), 36% (d16) and 81% (d24). It is believed that the emergence of strange odours and excessive softness were the main factors contributing to the rejection of the samples.

In the study of Villanueva-Carvajal et al. (2012), artisanal Oaxaca cheeses from 11 producers were evaluated by 14 trained judges. As a result, the flavour profile of Oaxaca cheeses was characterised by 14 sensory attributes, such as sweet, salty, sour, acid, creamy, fatty, buttery, fermented, smoked, burned milk, caramelised, watery, mouldy and cow. In addition, the Oaxaca cheeses presented low hardness and adhesiveness and high springiness, cohesiveness and chewiness using instrumental texture profile analysis (TPA).

9.3 Queso Chihuahua

Queso Chihuahua (Figure 9.2), also known as Chester or Mennonite cheese (González-Córdova et al. 2016), is a major cheese variety in Mexico. In 2017 it occupied fifth place in the ranking of cheese production, surpassing 40 000 tonnes. Its origin is closely related to the establishment in 1922 of the Mennonite group in the northwest region of Chihuahua (Sánchez-Gamboa et al. 2018a,b). Generally, this cheese is sold in blocks, which vary from 250 g to 10 kg, of semi-hard consistency (matured for three to six weeks), sliceable, easy to melt, light yellow in colour and spicy taste (Renye et al. 2011; Nájera-Domínguez et al. 2014; González-Córdova et al. 2016; Sánchez-Gamboa et al. 2018a,b).

Figure 9.2 Queso Chihuahua. *Source:* Barmalini/Adobe Stock.

Traditionally, the Queso Chihuahua is produced from raw cow's milk, without the addition of starter culture, by an artisanal manufacturing process similar to Cheddar cheese (Tunick et al. 2008). Due to food safety issues, a milk pasteurisation step for cheese production was implemented. However, small dairies claim that cheese produced with pasteurised milk does not have the same characteristics as cheese made with raw milk (Sánchez-Gamboa et al. 2018a,b). The Queso Chihuahua made with raw milk has a softer texture and a more intense flavour (Van Hekken et al. 2006), due to the growth of specific types of lactic acid bacteria found in the raw milk (Georgina Venegas-Ortega et al. 2020). A promising alternative to maintaining these sensory characteristics is the use of native lactic acid bacteria isolated from raw milk cheese (Nájera-Domínguez et al. 2014). According to Official Mexican Standards, the final product must contain a minimum of 22% protein and 25% fat, and a maximum of 45% moisture and 3% salt (Van Hekken et al. 2007).

Queso Chihuahua is generally known to have a spicy taste, smooth texture and intense flavour (Van Hekken et al. 2006). It is known that the sensory characteristics of this product can be influenced by the geographic region in which it was produced, seasonality and the type of milk (raw or pasteurised) used in the production of the cheese (Villalobos-Chaparro et al. 2018). However, relatively little effort has been made to assess the sensory profile of this cheese.

Recently, a quantitative descriptive analysis was carried out by eight trained judges to characterise Chihuahua cheese produced with raw cow's milk. The product was a semi-matured cheese with a bitter flavour, slightly salted and with a cream flavour, with aroma notes associated with whey and sour (Villalobos-Chaparro et al. 2018). Similarly, in the study of López-Díaz and Martínez-Ruiz (2018), 10 trained judges evaluated the taste, odour and mouthfeel attributes of 10 samples of Queso Chihuahua. In addition, an acceptance test was carried out with 120 untrained consumers. The Queso Chihuahua was characterised by a medium to medium-high odour of cooked milk, fresh milk and melted butter, with low-medium intensity of taste, firmness, moistness and fat character. Furthermore, cheeses with higher protein, fat, mineral and carbohydrate contents were preferred in the acceptance test. Likewise, in the work of Gamboa Alvarado et al. (2013), a sensory panel of 12 trained judges evaluated Queso Chihuahua for 21 attributes; acid, creamy, hard in the mouth, soft in the mouth, dry and yellow were the words most used by the evaluators.

In an attempt to assess the influence of the brand on the sensory characteristics of the Queso Chihuahua, Paul et al. (2014) compared four commercial brands of Queso Chihuahua produced in Mexico (Queseria America labelled W, Queseria Blumen labelled X, Queseria Lago Grande labelled Y and Queseria Super Fino labelled Z). The panellists identified flavours such as cooked, whey, milk fat and diacetyl consistently in all four samples. In addition, the flavours sour, bitter, salty and astringent were also identified in all four cheeses. As expected, the X cheese, which displayed the highest salt content, was associated with a maximum salty taste.

On the other hand, Trancoso-Reyes et al. (2014) investigated the effect of adding phospholipase A1 (PL-A1) and exopolysaccharide (EPS) producing bacteria (separately and in combination) on the sensory profile of Chihuahua Queso. Four different cheeses were manufactured: cheese made with PL-A1; cheese made with EPS-producing bacteria; cheese with both PL-A1 and EPS-producing bacteria and control cheese without PL-A1 or EPS-producing bacteria. The sensory analysis was carried out with 50 untrained participants, all regular consumers of Queso Chihuahua. All cheeses were equally rated for flavour, except the cheese manufactured with both PL-A1 and EPS-producing bacteria, which had the highest flavour preference. One concern in cheeses manufactured with phospholipase is the possible development of rancid flavours produced by hydrolysis of triglycerides. Nevertheless, and according to the comments of the judges, these cheeses did not have a rancid flavour. On the other hand, the texture of the cheese manufactured with EPS-producing bacteria was perceived as unpleasant by most of the judges. This alteration in texture may have been caused by the excessive water retention and modification of rheological properties.

Sánchez-Gamboa et al. (2018a) investigated the influence of season on the textural parameters of Queso Chihuahua from five artisanal cheese factories in Mexico. These authors concluded that none of the samples displayed fracturability, and there were no differences in hardness, cohesiveness and adhesiveness between the dairies or the seasons. Regarding springiness, there were differences between the seasons (winter, summer, autumn), but not between the dairies. The only parameter that presented a highly significant difference among seasons (winter, summer > autumn) and dairies was chewiness. Lastly, Nájera-Domínguez et al. (2014) produced Queso Chihuahua with lactic acid bacteria isolated from vegetables and Queso Chihuahua made with raw milk. The authors concluded that cheese manufactured with strains of *L. lactis* isolated from vegetables did not produce harder cheese than commercial strains or wild strains isolated from raw milk cheese. Conversely, the plant-derived strain isolated from lucerne (Alf-2) made cheese with the softest texture.

9.4 Mozzarella-type pizza topping cheese

Mozzarella cheese (Figure 9.3) is a cheese of Italian origin, more specifically from the southern region of Italy, and is currently one of the most consumed and produced cheeses in the world (Smith et al. 2018; Faccia et al. 2021). In terms of world consumption, the United States of America (USA) leads with 54.5% of the volume, followed by Brazil (6.6%), Italy (5.2%), Germany (5.2%) and Canada (3.7%). Mozzarella is

Figure 9.3 Mozzarella cheese. *Source:* Moving Moment/Adobe Stock.

American's favourite cheese accounting for nearly a third of consumption. Due to its melt and stretch capacity, it is one of the main ingredients of pizza (Jana and Tagalpallewar 2017) and more than 70% of all Mozzarella consumed in the United States is used in pizzas. Traditional Mozzarella cheese is made with buffalo milk. However, most of the Mozzarella produced today is made from partially skimmed pasteurised cow's milk (Smith et al. 2018). This cheese belongs to the cheese family of *pasta filata* and is characterised as soft, white, unripened, which may be consumed shortly after manufacture (Smith et al. 2018; Gonçalves and Cardarelli 2020).

There are four types of Mozzarella cheese in the United States, which are differentiated by their moisture and fat contents (Smith et al. 2018), as shown in Table 9.1. Its use in cooking depends on the moisture content of the product. For example, low-moisture Mozzarella is typically used for cooking procedures such as dressing pizza, while the high-moisture Mozzarella is mainly used as table cheese (Cincotta et al. 2021). During the production of Mozzarella cheese, mesophilic (*L. lactis* ssp. *lactis* and *L. lactis* ssp. *cremoris*) and thermophilic (*S. salivarius* ssp. *thermophilus*, *Lactobacillus delbrueckii* ssp. *bulgaricus* and/or *Lactobacillus helveticus*) cultures can be used (Gonçalves and Cardarelli 2021). Generally, the cheese used in pizzas is produced with a combination of thermophilic cultures, such as *Lactobacillus* ssp. and *S. thermophilus*. On the other hand, no *Lactobacillus* strains are added to table cheese since the addition of *Lactobacillus* is used to achieve the higher acidification rate needed in Mozzarella for pizza in order to reach the required moisture content (Smith et al. 2018).

The physicochemical composition of Mozzarella cheese reported in the literature is quite variable, as it is influenced by the type of milk used in cheese production, the starter cultures, process conditions and geographic region where it was produced. Table 9.2 shows the approximate physicochemical composition of Mozzarella cheese sold in different regions of the Americas.

Table 9.1 Compositional standards for Mozzarella cheese in the USA.

Mozzarella type	Moisture range (%)	Fat in dry matter (%)
Mozzarella	52–60	≥45
Low-moisture Mozzarella	45–52	≥45
Partially skimmed Mozzarella	52–60	30–45
Low-moisture partially skimmed Mozzarella	45–52	30–45

Table 9.2 Physicochemical composition of Mozzarella cheese.

Constituents	Silva et al. (2020)[a]	Ozturk et al. (2018)[b]	Paz et al. (2017)[c]
Moisture (%)	56.65	48.60	49.35
Fat (%)	22.75	22.50	42.44
Total protein (%)	17.54	24.50	25.70
Ash (%)	1.33	–	2.64
Salt (%)	–	1.00	–
pH	5.60	5.23	5.42

[a] Brazil – Cheese produced with buffalo's milk;
[b] USA – Cheese made with cow's milk;
[c] Argentina – Cheese produced with goat's milk.

Mozzarella cheese has a white colour, soft texture, mild (slightly acid) taste and buttery flavour and may be consumed shortly after manufacture. Furthermore, it is well known in the culinary industry for its excellent melt and stretch properties. However, these sensory characteristics can be influenced by several factors, such as the type of milk used in cheese production, the bacterial cultures used and the processing conditions.

Recently published research shows a trend towards offering Mozzarella cheese to consumers with an enhanced sensory profile and sometimes with a functional appeal for the consumer's health. The influence of some variables on the sensory profile of Mozzarella cheese has been studied, for example; the effect of the stretching temperature (Gonçalves and Cardarelli 2020), the use of autochthonous bacteria (Moynihan et al. 2014; Silva et al. 2020), the effect of high-pressure treatment (Ozturk et al. 2018), reducing the sodium content (Arboatti et al. 2014; Ganesan et al. 2014; Rodrigues et al. 2014; Ozturk et al. 2018), the influence of compression parameters (Fogaça et al. 2017), reducing the fat content (Wadhwani et al. 2011; Oberg et al. 2015; Fang et al. 2016), the influence of the cows' diet (Corradini et al. 2013; Oeffner et al. 2013) and using goat's milk (Paz et al. 2017).

Silva et al. (2020) investigated the influence of *S. thermophilus* (commercial and autochthonous) on the texture properties of Mozzarella cheeses produced with buffalo milk. The authors concluded through the TPA (TA.XT/Plus/50 texture analyser) that Mozzarella cheese made with autochthonous cultures had similar hardness, cohesiveness, springiness and chewiness to commercial culture cheese. Furthermore, they emphasised the technological potential of using the *S. thermophilus* SJRP107 autochthonous culture in Mozzarella cheese production due to the fast acidification and high production of organic acids. Fang et al. (2016) studied the texture properties of Mozzarella cheese and light Mozzarella cheese, with fat contents of 20% and 14%, respectively. The authors verified that cheese's springiness was not significantly different due to their similar protein network density. However, the hardness and the cohesiveness of light Mozzarella were higher than those of conventional Mozzarella cheese. Fat globules contribute to softening of the cheese texture because they interfere with the integrity of the protein network, and a harder texture is associated with a lower fat content.

On the other hand, Paz et al. (2017) evaluated the technological potential of using goat's milk to produce Mozzarella cheese. Mozzarella cheese produced with goat's milk had a more intense white colour than cheese made with cow's milk. The authors reported that cheeses made from goat milk are generally whiter because goats are capable of converting β-carotene into vitamin A and produce milk fat globules that are smaller in diameter than those in cow milk.

In an attempt to produce reduced-fat Mozzarella cheeses, Oberg et al. (2015) made cheeses with 0.25%, 0.5%, 1.0% and 1.5% of added xanthan gum. The incorporation of 1% xanthan gum in commercial Mozzarella cheese made with partially skimmed cow's milk produced Mozzarella cheese with greater stringiness and a similar flavour and texture to the control sample. Also, they tended to get softer with additional storage. Similarly, Arboatti et al. (2014) investigated the effect of salt reduction on the sensory properties of Mozzarella cheeses. Three samples (C: control cheese; S_1: cheese with 60% salt reduction and S_2: cheese with 35% salt reduction) were evaluated by a trained sensory panel of eight judges. The salt concentration significantly affected aroma, colour, overall flavour, creaminess and saltiness. In general, the scores observed for the control cheese (C) were slightly higher than for cheeses S_1 and S_2. Mozzarella cheese has a delicate, mild and milky flavour, and therefore, bitterness or acidity can limit cheese acceptability if those attributes are too intense. However, in their study, the attributes bitterness, acidity and residual flavour were not significantly affected by the salt concentration.

The composition of the milk used in cheese production is another factor that can influence the sensory profile of the final product. Corradini et al. (2013) studied the effect of dietary supplementation with palm oil and coconut oil on the physicochemical characteristics of milk from 23 healthy cows. Milk collection was performed at 21 and 36 days after the introduction of the diets. Three Mozzarella cheeses were produced from the milk of control, palm oil and coconut oil supplemented cows. These cheeses were submitted to a sensory acceptance test carried out with 50 untrained judges, as well as instrumental TPA. The judges accepted all samples; however, cheese made from milk of the coconut oil supplemented cows was the most preferred. Furthermore, according to the TPA, at both periods (21 and 36 days), Mozzarella cheeses produced with the milk of cows fed coconut oil were firmer.

9.5 Quark

Quark is a fresh cheese, considered typical of the Central and Eastern Europe region, produced with whole milk, skim or partly skimmed milk, generally cow's milk (Figure 9.4) (Lücke and Zangerl 2014). The consumption of this type of cheese has been growing in the Americas. In some cases, a starter culture that produces lactic acid is

Figure 9.4 Quark cheese. *Source:* Baibaz/Adobe Stock.

used, such as *Streptococcus cremoris* and *Leuconostoc citrovorum*. Other cultures can also be used to produce aromas such as *S. diacetylactis* (Sohal et al. 1988).

The sensory characteristics of quark cheese are related to the type of milk and starter culture used for its production. However, it is common to find in quark cheese a slightly acidic flavour and a mild bitterness due to the use of rennet. In addition, quark cheese has a smooth texture and short shelf life due to its high-water content (>72%, w/w) (Schulz-Collins and Senge 2004; Yadav et al. 2019). The chemical composition criteria are also influenced by the origin of the milk used for production. Table 9.3 shows the chemical composition of quark cheese depending on the chemical composition of milk. Another observation about the chemical composition is a low amount of calcium, and this is due to the more significant calcium loss in the whey (Schulz-Collins and Senge 2004).

Quark cheese is produced from acid coagulation that occurs below pH 4.6, leading to the precipitation of casein protein. In some cases, a small amount of rennet is added, with a function to make the curd firmer. The acid clot is broken by stirring, and the curd is recovered by centrifugation, in an industrial process, or by muslin cloth draining in a traditional method (Yadav et al. 2019; Dank et al. 2021).

Fresh cheeses are known for their high nutritional value and are appreciated for their unique sensory characteristics of taste, appearance and texture. This uniqueness results from factors such as milk composition and quality, ripening conditions, the coagulation process and microorganisms present in the cheese (González et al. 2017). Acidity is an essential factor in fresh cheeses like quark cheese. Lactic acid, hexanal, nonanal and decanal are responsible for the freshness characteristic. High moisture is also characteristic of this class of cheese (Litopoulou-Tzanetaki 2007). Quark cheese has a fresh, mildly acidic flavour and a soft, smooth pasty texture, and cream, fruit and seasonings can be added to create line extensions. It is a cheese rich in proteins, with a milky flavour, and the colour is usually white or slightly yellow. The addition of cream provides more smoothness to the cheese (Lucey 2011; Yadav et al. 2019).

Currently, there is a new product development preoccupation with changes in the process, improving the yield or improving nutritional attributes, and these changes can affect the sensory characteristics. Given the above, it is possible to find publications on the production of quark cheese with added probiotics (Kim et al. 2019), the utilisation

Table 9.3 Chemical composition of quark cheese.

		Low fat	Half fat	Whole fat
Specific criteria	Fat content [% dry matter]	<10	>20	>40
	Minimum content of dry matter [%]	18	20	24
	Minimum protein content [%]	12	10.5	8.7
Chemical composition	Energy (Kcal/100 g)	75.3	109.5	158.7
	Protein (g/100 g)	13.5	12.5	11.1
	Fat (g/100 g)	0.2	5.1	11.4
	Carbohydrate (g/100 g)	4	2.7	2.6
	Water (g/100 g)	80.7	78.1	73.8
	Calcium (mg/100 g)	114	85	95

Source: Drusch and Einhorn-Stoll (2016).

of by-products (Ozturkoglu-Budak et al. 2021) and use of different starter culture strains (Skryplonek et al. 2019). The impact of a variation on pasteurisation temperature on sensory attributes was reported (Miloradovic et al. 2018).

The addition of ginseng to quark cheese was investigated by Kim et al. (2019). Ginseng is a traditional product popular in northeast Asian countries and rich in bioactive compounds. During the production of quark cheese, ginseng extract was added, and the antioxidant activity, physicochemical and sensory properties were evaluated. Qualitative descriptive analysis was used for sensory evaluation. A panel composed of 22 panellists (12 females, 10 males, age 26–30 years) and a continuous scale (0–8) was used to measure characteristics, such as odour, taste, aftertaste and overall quality. The ginseng flavour, aroma and aftertaste increased significantly ($p<0.05$) with increasing concentration of the ginseng in cheese (>0.5% w/w). In contrast, the sensory quality was not significantly affected by ginseng extract at a concentration of less to 0.5%. The authors also reported increased antioxidant activity, using the ABTS method, with radical-scavenging activity ($r = 0.803$, $p<0.01$). The final colour of quark cheese was affected by increasing the extract concentration leading to a dark brown colour at the higher levels of addition of extract. Regarding texture, there was an increase in hardness with increasing concentration of the extract.

Skryplonek et al. (2019) utilised buttermilk for the production of quark cheese. Buttermilk is a by-product of butter production and is used in several dairy products to improve texture, elasticity, moisture retention and sensory properties. For sensory analysis, the sensory panel was composed of five trained panellists. Taste, smell, structure, consistency and colour were evaluated using a scale from 1 to 5. Quark cheese produced from buttermilk had a low-titratable acidity. The authors assumed this could be related to the low-protein content of the buttermilk compared with the sample made with milk. The authors also observed that buttermilk was responsible for texture modification in the quark cheese. Buttermilk with the highest water content had the lowest hardness, but the relationship needed more investigation. The taste and smell were categorised as particularly good, clean, mildly acidic and aromatic. Uniform structure and consistency were also observed, the sample with the lowest hardness had an inferior consistency.

The impact of starter culture substitution in the production of quark cheese was reported by Ozturkoglu-Budak et al. (2021). Kefir was used as a starter culture alternative, and two coagulation temperatures – time combinations was also examined, 35 °C and 100 °C by 16 hours and 2 hours, respectively. Kefir is a fermented dairy product that contains lactic acid bacteria, acetic acid bacteria and yeasts. A panel of seven trained panellists was used for sensory analyses over a 14-day period using a scoring card scale from 0 to 10. Characteristics such as colour, appearance, body texture, odour and flavour were evaluated. Proteolysis was identified in the samples, probably due to the presence of several microorganisms in the kefir. The cheese heated to a temperature of 100 °C had the highest texture, flavour and colour acceptability (scored >8). This preference may be related to the inhibition of microorganisms at the higher temperature used for coagulum formation. Smaller peptides resulting from protein hydrolysis by enzymes produced by the different microorganisms were found in samples heated at 35 °C.

9.6 Cottage cheese

Cottage cheese is defined as a soft unripened, cheese, produced by acid coagulation of skim milk, generally by the addition of starter culture and a certain amount of rennet (Drake et al. 2009), as illustrated in Figure 9.5. Cow's milk, buffalo milk or mixtures can be used (Chakraborty et al. 2021). The cultures used usually contain bacteria such as *L. lactis* subsp. *lactis* and *Lactococcus. lactis* subsp. *cremoris* (Araújo et al. 2016). The chemical composition of Cottage cheese can be modified according to the type of milk used. Figure 9.6 shows a variation in cheese composition made with different sources of milk.

The Cottage cheese process has similarities with quark cheese production. The steps include fermentation, cutting, stirring and draining. However, for Cottage cheese production two steps of curd rinsing are necessary. The first rinse occurs at room temperature and the second rinse at 3–4 °C, both with water. Acidification with starter culture requires a longer incubation time. A range of different fermentation times are used, such as a short incubation time of five hours, medium of eight hours and long of 12–16 hours. Like quark cheese, rennet can be used to improve the dry matter yield and the firmness of the curd particles due to the contraction of the cheese matrix (Guinee et al. 1993).

Cottage cheese is white in colour, has a slightly acidic and salty taste with a delicate diacetyl or creamy flavour and aroma. Cottage cheese has a granular texture, with curd granules of on average 3–12 mm in diameter and has a typical shelf life of three to four weeks under refrigerated conditions (Lucey 2011; Araújo et al. 2016; Jain et al. 2019). Several studies have been reported regarding modifying the process and evaluating how the changes can affect sensory aspects, the nutritional value and process efficiency or yield. In Latin America, there a few works focused on the sensory analysis of Cottage cheese. On the other hand, the United States has the highest volumes of

Figure 9.5 Cottage cheese. *Source:* Pixabay.

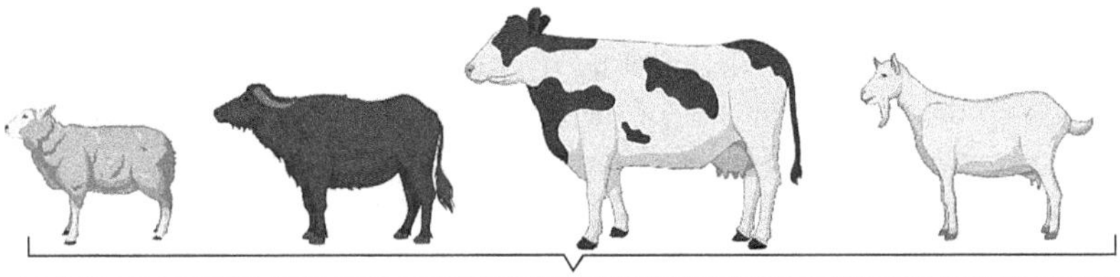

	Chemical composition			
Solid no fat [%]	9.51	8.78	7.72	7.06
Fat [%]	6.91	6.20	3.55	3.03
Protein [%]	4.77	4.10	3.41	3.37
Lactose [%]	4.93	4.76	4.69	4.33
Ash [%]	0.88	0.88	0.76	0.72
Water [%]	6.28	6.07	6.05	5.73

Figure 9.6 The chemical composition of Cottage cheese can vary according to the type of milk used. *Source:* Rasheed et al. (2016).

publications. Studies on Cottage cheese production and sensory analysis in the last 10 years will be considered here.

Brito et al. (2006) made Cottage cheese with homogenised and unhomogenised cream and different starter cultures. The authors carried out sensory analysis with eight panellists who used a 9-point hedonic scoring scale. No statistical difference in sensory attributes such as taste, aroma, texture, colour and general appearance was found between the treatments. The values for sour taste, characteristic aroma and overall flavour were within the normal range of 4 to 6 points. The appearance attributes analysed in the study were: the presence of free whey (2.1–2.3), colour (7.0–7.1), cream perception and grain size uniformity, and all attributes evaluated were within the normal range, which ranged from 1 to 3 for the first parameter and 7–9 points for the second parameter. For grain colour and cream perception attributes, there was a statistically significant difference ($p < 0.05$) between homogenised and non-homogenised cream treatments. However, both treatments were within the same colour range ranging from white to cream. In general acceptance, the scores obtained in Cottage cheese are close to 7.0, which corresponds to moderately liking, with no statistical difference between treatments.

Mónica et al. (2010) studied the probiotic (*Lacticaseibacillus casei*, *Lactobacillus acidophilus* La-5 and *Bifidobacterium* BB-12) viability in Cottage cheese during shelf life. A sensory panel with eight trained judges familiar with the typical characteristics of Cottage cheese was used. A 5-point hedonic scale was used. *Bifidobacterium* counts were greater than 1×10^6 CFU/g up to 14 days, and *L. acidophilus* and *L. casei* count were greater than 1×10^6 CFU/g up to 21 days shelf life. However, a bigger decline in *L. casei* numbers was observed during storage. No statistical differences in physicochemical characteristics were found between samples. On the other hand, sensory attributes were affected by the microorganisms added – statistical differences were found between samples for aroma and watery feeling. The authors obtained more aromatic cheeses when they used the microorganisms of the *Bifidobacterium* strain. They justified that this can be explained due to the heterofermentative characteristics of the strain, which can produce several aromatic compounds, mainly through citrate fermentation.

9.7 Queso Fresco

Queso Fresco (Figure 9.7) is a type of fresh Mexican cheese. It is traditionally made from raw bovine, ovine or caprine milk. It is considered the most popular Hispanic cheese found in the southern and western United States and Mexico. However, the traditional method of producing Queso Fresco with raw milk is not allowed in the USA due to a number of documented food poisoning outbreaks associated with the presence of *Listeria monocytogenes* and *Salmonella typhimurium* in the cheese (Sandra et al. 2004).

The composition of Queso Fresco can change due to differences in regional manufacturing protocols. In general, its physicochemical composition is as follows: 46–57% moisture, 18–29% fat, 17–21% protein and 1–3% salt (Sandra et al. 2004; Guo et al. 2011; Holle et al. 2018). The crumbly texture of Queso Fresco is the result of milling before salting during cheesemaking. Traditionally made Queso Fresco does not melt, and because of this, it is sometimes eaten fried (Holle et al. 2018). Queso Fresco may be produced with whole, partially skimmed or skim milk. It is moulded into a cylindrical shape, and its weight can vary from 200 g to 1 kg (González-Córdova et al. 2016).

Queso Fresco has high-moisture content, soft and crumbly texture, mild fresh-milk flavour, low-salt and near-neutral pH (Clark et al. 2001). Due to the traditional method of Queso Blanco production with raw milk, several researchers have examined alternative processing procedures to reduce the risk of pathogenic bacteria but without changing its sensory characteristics. A combination of the high-pressure processing (100, 200 and

Figure 9.7 Queso Fresco. *Source:* Angel Simon/Adobe Stock.

300 MPa) and pasteurisation was studied as an alternative processing method for Queso Fresco cheese production by Escobar et al. (2011). High pressure allows the inactivation of pathogens organisms under mild temperature conditions, for example, room temperature. For sensory analysis, 16-point hedonic scale, where 0 = none, 3 = slight, 7 = moderate and 15 = extreme was utilised. A group of nine students was selected and trained during six sessions. Six texture attributes were analysed for: firmness, springiness, crumbliness, stickiness, cohesiveness and oiliness/greasiness residual after crumbling. The combination of heat treatment and high pressure (300 MPa) provided a higher yield of cheese. Comparing the production of cheese from raw and pasteurised milk submitted to 300 MPa, the authors observed that the cheeses produced with raw milk had a lower moisture content and higher density. According to the authors, this may be related to protein denaturation caused by the thermal treatment. Protein denaturation was lower in the micelles of raw milk at the high pressure and due to dissociation and reorganisation formed dense agglomerates. With high-pressure-treated pasteurised milk, a better score for crumbling was observed. Components such as tension and cohesiveness showed a negative correlation with yield and moisture content. Sensory fragility, yield and moisture content were strongly correlated with each other and negatively correlated with variables such as wet protein, yield stress and sensory cohesiveness. The authors concluded the combined effects result in Queso Fresco containing a thin casein–whey protein matrix structure able to better retain sweet whey. The combined use of HPH and thermal processing of milk offers the potential to increase yield, moisture content and textural quality of Queso Fresco.

The incorporation of antibacterial substances in Queso Fresco was proposed by Van Tassell et al. (2015). Nisin and ferulic acid alone and in combination were used to prevent the growth of *Listeria innocua* and *L. monocytogenes*. An affective test performed by an untrained panel of 28 males and 32 females, 18–46 years old, was conducted for sensory analysis. A hedonic scale with 9 points was used (1 = dislike extremely, 5 = neither like nor a dislike and 9 = like extremely). Attributes such as: appearance, colour, aroma and texture were evaluated. No significant differences in consumer acceptance were detected between the control group and fresh cheese samples treated with ferulic acid. Nisin was not stable enough in the fresh cheese. The authors suggest that this may be related to its nearly neutral pH and difficulty staying active for long periods. Consequently, sensory tests were only performed with samples of ferulic acid. Texture attributes such as elasticity and chewiness were significantly lower compared to the control group. None of the samples treated with ferulic acid could be evaluated for taste, as purified food-grade ferulic acid was not used for sample preparation.

The use of dry bromelain extract (5%, 10% and 15% w/w) as a coagulating agent for Queso Fresco was proposed by Vergara Alvarez et al. (2019). For the study, raw and pasteurised milk was used to produce fresh cheese. A descriptive quantitative test was used for sensory analysis with a panel consisting of five trained judges. A 10-point hedonic scale was used for colour, taste, texture and odour/aroma. The physicochemical characteristics pH and acidity were also measured. The lowest concentration (5% w/w) of bromelain had the highest sensory acceptance. At low concentrations, no statistical differences were found for colour, smell and texture attributes between samples produced from raw and pasteurised milk.

9.8 Queso Blanco

Queso Blanco or Latin American White cheese is a fresh and soft cheese with a short shelf life produced by acidification of milk with acetic, lactic or citric acid (Figure 9.8). Its texture and structure resemble Cheddar cheese with a high-moisture content, and it is made traditionally from cow's milk. Its physicochemical composition depends on the raw milk, cheesemaking procedure and coagulation method (Torres and Chandan 1981; Farkye et al. 1995; Gurdian et al. 2017a). Queso Blanco contains: 51–53% of moisture, 19–20.7% of fat, 19.85–21.82% of protein, 1.79–1.94% of salt, presents a pH of about 5.27–5.28 and yield of 13.54–14.20% (Farkye et al. 1995).

Queso Fresco and Queso Blanco have similar characteristics but Queso Fresco is usually made with rennet addition, while Queso Blanco is usually made without. Queso Blanco curds are generally cut to a larger size before salting, providing a firmer texture compared to fresh cheese.

Queso Blanco cheeses are usually creamy, highly salty and acid in flavour. The texture and body resemble those of unripened high-moisture cheese (Litopoulou-Tzanetaki 2007). The incorporation of antioxidants and bioactive compounds into Queso Blanco to extend its shelf life or enhance its nutrition composition are areas that have attracted research interest in the recent years. Adding these compounds to Queso Blanco without modifying its sensory characteristics to maintain the identity and quality standards is especially important for the development of new products.

The addition of antioxidant oils to reduce lipid oxidation during their low-temperature storage has been investigated. However, it is noteworthy that the addition of antioxidant oils in white cheese needs to be evaluated by regulatory agencies before marketing. The addition of flaxseed oil at different stages (homogenisation, curd and salting) of the Queso Blanco was studied (Gurdian et al. 2017b). Instrumental texture profiling and colour analysis were the parameters analysed. The colour was determined using a Lab Scan XE colorimeter, and TPA was conducted with an Instron Universal model 5544 test instrument (Instron Co. Canton, MA). Other parameters such as moisture and pH

Figure 9.8 Queso Blanco. *Source:* Istetiana/Adobe Stock.

were also analysed. The storage period was 30 days. When the oil was added during salting, samples had a lower hardness and higher elasticity. The addition of oil to the curd resulted in a more reddish colour, which may have been due to the growth of fungi and yeasts.

The incorporation of microcapsules with tomato extract (2.0%) into Queso Blanco was studied by Jeong et al. (2017). Tomato extract is rich in lycopene which is an effective antioxidant. Ten trained sensory panellists evaluated texture, appearance and taste. Before evaluating, the panellists were trained for one hour. A hedonic scale was used for texture and taste with 7 points (1 = very weak and 7 = very strong). An instrumental colorimeter was used to measure colour. Instrumental texture properties were also evaluated utilising the TMS-Pro texture analyser. The tests were carried out over a 60-day storage period. Higher concentrations of microcapsules lead to an increase in the yellow colour, astringency and firmness of the cheese. Samples supplemented with 2% (m/m) of lycopene microcapsules, had a mean score of 6.11 for acidity, while the control sample had a mean score of 1.25. Differences were also observed for astringency (scored 3.25), firmness (5.92) and texture (5.54); the scores for the control group were 1.38, 1.34 and 4.25, respectively.

9.9 Cotija cheese

Cotija cheese (Figure 9.9) has been produced in Mexico for more than 400 years in the western hills of the country which are home to deciduous forest vegetation. However, a similar version of this cheese can be found in other regions of the country. It is a cheese

Figure 9.9 Cotija Cheese. *Source:* Marcos/Adobe Stock.

aged under environmental conditions for few months, with a maturation time of at least three months, which can be extended to years. Factors such as storage temperature and humidity, strongly influence its sensory characteristics (González-Córdova et al. 2016; Chombo-Morales et al. 2016; Escobar-Zepeda et al. 2016; Murugesan et al. 2018).

Cotija cheese is produced from raw cow's milk, without any thermal treatment or starter culture addition. Cheesemaking steps such as cutting of the curd, adding salt, pre-setting and the ripening process can vary among rural producers. Traditionally, it was only made during the summer, has a cylindrical shape of 40 cm in diameter and 20 cm in height, and its typical weight is approximately 20 kg (Chombo-Morales et al. 2016; Escobar-Zepeda et al. 2016; Murugesan et al. 2018). Cotija cheese has a chemical composition of approximately 46% moisture, 15% ash in dry matter, 36% protein and 37% fat in dry matter (Murugesan et al. 2018).

Mesophilic and thermophilic starter cultures and high-cooking temperatures are used for Cotija cheese production. Exogenous lipases are typically added to enhance lipolysis.

Cotija cheese is a semi-hard cheese, and its sensory attributes, such as smell, texture and taste, depend on environmental conditions and the production method (Murugesan et al. 2018). There are no recent studies about the sensory properties of Cotija cheese. The work described in the literature identifies the primary microbial populations contained in the cheese (Flores-Magallón et al. 2011; Chombo-Morales et al. 2016; Escobar-Zepeda et al. 2016). Bacteria of the genera *Lactobacillus*, *Weissella* and *Enterococcus* and yeasts of the genera *Candida*, *Pichia*, *Moniliella*, *Kluyveromyces*, *Rhodotorula* and *Galactomyces* were present in most of the cheeses analysed. Bacteria of the genus *Leuconostoc* and *Weissella* were also found by Escobar-Zepeda et al. (2016), although they are not generally associated with Cotija cheese. The prevalence of these genus (*Leuconostoc* and *Weissella*) may be related to the geographic region where the cheese was produced. However, the genus *Leuconostoc* and *Weissella* were also moderately abundant in the cheese analysed by Murugesan et al. (2018).

9.10 Mexican Manchego

Manchego cheese is produced in Mexico's Queretaro area and manufactured with pasteurised cow's milk and rennet. It is a semi-hard cheese, pressed, uncooked and slightly ripened (for between 14 and 30 days). The addition of starter cultures is optional. Mexican Manchego cheese has a chemical composition of approximately 26% fat, 40% moisture, 1.5% salt and a pH of 5.8. Traditional Manchego is made from raw milk (no starter cultures added). In contrast, commercial Manchego cheese is made with pasteurised milk inoculated with mesophilic starter cultures. The steps of production have been described by Ibarra-Sánchez et al. (2017).

Mexican Manchego cheese have different characteristics to the Manchego cheese produced in Spain. Mexican Manchego cheese is closer to Gouda or Monterey Jack in body and flavour (Solano-López and Hernández-Sánchez 2000; Lobato-Callero et al. 2001; González-Córdova et al. 2016). Chihuahua cheese and Mexican Manchego cheese have similar characteristics (Lluis-Arroyo et al. 2014). Mezo-Solís et al. (2020)

evaluated the physicochemical, sensory and texture attributes of Mexican Manchego cheese made from sheep's milk. For sensory analysis, consumer tests were carried out with 75 judges (27 female and 48 male) using a 9-point hedonic scale with (where 1 = extremely dislike; 5 = neither dislike nor like and 9 = remarkably like). The attributes evaluated were colour, texture, aroma and overall acceptability. Instrumental textural attributes of the cheese were measured by the uniaxial compression test and related to sensory evaluation. Samples were evaluated after ripening for 1, 30, 60, 90, 120, 150 and 180 days. The authors observed a decrease in moisture after 60 days. No significant change in pH, which ranged from 5.2 to 5.4, occurred during ripening. After 120 days of ripening, only a decrease in whiteness was exhibited. The firmness decreased after 30 days of ripening and remained constant until 150 days. On the first day of ripening, all the sensory attributes had high values – in the range of 7–8 points. However, after 180 days of ripening, the scores for those attributes had decreased significantly and ranged from 6 to 7 points. Consumer sensory evaluation showed that the cheese manufacturing protocol used in this study was not suitable for the production of an acceptable aged cheese. Through the sensory evaluation, the authors also observed that Latin American consumers prefer fresh and mild cheeses made from cow's milk to cheese made from goat and/or sheep's milk.

9.11 Minas Frescal

According to Bemfeito et al. (2016) and Margalho et al. (2021), some regions are recognised as ideal for cheese production due to a suitable climate, native pastures and traditional production techniques, resulting in products with unique sensory characteristics. Minas Frescal cheese (Figure 9.10) is one of the main dairy products consumed in Brazil, Minas Gerais being considered the prominent cheese producing state (Bemfeito et al. 2016; Souza et al. 2021). Two different kinds of cheese are made in the state: Minas

Figure 9.10 Minas Frescal cheese. *Source:* Leonidas/Adobe Stock.

Frescal cheese, a fresh unripened cheese, and Minas Padrão cheese, a maturated cheese with a more yellowish appearance. Cheese produced in this region won international awards in 2017 and 2019 at the Mondial du Fromage et des Produits Laitiers in France. According to Souza et al. (2021), this award demonstrates the quality of the cheese, as this contest is one of the most important in the cheese production world.

Production of Minas Frescal cheese is among the highest for dairy products in Brazil, ranking in third place right after Mozzarella and Prato-type cheeses (Gomes et al. 2011). It is characterised as a white cheese, with a soft texture, low-salt content (approximately 1.5% v/v NaCl), high-moisture content (consequent a high-water activity), pH higher than 5.0 and slight lactic acid taste which is produced by the enzymatic coagulation of pasteurised milk with rennet or other appropriate coagulating enzymes. This cheese is not matured and is a highly perishable product with a short shelf life, even when stored under refrigeration (Gomes et al. 2011; Felicio et al. 2015; Dantas et al. 2016). The industry's high level of interest in this cheese is understandably related to its high-yield, simple manufacturing process, low-production cost and popularity with most consumers. Also, due to the absence of a maturation period, producers have a quick return on investment (Sant'Ana et al. 2013; Magenis et al. 2014).

According to Brazilian legislation (Brasil 1997), when determining the characteristics of the Minas Frescal cheese, only two physicochemical characteristics are prescribed, a defined limit for moisture of 55 g/100 g and a fat in a dry matter between 25 and 45 g/100 g. This regulation specifies that Minas Frescal cheese must have a cylindrical shape and weigh between 0.3 and 5.0 kg. The absence of a manufacturing protocol, several parameters being defined as optional and only two physicochemical characteristics specified by the legislation has resulted in physicochemical and sensory variations in Minas Frescal cheese around the country (Oliveira et al. 2017). Verruck et al. (2015) reported that changes in the raw material and variations in processing and storage could affect the physicochemical characteristics and sensory parameters of the cheese such as: colour, flavour, structure and rheological properties. Consequently, the acceptability may be affected.

The sensory characteristics of Minas Frescal cheese can define the acceptability of the product to consumers (Bemfeito et al. 2016). Sensory analysis is essential for determining the opinions of consumers and from a description of the sensory characteristics expected by consumers, developing recommendations for formulation and manufacture (Oliveira et al. 2017).

Minas Frescal cheese is a product influenced mainly by extrinsic factors, such as climate, native pastures, altitude, temperature, soil type and humidity (Bemfeito et al. 2016; Souza et al. 2021). The presence of several microorganisms during the production of this cheese results in several biochemical changes in lactose, casein and lipids. As a result of these reactions, several volatile compounds are produced which contribute to its characteristic taste and aroma (Souza et al. 2021).

According to Oliveira et al. (2017), studies on the sensory characteristics and acceptability of Minas Frescal cheese are scarce. Knowledge of consumer preferences is essential in order to improve sensory quality and acceptability. Table 9.4 summarises some studies that evaluated the sensory characteristics of Minas Frescal cheese. It is important to highlight the changes in the sensory characteristics of the products by these

Table 9.4 Sensory characteristics of Minas Frescal cheese produced by traditional methods or with variations.

Method of production	Colour	Aroma/Flavour	Texture	Consumer acceptance	References
Incorporation of *Lactobacillus acidophilus*	–	The incorporation of *Lactobacillus* resulted in a more acidic cheese	No significant difference between the samples containing probiotic culture and control, described as a brittle texture	Described as pleasant	Buriti et al. (2005)
Addition of lactic acid and Bifidobacterium BB-12	Slightly green colouration after storage time	–	Cheese became harder during storage and consequently chewier	Majority of consumers rated as liked or liked significantly	Fritzen-Freire et al. (2010)
Addition of soy derivatives	Predominant cream colour	Soy aroma in some samples. No soy flavour was detected	Very soft texture	Majority of consumers rated as liked and liked extremely	Piazzon-Gomes et al. (2010)
Addition of *Lactobacillus casei Zhang*	Pale colour with yellow tone	–	The addition of *Lactobacillus* decreased the rigidity, resulting in softer and less elastic cheeses	Poor acceptability related to off-flavours	Dantas et al. (2016)
Low-sodium Minas Frescal cheese	–	Disagreeable odour due to the arginine concentration	Less brittle and softer cheese	All the samples were classified as 'neither like nor dislike'	Felicio et al. (2016)
Commercial brands of Minas Frescal cheese	Yellow colour	Accentuated bitter and acid taste with sweet aromas	Fragmented texture	Majority described as 'I liked it slightly'	Oliveira et al. (2017)
A2 milk	–	Some consumers rated the taste as acidic	Softer and creamier	–	Mendes et al. (2019)
Milk subjected to ohmic heating	White colour	Less bitter taste, milk flavour and butter aroma	More homogeneous mass and elastic consistency	Majority described as 'like'	(Rocha et al. 2020)

factors and to study these changes thoroughly, in order to assure high-consumer acceptability. In general, Minas Frescal cheese with a yellow colour, acid taste and soft texture, has high-consumer acceptability.

The wheying-off defect of fresh cheese can be eliminated by using ultrafiltration to concentrate the protein content of the cheese milk (Ribeiro et al. 2017). The adoption of the UF process has resulted in a significant increase in Minas Frescal cheese production over the last few years. Consumer acceptability of this product is high even though it has a considerably different texture when compared with the original product. To avoid hard and crumbly texture and maintain the texture as close as possible to conventional cheese, low-concentration retentates are used (Cunha et al. 2006; Catarino et al. 2013).

The cheese manufacturing process comprises a series of chemical, biochemical and microbiological changes that can affect the final sensory characteristics of the product. An essential factor to be considered is that casein hydrolysis by rennet results in the release of many peptides. These peptides in turn are substrates for peptidase enzymes which release smaller peptides and free amino acids, contributing directly to the cheese flavour and texture (Faion et al. 2019). In ultrafiltered cheese, casein hydrolysis is even more critical since there is no whey-drainage step, and the coagulants remain entrapped on the cheese curd and can cause sensory defects during storage. In addition, even under refrigeration, there is a residual activity that can cause a bitter taste in cheese (Ribeiro et al. 2018).

Hydamaka et al. (2001) studied the manufacturing of direct acidified cheese from UF retentates and reported higher values of hardness, gumminess and chewiness when produced with high-concentration ultrafiltered milk. This behaviour can be related to the moisture content of the sample. These authors also performed a sensory evaluation and reported that ultrafiltered cheese scored higher than cheese from whole milk. Flavour, body/texture and colour/appearance were evaluated and the UF cheese received a total score of 91 compared to 72 for whole milk cheese.

Ultrafiltered Minas Frescal cheese has been reported to be more resistant to rupture, firmer and more elastic than the traditional cheese. The only parameter that was not altered was crumbliness (Cunha et al. 2006). Flavour was the only sensory parameter with a similar score for both products. Retentate concentration factor influenced the preference of the judges, and concentration factors of 1.2 and 1.5 were preferred over 1.8. The authors related this preference due to greater firmness and elasticity (Cunha et al. 2006).

Ribeiro et al. (2018) evaluated the texture and colour of fresh cheese made with ultrafiltered milk. These authors reported that the cheeses had a high-moisture content and was classified as soft. Colour values demonstrated that the cheese presented high luminosity and yellow colour. The low number of papers evaluating the sensory characteristics of Ultrafiltered Fresh cheese is a shortcoming and further research is needed.

9.12 Coalho cheese

The northeastern region of Brazil is well known for producing Coalho cheese, which has been made for over 150 years. This product is a traditional Brazilian cheese produced mainly by small-scale producers and has social and economic relevance for the

Figure 9.11 Coalho cheese. *Source:* TiagoCaldas/Adobe Stock.

region, being the primary income source for many local families. In recent years, the consumption of Coalho cheese has grown in other regions of the country, with a household penetration rate of around 52% (Silva et al. 2012). Coalho cheese (Figure 9.11) has high commercial value, good return and high-consumer acceptability (Cavalcante et al. 2007; Queiroga et al. 2013; Bruno et al. 2017; Costa et al. 2020). Although produced for many years, Coalho cheese is still a product with significant variability in physicochemical, technological and sensory properties (Cavalcante et al. 2007; Queiroga et al. 2013).

The cheese is made from raw or pasteurised cow's milk, coagulated using rennet or other coagulating enzymes and acidified with selected lactic acid bacteria when necessary. According to the chosen protocol for cheese production, it is characterised as type A or B, the former made from pasteurised milk and the latter from raw milk (Araújo et al. 2020). Pasteurisation reduces the pathogenic and endogenous flora, which can affect the sensory characteristics of the cheese. The main characteristics of Coalho cheese are: a semi-hard body, medium- to high-moisture content, fat content in the range 35–60% and a rectangular shape, with the weight varying between 1 and 5 kg. It has a compact structure, with or without small eyes and a thin rind. It is lightly salted (around 2%), with an acid flavour and white uniform colour (Cavalcante et al. 2007; Oliveira et al. 2014; Bruno et al. 2017; Soares et al. 2017; Costa et al. 2020).

One factor of great importance for Coalho is resistance to heat without melting, which is a feature in several applications, such as the popular 'roast cheese' one (Oliveira et al. 2014). Coalho is widely commercially available, not only in the northeastern region but especially on beaches all around the country where it is served roasted. This is often the first point of contact through which the cheese is introduced to tourists from several countries (Machado et al. 2021).

According to Costa et al. (2020), cheese spoilage during storage and reduced shelf life are an issue. On the other hand, Coalho cheese has attracted the attention of researchers since it can be an excellent carrier of probiotic microorganisms. The use of probiotic cultures in Coalho cheese production to enhance the nutritional value and sensory quality has been recommended by Bezerra et al. (2017a). The current legislation in Brazil stipulates that Coalho cheese must be made with pasteurised milk. This ensures food safety but, as mentioned above, reduces the natural milk microflora that contributes to the desirable sensory characteristics of the cheese (Cruz et al. 2020). Also, this legislation specifies some sensory characteristics for the cheese such as: elastic consistency with soft texture, white-yellow colour, slightly acidic and salty taste, and a slightly acid odour (Brasil 2001).

It is well known that during cheese manufacturing, the milk undergoes various chemical and biochemical reactions simultaneously. These reactions determine the quality of the Coalho cheese since they result in the aroma and flavour development, the sensory criteria that determine the consumer acceptability (Bezerra et al. 2017b).

According to Soares et al. (2017), the evolution of consumer perception of a food product is an essential insight for researchers as it leads to a better understanding of consumer behaviour and preferences. Some descriptors commonly used to characterise Coalho cheese are: leakage of whey (appearance alteration), butter or milk flavour, butter taste and rubbery texture (Cavalcante et al. 2007; Queiroga et al. 2013). Table 9.5 shows some references that describe the sensory characteristics of Coalho cheese resulting from different production methods and different compositions. The evaluation of consumer acceptance is of great importance, since the maintenance or improvement of sensory acceptability is a challenge for the dairy industry. Some essential descriptors for Coalho cheese must be considered by dairy processors during cheesemaking: yellow colour, the presence of eyes, brightness, Coalho cheese aroma, salty taste, bitter taste, acidic taste and rubbery texture (Machado et al. 2021). It is important to emphasise that, in a general way, most of the research papers found in the literature deal with the impact of altering the composition or the manufacturing process on Coalho cheese characteristics. There is a need for more studies on traditional Coalho cheese to better understand consumer perception of the product.

9.13 Conclusions

This chapter summarises the literature on the sensory characteristics of some of the most popular Pan-American cheese varieties including the most recent studies on Cottage cheese, Cotija cheese, Quark cheese, Manchego cheese, Queso Blanco, Queso Fresco, Mozzarella cheese, Oaxaca cheese, Queso Chihuahua, Minas Frescal cheese and Coalho cheese. Mozzarella cheese stands out in this context because of its higher production and consumption levels and has been the subject of more studies related to the description of sensory characteristics. On the other hand, only a few comparative studies could be incorporated in this chapter focusing on the sensory evaluation of traditional Pan-American fresh and soft cheeses. Thus, more studies on the traditional Pan-American cheeses must be undertaken to describe the unique sensory characteristics of these cheese varieties.

Table 9.5 Sensory characteristics of Coalho cheese produced by traditional methods or with variations.

Method of production	Colour	Aroma/Flavour	Texture	Consumer perception	References
Pasteurised milk and endogenous culture	Yellow colour	Pleasant aroma and flavour	–	Moderate acceptability	Cavalcante et al. (2007)
Mixture of goat and cow's milk	White-yellow	–	–	Goat cheese had higher acceptability	Santos et al. (2011)
Produced from goat's milk with added probiotic lactic acid bacteria	White-yellow	–	Increase in hardness	Acceptability ranged from 5 to 7 on a 9-point scale	de Oliveira et al. (2012)
Goat's milk, cow's milk and mixtures	White-yellow	Goat flavour, bitter taste	Cheese made from goat's milk was harder	Good acceptability	Queiroga et al. (2013)
Influence of seasonality and somatic cell counts	–	–	–	Slightly to moderately accepted	de Lima et al. (2019)
Traditional Coalho cheese	Yellow colour, white colour	Alcoholic flavour, salty, bitter, acidic taste, milk flavour	Coalho cheese aroma, fermented aroma	Rubbery, firm texture, creaminess, crumbly	Soares et al. (2020)
Made from goat's milk with added quince	Yellow colour	No aroma and flavour of goat milk	–	Slightly to moderately accepted	Alves et al. (2020)
Milk with a high somatic cell count	–	–	–	Acceptability around 80% for milk with low-somatic cell count	Bezerra et al. (2020)
Traditional, lactose-free milk and with added probiotics	–	Cheese made from Lactose-free milk had a sweeter taste	–	Acceptability ranged from 5 to 7 on a 9-point scale	Leite et al. (2021)
Probiotic strains of *Enterococcus faecium*	White-yellow colour	–	–	'Liked slightly' or 'liked very much'	Machado et al. (2021)
Goat Coalho cheese seasoned with an alcoholic beverage	Tendency to white colour	–	Softer cheese	'Liked slightly' or 'moderately liked'	Silva et al. (2021)

References

Alves, R.N., Lima, T.L.S., Cavalcanti, M.T., and Goncalves, M.C. (2020). Cheese Coalho goat condimented with marmeleiro shelf life study, market profile and sensory acceptance. *Research, Society and Development* 9: e49932330.

Araújo, E.A., Pinto, M.S., dos Santos Pires, A.C., and de Carvalho, A.F. (2016). Chapter 15. Cottage cheese: fundamentals and technology. In: *Handbook of Animal-Based Fermented Food and Beverage Technology*, 2e (ed. Y.H. Hui and E.Ö. Evranuz), 257–267. CRC Press.

de Araújo, L., A., Marinho, M.E., Melo, A.P. et al. (2020). Type B Coalho cheese with the addition of aqueous coriander extract. *The Research, Society and Development Journal* 53: 1689–1699.

Arboatti, A.S., Olivares, M.L., Sabbag, N.G. et al. (2014). The influence of sodium chloride reduction on physicochemical, biochemical, rheological and sensory characteristics of mozzarella cheese. *Dairy Science & Technology* 94: 373–386. https://doi.org/10.1007/s13594-014-0169-2.

Bemfeito, R.M., Rodrigues, J.F., Silva, J.G.E., and Abreu, L.R. (2016). Temporal dominance of sensations sensory profile and drivers of liking of artisanal Minas cheese produced in the region of Serra da Canastra, Brazil. *Journal of Dairy Science* 99: 7886–7897. https://doi.org/10.3168/jds.2016-11056.

Bezerra, T.K.A., de Oliveira Arcanjo, N.M., de Araújo, A.R.R. et al. (2017a). Volatile profile in goat Coalho cheese supplemented with probiotic lactic acid bacteria. *LWT – Food Science and Technology* 76: 209–215. https://doi.org/10.1016/j.lwt.2016.03.041.

Bezerra, T.K.A., de Oliveira Arcanjo, N.M., Garcia, E.F. et al. (2017b). Effect of supplementation with probiotic lactic acid bacteria, separately or combined, on acid and sugar production in goat 'Coalho' cheese. *LWT – Food Science and Technology* 75: 710–718. https://doi.org/10.1016/j.lwt.2016.10.023.

da Bezerra, J., S., Sales, D.C., de JPF, O. et al. (2020). Effect of high somatic cell counts on the sensory acceptance and consumption intent of pasteurized milk and coalho cheese. *Food Science and Technology* 2061: 1–9. https://doi.org/10.1590/fst.21620.

Brasil (1997). Technical Regulation on Identity and Quality of Minas Fresca Cheese (Regulamento Técnico de Identidade e Qualidade de Queijo Minas Frescal). Ministério da Agricultura, Pecuária e Abastecimento.

Brasil (2001). Technical Regulation of Identity and Quality of Queijo de Coalho (Regulamento Técnico de Identidade e Qualidade de Queijo de Coalho). Ministério da Agricultura, Pecuária e Abastecimento.

Brito, C., Pino, F.M., Molina, C.L.H. et al. (2006). Cottage cheese Made with redi-set and DVS lactic cultivation, using homogenized milk cream without homogenizing (Queso cottage Elaborado con cultivo láctico redi-set y DVS, usando crema Lactea Homogeneizada y sin homogeneizar). *Revista Chilena de Nutricion* 33: https://doi.org/10.4067/S0717-75182006000100008.

Bruno, L.M., Briggiler Marcó, M., Capra, M.L. et al. (2017). Wild lactobacillus strains: technological characterisation and design of Coalho cheese lactic culture. *International Journal of Dairy Technology* 70: 572–582. https://doi.org/10.1111/1471-0307.12360.

Buriti, F.C.A., Da Rocha, J.S., and Saad, S.M.I. (2005). Incorporation of *Lactobacillus acidophilus* in Minas fresh cheese and its implications for textural and sensorial properties during storage. *International Dairy Journal* 15: 1279–1288. https://doi.org/10.1016/j.idairyj.2004.12.011.

Caro, I., Quinto, E.J., Fuentes, L. et al. (2020). Characterization of Lactococcus strains isolated from artisanal Oaxaca cheese. *LWT* 122: 109041. https://doi.org/10.1016/j.lwt.2020.109041.

Carvajal-Moreno, M., Vargas-Ortiz, M., Hernández-Camarillo, E. et al. (2019). Presence of unreported carcinogens, aflatoxins and their hydroxylated metabolites, in industrialized Oaxaca cheese from Mexico City. *Food and Chemical Toxicology* 124: 128–138. https://doi.org/10.1016/j.fct.2018.11.046.

Catarino, I., Martins, A.P.L., Duarte, E. et al. (2013). Rennet coagulation of sheep milk processed by ultrafiltration at low concentration factors. *Journal of Food Engineering* 114: 249–254. https://doi.org/10.1016/j.jfoodeng.2012.08.013.

Cavalcante, J.F.M., de Andrade, N.J., Furtado, M.M. et al. (2007). Processing of regional coalho cheese using pasteurized milk and endogenous lactic culture (Processamento do queijo coalho regional empregando leite pasteurizado e cultura lática endógena). *Ciência e Tecnol Aliment* 27: 205–214. https://doi.org/10.1590/s0101-20612007000100036.

Chakraborty, P., Shivhare, U.S., and Basu, S. (2021). Effect of milk composition on sensory attributes and instrumental properties of Indian cottage cheese (Chhana). *NFS Journal* 23: 8–16. https://doi.org/10.1016/j.nfs.2021.02.002.

Chombo-Morales, P., Kirchmayr, M., Gschaedler, A. et al. (2016). Effects of controlling ripening conditions on the dynamics of the native microbial population of Mexican artisanal Cotija cheese assessed by PCR-DGGE. *LWT - Food Science and Technology* 65: 1153–1161. https://doi.org/10.1016/j.lwt.2015.09.044.

Cincotta, F., Condurso, C., Tripodi, G. et al. (2021). Comparison of lactose free and traditional mozzarella cheese during shelf-life by aroma compounds and sensory analysis. *LWT* 140: 110845. https://doi.org/10.1016/j.lwt.2020.110845.

Clark, S., Warner, H., and Luedecke, L. (2001). Acceptability of Queso Fresco cheese by traditional and nontraditional consumers. *Food Science and Technology International* 7: 165–170. https://doi.org/10.1177/108201320100700210.

da Corradini, S.A., S., Madrona, G.S., de Souza, N.E. et al. (2013). Sensory characteristics and fatty acids of mozzarella cheese from the milk of crossbred cows fed palm oil and coconut fat (Características sensoriais e ácidos graxos de queijo mussarela do leite de vacas mestiças alimentadas com óleo de palma e gordura de coco). *Acta Scientiarum Technology* 35: 789–795. https://doi.org/10.4025/actascitechnol.v35i4.20158.

Costa, R.G.B., Alves, R.C., da Cruz, A.G. et al. (2018). Manufacture of reduced-sodium Coalho cheese by partial replacement of NaCl with KCl. *International Dairy Journal* 87: 37–43. https://doi.org/10.1016/j.idairyj.2018.07.012.

Costa, C.F., Fusieger, A., Andretta, M. et al. (2020). Short communication: potential use of passion fruit (*Passiflora cincinnata*) as a biopreservative in the production of Coalho cheese, a traditional Brazilian cheese. *Journal of Dairy Science* 103: 3082–3087. https://doi.org/10.3168/jds.2019-17791.

Cunha, CR Da, Viotto, W.H., and Viotto, L.A. (2006). Use of low concentration factor ultrafiltration retentates in reduced fat "Minas Frescal" cheese manufacture: effect on composition, proteolysis, viscoelastic properties and sensory acceptance. *International Dairy Journal* 16: 215–224. https://doi.org/10.1016/j.idairyj.2005.03.004.

Cruz, G.A., Bruno, L.M., de Gadelha, G.B., 0. et al. (2020). Effect of whey protein concentrate and endogenous lactic acid bacteria in low-fat Coalho cheese. *The Research, Society and Development Journal* 9: 1689–1699.

Dank, A., van Mastrigt, O., Yang, Z. et al. (2021). The cross-over fermentation concept and its application in a novel food product: the dairy miso case study. *LWT* 142: 111041. https://doi.org/10.1016/j.lwt.2021.111041.

Dantas, A.B., Jesus, V.F., Silva, R. et al. (2016). Manufacture of probiotic Minas Frescal cheese with *Lactobacillus casei* Zhang. *Journal of Dairy Science* 99: 18–30. https://doi.org/10.3168/jds.2015-9880.

De Oca-flores, E.M., Castela, O.A., Estrada-flores, J.G., and Lica Espinoza-ortega, A. (2009). Oaxaca cheese: manufacture process and physicochemical characteristics. *International Journal of Dairy Science* 62: 535–540. https://doi.org/10.1111/j.1471-0307.2009.00533.x.

Drake, S.L., Lopetcharat, K., and Drake, M.A. (2009). Comparison of two methods to explore consumer preferences for cottage cheese. *Journal of Dairy Science* 92: 5883–5897. https://doi.org/10.3168/jds.2009-2389.

Drusch, S. and Einhorn-Stoll, U. (2016). Chapter 3: Quark: a traditional German fresh cheese. In: *Modernization of Traditional Food Processes and Products* (ed. A. McElhatton and M.M. El Idrissi), 21–31. US, Boston, MA: Springer.

Escobar, D., Clark, S., Ganesan, V. et al. (2011). High-pressure homogenization of raw and pasteurized milk modifies the yield, composition, and texture of queso fresco cheese. *Journal of Dairy Science* 94: 1201–1210. https://doi.org/10.3168/jds.2010-3870.

Escobar-Zepeda, A., Sanchez-Flores, A., and Quirasco Baruch, M. (2016). Metagenomic analysis of a Mexican ripened cheese reveals a unique complex microbiota. *Food Microbiology* 57: 116–127. https://doi.org/10.1016/j.fm.2016.02.004.

Estela Hernández, C., Alejandra, R.-M., Manuel, V.-O. et al. (2016). Consumption data for the Main cheeses (Mexican-style fresh and Oaxaca) for dietary exposure assessment among the population of Veracruz city, Mexico. *Advances in Dairy Research* 04: 1000164. https://doi.org/10.4172/2329-888x.1000164.

Faccia, M., Natrella, G., Pietro, L.P. et al. (2021). Quality characteristics of mozzarella cheese manufactured with recycled stretchwater. *LWT* 147: 111554. https://doi.org/10.1016/j.lwt.2021.111554.

Faion, A.M., Becker, J., Fernandes, I.A. et al. (2019). Sheep's milk concentration by ultrafiltration and cheese elaboration. *Journal of Food Process Engineering* 42: 1–9. https://doi.org/10.1111/jfpe.13058.

Fang, X., Rioux, L.E., Labrie, S., and Turgeon, S.L. (2016). Disintegration and nutrients release from cheese with different textural properties during in vitro digestion. *Food Research International* 88: 276–283. https://doi.org/10.1016/j.foodres.2016.04.008.

FAO (2017). Agricultural Prospects (2017–2026).

Farkye, N.Y., Bhanu Prasad, B., Rossi, R., and Noyes, O.R. (1995). Sensory and textural properties of queso blanco-type cheese influenced by acid type. *Journal of Dairy Science* 78: 1649–1656. https://doi.org/10.3168/jds.S0022-0302(95)76789-3.

Felicio, B.A., Pinto, M.S., Oliveira, F.S. et al. (2015). Effects of nisin on *Staphylococcus aureus* count and physicochemical properties of Minas Frescal cheese. *Journal of Dairy Science* 98: 4364–4369. https://doi.org/10.3168/jds.2015-9520.

Felicio, T.L., Esmerino, E.A., Vidal, V.A.S. et al. (2016). Physico-chemical changes during storage and sensory acceptance of low sodium probiotic Minas cheese added with arginine. *Food Chemistry* 196: 628–637. https://doi.org/10.1016/j.foodchem.2015.09.102.

Flores-Magallón, R., Oliva-Hernández, A.A., and Narváez-Zapata, J.A. (2011). Characterization of microbial traits involved with the elaboration of the Cotija cheese. *Food Science and Biotechnology* 20: 997–1003. https://doi.org/10.1007/s10068-011-0137-z.

Fogaça, D.N.L., da Silva, W.S., and Rodrigues, L.B. (2017). Influence of compression parameters on mechanical behavior of mozzarella cheese. *Journal of Texture Studies* 48: 427–432. https://doi.org/10.1111/jtxs.12246.

Fritzen-Freire, C.B., CMO, M., Laurindo, J.B. et al. (2010). The effect of direct acidification on the microbiological, physicochemical and sensory properties of probiotic Minas Frescal cheese. *International Journal of Dairy Technology* 63: 561–568. https://doi.org/10.1111/j.1471-0307.2010.00617.x.

Fuentes, L., Mateo, J., Quinto, E.J., and Caro, I. (2015). Changes in quality of nonaged pasta filata Mexican cheese during refrigerated vacuum storage. *Journal of Dairy Science* 98: 2833–2842. https://doi.org/10.3168/jds.2014-8152.

Gamboa Alvarado, J.G., Rojas Almaraz, D., Ramón Canul, L.G., and Ramírez Rivera, E.J. (2013). Determination of the quality of cheese "Chihuahua" type: sensory and physicochemical approaches. *Emirates Journal of Food and Agriculture* 25: 409–417. https://doi.org/10.9755/ejfa.v25i6.12496.

Ganesan, B., Brown, K., Irish, D.A. et al. (2014). Manufacture and sensory analysis of reduced- and low-sodium cheddar and mozzarella cheeses. *Journal of Dairy Science* 97: 1970–1982. https://doi.org/10.3168/jds.2013-7443.

Georgina Venegas-Ortega, M., Carolina Flores-Gallegos, A., Noé Aguilar, C. et al. (2020). Multi-functional potential of presumptive lactic acid bacteria isolated from Chihuahua cheese. *Food* 9: 276. https://doi.org/10.3390/foods9030276.

Gomes, A.P., Cruz, A.G., Cadena, R.S. et al. (2011). Manufacture of low-sodium Minas fresh cheese: effect of the partial replacement of sodium chloride with potassium chloride. *Journal of Dairy Science* 94: 2701–2706. https://doi.org/10.3168/jds.2010-3774.

Gonçalves, M.C. and Cardarelli, H.R. (2020). Effect of the stretching temperature on the texture and thermophysical properties of mozzarella cheese. *Journal of Food Processing & Preservation* 44: e14703. https://doi.org/10.1111/jfpp.14703.

Gonçalves, M.C. and Cardarelli, H.R. (2021). Mozzarella cheese stretching: a minireview. *Food Technology and Biotechnology* 59: 82–91. https://doi.org/10.17113/ftb.59.01.21.6707.

González, M.L., Sánchez, H.C., Franco, F.M.J. et al. (2017). Physical, chemical and texture characteristics of Aro cheese. *Food Research* 2: 61–67. https://doi.org/10.26656/fr.2017.2(1).211.

González-Córdova, A.F., Yescas, C., Ortiz-Estrada, Á.M. et al. (2016). Invited review: artisanal Mexican cheeses. *Journal of Dairy Science* 99: 3250–3262. https://doi.org/10.3168/jds.2015-10103.

GTP (2020). Global cheese market hit record highs but is to lose momentum against the pandemic. Portal Glob. Trade. https://www.globaltrademag.com/global-cheese-market-hit-record-highs-but-is-to-lose-momentum-against-the-pandemic (accessed 5 Jun 2021).

Guinee, T.P., Pudja, P.D., and Farkye, N.Y. (1993). Fresh acid-curd cheese varieties. In: *Cheese: Chemistry, Physics and Microbiology*, vol. 2 (ed P.F. Fox), 363–419. US, Boston, MA: Springer.

Guo, L., Van Hekken, D.L., Tomasula, P.M. et al. (2011). Effect of salt on the chemical, functional, and rheological properties of Queso Fresco during storage. *International Dairy Journal* 21: 352–357. https://doi.org/10.1016/j.idairyj.2010.12.009.

Gurdian, C., Chouljenko, A., Solval, K.M. et al. (2017a). Application of edible films containing oregano (*Origanum vulgare*) essential oil on Queso Blanco cheese prepared with flaxseed (*Linum usitatissimum*) oil. *Journal of Food Science* 82: 1395–1401. https://doi.org/10.1111/1750-3841.13733.

Gurdian, C., Reyes, V., Kyereh, E. et al. (2017b). Incorporating flaxseed (*Linum usitatissimum*) oil into Queso Blanco at different stages of the cheese manufacturing process. *Journal of Food Processing & Preservation* 41: e13279. https://doi.org/10.1111/jfpp.13279.

Holle, M.J., Ibarra-Sánchez, L.A., Liu, X. et al. (2018). Microbial analysis of commercially available US Queso Fresco. *Journal of Dairy Science* 101: 7736–7745. https://doi.org/10.3168/jds.2017-14037.

Hydamaka, A.W., Wilbey, R.A., Lewis, M.J., and Kuo, A.W. (2001). Manufacture of heat and acid coagulated cheese from ultrafiltered milk retentates. *Food Research International* 34: 197–205. https://doi.org/10.1016/S0963-9969(00)00152-6.

Ibarra-Sánchez, L.A., Van Tassell, M.L., and Miller, M.J. (2017). Invited review: Hispanic-style cheeses and their association with *Listeria monocytogenes*. *Journal of Dairy Science* 100: 2421–2432. https://doi.org/10.3168/jds.2016-12116.

Jain, S., Gupta, S., Kumar, A. et al. (2019). Preparation and evaluation of cottage cheese using enzyme extracted from mustard oilseed cake. *International Journal of Current Microbiology and Applied Sciences* 8: 1202–1209. https://doi.org/10.20546/ijcmas.2019.812.148.

Jana, A. and Tagalpallewar, G.P. (2017). Functional properties of mozzarella cheese for its end use application. *Journal of Food Science and Technology* 54: 3766–3778. https://doi.org/10.1007/s13197-017-2886-z.

Jeong, H.-J., Lee, Y.-K., Ganesan, P. et al. (2017). Physicochemical, microbial, and sensory properties of Queso Blanco cheese supplemented with powdered microcapsules of tomato extracts. *Korean Journal for Food Science of Animal Resources* 37: 342–350. https://doi.org/10.5851/kosfa.2017.37.3.342.

Kamimura, B.A., Magnani, M., Luciano, W.A. et al. (2019). Brazilian artisanal cheeses: an overview of their characteristics, main types and regulatory aspects. *Comprehensive Reviews in Food Science and Food Safety* 18: 1636–1657. https://doi.org/10.1111/1541-4337.12486.

Kim, K.-T., Hwang, J.E., Eum, S.J., and Paik, H.-D. (2019). Physiochemical analysis, antioxidant effects, and sensory characteristics of quark cheese supplemented with ginseng extract. *The Food Science of Animal Resources* 39: 324–331. https://doi.org/10.5851/kosfa.2019.e26.

Lauzin, A., Pouliot, Y., and Britten, M. (2020). Understanding the differences in cheese-making properties between reverse osmosis and ultrafiltration concentrates. *Journal of Dairy Science* 103: 201–209. https://doi.org/10.3168/jds.2019-16542.

Leite, A.C.S., Cortez, N.M.D.S., de Albuquerque, S.S.M.C. et al. (2021). Development of stuffed Coalho cheese in the traditional, lactose-free and probiotic-added formulations. *Ciencia Rural* 51: https://doi.org/10.1590/0103-8478cr20200049.

de Lima, E.R., de Oliveira, J.P.F., de Bezerra, M., F. et al. (2019). Influence seasons of the year and somatic cell counts on physical-chemical and sensory parameters of milk and coalho cheese. *Acta Scientiarum – Animal Sciences* 41: 1–7. https://doi.org/10.4025/actascianimsci.v41i1.44666.

Litopoulou-Tzanetaki, E. (2007). Soft-ripened and fresh cheeses: Feta, Quark, Halloumi and related varieties. In: *Improving the Flavour of Cheese*, 474–493. Elsevier.

Lluis-Arroyo, D., Flores-Nájera, A., Cruz-Guerrero, A. et al. (2014). Effect of an exopolysaccharide-producing strain of *Streptococcus thermophilus* on the yield and texture of Mexican Manchego-type cheese. *International Journal of Food Properties* 17: 1680–1693. https://doi.org/10.1080/10942912.2011.599091.

Lobato-Callero, C., Robles-Martinez, J.C., Caballero-Perez, J.F. et al. (2001). Fat replacers in low-fat Mexican Manchego cheese. *Journal of Texture Studies* 32: 1–14. https://doi.org/10.1111/j.1745-4603.2001.tb01030.x.

López-Díaz, J.A. and del Martínez-Ruiz, N.R. (2018). Sensorial and physicochemical profile of Chihuahua cheese considering consumer preferences. *Agrociencia* 52: 361–378.

Lucey, J.A. (2011). Cheese | acid- and acid/heat coagulated cheese. In: *Encyclopedia of Dairy Sciences* (ed. J.W. Fuquay, P.F. Fox and P.L.H. McSweeney), 698–705. Elsevier.

Lücke, F.-K. and Zangerl, P. (2014). Food safety challenges associated with traditional foods in German-speaking regions. *Food Control* 43: 217–230. https://doi.org/10.1016/j.foodcont.2014.03.014.

de Machado, T., O.X., dos Santos Carvalho Mesquita, R.V., da Silva, V.O. et al. (2021). Effect of the application of probiotic strains of *Enterococcus faecium* on the physicochemical and sensory characteristics of Coalho cheese. *Semina-Ciencias Agrarias* 42: 167–178. https://doi.org/10.5433/1679-0359.2021v42n1p167.

Magenis, R.B., Prudêncio, E.S., Fritzen-Freire, C.B. et al. (2014). Rheological, physicochemical and authenticity assessment of Minas Frescal cheese. *Food Control* 45: 22–28. https://doi.org/10.1016/j.foodcont.2014.04.012.

Margalho, L.P., Kamimura, B.A., Pimentel, T.C. et al. (2021). A large survey of the fatty acid profile and gross composition of Brazilian artisanal cheeses. *Journal of Food Composition and Analysis* 101: 103955. https://doi.org/10.1016/j.jfca.2021.103955.

Maubois, J.L. (2002). Membrane microfiltration: a tool for a new approach in dairy technology. *Australian Journal of Dairy Technology* 57: 92.

Mehaia, M.A. (2002). Manufacture of fresh soft white cheese (Domiati-type) from ultrafiltered goats' milk. *Food Chemistry* 79: 445–452. https://doi.org/10.1016/S0308-8146(02)00195-4.

Mendes, M.O., de Morais, M.F., and Rodrigues, J.F. (2019). A2A2 milk: Brazilian consumers' opinions and effect on sensory characteristics of petit Suisse and Minas cheeses. *LWT* 108: 207–213. https://doi.org/10.1016/j.lwt.2019.03.064.

Mezo-Solís, J.A., Moo-Huchin, V.M., Sánchez-Zarate, A. et al. (2020). Physico-chemical, sensory and texture properties of an aged Mexican Manchego-style cheese produced from hair sheep milk. *Food* 9: 1666. https://doi.org/10.3390/foods9111666.

Miloradovic, Z., Miocinovic, J., Kljajevic, N. et al. (2018). The influence of milk heat treatment on composition, texture, colour and sensory characteristics of cows' and goats' quark-type cheeses. *Small Ruminant Research* 169: 154–159. https://doi.org/10.1016/j.smallrumres.2018.09.012.

Mónica, O.C., Brito, C.C.S., TRP, S. et al. (2010). Viability of probiotic microorganisms *Lactobacillus casei* 01, *Lactobacillus acidophilus* La-5, *Bifidobacterium* BB12 during cottage cheese storage (Viabilidad de los microorganismos probióticos *Lactobacillus casei* 01, *Lactobacillus acidophilus* La-5, *Bifidobacterium* BB12 durante el almacenamiento de queso cottage). *Vitae [online]* 17: 141–148.

Moynihan, A.C., Govindasamy-Lucey, S., Jaeggi, J.J. et al. (2014). Effect of camel chymosin on the texture, functionality, and sensory properties of low-moisture, part-skim mozzarella cheese. *Journal of Dairy Science* 97: 85–96. https://doi.org/10.3168/jds.2013-7081.

Murugesan, S., Reyes-Mata, M.P., Nirmalkar, K. et al. (2018). Profiling of bacterial and fungal communities of Mexican cheeses by high throughput DNA sequencing. *Food Research International* 113: 371–381. https://doi.org/10.1016/j.foodres.2018.07.023.

Nájera-Domínguez, C., Gutiérrez-Méndez, N., Aguirre-Gardea, K. et al. (2014). Texture properties of miniature Chihuahua-type cheese manufactured with different strains of *Lactococcus lactis* isolated from plants and raw milk cheese. *Journal of Texture Studies* 45: 487–494. https://doi.org/10.1111/jtxs.12092.

do Nascimento Alves, R., TLS, L., Cavalcanti, M.T., and Gonçalves, M.C. (2020). Cheese Coalho goat condimented with Marmeleiro shelf life study, market profile and sensory acceptance. *The Research, Society and Development Journal* 9: 1–9.

Oberg, E.N., Oberg, C.J., Motawee, M.M. et al. (2015). Increasing stringiness of low-fat mozzarella string cheese using polysaccharides. *Journal of Dairy Science* 98: 4243–4254. https://doi.org/10.3168/jds.2014-8733.

Oeffner, S.P., Qu, Y., Just, J. et al. (2013). Effect of flaxseed supplementation rate and processing on the production, fatty acid profile, and texture of milk, butter, and cheese. *Journal of Dairy Science* 96: 1177–1188. https://doi.org/10.3168/jds.2012-5941.

de Oliveira, M.E.G., Garcia, E.F., de Queiroga, R., and de Souza, E.L. (2012). Technological, physicochemical and sensory characteristics of a Brazilian semi-hard goat cheese (Coalho) with added probiotic lactic acid bacteria. *Science in Agriculture* 69: 370–379. https://doi.org/10.1590/S0103-90162012000600005.

de Oliveira, M.G., E., Fernandes Garcia, E., Vasconcelos de Oliveira, C.E. et al. (2014). Addition of probiotic bacteria in a semi-hard goat cheese (Coalho): survival to simulated gastrointestinal conditions and inhibitory effect against pathogenic bacteria. *Food Research International* 64: 241–247. https://doi.org/10.1016/j.foodres.2014.06.032.

Oliveira, E.W., Esmerino, E.A., Carr, B.T. et al. (2017). Reformulating Minas Frescal cheese using consumers' perceptions: insights from intensity scales and check-all-that-apply questionnaires. *Journal of Dairy Science* 100: 6111–6124. https://doi.org/10.3168/jds.2016-12335.

Ozturk, M., Govindasamy-Lucey, S., Jaeggi, J.J. et al. (2018). Investigating the properties of high-pressure-treated, reduced-sodium, low-moisture, part-skim mozzarella cheese during refrigerated storage. *Journal of Dairy Science* 101: 6853–6865. https://doi.org/10.3168/jds.2018-14415.

Ozturkoglu-Budak, S., Akal, H.C., and Türkmen, N. (2021). Use of kefir and buttermilk to produce an innovative quark cheese. *Journal of Food Science and Technology* 58: 74–84. https://doi.org/10.1007/s13197-020-04516-0.

Paul, M., Nuñez, A., Van Hekken, D.L., and Renye, J.A. (2014). Sensory and protein profiles of Mexican Chihuahua cheese. *Journal of Food Science and Technology* 51: 3432–3438. https://doi.org/10.1007/s13197-012-0868-8.

Paz, N.F., Gonçalvez De Oliveira, E., Villalva, F.J. et al. (2017). Effect of pH at drainage on the physicochemical, textural and microstructural characteristics of mozzarella cheese from goat milk. *Food Science and Technology* 37: 193–201. https://doi.org/10.1590/1678-457X.05116.

Piazzon-Gomes, J., Prudêncio, S.H., and da, dos Silva, R.S., S.F. (2010). Minas Frescal cheese with soy derivatives: Physical, chemical and sensory characteristics (Queijo tipo minas frescal com derivados de soja: Características físicas, químicas e sensoriais). *Cienc e Tecnol Aliment* 30: 77–85. https://doi.org/10.1590/s0101-20612010000500013.

de, do Queiroga, R., C.R., E., Santos, B.M., AMP, G. et al. (2013). Nutritional, textural and sensory properties of Coalho cheese made of goats', cows' milk and their mixture. *LWT - Food Science and Technology* 50: 538–544. https://doi.org/10.1016/j.lwt.2012.08.011.

Rasheed, S., Qazi, I.M., Ahmed, I. et al. (2016). Comparative study of cottage cheese prepared from various sources of milk. *Proceedings of the Pakistan Academy of Sciences* 269–282.

Renye, J.A., Somkuti, G.A., Van Hekken, D.L., and Guerrero Prieto, V.M. (2011). Short communication: characterization of microflora in Mexican Chihuahua cheese. *Journal of Dairy Science* 94: 3311–3315.

Ribeiro, L.R., de Leite Júnior, B.R., C., AAL, T., and Cristianini, M. (2017). Comparative study of the effect of high pressure processing on the residual activity of milk coagulants in buffer and in ultrafiltered cheese. *LWT – Food Science and Technology* 82: 1–7. https://doi.org/10.1016/j.lwt.2017.04.016.

Ribeiro, L.R., de Leite Júnior, B.R., C., and Cristianini, M. (2018). Effect of high-pressure processing on the characteristics of cheese made from ultrafiltered milk: influence of the kind of rennet. *Innovative Food Science and Emerging Technologies* 50: 57–65. https://doi.org/10.1016/j.ifset.2018.10.012.

Rocha, R.S., Calvalcanti, R.N., Silva, R. et al. (2020). Consumer acceptance and sensory drivers of liking of Minas Frescal Minas cheese manufactured using milk subjected to ohmic heating: performance of machine learning methods. *LWT* 126: 109342. https://doi.org/10.1016/j.lwt.2020.109342.

Rodrigues, J.F., Gonçalves, C.S., Pereira, R.C. et al. (2014). Utilization of temporal dominance of sensations and time intensity methodology for development of low-sodium mozzarella cheese using a mixture of salts. *Journal of Dairy Science* 97: 4733–4744. https://doi.org/10.3168/jds.2014-7913.

Sánchez-Gamboa, C., Hicks-Pérez, L., Gutiérrez-Méndez, N. et al. (2018a). Seasonal influence on the microbial profile of Chihuahua cheese manufactured from raw milk. *International Journal of Dairy Technology* 71: 81–89. https://doi.org/10.1111/1471-0307.12423.

Sánchez-Gamboa, C., Hicks-Pérez, L., Gutiérrez-Méndez, N. et al. (2018b). Microbiological changes during ripening of Chihuahua cheese manufactured with raw milk and its seasonal variations. *Food* 7: 1–12. https://doi.org/10.3390/foods7090153.

Sandoval-Copado, J., Orozco-Villafuerte, J., Pedrero-Fuehrer, D., and Colín-Cruz, M.A. (2016). Sensory profile development of Oaxaca cheese and relationship with physicochemical parameters. *Journal of Dairy Science* 99: 7075–7084. https://doi.org/10.3168/jds.2015-10833.

Sandra, S., Stanford, M.A., and Goddik, L.M. (2004). The use of high-pressure processing in the production of Queso Fresco cheese. *Journal of Food Science* 69: FEP153–FEP158. https://doi.org/10.1111/j.1365-2621.2004.tb06340.x.

Sant'Ana, A.M.S., Bezerril, F.F., Madruga, M.S. et al. (2013). Nutritional and sensory characteristics of Minas fresh cheese made with goat milk, cow milk, or a mixture of both. *Journal of Dairy Science* 96: 7442–7453. https://doi.org/10.3168/jds.2013-6915.

Santos, B.M., de MEG, O., de YRF, S. et al. (2011). Physicochemical and sensory characterization of coalho cheese produced with a mixture of goat's milk and cow's milk (Caracterização físico-química e sensorial de queijo de coalho produzido com mistura de leite de cabra e de leite de vaca). *Revista do Instituto Adolfo Lutz* 70: 302–310.

Schulz-Collins, D. and Senge, B. (2004). Acid- and acid/rennet-curd cheeses part A: quark, cream cheese and related varieties. In: *Cheese: Chemistry, Physics and Microbiology, Volume 2* (ed. P.F. Fox, P.L.H. McSweeney, T.M. Cogan and T.P. Guinee), 301–328. Academic Press.

Silva, R.A., Lima, M.S.F., Viana, J.B.M. et al. (2012). Can artisanal "coalho" cheese from northeastern Brazil be used as a functional food? *Food Chemistry* 135: 1533–1538. https://doi.org/10.1016/j.foodchem.2012.06.058.

Silva, L.F., Sunakozawa, T.N., Amaral, D.M.F. et al. (2020). Safety and technological application of autochthonous *Streptococcus thermophilus* cultures in the buffalo mozzarella cheese. *Food Microbiology* 87: 103383. https://doi.org/10.1016/j.fm.2019.103383.

da Silva, V.B., de Alves, B.P., O., dos Santos, B.C. et al. (2021). Physicochemical analysis, consumer profile and sensory analysis of goat Coalho cheeses seasoned with alcoholic beverages / Análise Fisico-Química, Perfil De Consumidor E Análise sensorial De Queijos De Coalho De Cabra Condimentados com Bebidas Alcoólicas. *Brazilian Journal of Development* 7: 18160–18180. https://doi.org/10.34117/bjdv7n2-451.

Skryplonek, K., Dmytrów, I., and Mituniewicz-Małek, A. (2019). The use of buttermilk as a raw material for cheese production. *International Journal of Dairy Technology* 72: 610–616. https://doi.org/10.1111/1471-0307.12614.

Smith, J.R., Carr, A.J., Golding, M., and Reid, D. (2018). Mozzarella cheese – a review of the structural development during processing. *Food Biophysics* 13: 1–10. https://doi.org/10.1007/s11483-017-9511-6.

Soares, E.K.B., Esmerino, E.A., Ferreira, M.V.S. et al. (2017). What are the cultural effects on consumers' perceptions? A case study covering Coalho cheese in the Brazilian northeast and southeast area using word association. *Food Research International* 102: 553–558. https://doi.org/10.1016/j.foodres.2017.08.053.

Soares, E.K.B., Silva, R., da Silva, W.P. et al. (2020). An intra-cultural investigation in Brazil using Coalho cheese and preferred attribute elicitation. *Journal of Sensory Studies* 35: 1–10. https://doi.org/10.1111/joss.12543.

Sohal, T.S., Roehl, D., and Jelen, P. (1988). Rennet as a cause of bitterness development in Quarg. *Journal of Dairy Science* 71: 3188–3196. https://doi.org/10.3168/jds.S0022-0302(88)79923-3.

Solano-López, C. and Hernández-Sánchez, H. (2000). Behaviour of *Listeria monocytogenes* during the manufacture and ripening of Manchego and Chihuahua Mexican cheeses. *International Journal of Food Microbiology* 62: 149–153. https://doi.org/10.1016/S0168-1605(00)00403-7.

Souza, T.P., Evangelista, S.R., Passamani, F.R.F. et al. (2021). Mycobiota of Minas artisanal cheese: safety and quality. *International Dairy Journal* 120: 105085. https://doi.org/10.1016/j.idairyj.2021.105085.

Torres, N. and Chandan, R.C. (1981). Latin American white cheese – a review. *Journal of Dairy Science* 64: 552–557. https://doi.org/10.3168/jds.S0022-0302(81)82608-2.

Trancoso-Reyes, N., Gutiérrez-Méndez, N., Sepulveda, D.R., and Hernández-Ochoa, L.R. (2014). Assessing the yield, microstructure, and texture properties of miniature Chihuahua-type cheese manufactured with a phospholipase A1 and exopolysaccharide-producing bacteria. *Journal of Dairy Science* 97: 598–608. https://doi.org/10.3168/jds.2013-6624.

Tunick, M.H., Van Hekken, D.L., Molina-Corral, F.J. et al. (2008). Original Research Queso Chihuahua: manufacturing procedures, composition, protein profiles, and microbiology. *International Journal of Dairy Technology* 61: 62–69.

Upreti, P., Mistry, V.V., and Acharya, M.R. (2011). Characterization of rennet coagulation of milk concentrated by vacuum condensing and ultrafiltration. *Dairy Science & Technology* 91: 383–395. https://doi.org/10.1007/s13594-011-0004-y.

Van Hekken, D.L., Drake, M.A., Corral, F.J.M. et al. (2006). Mexican Chihuahua cheese: sensory profiles of young cheese. *Journal of Dairy Science* 89: 3729–3738. https://doi.org/10.3168/jds.S0022-0302(06)72414-6.

Van Hekken, D.L., Tunick, M.H., Tomasula, P.M. et al. (2007). Mexican Queso Chihuahua: rheology of fresh cheese. *International Journal of Dairy Technology* 60: 5–12.

Van Tassell, M.L., Ibarra-Sánchez, L.A., Takhar, S.R. et al. (2015). Use of a miniature laboratory fresh cheese model for investigating antimicrobial activities. *Journal of Dairy Science* 98: 8515–8524. https://doi.org/10.3168/jds.2015-9967.

Vergara Alvarez, W.Y., Arteaga- Márquez, M., and Hernández-Ramos, E.J. (2019). Sensory acceptance and shelf life of fresh cheese made with dry bromelain extract as a coagulating agent. *DYNA* 86: 270–275. https://doi.org/10.15446/dyna.v86n210.76949.

Verruck, S., Prudêncio, E.S., Vieira, C.R.W. et al. (2015). The buffalo Minas Frescal cheese as a protective matrix of Bifidobacterium BB-12 under invitro simulated gastrointestinal conditions. *LWT - Food Science and Technology* 63: 1179–1183. https://doi.org/10.1016/j.lwt.2015.04.014.

Villalobos-Chaparro, S., Salas-Muñóz, E., Gutiérrez-Méndez, N., and Virginia Nevárez-Moorillón, G. (2018). Sensory profile of Chihuahua cheese manufactured from raw Milk. *International Journal of Food Science* 8494105. https://doi.org/10.1155/2018/8494105.

Villanueva-Carvajal, A., Esteban-Chávez, M., Espinoza-Ortega, A. et al. (2012). Oaxaca cheese: flavour, texture and their interaction in a Mexican traditional *pastaï filata* type cheese. *CyTA-Journal Food* 10: 63–70. https://doi.org/10.1080/19476337.2011.557840.

Wadhwani, R., McManus, W.R., and McMahon, D.J. (2011). Improvement in melting and baking properties of low-fat mozzarella cheese. *Journal of Dairy Science* 94: 1713–1723. https://doi.org/10.3168/jds.2010-3952.

Yadav, P., Chauhan, A.K., and Singh, R.B. (2019). Quark cheese: characteristics, preparation, and recent advances as a functional food. In: *The Role of Functional Food Security in Global Health* (ed. R.B. Singh, R.R. Watson and T. Takahashi), 395–408. Elsevier.

10 Sensory Characteristics of Cheddar and Related Cheeses Varieties

Maurice G. O'Sullivan

10.1 Introduction

A range of different approaches exist to gain information on the sensory character of Cheddar cheese and these can be separated in to four different areas: (i) expert cheese grading, (ii) difference methods, (iii) descriptive methods and (iv) affective methods. The choice of sensory approach depends upon factors, such as the complexity of information required, application, cost and time. Additionally, flavour chemistry, using techniques like gas chromatography mass spectrometry (GCMS), can be used to quantify the specific flavour chemicals present in these cheese varieties (O'Sullivan 2017). GCMS separates volatiles in a column as they are forced along by a carrier gas, and the analytes are labelled and quantified by a detector. The flame ionisation detector (FID) or mass spectrometry (MS) are the most often used detectors (Kong and Singh 2011). GC-olfactometry is a method where trained panellists, using a sniffer port, interface to qualitatively and quantitatively evaluate the odour for each analyte leaving the chromatographic column, in order to understand the sensory properties of a certain compound at a given concentration (Plutowska and Wardencki 2008; Kilcawley 2016).

Expert Grading is used by specialists in the commercial cheese sector to evaluate cheese quality. As Cheddar cheese is aged, it matures which is manifested as changes to appearance, aroma/flavour and textural properties and also potentially the development of off-flavours. The grading method utilises a negative marking system that deducts points where defects are detected. This method allows expert graders to direct different batches of Cheddar to specific markets based on their sensory properties and grade obtained, or hold the cheese that meet specific criteria for special attention where the batch matures further for premium matured or extra mature Cheddar products. Alternatively, low-grade Cheddar can be diverted early for processing, thus saving companies storage and maturation costs (O'Sullivan 2017).

Difference methods are relatively simple procedures whereby cheese samples are compared directly to other Cheddar samples and assessors are asked to determine if they are the same or different. Difference testing methods include 'paired comparison', 'duo tri', 'triangle' and 'tetrad' methods. These methods are routinely used to monitor cheese quality during production and maturation but give little information regarding specific flavour aspects and thus will not be discussed in the present chapter.

Sensory Profiling of Dairy Products, First Edition. Edited by John J. Tuohy.
© 2023 John Wiley & Sons Ltd. Published 2023 by John Wiley & Sons Ltd.

However, descriptive methods are fundamental to quantitatively determine the sensory attributes in a cheese sample and involve the training of specially screened panellists, usually 8–12. The assessors are trained to measure the attributes associated with the relevant sensory modalities of cheese including: appearance, aroma, flavour, texture, taste and aftertaste. The language is descriptive and non-hedonic, in that assessors are not asked, for example, how much they rate or like the product being tested. The different methods for descriptive profiling include the flavour profile, quantitative descriptive analysis (QDA), spectrum, free choice profiling and texture profile methods. Some of these are widely used for cheese product development and research purposes, while others appear to be academic curiosities not very widely used (Murray et al., 2001). This book chapter will focus specifically in the two main methods used. Spectrum, principally in the USA, and QDA mainly in the rest of the world. In the QDA method, experts with cheese knowledge can evaluate Cheddar cheese and suggest descriptive terms that specifically describe them and the sensory dimension to be examined to produce an initial 'meta' sensory list. This meta-list is then used to commence sensory training and in conjunction with sensory references and using chemometric statistical methods is reduced to provide a stable profiling lexicon for subsequent analysis. These attributes/lexicons are available for various Cheddar cheese products (Drake et al. 2005; O'Sullivan 2017),

Descriptive sensory methods, although often the benchmark for quantitative analysis can be expensive and time consuming because of the requirement to train and profile individual panellists over extended periods of time (O'Sullivan et al. 2011). However, affective methods are restricted to hedonic (liking) assessments, including acceptability or preference, but do not describe the product. Traditionally, this issue was solved by combining descriptive data and hedonic data using data statistical tools involving predominantly chemometrics in a method called 'preference mapping' (Kilcawley and O'Sullivan 2017). Affective methods use hedonics or ask untrained assessors their opinion of cheese products. These might include rating their liking of appearance, aroma, flavour, texture and their overall impression, or the overall acceptability of a product. Typical affective methods include the classical 'sensory acceptance test' and the 'consumer test'. Sensory acceptance testing can be employed during the development and optimisation processes as a means of assessing variant suitability in a hedonic fashion and involves anywhere from 25 to 75 individuals and where possible analysis should be performed in duplicate (Stone et al. 2012a, b; Stone and Sidel 2004). Whereas consumer testing is generally involved after the various stages of product development is complete as a means of final product validation before product launch and involves large numbers of consumers (>100). Overall, the greater the number of assessors used for affective analysis the greater the statistical reliability of the data that is obtained (O'Sullivan 2017).

10.2 Cheddar and related varieties

10.2.1 *Cheddar*

There are over 500 different varieties of commercially available cheese, which have a wide range of visual, physical and flavour attributes. The flavour of cheese is governed by three main biochemical pathways: glycolysis, lipolysis and proteolysis. In general terms, the extent of each of these processes is characteristic of the individual cheese variety (Kilcawley and O'Sullivan 2017). Cheddar cheese is the most consumed cheese in the world and is extensively used as both a table cheese and as an ingredient in various culinary dishes in food service and in formulated foods products such as soups and sauces (Guinee 2016). The majority of Cheddar cheese flavour development occurs during ageing or ripening (Walstra et al. 1999; Singh et al. 2003). Cheddar is a relatively hard cheese that varies extensively in consistency depending on how long it has been aged. Young Cheddar is mild flavoured and slightly soft and pasty becoming more crumbly in texture and more strongly flavoured as it matures during ripening. Cheddar originated from the village of Cheddar the south-west of Britain and is now among the most important cheeses made worldwide. It is made from pasteurised, standardised cow's milk using mesophilic starter cultures and in large cheese factories, defined-strain starter systems are common (McSweeney 2007). Cheddar has also been made from buffalo milk. Murtaza et al. (2013) investigated Cheddar cheese prepared from cow and buffalo milk using indigenous and commercial cultures. Sensory scores for odour, flavour and texture attributes obtained for buffalo milk cheese were significantly higher.

10.2.2 *Washed-curd cheeses*

The closed-texture semi-hard cheeses typified by Gouda and Edam are made by a process which includes a curd 'washing' stage to replace some of the whey with water and reduce acidity (Law 2010). Gouda is generally produced from pasteurised milk acidified with a mesophilic starter of multiple strains of lactic acid bacteria (LAB) and is brine salted (Van den Berg et al. 2004). Gouda is matured for two to three months and sometimes up to two years at approximately 13–15 °C. During ripening, the microorganism-rich substrate is changed by endogenous milk enzymes, milk-clotting enzymes and enzymes of the microbial communities of the cheese (particularly enzymes from lactococci and lactobacilli) (Ruyssen et al. 2013). The brining of Gouda cheese is used to control the maturation process and flavour of the cheese (Cruz et al. 2011). Edam-type cheese (Dutch-type cheese) is one of the most widespread groups of cheeses. It is a semi-hard cheese and is named after the town of Edam in the province of North Holland. Edam is traditionally sold in flat-ended spheres with a pale-yellow interior and a coat, or rind, of red paraffin wax. Optimum ripening time of these products is 6–10 weeks, usually at a temperature of 10–14 °C (Fox et al. 2000). Edam cheese generally has a very mild flavour, slightly salty or nutty and almost no smell when compared to other cheeses. As the cheese is aged, its flavour sharpens, and it becomes firmer (Jenkins 1996).

10.2.3 *Monterey Jack*

Monterey Jack is a light yellow coloured semi-hard cheese that melts easily and is often flavoured. Originating in Monterey California, this is one of the most famous American cheeses. Mexican Franciscan friars developed this semi-firm, creamy, mild flavoured cheese sometime before the 1850s (Hyde 2014).

10.3 Cheddar cheese grading methods

Grading has quite a special and long traditional association with Cheddar quality assessment by cheese producers. During grading, large numbers of cheeses can be rapidly scored for overall flavour and texture quality, based on an idealised concept of the perfect Cheddar or graded for a specific market (Kilcawley and O'Sullivan 2017). Grading typically involves defect-based judgements with graders looking for negative attributes and often have a predetermined list of defects that have been developed for Cheddar cheese through experience, which can be either formal or informal (Delahunty and Murray 1997; Delahunty and Drake 2004; Drake 2007; Kilcawley 2016). Experienced factory graders can diagnose problems and provide feedback to those in production to aid improvement and consistency of manufacture (Muir 2010). Some experienced graders also use compositional data to get a better overall assessment of quality, but also potentially gain a better understanding of predicated ripening characteristics (Kilcawley 2016). Grading systems can thus determine the reliability and reproducibility of the plant or process; contribute to quality control and guide to optimum storage and marketing strategies. For the latter, the strategy determines what cheeses are selected for specific markets, for national and multinational retailers, or for individual cheesemongers (Muir 2010; Kilcawley 2016). A Cheddar cheese grader can assess in excess of a hundred cheeses in a session but does not always provide a score just an overall comment as an assessment of quality. Graders make rapid visual assessments of large Cheddar blocks and then make further assessments by taking a sample cheese plug using a cheese trier. Graders are assessing how cleanly the plug comes out, along with its appearance, colour and adhesiveness to the trier. A small sample of the plug is repeatedly manipulated between the index finger and thumb. The aroma of this warm cheese is inhaled and finally the cheese is rolled within the mouth for aroma and taste (Kilcawley 2016). For Cheddar cheese production, with potential maturation times of up to two years, the grader plays a vital role in monitoring on-going quality development (O'Sullivan 2017). Cheese which meets the quality grade or criteria is held for maturation while inferior graded cheese may be sent to processing early as the experienced grader will determine that it cannot obtain the desired quality if held for longer. In many cases, they will try to identify a specific time that the cheese should be retailed at, or if it should be re-evaluated in the future to ensure that it is following a predicted quality route (Kilcawley 2016). Figure 10.1 depicts a typical 25 g Cheddar cheese block prior to storage or retail sectioning.

10.4 Sensory profiling methods for Cheddar cheese

Descriptive methods involve the training of panellists to quantitatively determine the sensory attributes in a selection of samples. Descriptive analysis is the most powerful sensory tool in cheese flavour research. It can be used to differentiate cheeses based on a full complement of sensory characteristics/attributes and determine a quantitative description of all the sensory aspects that can be identified (Singh et al. 2003). Some of these are tests are widely used for product development and for research purposes, while others appear to be of less practical use (Kilcawley and O'Sullivan 2017). Descriptive

Figure 10.1 20 kg Cheddar cheese block (polypropylene wrap).

analysis is a method where defined sensory terms are quantified by sensory panellists. Detailed descriptions of sensory terminology and procedural guidelines for the identification and selection of descriptors for establishing a sensory profile by a multidimensional approach have been described in ISO (1992) and ISO (1994). A list of descriptive terms is determined initially and is referred to as a 'lexicon' or 'descriptive vocabulary' and describe the specific sensory attributes in the cheese samples and can be used to evaluate the changes in these attributes. The two most commonly used methods are the QDA and the spectrum methods. For the QDA method, the assessors are trained to measure the attributes associated with the relevant sensory modalities of appearance, aroma, flavour, texture, taste and aftertaste. In the QDA method, experts with product knowledge can evaluate a sample set of the products to be profiled in the laboratory and suggest descriptive terms that specifically describe the cheese to be tested and the sensory dimension to be examined to produce an initial 'meta' sensory term list. This vocabulary can be based on terms suggested by the panellists themselves in discussions under supervision of the panel-leader. The language is descriptive and non-hedonic, in that assessors are not asked for example how much they rate or like the product being tested. The method follows a strict protocol whereby specially screened assessors are firstly selected based on familiarity with the product and their sensory acuity. Training involves a sensory term reduction process where a lexicon of relevant sensory terms is calibrated with the assessors using sensory references defining the specific term. Whether a term is included or not depends on its statistical relevance measured using chemometrics or multivariate data analysis. Only calibrated and trained assessors and specific sensory terms which absolutely define the cheese sample set will ultimately be used for the sensory profile. Sensory lexicons can also be provided which consist of lists of sensory terms describing appearance, aroma, flavour, texture, taste and aftertaste attributes. Lexicons are available for Cheddar cheese (Drake et al. 2005). The QDA method was first proposed by Stone et al. (1974) and relies heavily on statistical analysis to determine the appropriate terms, procedures and panellists to be used for the analysis of a specific product. Each descriptive test has three stages (i) selecting a panel, (ii) establishing the language or vocabulary to describe the cheese and (iii) quantify the sensory results (Delahunty and Drake 2004). A panel of typically 8–12 individuals are used to assess products in the controlled environment of the sensory laboratory. ISO (2007) describes the requirements to set up a test room comprising a testing area, a preparation area and an office, specifying those that are essential or those that are merely desirable (ISO 2007). Assessors are trained to identify and quantify a wide range of specific sensory attributes. Such attributes relate to flavour, aroma and texture and in some cases appearance (Kilcawley 2016). Initially, this statistical analysis is used in the sensory term reduction process during training, and this training of the QDA panel requires the use of product references to stimulate the generation of sensory terms. These references help the panellist define and quantify the attribute they are assessing and greatly assist in the training process. Sample presentation order must be unbiased with samples presented in a randomised order to prevent first order and carry-over effects (MacFie et al. 1989). The panel leader acts as moderator and facilitator, without directly influencing the group. Panellists cannot discuss data, terminology or samples after each taste session, but must rely on the discretion of the panel leader for any information on their performance. Feedback is provided by the facilitator based on the statistical analysis of the taste session data (Lawless and Heymann 1998; Meilgaard et al. 1999). From here, training involves sensory term reduction followed by sensory profiling. Panels act as a highly trained instrument and panels require regular 'calibration' or training to ensure accuracy. These panels are costly to operate mainly because panels operate best when regularly active (Kilcawley 2016).

The spectrum method is similar to QDA but more complex. It was developed in the 1970s (Civille and Szczesniak 1973) and is a method which uses a strict technical sensory vocabulary linked to defined sensory reference materials. The scales are anchored using these reference points which correspond to food reference samples that reduces panel variability. Panellists develop their list of attributes by evaluating a large array of products within the category. Products may be described in terms of only one attribute (e.g. 'appearance' or 'aroma'), or they may be trained to evaluate all attributes (Murray et al. 2001). Its principal characteristic is that the panellist scores the perceived intensities with reference to pre-learned 'absolute' intensity scales (Murray et al., 2001). The purpose is to make the resulting profiles universally understandable. The method provides an array of standard attribute names ('lexicons'), each with its set of standards that define a scale of intensity (Muñoz and Civille 1992; Meilgaard et al. 1999). These descriptive terms have been developed and employed by a number of authors for the sensory evaluation of Cheddar cheese e.g. Murray and Delahunty (2000a, b); Drake et al. (2005). Table 10.1 presents some hedonic and descriptive terms typically used for the sensory evaluation of Cheddar cheese. One of the drawbacks is that extensive training (of panellists) is required when using the spectrum method. Other potential drawbacks include cultural differences of panels and the difficulty of quantifying an attribute over a range of different products (Murray et al. 2001).

Table 10.1 Sensory terms for the affective and descriptive evaluation of Cheddar Cheese.

Attribute	Definition	Scale
Hedonic		
Appearance-Liking	The liking of appearance	0 = extremely dislike, 10 = extremely like
Flavour-Liking	The liking of flavour	0 = extremely dislike, 10 = extremely like
Aroma-Liking	The liking of aroma	0 = extremely dislike, 10 = extremely like
Texture-Liking	The liking of appearance	0 = extremely dislike, 10 = extremely like
Overall acceptability	The acceptability of the product	0 = extremely unacceptable 10 = extremely acceptable
Intensity		
Appearance-colour	Appearance-Ivory to orange colour	0 = ivory 10 = orange
Creamy aroma	The smell associated with creamy/milky products	0 = none, 10 = extreme
Oxidised aroma	The smell associated with oxidised dairy products	0 = none, 10 = extreme
Barnyard aroma	The smell associated the farm, barnyard, ox tail	0 = none, 10 = extreme
Sweaty/sour aroma	The aromatics reminiscent of perspiration, foot odour. Sour, stale, slightly cheesy, moist, stained or odorous with sweat	0 = none, 10 = extreme
Firmness in the mouth	Firm texture in the mouth	0 = none, 10 = extreme
Crumbly	Crumbly texture in the mouth	0 = none, 10 = extreme
Pasty	Pasty texture in the mouth	0 = none, 10 = extreme
Sweet taste	Fundamental taste sensation of which sucrose is typical	0 = none, 10 = extreme
Salt taste	Fundamental taste sensation of which sodium chloride is typical	0 = none, 10 = extreme
Sour	Fundamental taste sensation of which lactic acid is typical	
Bitter taste	Fundamental taste sensation of which caffeine or quinine in soda water is typical	0 = none, 10 = extreme
Cheddar flavour	Intensity of Cheddar cheese flavour	0 = none, 10 = extreme
Cream flavour	The flavour associated with creamy/milky products	0 = none, 10 = extreme
Dairy sweet flavour	The flavours associated with sweetened cultured dairy products such as fruit yoghurt	0 = none, 10 = extreme
Dairy fat flavour	Intensity of fat flavour	0 = none, 10 = extreme
Off-flavour	Off-flavour (Rancid)	0 = none, 10 = extreme
Oxidised flavour	The flavour associated with rancid or oxidised products	0 = none, 10 = extreme
Barnyard flavour	The flavour associated with the farm, barnyard, ox tail	0 = none, 10 = extreme
Fruity/Estery flavour	The flavours associated with fatty acid ethyl esters	0 = none, 10 = extreme

Source: From O'Sullivan (2017) / With permission of Elsevier.

10.5 Origin of Cheddar flavour and texture development

During the production of these cheese varieties, milk is acidified, typically through the action of an added starter culture of bacteria (LAB) which convert lactose into lactic acid. Rennet is then added which is a complex mixture of enzymes containing chymosin, the main protease, extracted from the stomach of calves which curdles the casein in milk. The curds and whey are then separated with the curds typically pressed into moulds and ripened to make cheese. Cheddar sensory properties can be categorised as taste, texture and aroma/flavour properties, but colour is also important for some varieties (Kilcawley and O'Sullivan 2017). With regard to taste, salt (NaCl) is of particular importance as it directly impacts on taste and acts as a flavour enhancer (FE) and influences structure and rheological properties of cheese. The extent of the impact of salt in cheese flavour depends upon its concentration, the composition and the age of the cheese. Salt reduction is a recent trend in cheese manufacture undertaken to make the product healthier as NaCl is associated with certain civilisation diseases such as heart disease and high blood pressure. For this reason, and the popularity of nutritionally optimised varieties of Cheddar cheese, this topic will be fully explored in the following section on salt reduction.

Cheese is a dynamic product, with many varieties having up to 100 billion bacteria per gram, which metabolise carbohydrates, lipids and/or proteins to create the volatile and non-volatile compounds that interact to create the particular cheese flavour. The biochemical reactions in cheese occur by the addition of microbial populations and/or exogenous enzymes during production. These biochemical reactions are also influenced by the milk (type, quality and treatment), production equipment/processes, indigenous/exogenous microbial populations (selection and concentration), indigenous/exogenous enzymes (selection and concentration), salting (dry and brine), production processes and ripening regimes (time, temperature and humidity), all of help contribute to the wide variety of cheeses available (Kilcawley and O'Sullivan 2017). Standard Cheddar cheese usually has 25.5% protein, which is predominantly casein (O'Callaghan et al. 2000).

LAB are Gram-positive, non-motile and non-spore forming, grow anaerobically, are obligate carbohydrate fermenters which metabolise lactose to lactic acid, which aid in curd formation (post whey drainage) and in the flavour development in the early stages of Cheddar cheese ripening (Kilcawley and O'Sullivan 2017; Walstra et al. 2006). Lactic acid produced during fermentation has a slightly tart taste, but also influences the final pH of cheese and overall cheese flavour perception. Lactate, diacetyl, acetoin, acetaldehyde and ethanol are all important Cheddar cheese flavour compounds. Acetaldehyde has a characteristic yoghurt aroma, while diacetyl (2,3-butanedione) and acetoin (3-hydroxy-2-butanone) impart a very characteristic creamy buttery aroma, with acetic acid providing a sharp vinegar note (Curioni and Bosset 2002; Smit et al. 2005). Ethanol likely has a minimal direct contribution to Cheddar flavour as is described as having a dry dust aroma, but is much more important in relation to the formation of ethyl esters with free fatty acids, via esterification or alcohol lysis as these compounds are very odour active and are responsible for fruity flavours in many cheeses (Curioni and Bosset 2002). Diacetyl, acetoin and 2,3-butanediol are also commonly found in cheese, but their production is primarily due to amino acid metabolism (McSweeney 2004).

Lipolysis is the hydrolysis of free fatty acids from tri-, di- and mono-acylglycerides and carried out by two hydrolytic enzymes: esterases and lipases. These enzymes catalyse the hydrolysis of the ester bond of a glyceride yielding a fatty acid and an alcohol (glycerol), but operate in different environments and have different specificities (Kilcawley and O'Sullivan 2017). The source of lipolytic enzymes in cheese can originate from a number of different sources: (i) the indigenous milk lipase (lipoprotein lipase), (ii) pregastric esterases/rennet pastes, (iii) starter bacteria, (iv) adjunct starter bacteria, (v) non-starter lactic acid bacteria (NSLAB), (vi) yeasts and moulds and (vii) the addition as exogenous lipases (Deeth and Fitz-Gerald 1995; Fox and Wallace 1997; McSweeney and Sousa 2000). Cheddar cheeses contain extremely high levels of free fatty acids. Free fatty acid metabolism forms methyl ketones, secondary alcohols, esters (methyl esters, ethyl esters and propyl esters) and lactones which are odour active and contribute to cheese flavour. Secondary alcohols in cheese form through the reduction of their corresponding methyl ketones (Engels and Visser 1997). Esterification forms esters from alcohols and carboxylic acids, whereas alcoholysis is the production of esters from alcohols and acylglycerols or from alcohols and fatty acyl-CoAs derived from the metabolism of fatty acids, amino acids and/or carbohydrates. Ethyl acetate, ethyl butyrate, ethyl hexanoate, ethyl octanoate and ethyl decanoate are found in most cheese varieties (Liu et al. 2004). Ethyl esters are associated with different and diverse fruity aromas. Liu et al. (2004) specified these aromas as follows: ethyl acetate (solvent, fruity, pineapple), ethyl butanoate (apple, banana, sweet, fruity, fragrant), ethyl hexanoate (banana, pineapple, sweet, fruity, wine-like, brandy, powerful), ethyl octanoate (pear, sweet, fruity, banana, pineapple, apricot, wine, floral), ethyl decanoate (apple, brandy, grape-like, fruity, oily). However, fruity notes can be considered a defect in Cheddar cheese (McSweeney and Sousa 2000). In general, ester formation is minimal in cheeses such as Gouda and Cheddar in comparison to Italian-type cheeses, where they are important characteristic aroma compounds. Alewijn et al. (2005) reported that lactones contribute to the flavour of Gouda cheese, although their relatively high flavour threshold may limit their overall contribution to cheese flavour (Qian and Burbank 2007).

Proteolysis assists in the softening of cheese texture during ripening due to the hydrolysis of caseins within the curd and through a decrease in water activity (McSweeney 2004). This contributes directly to cheese flavour through the hydrolysis of caseins to small peptides and amino acids, and via the metabolism of amino acids by starter bacteria and secondary cultures (Kilcawley and O'Sullivan 2017). Milk contains indigenous proteolytic enzymes (plasmin and cathepsin D) that can play an active role in primary proteolysis. Direct associations between proteolysis and the development of mouthfeel, texture, taste and aroma of maturing cheese have been reported (Pripp et al. 2006). Proteolysis is the most complex of the biochemical pathways involved in Cheddar cheese ripening because of the diversity of potential enzymatic and chemical reactions involved. Proteolysis is initiated primarily by the coagulant action on casein and subsequently by the action of cell wall proteinases of LAB and intracellular metabolic activity. Bitterness in Cheddar cheese is due to the accumulation of small hydrophobic peptide end sequences and the presence of certain amino acids at the carboxy or amino terminal. The main bitter amino acids are isoleucine, phenylalanine, leucine, methionine, proline, valine, although lysine, tryptophan, tyrosine, histidine and arginine can impart bitterness is some circumstances (Kilcawley 2016). On the other hand, some amino acids are associated with sweetness (proline, lysine, alanine, glycine, serine and threonine), sourness (histidine, asparagine and glutamic acid) and umami flavour (leucine, tyrosine, asparagine and glutamic acid), (McSweeney 1997). The most important aroma compounds resulting from amino acid metabolism are aldehydes, alcohols, carboxylic acids, thiols and thio-esters (Kilcawley 2016).

10.6 Reduced-salt Cheddar

Salt (NaCl) is added as an important process of Cheddar cheese production and directly impacts taste, acts as a flavour enhancer and influences structure and rheological properties of cheese. Salt affects the compositional, water activity, microbial growth, enzymes activity, curd syneresis, solubility and hydration of protein during ripening, leading to the synchronised improvement in flavour and texture of the final product (Rulikowska et al. 2013). The resulting Cheddar cheese flavour depends upon salt concentration, cheese composition and the age of the cheese. Salting (dry and brine), production processes and ripening regimes (time, temperature and humidity) all of help contribute to the wide variety of cheeses available. Dry salting involves the addition of salt to fresh curd for cheese such as Cheddar or Monterey Jack. Salt can be applied manually by the cheesemaker while mixing the curd whereas automated production lines utilise salting belts where the salt is accurately dosed onto the curd. Brining involves the immersion of cheese in a brine for varieties such as Mozzarella, Edam and Gouda. The brine is absorbed by the cheese while water is released thus diluting the brine over time and it is important to top-up the salt in the brine to ensure appropriate salt addition (Kilcawley and O'Sullivan 2017). Salt influences cheese composition, microbial growth, enzymatic activities, biochemical changes and casein hydration during ripening, and cheese quality (Guinee and Sutherland 2020).

Salt is a component of foods that has come under considerable scrutiny over the last couple of decades because of its association with the civilisation diseases such as coronary heart disease (CHD), high blood pressure or hypertension being the most prominent. Increasing consumption of Cheddar cheese worldwide has concurrently resulted in an increase in the contribution to total dietary sodium (Khetra et al. 2019). Cheddar cheese contains 6 g Na/kg cheese and thus is considered as a high sodium containing food (Beardsley 2017). Research in to reduced-salt Cheddar cheese has been a common endeavour with the objective of producing quality characteristics (flavour, texture and cooking properties) as close as possible to traditional Cheddar (McCarthy et al. 2016). Low salt containing cheeses are normally pasty and have off-flavour (Fox 1975). Thus, salt reduction in Cheddar cheese is a major technological challenge as it contributes to flavour and affects microbial growth, enzyme activity and syneresis, which in turn affects the cheese composition (Guinee and Fox 2004). Salt reduction results in increased moisture content, water activity, rate of lactose metabolism, extent of proteolysis and a lower rate of starter cell autolysis. The fracture force and firmness of raw Cheddar cheese is increased and the flow and strength of the heated Cheddar is altered depending on the level of salt reduction (McCarthy et al. 2015). Reduction of salt also accelerates protein hydration which has a major influence on the physical properties and quality of Cheddar cheese due to their influence on microstructure, rheology and texture (McSweeney and Fox 2009).

Salt replacers such as KCl and $MgCl_2$ may compensate for sensory and textural losses in cheese but, similar to applications in processed meats, have issues with metallic and bitter aftertastes in the final product (Grummer et al. 2012). Bitter and metallic flavours as well as soft and pasty body are common defects of KCl applications for sodium reduction in Cheddar cheese (Cruz et al. 2011). Khetra et al. (2019) substituted 75% of sodium chloride (NaCl) with potassium chloride (KCl) in Cheddar cheese and addressed the bitterness associated with KCl by incorporating hydrolysed vegetable protein (HVP) as a flavour enhancer and adenosine-50-monophosphate (AMP) as bitter blocker (BB) in experimental variants. They found that the addition of this combination of ingredients significantly improved the flavour, saltiness and bitterness characteristics of the developed products. This study validated masking as a strategy for compensation for bitterness by using KCl in low or reduced sodium cheese. Grummer et al. (2012) used $CaCl_2$ and $MgCl_2$ to reduce sodium in Cheddar-style cheese. Reduced-salt variants had undesirable flavours and tastes such as bitter, metallic, earthy, unclean, soapy and numbing which would not be desirable in high-quality Cheddar cheese. Murtaza et al. (2014) investigated texture, flavour and sensory quality of buffalo milk Cheddar cheese as influenced by reducing the sodium salt content. They concluded that texture, hardness, toughness and crumbliness lessened upon reducing the salt content during the course of ripening which effected cheese composition, proteolysis, increased the aw, volatile flavour compounds content, but decreases the sensory quality (Murtaza et al. 2014).

10.7 Reduced-fat Cheddar

Low-fat dairy products are in demand by those who wish to reduce their calorie intake, particularly saturated fat, and thus have driven the market to develop healthier reduced-fat variants of Cheddar cheese. Fat has an important sensory role in Cheddar cheese contributing to both flavour and texture and thus technological difficulties can be encountered when levels are reduced (Kilcawley and O'Sullivan 2017). Reducing the fat content of Cheddar leads to higher concentrations of protein, moisture, pH changes and a reduction in moisture-in-non-fat substances (McCarthy et al. 2016). Extensive research has also been undertaken on reducing fat in Cheddar cheese, as a means of providing insights for the development of reduced-fat cheeses with quality characteristics (flavour, texture and cooking properties) as close as possible to full-fat Cheddar cheese (McCarthy et al. 2016). In Europe, a reduction of 30% of the fat content relative to the reference full-fat cheese is required to use the term 'reduced-fat' in the description of the product, according to European Council regulation No. 1924/2006 (Deegan et al. 2014). In the USA, the FDA regulation mandates that food products claiming to be low fat must not contain more than 3 g of fat per reference amount (50 g), whereas reduced-fat labelling can be used for food that contains 25% less fat than the regular version (FDA-DHHS 2002) (Amelia et al. 2013).

Cheddar becomes firmer and more rubbery as fat is reduced, which increases the corresponding protein content. Flavour is particularly affected as the fat/protein matrix changes and affects flavour perception deleteriously. The greater the fat reduction the more negative the sensory flavour and texture quality. Some of the flavour defects mentioned included meaty, brothy, burnt, bitterness, low-flavour intensity and milk fat flavour (Amelia et al. 2013; Drake and Swanson 1995; Kilcawley and O'Sullivan 2017).

Fenelon et al. (2002) used the adjunct *Lactobacillus helveticus* in combination with *Leuconostoc cremoris*, *Lactococcus lactis* var. diacetylactis and Streptococcus thermophiles to produce 50% fat reduced Cheddar cheese. The 50% fat reduced Cheddar had a higher preference score than a similar cheese made without the adjunct culture although the flavour profile was clearly different from regular Cheddar as burnt off-flavour notes were produced. Curd washing has also been investigated for the sensory optimisation of reduced-fat cheese (Johnson et al. 1998).

Nelson and Barbano (2004) studied flavour optimisation of reduced-fat Cheddar cheese by only removing the fat from regular aged Cheddar once the flavour had fully developed. The resulting products retained the same flavour intensity as full-fat Cheddar but the texture was quite different. Ritvanen et al. (2005) studied sensory characteristics, chemical composition and consumer acceptance of full-fat and reduced-fat curd washed cheeses (Havarti-type, Edam and Emmental) on the Finnish market. These authors found almost all cheeses were considered pleasant, but differences in flavour and texture to full-fat and reduced-fat cheeses were detected.

References

Alewijn, M., Sliwinski, E.L., and Wouters, J.T.M. (2005). Production of fat-derived (flavour) compounds during the ripening of Gouda cheese. *International Dairy Journal* 15: 733–740.

Amelia, I., Drake, M.A., Nelson, B., and Barbano, D.M. (2013). A new method for the production of low-fat Cheddar cheese. *Journal of Dairy Science* 96: 4870–4884.

Beardsley, R.J. (2017). Growth of *E. coli* in reduced salt cheddar cheese. Doctoral dissertation. Pretoria, South Africa: University of Pretoria.

Civille, G.V. and Szczesniak, A.S. (1973). Guidelines to training a texture profile panel. *Journal of Texture Studies* 4: 204–223.

Cruz, A.G., Faria, J.A., Pollonio, M.A. et al. (2011). Cheeses with reduced sodium content: effects on functionality, public health benefits and sensory properties. *Trends in Food Science and Technology* 22: 276–291.

Curioni, P.M.G. and Bosset, J.O. (2002). Key odorants in various cheese types as determined by gas chromatography-olfactometry. *International Dairy Journal* 12: 959–984.

Deegan, K.C., Holopainen, U., McSweeney, P.L.H. et al. (2014). Characterisation of the sensory properties and market positioning of novel reduced-fat cheese. *Innovative Food Science and Emerging Technologies* 21: 169–178.

Deeth, H.C. and Fitz-Gerald, C.H. (1995). Lipolytic enzymes and hydrolytic rancidity in milk and milk products. In: *Advanced Dairy Chemistry, Lipids*, vol. 2 (ed. P.F. Fox), 247–308. London: Chapman & Hall.

Delahunty, C.M. and Drake, M.A. (2004). Sensory character of cheese and its evaluation. In: *Cheese: Chemistry, Physics and Microbiology*, General Aspects, vol. 1 (ed. P.F. Fox, P.L.H. McSweeney, T.M. Cogan and T.P. Guinee), 455–487. London, UK: Elsevier.

Delahunty, C.M. and Murray, J.M. (1997). Organoleptic evaluation of cheese. In: *Proceedings of the 5th Cheese Symposium* (ed. T.M. Cogan, P.F. Fox and R.P. Ross), 90–97. Teagasc, Dublin 4.

Drake, M.A. (2007). Invited review: sensory analysis of dairy foods. *Journal of Dairy Science* 90: 4925–4937.

Drake, M.A. and Swanson, B.G. (1995). Reduced and low-fat cheese technology. A review. *Trends in Food Science and Technology* 6: 366–369.

Drake, M., Yates, M.D., Gerard, P.D. et al. (2005). Comparison of differences between lexicons for descriptive analysis of Cheddar cheese flavour in Ireland, New Zealand, and the United States of America. *International Dairy Journal* 15 (5): 473–483. https://doi.org/10.1016/j.idairyj.2004.09.006.

Engels, W.J.M. and Visser, S. (1997). A comparative study of volatile compounds in the water-soluble fraction of various types of ripened cheese. *International Dairy Journal* 48: 127–140.

FDA-DHHS (2002). *21CFR101.62b: Nutrient content claims for fat, fatty acid, and cholesterol content of foods*. Food and Drug Administration–Department of Health and Human Services, Washington, D https://www.accessdata.fda.gov/scripts/cdrh/cfdocs/cfcfr/cfrsearch.cfm?fr=101.62 (accessed July 2020).

Fenelon, M.A., Beresford, T.P., and Guinee, T.P. (2002). Comparison of different bacterial culture systems for the production of reduced-fat Cheddar cheese. *International Journal of Dairy Technology* 55: 194–203.

Fox, P.F. (1975). Influence of cheese composition on quality. *Irish Journal of Agricultural and Food Research* 14: 33–42.

Fox, P.F. and Wallace, J.M. (1997). Formation of flavour compounds in cheese. *Advances in Applied Microbiology* 45: 17–85.

Fox, P.F., Guinee, T.P., Cogan, T.M., and McSweeney, P.L.H. (2000). *Fundamentals of Cheese Science*. Aspen Publication 638 pp.

Grummer, J., Karalus, M., Zhang, K. et al. (2012). Manufacture of reduced sodium Cheddar-style cheese with mineral salt replacers. *Journal of Dairy Science* 95: 2830–2839.

Guinee, T.P. (2016). Protein in cheese products: structure-function relationships. In: *Advanced Dairy Chemistry*, Proteins, Applied Aspects, vol. 1B (ed. P.L.H. McSweeney and S.A. O'Mahony), 347–415. New York, NY, USA: Springer Science+Business Media.

Guinee, T.P. and Fox, P.F. (2004). Salt in cheese: physical, chemical and biological aspects. In: *General Aspects*, Cheese: Chemistry, Physics, and Microbiology, 3e, vol. 1 (ed. P.F. Fox, P.L.H. McSweeney, T.M. Cogan and T.P. Guinee), 347–360. London, UK: Elsevier Academic Press.

Guinee, T.P. and Sutherland, B.J. (2020). Salting of cheese. In: *Encyclopedia of Dairy Sciences*, 3e (ed. P.L.H. McSweeney and J.P. McNamara), 321–335. London, UK: Academic Press.

Hyde, V. (2014). Regional Cheese Cuisine: Monterey Jack. Culturecheesemag.com https://culturecheesemag.com/blog/regional-cheese-cuisine-monterey-jack (accessed 11 November 2022).

ISO (1992). *International Standard 5496. Sensory Analysis Methodology. Initiation and Training of Assessors in the Detection and Recognition of Odours*. Ref. No. ISO 5496:1992 €. Genève, Switzerland: International Organization for Standardization.

ISO (1994). *International Standard. 11035. Sensory Analysis-Identification and Selection of Descriptors Establishing a Sensory Profile by a Multidimensional Approach*. Ref. no. ISO 11035:1994 (E). Genève: International Organization for Standardization.

ISO (2007). *International Standard 8589. Sensory Analysis — General Guidance for the Design of Test Rooms*. Genève: International Organization for Standardization.

Jenkins, S.W. (1996). *Cheese Primer*, 373. Workman Publishing Company ISBN: 0-89480-762-5.

Johnson, M.E., Steele, J.L., Broadbent, J., and Weimer, B.C. (1998). Manufacture of Gouda and flavour development in reduced-fat Cheddar cheese. *Australian Journal of Dairy Technology* 53 (2): 67–69.

Khetra, Y., Kanawjia, S.K., Puri, R. et al. (2019). Using taste-induced saltiness enhancement for reducing sodium in Cheddar cheese: effect on physico-chemical and sensorial attributes. *International Dairy Journal* 91: 165–171.

Kilcawley, K. (2016). Cheese Flavour. Chapter 13. In: *Fundamentals of Cheese Science*, 2e (ed. P.F. Fox, T.P. Guinee, T.M. Cogan and P.L.H. McSweeney). New York: Springer.

Kilcawley, K.N. and O'Sullivan, M.G. (2017). Cheese flavour development and sensory characteristics. In: *Global Cheesemaking Technology: Cheese Quality and Characteristics* (ed. P. Papademas and T. Bintsis). Chichester: Wiley.

Kong, F. and Singh, R.P. (2011). Advances in instrumental methods to determine food quality deterioration. In: *The Stability and Shelf-Life of Food* (ed. D. Kilcast and P. Subramaniam). Cambridge, UK: Woodhead Publishing Limited.

Law, B.A. (2010). Chapter 7. Cheese-ripening and cheese flavour technology. In: *Technology of Cheesemaking*, 2e (ed. B.A. Law and A.Y. Tamine), 231–259. Oxford: Wiley-Blackwell Publication.

Lawless, H.T. and Heymann, H. (1998). Descriptive analysis. In: *Sensory Evaluation of Food, Principles and Practices* (ed. H.T. Lawless and H. Heymann), 117–138, 341–378. New York: Chapman and Hall.

Liu, S.Q., Holland, R., and Crow, V. (2004). Esters and their biosynthesis in fermented dairy products: a review. *International Dairy Journal* 14: 923–945.

MacFie, H.J., Bratchell, N., Greenhoff, K., and Vallis, L.V. (1989). Designs to balance the effect of order of presentation and first-order carry-over effects in hall tests. *Journal of Sensory Studies* 4: 129–148.

McCarthy, C.M., Wilkinson, M.G., Kelly, P.M., and Guinee, T.P. (2015). Effect of salt and fat reduction on the composition, lactose metabolism, water activity and microbiology of Cheddar cheese. *Dairy Science and Technology* 95: 587–611.

McCarthy, C.M., Wilkinson, M.G., Kelly, P.M., and Guinee, T.P. (2016). Effect of salt and fat reduction on proteolysis, rheology and cooking properties of Cheddar cheese. *International Dairy Journal* 56: 74–86.

McSweeney, P.L.H. (1997). The flavour of milk and dairy products: III. Cheese: taste. *International Journal of Dairy Technology* 50: 123–128.

McSweeney, P.L.H. (2004). Biochemistry of cheese ripening. *International Journal of Dairy Technology* 57 (2–3): 127–144.

McSweeney, P.L.H. (2007). *Cheese Problems Solved*, 214–229. Woodhead Publishing Series in Food Science, Technology and Nutrition.

McSweeney, P.L.H. and Fox, P.F. (2009). *Advanced Dairy Chemistry*, Lactose, Water, Salts and Minor Constituents, vol. 3. New York: Springer-Verlag.

McSweeney, P.L.H. and Sousa, M.J. (2000). Biochemical pathways for the production of flavour compounds in cheese during ripening. *Le Lait* 80: 293–324.

Meilgaard, M.C., Civille, G.V., and Carr, B.T. (1999). *Sensory evaluation techniques*, 3e, Chapter 5, 54–55. Florida: Academic Press.

Muir, D.D. (2010). The grading and sensory profiling of cheese. Chapter 13. In: *Technology of Cheesemaking*, 2e (ed. B.A. Law and A.Y. Tamime), 440–472. UK: Wiley.

Muñoz, A.M. and Civille, G.V. (1992). The spectrum descriptive analysis method. In: *ASTM Manual on Descriptive Analysis* (ed. R.C. Hootman), 22–34. Pennsylvania: American Society for Testing and Materials.

Murray, J.M. and Delahunty, C.M. (2000a). Description of Cheddar cheese packaging attributes using an agreed vocabulary. *Journal of Sensory Studies* 15 (2): 201–218.

Murray, J.M. and Delahunty, C.M. (2000b). Selection of standards to reference terms in a Cheddar-type cheese flavor language. *Journal of Sensory Studies* 15 (2): 179–199.

Murray, J., Delahunty, C., and Baxter, I. (2001). Descriptive sensory analysis: past, present and future. *Food Research International* 34 (6): 461–471. https://doi.org/10.1016/S0963-9969(01)00070-9.

Murtaza, M.A., Rehman, S.U., Anjum, F.M., and Huma, N. (2013). Descriptive sensory profile of xo wans buffalo milk cheddar cheese prepared using indigenous cultures. *Journal of Dairy Science* 96: 1380–1386.

Murtaza, M.A., Huma, N., Sameen, A. et al. (2014). Texture, flavor, and sensory quality of buffalo milk Cheddar cheese as influenced by reducing sodium salt content. *Journal of Dairy Science* 97: 6700–6707.

Nelson, B.K. and Barbano, D.M. (2004). Reduced-fat Cheddar cheese manufactured using a novel fat removal process. *Journal of Dairy Science* 87: 841–853.

O'Callaghan, Y.C., O'Conner, T.P., and O'Brien, N.M. (2000). Nutritional aspects of cheese. Chapter 21. In: *Fundamentals of Cheese Science*, 2e (ed. P.F. Fox, T.P. Guinee, T.M. Cogan and P.L.H. McSweeney), 504–512. Gaithersburg, MD: Aspen Publishers.

O'Sullivan, M.G. (2017). Sensory properties of dairy products, Chapter 12. In: *A Handbook for Sensory and Consumer Driven New Product Development: Innovative Technologies for the Food and Beverage Industry*, 259–280. United Kingdom: Woodhead Publishing Ltd.

O'Sullivan, M.G., Kerry, J.P., and Byrne, D.V. (2011). Use of sensory science as a practical commercial tool in the development of consumer-led processed meat products, Chapter 7. In: *Processed Meats* (ed. J.P. Kerry and J.F. Kerry), 156–182. United Kingdom: Woodhead Publishing Ltd.

Plutowska, B. and Wardencki, W. (2008). Application of gas chromatography-olfactometry (GC-O) in analysis and quality assessment of alcoholic beverages ± a review. *Food Chemistry* 107 (1): 449±463.

Pripp, A.H., Skeie, S., Isaksson, T. et al. (2006). Multivariate modelling of relationships between proteolysis and sensory quality of Prast cheese. *International Dairy Journal* 16: 225–235.

Qian, M.C. and Burbank, H.M. (2007). Hard Italian cheeses: Parmigiano-Reggiano and Grana Padano, Chapter 19. In: *Improving the Flavour of Cheese* (ed. B.C. Weimer), 421–443. Cambridge England: Woodhead Publishing Ltd.

Ritvanen, T., Lampolahti, S., Lilleberg, L. et al. (2005). Sensory evaluation, chemical composition and consumer acceptance of full fat and reduced fat cheeses in the Finnish market. *Food Quality and Preference* 16: 479–492.

Rulikowska, A., Kilcawley, K.N., Doolan, I.A. et al. (2013). The impact of reduced sodium chloride content on Cheddar cheese quality. *International Dairy Journal* 28: 45–55.

Ruyssen, T., Janssens, M., Van Gasse, B. et al. (2013). Characterisation of Gouda cheeses based on sensory, analytical and high-field 1H nuclear magnetic resonance spectroscopy determinations: effect of adjunct cultures and brine composition on sodium-reduced Gouda cheese. *International Dairy Journal* 33: 142–152. https://doi.org/10.1016/j.idairyj.2013.04.009.

Singh, T.K., Drake, M.A., and Cadwallader, K.R. (2003). Flavor of Cheddar cheese: a chemical and sensory perspective. *Comprehensive Reviews in Food Science and Food Safety* 2: 139–162.

Smit, G., Smit, B.A., and Engles, W.J.M. (2005). Flavour formation by lactic acid bacteria and biochemical flavour profiling of cheese products, FEMS. *Microbiology Reviews* 29: 591–610.

Stone, H. and Sidel, J.L. (2004). Affective testing. Chapter 7. In: *Sensory Evaluation Practices*, Food Science and Technology, International Series, 3e (ed. H. Stone and J.L. Sidel), 247–277. USA: Academic Press/Elsevier.

Stone, H., Sidel, J., Oliver, S. et al. (1974). Sensory evaluation by quantitative descriptive analysis. *Food Technology* 28: 24–34.

Stone, H., Bleibaum, R.N., and Thomas, H.A. (2012a). Affective testing. In: *Sensory Evaluation Practices*, 4e, vol. 4 (ed. H. Stone, R.N. Bleibaum and H. Thomas), 291–325. USA: Elsevier Academic Press.

Stone, H., Bleibaum, R., and Thomas, H. (2012b). Test strategy and design of experiments. In: *Sensory Evaluation Practices*, 4e, vol. 4 (ed. H. Stone, R.N. Bleibaum and H.A. Thomas), 117–157. USA: Elsevier Academic Press.

Van den Berg, G., Meijer, W.C., Düsterhöft, E.-M., and Smit, G. (2004). Gouda and related cheeses. Chapter 5. In: *Major Cheese Groups*, Cheese: Chemistry, Physics and Microbiology, vol. 2 (ed. P.F. Fox), 103–135. London, UK: Elsevier Academic Press.

Walstra, P., Geurts, T.J., Noomen, A. et al. (1999). *Cheese Varieties*, 2e. New York: Marcel Decker.

Walstra, P., Wouters, J.T.M., and Geurts, T.J. (ed.) (2006). Lactic fermentations. In: *Dairy Science and Technology*, 2e, 357–397. Boca Raton, London, New York: Taylor & Francis Group.

11 Sensory Characteristics of Swiss-type Cheese Varieties

Barbara Guggenbühl Gasser, Pascal Fuchsmann and Marie-Therese Fröhlich-Wyder

11.1 Introduction

Swiss-type cheeses were originally manufactured in Switzerland in the Emmen valley – Emmental in Swiss German; their precursors were mountain cheeses dating from the twelfth century (Imhof 2013). Nowadays, they are manufactured in many countries using a wide range of production technologies resulting in cheeses often quite different from the original Emmentaler. The application of various terms describing these cheeses often leads to confusion and to false classifications (Fröhlich-Wyder et al. 2017). For this reason, the term 'Swiss-type cheese' is used in the present chapter for types of cheese which undergo a propionic acid fermentation.

Swiss-type cheeses are always cooked and ripened. Common to all Swiss-type cheeses is the propionic acid fermentation by *Propionibacterium freudenreichii* which is usually achieved by adding a culture of selected *P. freudenreichii* strains. However, the natural propionibacterial community in raw milk is still diverse and may prompt spontaneous propionic acid fermentation in cheeses made therefrom (Turgay et al. 2011).

Hard Swiss-type cheeses, especially traditional varieties such as Emmentaler PDO, Allgäuer Emmentaler PDO or Emmental de Savoie PGI are often made from untreated cows' raw milk. The bacteriological requirements are stringent. Typically, thermophilic, homofermentative lactic acid bacteria are used as mixed cultures containing lactobacilli (*Lactobacillus helveticus*, *Lactobacillus delbrueckii* ssp. *lactis*) and streptococci (*Streptococcus salivarius* ssp. *thermophilus*). Facultatively heterofermentative non-starter lactobacilli may be present as indigenous microflora or may be added purposely.

The semi-hard Swiss-type cheeses such as Jarlsberg, Maasdam, Leerdammer, Samsoe, Felsberg, Grevé and Fol Epi are usually made from pasteurised cows' milk which is often either bactofuged or supplemented with suitable preservatives to prevent butyric acid fermentation (Daly et al. 2010; Rehn et al. 2011; Rodriguez et al. 2011). These so-called 'Goutalers', a cross between Gouda and Emmentaler, are made with mesophilic starter cultures containing lactic acid cocci, namely *Leuconostoc* spp. and *Lactococcus lactis* ssp. *lactis* biovar *diacetylactis* (Mikš-Krajnik et al. 2013).

Sensory Profiling of Dairy Products, First Edition. Edited by John J. Tuohy.
© 2023 John Wiley & Sons Ltd. Published 2023 by John Wiley & Sons Ltd.

In the last decade, several in-depth contributions and reviews on Swiss-type cheeses and propionic bacteria had been published (Bachmann et al. 2011; Fox et al. 2017; Fröhlich-Wyder et al. 2017; Hartmann et al. 2017; Turgay et al. 2021). They all give a comprehensive overview on how to manufacture and ripen a Swiss-type cheese.

Therefore, the focus of this chapter lies on a topic which had not been discussed in detail in the aforementioned contributions, the 'sensory evaluation of Swiss-type cheeses'.

11.2 Sensory evaluation methods

The human senses have been used to measure food quality for centuries.

Today, there is a broad collection of test methods which allow us to address a wide range of sensory-related questions and to systematically and accurately evaluate the human responses to specific stimuli. Thereby, sensory evaluation methods are often grouped according to the type of panellists used for the sensory judgements.

11.2.1 *Grading and quality scoring by cheese experts*

Especially in artisanal cheese production, typical of countries such as Switzerland or France, sensory evaluation by cheese experts has a long tradition and is still widely applied. This type of sensory judgement is suitable for experts testing and judging Swiss-type cheeses on a daily basis. Thus, they have a lot of experience and can compare the perceived intensities of specific sensory characteristics to a predetermined, desired sensory quality level.

The main focus of this type of sensory evaluation is on cheese defects, and assigning overall scores for quality classification often relevant for product pricing (Federal Office for Agriculture FAOG 2018).

As an example, quality scoring of Emmentaler PDO is given in Table 11.1.

The criteria eyes, cheese body, texture and aroma are assessed according to a 5-point scale by a group of at least three experts consisting of a representative from the consortium Emmentaler Switzerland, from the cheese production plant and the corresponding trading company. Only cheese which meets the quality requirements and is not declassified during the further ripening period may be labelled as Emmentaler PDO.

11.2.2 *Descriptive profiling methods (quantitative descriptive tests)*

The best known and widely applied approach to sensory evaluation is a neutral overall description of the perceived characteristics of a cheese whereby trained panellists are used as objective analytical instruments. Descriptive profiling methods allow the determination of an overall sensory cheese profile by generating a list of specific descriptors and subsequent measurement of their perceived intensity. The different methods can be grouped in two categories. The classic form, namely methods summarised under the

Table 11.1 Quality scoring according to the specification for Emmentaler PDO.

Classification criteria	
Eyes Cheese body Aroma Texture	
Quality grades	
Emmentaler quality grade 1:	Emmentaler with at least 18 points, whereby no criterion may have less than 4 points. Is intended for direct sale by retailers and for pre-packaging.
Emmentaler quality grade 2A	Emmentaler with at least 17 points, whereby the criterion 'aroma' must have at least 4 points. Not allowed for direct sale by retailers and pre-packaging. Processing must take place in a Swiss cheese melting plant. The designation Emmentaler may be used for the final products.
Emmentaler quality grade 2B	Cheeses whose content is outside the standards of the specifications, which have less than 17 points or which have less than 4 points in the aroma criterion, have to be processed in a melting plant. It is not allowed to use the designation Emmentaler for the final products.
Emmentaler quality grade 3	Cheeses with a quality grade 3 must be put to a special use (e.g. feeding of pigs). It may be used under no circumstances for human consumption.

Source: Adapted from Federal Office for Agriculture (2018).

term quantitative descriptive analysis, is carried out by trained judges. In contrast, untrained naïve persons can be used for less time consuming and thus cheaper methods such flash profiling, projective mapping or napping (Delarue et al. 2016).

Especially for the conventional methods, the attributes used to describe specific sensory sensations is a key factor. In order to be able to characterise and especially compare the sensory aspects of the broad range of different Swiss-type cheeses in an objective and reproducible way, it is important to use a standardised sensory language and if possible, reference material in order to train panellists. Although in general descriptive panels create their own list of descriptors, existing lexicons are helpful to generate a suitable list of adequate qualitative sensory aspects of Swiss-type cheeses. Compilations of well-defined terms together with indications for reference material are given by Drake et al. (2007), Drake and Civille (2003), Drake and Delahunty (2017), Nielsen et al. (1998) Rétiveau et al. (2005) and Talavera-Bianchi and Chambers (2008).

Methods such as temporal dominance of sensation (TDS) and time-intensity (TI) allow information on qualitative and quantitative changes in temporal perception of sensory characteristics to be obtained.

Although descriptive profiling methods allow the sensory evaluation of Swiss-type cheeses in a reproducible way, one has to keep in mind that the sensory profiles are influenced by the combination of cheese samples tested within a test session, the vocabulary used for the description, as well as the test method itself. Moreover, the perceived quality can change depending on the concentration of a compound. For example, pentane was described as 'beany' at low-concentration levels (1 ppm), 'sweaty' at intermediate concentrations (10–5000 ppm) and at concentrations between 10 000 and 100 000 ppm as manure/barnyard like (Vara-Ubol et al. 2004).

11.2.3 *Consumer testing*

A third possibility to get information about sensory characteristics of Swiss-type cheeses is the use of naïve consumers. In general, data collected with consumers are related to hedonic aspects such as acceptance, overall liking or subjective intensity ratings of selected descriptors by just about right (JAR) scales. For this type of sensory evaluation, generally large numbers of evaluators are needed to get reliable product information.

In recent years, sensory perspectives as perceived by consumers have become more important. In response, new more time efficient and less expensive sensory methods were developed and are widely used nowadays. Among these rapid methods, verbal-based tests, such as check-all-that-apply (CATA) and rate all that apply (RATA), flash profiling, several methods based on similarity, such as different sorting tasks and projective mapping, as well as reference-based methods such as polarised sensory positioning, polarised projective mapping or pivot profiling, have to be mentioned (Delarue et al. 2016; Thuillier et al. 2015).

11.3 Sensory characteristics of Swiss-type cheese varieties

As for any food, the sensory characteristics of Swiss-type cheeses result from complex interactions between chemical ingredients and physical properties of the cheese and can be described with attributes belonging to the three sensory modalities: appearance, texture and flavour.

Important factors that contribute to the sensory profile of Swiss-type cheeses are: raw milk quality, composition of bacterial cultures, processing parameters and ripening conditions.

According to the specifications of Emmentaler PDO, as an important representative of Swiss-type cheeses, the variety has to show round to slightly oval eyes of a size between 2 and 4 cm, an elastic, non-adhesive texture/body, a typical nutty flavour due to propionic acid fermentation including 'sweet' and 'spicy' as well as 'slightly salty' and 'slightly acidic' notes (Federal Office for Agriculture 2018).

However, many of these qualitative descriptors are not unique to Swiss-type cheeses and could be used also for the sensory characterisation of other cheese varieties.

Therefore, the question arises, which key factors are responsible for the uniqueness of Swiss-type cheeses?

11.3.1 *Role of propionic acid fermentation*

The propionic acid fermentation common to all Swiss-type cheeses is considered one of the main factors contributing to the sensory profile of this cheese variety. In fact, the most salient and famous visual sensory aspect, the big holes or eyes, as well as the sweet and nutty flavour notes often associated with Swiss-type cheeses are known to be related to the propionic acid fermentation.

The majority of these cheeses, either made from raw milk or from pasteurised milk, is manufactured with the application of a culture consisting of selected *P. freudenreichii* strains. Only a few traditional cheeses made from raw milk, such as Comté PDO or Fontina PDO, may undergo a spontaneous propionic acid fermentation. The main emerging products are propionic and acetic acid, CO_2 and mostly also succinic acid. As a preliminary step in propionic acid fermentation, lactic acid is oxidised to pyruvic acid, which is further oxidised in a one-third share to acetic acid and CO_2. The remaining two-thirds of pyruvic acid are reduced to propionic acid via the Wood-Werkman cycle (Turgay et al. 2021). However, various strains of *P. freudenreichii* co-metabolise aspartic acid and lactic acid. Aspartic acid is deaminated to fumaric acid and further reduced to succinic acid. To maintain cell redox balance, more pyruvic acid is oxidised to acetic acid and CO_2 than to propionic acid. The use of *P. freudenreichii* strains with high-aspartase activity may contribute to a shorter ripening period primarily as a result of the flavour enhancing products succinic acid and ammonia (Drake et al. 2007; Wyder et al. 2001).

The technological operations during manufacture of a Swiss-type cheese are aimed at creating ideal conditions for the propionic acid fermentation. Many factors influence the growth of *P. freudenreichii*. Important factors are the temperature during the production and ripening period, the pH of the cheese and the NaCl concentration. *P. freudenreichii* grows optimally at a temperature of 25–35 °C, but endures temperatures of up to 55 °C (Turgay et al. 2021). Thus, the species can be applied not only in semi-hard Swiss-type cheeses with scalding temperatures of 35–40 °C but also in hard Swiss-type cheeses with scalding temperatures of 50–53 °C. Furthermore, it is advantageous to place the cheeses after manufacture for a certain period of time in a warm ripening room (17–24 °C) in order to promote the growth of *P. freudenreichii*.

The species is quite sensitive to low pH values and is unable to grow at pH values below 5.0. Thus, a fundamental step in the manufacture of e.g. Emmentaler PDO, Jarlsberg, Maasdam and Grevé is the reduction of the lactose concentration in the milk e.g. by adding water to the milk and/or curd (12–30%). This leads to a relatively high pH value after lactic acid fermentation (pH of 5.20–5.35), which promotes propionic acid fermentation.

The salt sensitivity of propionibacteria is strain dependent and can be addressed by using lower salt contents in cheeses, such as Emmentaler PDO (0.5% w/w) or French Emmental (0.4–0.7%). It is also possible to select less salt-sensitive strains for cheeses with higher salt contents, such as Fol Epi, Grevé, Leerdammer or Maasdam (1.1–1.6%).

Addressing all of these factors during cheese manufacture can guarantee an optimal propionic acid fermentation.

11.3.2 *Appearance*

The most distinct, desired visual characteristic of Swiss-type cheeses are the holes or eyes (Figure 11.1) which many consumers all over the world immediately associate with a Swiss-type cheese. As mentioned earlier in this chapter, for Emmentaler PDO the specification of the form and size of the holes are defined as round to slightly oval of 2–4 cm in diameter (Federal Office for Agriculture 2018).

Figure 11.1 Desired visual appearance of Swiss-type cheese varieties.

In general, eye formation in cheese depends mainly on time, intensity of CO_2 production and its diffusion, the presence of nuclei as well as structure and temperature of the cheese body. Eye quantity and distribution are strongly influenced by technological aspects of milk treatment such as centrifugation, bactofugation, microfiltration and heating. In Swiss-type cheeses, eye formation results mainly from CO_2 production during propionic fermentation (Fröhlich-Wyder et al. 2017; Guggisberg et al. 2015). The number, size, shape and distribution are crucial quality factors especially for producers and retailers but also for consumers.

Traditionally, the monitoring of eye formation is still carried out by listening to the sound pattern elicited by tapping on the cheese surface with the handle of a cheese trier. Depending on the sound, the cheesemaker decides on possible adaptations of the further ripening process. Visual inspection of small cheese cylinders sampled with a cheese trier or cutting cheese loafs in half is another possibility to assess eye formation. However, both of these methods are highly dependent on the experience of the test person. Nowadays, newer non-destructive instrumental methods such as X-rays, ultrasound, X-ray computed tomography (CT) and magnetic resonance imaging (MRI) are used especially either in large-scale production or in research (Fröhlich-Wyder et al. 2017; González et al. 2020; González et al. 2019).

It is known that consumers associate certain colours with a specific flavour. In consequence, when colour differs from an expected standard, the effect of colour on liking increases because of a more accentuated interaction between colour and flavour (Roth et al. 1988). There is little information on the influence of colour on the perceived quality of Swiss-type cheese. Although colour is explicitly mentioned quality criterion for example in the quality requirements for Emmental PDO cheese, no scientific studies about the importance of the colour of Swiss-type cheeses for consumers were found in the literature.

11.3.3 *Texture*

Textural aspects together with flavour and appearance strongly contribute to the overall quality perception of cheese (Foegeding and Drake 2007). Physical structure of the cheese body is supposed to be one of the main factors influencing the perception of cheese texture as well as eye formation (Guggisberg et al. 2015). Dynamic interactions between the cheese matrix and the surface of the oral cavity, breakdown patterns and bolus formation due to manipulation of cheese in the mouth are thought to be crucial for texture perception such as elasticity, creaminess and crumbliness (Rogers et al. 2009; Saint-Eve et al. 2015).

In addition, the structure of the cheese body is supposed to play an important role in the release of flavour active substances. Strong interactions between composition, especially fat content and texture of cheese, were found which in turn influenced individual chewing behaviour and consequently flavour release (Tarrega et al. 2008). Generally, weaker gel networks and more porous structures are associated with a higher flavour release. Since proteolysis results in a weakening of the protein network in cheese, it can be assumed that the release of aroma compounds and tastants such as NaCl during mastication is facilitated by proteolytic activity of the bacteria present in the cheese (Lamichhane et al. 2018).

The texture of Swiss-type cheeses changes with ripening time. Young cheeses (up to approximately three months) show a soft and elastic body which is essential for the formation of regular, round shaped eyes. Elasticity of the cheese body is strongly related to the amount of associated calcium, which itself is influenced by the pH after whey separation and after lactic acid fermentation (Fröhlich-Wyder et al. 2017). Calcium content in Swiss-type cheeses is generally higher than in other cheese varieties since an elastic texture is a prerequisite for an optimal eye formation. With increasing ripening duration, Swiss-type cheeses become more solid and crumbly (Bachmann et al. 1999). These textural changes of aged cheese are commonly attributed to changes in the structure of the protein matrix resulting in changes of specific breakdown patterns during oral manipulation (Rogers et al. 2009).

11.3.4 *Flavour*

Flavour is considered to be one of the most important factors influencing consumer choice and acceptance and thus, the success of a product in the market. A flavour profile matching personal expectations is crucial when consumers are exposed to a food such as cheese for the first time. Therefore, knowledge about the relationship among the estimated more than 200 different volatile compounds found in Swiss-type cheeses and the impact of cross modal sensory interactions on the perception of the flavour is of high interest in order to influence and promote the formation of desirable aroma compounds during production and ripening of Swiss-type cheeses.

Moreover, it is well accepted that knowledge on odour activity values (OAV) is a key factor for flavour perception. The OAV corresponds to the concentration of a compound divided by its threshold value. It is assumed that only compounds showing OAVs greater than 1 significantly contribute to the flavour (Qian and Reineccius 2003). Thus, the ratio of volatiles relative to their odour threshold is the key factor for differences in flavour profiles between cheeses including Swiss-type cheeses (Taylor et al. 2013).

A summary of volatile and non-volatile compounds detected in Swiss-type cheese in different studies is given in Table 11.2. It has to be mentioned that the set of flavour compounds described in the literature contributing to the characteristic sensory

Table 11.2 Volatile and non-volatile compounds considered to contribute to the flavour of Swiss-type cheeses.

Compound	Descriptor	References
(E)-2-nonenal	Green, soapy, cucumber, melon-like	Thierry and Maillard (2002), Thierry et al. (2004), Warmke et al. (1996)
1-propanol	Earthy, fermented, nutty, fruity, bubble gum	Castada et al. (2019), Pillonel et al. (2002), Taylor et al. (2013), Agroscope (2020)
2-propanol	Green, spicy, floral	Pillonel et al. (2002), Taylor et al. (2013)
1-butanol	Fruity	Castada et al. (2019), Pillonel et al. (2002), Taylor et al. (2013)
2-butanol	Fruity, banana	Pillonel et al. (2002), Agroscope (2020)
2,3 butanediol	Odourless, sweet taste	Agroscope (2020)
butan-2-one	Sharp, sweet	Castada et al. (2019), Pillonel et al. (2002)
2,3 butanedione	Buttery, creamy	Castada et al. (2014),Langler et al. (1967),Pillonel et al. (2002), Taylor et al. (2013), Thierry and Maillard (2002), Thierry et al. (2004), Warmke et al. (1996), Agroscope (2020)
2,3-dimethyl-5-methylpyrazine	Nutty	Warmke et al. (1996)
1-methoxy-2-propanol	Sweetish, ethereal, bitter	Agroscope (2020)
2-methylpropanol (isobutyl alcohol)	Sweetish, fruity	Preininger et al. (1996), Taylor et al. (2013), Thierry et al. (2004), Agroscope (2020)
methyl acetate	Nail polish remover	Castada et al. (2019)
2-methylpropanal	Pungent	Castada et al. (2019), Preininger et al. (1996), Taylor et al. (2013), Thierry and Maillard (2002), Warmke et al. (1996)
2-methylbuytraldehyde	Chocolate like	Castada et al. (2019)
2-methylbutanal	Malty, green	Preininger et al. (1996), Taylor et al. (2013), Thierry and Maillard (2002), Warmke et al. (1996)
3-methylbutanal	Malty, green	Castada et al. (2014), Langler et al. (1967), Preininger et al. (1996), Taylor et al. (2013), Thierry and Maillard (2002), Thierry et al. (2004), Warmke et al. (1996), Agroscope (2020)
2-methyl-1-butanol	Fermented	Pillonel et al. (2002), Preininger et al. (1996), Taylor et al. (2013)
3-methyl-1-butanol	Fermented	Preininger et al. (1996), Taylor et al. (2013)
2-methylbutanoic acid (isovaleric acid)	Fruity, waxy, sweaty	Preininger et al. (1996)
3-methylbutanoic acid	Rancid, cheesy, sweaty socks, faecal	Castada et al. (2014), Langler et al. (1967), Preininger et al. (1996), Taylor et al. (2013), Thierry and Maillard (2002), Thierry et al. (2004), Warmke et al. (1996)

Table 11.2 (Continued)

Compound	Descriptor	References
2-methylbutyl acetate	Sweet, banana, fruity, tropical juicy fruit	Taylor et al. (2013)
3-methylbutyl acetate	Pear, banana	Castada et al. (2019), Preininger et al. (1996), Taylor et al. (2013)
2-methylbutyl propanoate	Sweet, fruity	Taylor et al. (2013)
3-methylbutyl propanoate	Apricot, pineapple	Preininger et al. (1996), Taylor et al. (2013)
2-methylpropanoic acid	Sweet, apple like, cheesy	Pillonel et al. (2002), Preininger et al. (1996), Thierry et al. (2004)
3-methylindole (skatole)	Faecal	Thierry and Maillard (2002), Thierry et al. (2004)
4-methylpentanoic acid	Cheese like	Preininger et al. (1996)
2-butenal	Pungent	Preininger et al. (1996)
2-propanone	Sweet, fruity	Langler et al. (1967), Preininger et al. (1996)
2-butanone	Chemical, fruity, green	Preininger et al. (1996), Agroscope (2020)
2-methyl-2-butanone	[a]	Taylor et al. (2013)
2-pentanol	Fermented, fruity, oily	Castada et al. (2019), Pillonel et al. (2002), Agroscope (2020)
2-pentanone	Sweet, banana, fermented	Castada et al. (2019), Taylor et al. (2013), Agroscope (2020)
2,3 pentandione	Buttery, creamy	Castada et al. (2019), Taylor et al. (2013), Agroscope (2020)
3-methyl pentane	Odourless	Pillonel et al. (2002)
2-methyl-3-pentanol	Alliaceous, brothy, meaty, burned, roasted, sulphurous	Agroscope (2020)
4-methyl-2 pentanone	Sharp solvent-like with green, herbal, fruity and dairy nuances (3)	Taylor et al. (2013)
3-methyl-2 pentanone	Peppermint like	Taylor et al. (2013)
2-phenyl-2-methylbutane	[a]	Castada et al. (2019)
2-heptanol	Fruity, green (2)	Agroscope (2020)
2-hexen-1-ol	Fruity, lavender, geranium, mint	Castada et al. (2019)
methyl hexanoate	Fruity	Castada et al. (2019)
methyl octanoate	Green, waxy	Castada et al. (2019)
2-hexanone	Fruity, fungal, meaty	Taylor et al. (2013)
2-heptanone	Spicy, forest	Castada et al. (2019), Taylor et al. (2013), Agroscope (2020)
3-hydroxy2-butanone (acetoin)	Buttery, creamy	Pillonel et al. (2002)
6-methyl-2-heptanone	Campheorous	Taylor et al. (2013)
5-methyl-2-heptanone	Herbal, fruity	Taylor et al. (2013)
2-pentadecanone	Floral, fatty	Castada et al. (2019)
2-phenylethanol	Rose	Castada et al. (2019)
2-tridecanone	Waxy, fatty	Castada et al. (2019)
acetone	Nail polish remover	Castada et al. (2019)
acetic acid	Vinegar	Castada et al. (2014), Langler et al. (1967), Taylor et al. (2013), Thierry and Maillard (2002), Thierry et al. (2004), Warmke et al. (1996), Agroscope (2020)

(Continued)

Table 11.2 (Continued)

Compound	Descriptor	References
acetic acid-3-methylbutyl ester (isoamyl acetate)	Banana, pear	Pillonel et al. (2002)
acetaldehyde	Yoghurt	Castada et al. (2019)
acetophenone	Pungent, medicinal, almond like,	Agroscope (2020)
ammonia	Sharp, suffocating	Thierry and Maillard (2002), Thierry et al. (2004), Warmke et al. (1996)
benzaldehyde	Bitter almond	Castada et al. (2019), Preininger et al. (1996), Taylor et al. (2013), Agroscope (2020)
benzothiazole	Meaty, sulphurous	Castada et al. (2019)
butanal	Pungent	Thierry and Maillard (2002), Thierry et al. (2004), Warmke et al. (1996)
2-butenal	Sweet, fruity, cherry, nutty	Taylor et al. (2013)
butanoic acid (butyric acid)	Sweaty, rancid	Castada et al. (2014), Langler et al. (1967), Taylor et al. (2013), Thierry and Maillard (2002), Thierry et al. (2004), Warmke et al. (1996), Agroscope (2020)
butyl formate	Plum, fruity	Agroscope (2020)
butyl acetate	Fruity	Castada et al. (2019), Taylor et al. (2013)
caproic acid	Cheesy, acidic	Castada et al. (2014), Langler et al. (1967)
caprylic acid	Goaty	Castada et al. (2014), Langler et al. (1967)
capric acid	Creamy, heavy cream	Castada et al. (2014), Langler et al. (1967)
m-cresol	Phenolic, animalic, woody	Agroscope (2020)
p-cresol	Phenolic, animalic (2)	Agroscope (2020)
decanoic acid	Rancid	Agroscope (2020)
diethyl sulphide	Cooked, sulphury	Thierry and Maillard (2002), Thierry et al. (2004)
dimethyl disulphide (DMDS)	Cauliflower, garlic, onion	Preininger et al. (1996), Taylor et al. (2013), Thierry and Maillard (2002), Thierry et al. (2004), Warmke et al. (1996), Agroscope (2020)
dimethyl sulphide (DMS)	Cooked cabbage, onion	Castada et al. (2019), Preininger et al. (1996), Thierry and Maillard (2002), Thierry et al. (2004), Warmke et al. (1996), Agroscope (2020)
dimethyl trisulphide (DMTS)	Cabbage, onion	Castada et al. (2019), Preininger et al. (1996), Thierry and Maillard (2002), Thierry et al. (2004), Warmke et al. (1996), Agroscope (2020)
dimethyl sulphone	Sulphury, burned (2)	Agroscope (2020)
dimethyl benzene	Hazardous	Castada et al. (2019)

Table 11.2 (Continued)

Compound	Descriptor	References
dodecane	Mild	Castada et al. (2019)
dodecanoic acid (lauric acid)	Fatty, coconut, bay oil	Castada et al. (2019), Castada et al. (2014)
ethanol	Ethanol like	Castada et al. (2019), Taylor et al. (2013), Thierry and Maillard (2002), Thierry et al. (2004), Warmke et al. (1996)
ethyl acetate	Fruity	Taylor et al. (2013)
ethyl 3-methylbutanoate	Sweet, fruity, pineapple, apple, orange	Castada et al. (2014), Langler et al. (1967), Preininger et al. (1996)
ethyl butanoate	Fruity	Castada et al. (2019), Castada et al. (2014), Langler et al. (1967), Pillonel et al. (2002), Taylor et al. (2013), Thierry and Maillard (2002), Thierry et al. (2004), Warmke et al. (1996)
ethyl hexanoate	Fruity, pineapple	Castada et al. (2019), Castada et al. (2014), Langler et al. (1967), Taylor et al. (2013), Thierry and Maillard (2002), Thierry et al. (2004), Warmke et al. (1996), Agroscope (2020)
ethyl decanoate	Waxy	Castada et al. (2019)
ethyl dodecanoate	Soapy, waxy, sweet	Castada et al. (2019)
ethyl methyl sulphide	Sulphurous, cabbage, meaty	Thierry and Maillard (2002), Thierry et al. (2004)
ethyl octanoate	Waxy, sweet, musty, fruity, pineapple	Castada et al. (2019)
ethyl propanoate (= ethyl propionate)	Pineapple	Castada et al. (2019), Taylor et al. (2013), Agroscope (2020)
formic acid	Acetic acid like	Castada et al. (2019)
Furaneol	Caramel	Pillonel et al. (2002), Taylor et al. (2013), Thierry and Maillard (2002), Thierry et al. (2004)
γ-decalactone	Coconut	Castada et al. (2019), Langler et al. (1967), Thierry and Maillard (2002), Thierry et al. (2004), Agroscope (2020)
γ-dodecalactone	Fruity fatty creamy buttery peach (3)	Castada et al. (2019), Agroscope (2020)
δ-dodecalactone	Creamy, buttery	Castada et al. (2019)
glutamic acid	Umami, savoury	Langler et al. (1967)
heptanol	Solvent-like, fermented, nutty, fatty	Agroscope (2020)
hexan-1-ol	Herbal, green	Pillonel et al. (2002)
hexanal	Green, grassy, woody, vegetative	Pillonel et al. (2002)
2-hexanone	Fruity	Castada et al. (2019)
hexanoic acid	Cheesy, goaty	Taylor et al. (2013), Agroscope (2020)
heptanoic acid	Waxy, cheesy, fruity	Agroscope (2020)

(Continued)

Table 11.2 (Continued)

Compound	Descriptor	References
homofuraneol	Sweet caramelic, burned sugar, fruity	Thierry and Maillard (2002), Thierry et al. (2004), Warmke et al. (1996)
hydrogen sulphide	Rotten eggs	Thierry and Maillard (2002), Thierry et al. (2004)
isopropyl propanoate	Fruity, berry, bubble gum	Taylor et al. (2013)
isobutyric acid	Cheesy, rancid	Agroscope (2020)
methanethiol (methyl mercaptan)	Cooked cabbage, garlic, alliaceous	Preininger et al. (1996), Thierry and Maillard (2002), Thierry et al. (2004), Warmke et al. (1996)
methional	Boiled potato, potato, skin yeasty	Langler et al. (1967), Preininger et al. (1996), Thierry and Maillard (2002), Thierry et al. (2004), Warmke et al. (1996)
methionol	Onion- and garlic-like	Thierry and Maillard (2002), Thierry et al. (2004), Warmke et al. (1996)
methyl butanoate (= 3-methyl butanoic acid?)	Pungent, ethereal, fruity, perfumey	Taylor et al. (2013), Agroscope (2020)
methyl propanoate (propionic ester)	Sweet, fruity, black currant	Taylor et al. (2013), Agroscope (2020)
methyl decanoate	Fermented, fatty	Castada et al. (2019)
nonanoic acid	Fatty, waxy, cheesy	Agroscope (2020)
octanal	Green, citrus peel	Taylor et al. (2013)
octanoic acid	Rancid, soapy, cheesy, fatty	Agroscope (2020)
1-octene	Gasoline	Castada et al. (2019), Pillonel et al. (2002)
δ-octalactone	Coconut	Castada et al. (2019)
1-nonene	Grassy, onion like	Castada et al. (2019)
8-nonen-2-one	Fruity, baked	Agroscope (2020)
2-nonanone	Green, fruity	Castada et al. (2019), Agroscope (2020)
pentane	Petroleum like	Pillonel et al. (2002)
pentadecane	Waxy	Castada et al. (2019)
pentanoic acid	Cheesy, acidic	Agroscope (2020)
phenol	Distinct aromatic, sweet	Agroscope (2020)
phenyl ethanol	Rose like, floral	Preininger et al. (1996), Agroscope (2020)
phenylacetate aldehyde	Rose like	Castada et al. (2019),Preininger et al. (1996), Thierry and Maillard (2002), Thierry et al. (2004), Agroscope (2020)
pinene	Fresh pine, woody, resinous	Castada et al. (2019)
propyl acetate	Fruity	Pillonel et al. (2002)
propionic acid	Rancid	Langler et al. (1967), Taylor et al. (2013), Thierry et al. (2004), Warmke et al. (1996), Agroscope (2020)
skatole	Animal faecal naphthyl, civet	Agroscope (2020)
S-methyl thioacetate	Cauliflower, garlic, cheesy	Preininger et al. (1996)

Table 11.2 (Continued)

Compound	Descriptor	References
sotolon	Fenugreek seeds	Thierry and Maillard (2002)
succinic acid	Sour, umami	Langler et al. (1967)
tetradecalactone	Waxy fatty, cheesy	Agroscope (2020)
tetradecanoic acid (Myristic acid)	Waxy, fatty, cheesy	Agroscope (2020)
tetramethylpyriazine	Nutty, weedy, earthy, asparagus	Thierry and Maillard (2002), Thierry et al. (2004), Warmke et al. (1996)
toluene	Sweet pungent benzene-like, burned	Castada et al. (2019), Pillonel et al. (2002), Agroscope (2020)
2,3,4-trimethylpentane	Gasoline like	Pillonel et al. (2002)
2-undecanone	Fruity, fatty	Castada et al. (2019), Agroscope (2020)
2-tridecanone	Fatty, cheesy, earthy	Agroscope (2020)
3-hydroxybutan-2-one (acetoin)	Creamy, milky, buttery	Agroscope (2020)
2-undecanone	Floral, fatty, pineapple	Castada et al. (2019)
γ valerolactone	Herbal, tonka bean	Langler et al. (1967)

[a] No descriptor found.

flavour of Swiss-type cheeses might be determined by the technique chosen for the analysis, since the type of extraction as well as the material of GC column stationary phase were shown to greatly influence the palette of detectable volatiles (Bertuzzi et al. 2018).

Key volatile compounds

The typical flavour of Swiss-type cheeses is often described by distinct sweet and nutty notes. Especially the sweet flavour note is often more pronounced compared to other hard cheese varieties. Already more than 80 years ago, it was proposed that the flavour characteristic of Emmental cheese is related to the content of propionic acid (Langsrud and Reinbold 1973).

One of the first systematic studies on volatiles contributing to the flavour of Swiss-type cheese identified 66 different compounds encompassing alcohols, aldehydes, esters, hydrocarbons, lactones and methyl ketones as well as diacetyl (2,3-butanedione) and dimethyl sulphide (Langler et al. 1967). Results of an early extensive study on volatile compounds found in Swiss-type cheese and other cheese varieties are given by Langsrud and Reinbold (1973).

Besides propionic acid, acetic acid, lactic acid, succinic acid and glutamic acid either in their free form and/or as ammonium, sodium, potassium, magnesium or calcium salts, as well as the corresponding chlorides and phosphates of these cations are considered key compounds for the typical flavour of Swiss-type cheese. Magnesium and calcium propionate were supposed to be the main compounds responsible for the sweet flavour note (Warmke et al. 1996). Moreover, 2,3 butandione and 3-methylbutanal, together with homofuraneol, furaneol, ethyl butanoate, ethyl hexanoate and methional were identified as high-impact compounds to the overall flavour of Swiss-type cheese (Preininger et al. 1996). Methional with its potato-like odour, and characterised by a quite low-odour threshold is considered to highly contribute to the flavour of Swiss-type cheeses (Harper et al. 2011). To a smaller extent also butanoic and caproic acid as well as leucine were considered as important flavour compounds (Preininger et al. 1996; Warmke et al. 1996).

In addition to propionic and acetic acid formed during propionic fermentation, FFA derived from lipolysis and its metabolites derived from catabolism during cheese ripening are considered to have a high impact on the typical flavour notes of Swiss-type cheeses. Compared to lactic acid bacteria generally showing a low-lipolytic activity, propionic acid bacteria (PAB) have a 10–100 time higher lipolytic activity (Chamba and Perreard 2002; Fröhlich-Wyder et al. 2017; Turgay et al. 2021). Lipolysis results in the formation FFA of medium and high-carbon chain lengths. The amount of FFA is considered to be directly linked to the cheese flavour (Bertuzzi et al. 2018). FFAs are supposed to influence cheese flavour also indirectly by acting as precursors of many odour active molecules such as methyl ketones, aldehydes, lactones or esters (Thierry et al. 2017).

As mentioned before, volatile compounds formed due to the activity of the PAB are supposed to be one of the key factors responsible for the characteristic flavour of Swiss-type cheeses. However, although research on the volatile odour active compounds of Swiss-type cheese started almost a century ago, only a few studies have tried to evaluate the direct impact of volatile compounds formed by PAB on the perceived flavour of Swiss-type cheeses.

Thierry and Maillard (2002) studied the formation of volatile flavour compounds deriving from the catabolism of amino acids by *P. freudenreichii in vitro* and in cheese. Data showed that PBA strains produced several branched-chain fatty acids such as 2-methylpropanoic acid, 2-methylbutanoic, 3-methylbutanoic acid and 4-methylpentanoic acid in different culture media. The formation of sulphur containing compounds such as dimethyl sulphide (DMS), dimethyl disulphide (DMDS), dimethyl trisulphide (DMTS) and methanethiol in culture media was strain dependent.

In mini Emmental cheese produced from microfiltered milk according to a standardised protocol, isovaleric acid, derived by conversion of the amino acids leucine and isoleucine, was the volatile compound with the highest potential impact on cheese flavour formed by PAB followed by 2-methybutanal having an OAV of approximately 1. Branched-chain alcohols showed OAVs smaller than 1. The addition of PAB also resulted in significantly higher amounts of several branched-chain esters, aldehydes and alcohols. However, no effect on the amount of sulphur containing compounds was observed (Thierry and Maillard 2002). In contrast, in young cheese (30 days), the formation of dimethyl sulphide as well as methanethiol (methyl mercaptan) seemed to be positively related to an increase in the propionic to acetic acid ratio (Harper et al. 2011).

Further investigations on cheese juice (aqueous phase) and mini Emmental cheeses produced with and without addition of *P. freudenreichii* revealed a pronounced influence on the profile of volatiles in both matrices (Thierry et al. 2004). In the aqueous cheese juices, 57 of the 69 identified volatiles significantly differed in the samples fermented with and without PAB. The biggest differences were observed for the group of esters and alcohols. The addition of PAB to mini Emmental cheeses resulted in significantly higher amounts of short chain carboxylic acids such as acetic, propanoic, butanoic, hexanoic and 3-methylbutanoic acid (isovaleric acid) as well as of some ketones, whereas most aldehydes were found in lower quantities. However, no conclusions could be drawn on the contribution of PAB to the perceived Emmental flavour, since the amount and composition of volatiles of mini cheeses and Emmental cheese were not directly comparable (Thierry et al. 2004).

Evaluating the production of volatiles by 76 strains *of P. freudenreichii*, *Propionibacterium jensenii*, *Propionibacterium thoenii* and *Propionibacterium acidipropionici* distinct interspecies and intraspecies differences were observed (Yee et al. 2014). Multivariate analysis of the 36 compounds showing differences among species revealed that *P. freudenreichii* strains were associated with higher amounts of FFA, ethyl esters derived from the latter as well as metabolic products from amino acid degradation such as branched-chain acids and alcohols. *P. acidipropionici* produced more sulphur containing compounds deriving from the methionine catabolism, whereas selected strains of *P. jensenii* and *P. thoenii* were positively related to the formation of benzaldehyde and branched-chain aldehydes. In addition, enormous strain differences in the amount of flavour compounds produced, varying by up to 500-fold, were observed within species.

A recent study on the flavour of Swiss cheese produced in five different sites in the United States revealed a distinct fingerprint of high-impact volatile compounds for each of the five factories (Taylor et al. 2013). Concentration differences among production sites were found for the majority of the 24 compounds analysed. Sixteen of these substances showed an OAV greater than 1, whereas acetic acid, ammonia, butanal, methionol and 2,3,5,6-tetramethylpyrazine, all showing OAV's < 1, were considered having little effect on the flavour of Swiss cheese. A different set of key odour compounds was determined for each of the five production sites. Among the compounds with the most discriminating power were methional, diacetyl, butanoic acid, methylbutanals (2- and 3-methylbutanal) and furaneol (Taylor et al. 2013).

More recently, volatile compounds with high-flavour potential found in eight Swiss-type cheese varieties were investigated by GC–MS analysis (Agroscope, internal report 2020). The most abundant compounds were grouped into eight chemical families, namely sulphur compounds, aldehydes, esters, ketones, indoles and derivatives, carboxylic acids, alcohols and lactones (Figure 11.2).

For all families of volatiles, significant differences were found among the eight analysed Swiss-type cheeses. As could be expected, especially the ripening time of the cheeses had a pronounced influence on the abundance of the analysed volatiles.

Ketones, represented by butan-2-one, diacetyl, pentane-2,3-dione, heptan-2-one, 3-hydroxybutan-2-one (acetoin), nonan-2-one, 8-nonen-2-one, undecan-2-one, acetophenone and tridecan-2-one were found in significantly greater quantities in the younger cheeses such as Leerdammer, Fol Epi and Switzerland Swiss. It was assumed that diacetyl and acetoin, characterised by a distinct buttery note, were not formed by PAB but rather by the starter culture from the metabolism of citrate. As for the ketones, also compounds of the lactone family were present in significantly higher amounts in the young cheeses (Leerdammer, Fol Epi, Switzerland Swiss). Represented by δ-decalactone, δ-dodecalactone and δ-tetradecalactone, this family was supposed to contribute to the milky and buttery note of the Swiss-type cheeses studied.

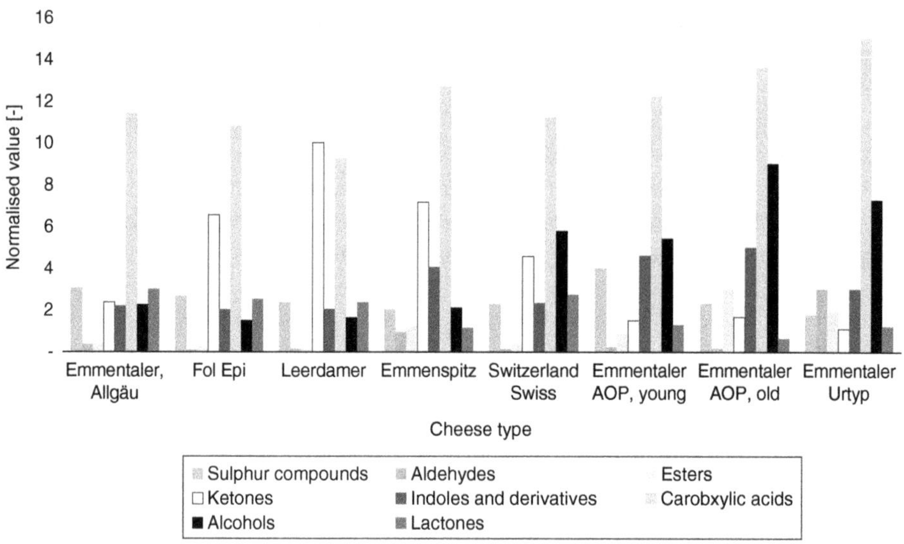

Figure 11.2 Aroma profile of eight Swiss-type cheeses. The volatile compounds were classified into eight chemical families (sulphur compounds, aldehydes, esters, ketones, indoles and derivatives, carboxylic acids, alcohols and lactones). Mass spectrometer signals (TIC) were normalised to the most intense value and divided by the number of compounds detected for each family.

By contrast, the quantities of the 15 carboxylic acids identified in the eight Swiss-type cheeses significantly increased with longer maturation periods of the cheeses. The highest amounts of these acids, mainly characterised by a cheesy, fatty flavour note, were determined for Emmentaler Urtyp and Emmentaler PDO, ripened for 18 months (Emmentaler PDO, old).

Moreover, Emmentaler Urtyp contained approximately 10 times more of the identified three aldehydes which are described as malty and bitter almond like compared to the other cheeses examined. This finding was most likely not related to PAB activity but rather the result of the maturation parameters of Emmentaler Urtyp. The latter is ripened at high relative humidity (>90% R.H.), whereas Swiss-type cheeses usually are matured at distinctly lower humidity values or else in plastic films. Thus, it can be assumed that duration of ripening and ripening parameters mainly contributed to the observed differences in aldehyde content.

Also, the compounds belonging to the ester, alcohol and indole families were found in significantly higher levels in the cheeses ripened for a longer period, namely Emmentaler PDO, old and Urtyp. The group of esters represented by ethyl propionate, ethyl hexanoate and butyl formate are known for their fruity notes. The aroma notes of the nine alcohols identified in the eight Swiss-type cheeses can be described as fruity, fermented, as well as alcoholic. The indole family and derivatives, encompassing toluene, phenol, p-cresol, m-cresol and skatole, contribute what are described as animal notes.

The sulphur containing compounds identified in the eight Swiss-type cheeses included dimethyl sulphide, dimethyl disulphide, dimethyl trisulphide and dimethyl sulphone. The latter is odourless and does not contribute to the flavour of Swiss cheese. These highly volatile and reactive compounds are usually characterised by onion, cabbage or garlic-like notes. Especially the concentration of dimethyl sulphide was considered already decades ago to have a high impact on perceived flavour of Swiss-type cheeses (Langsrud and Reinbold 1973). Compared to the alcohols, esters and indoles, the analysis of the sulphur family of the analysed cheeses showed a slightly less pronounced trend towards higher quantities with increasing maturation period of the cheeses.

Non-volatiles

The perceived saltiness of Swiss-type cheeses is mainly a result of the addition of sodium chloride during the brining process and salting during the ripening process (Bisig et al. 2019). Compared to other cheese varieties, Swiss-type cheeses usually show a low-to-medium salt concentration ranging from 0.2 to about 1.6% of sodium chloride (Hartmann et al. 2017) and thus generally show a low perceived saltiness. Sodium chloride concentration of Emmentaler PDO was found to be linked to the intensity of specific flavour characteristics by enhancing perceived nutty and fruity notes as well as by suppressing the metallic note or bitterness, both considered as undesirable flavour defects of Emmentaler cheese (Agroscope, internal report 2020).

Furthermore, other compounds showing a pronounced taste such as the salts of glutamate responsible for the sensory characteristic known as ‘umami’ play an important role in Swiss-type cheese flavour (Drake et al. 2007). Perceived umami intensity was found to be positively related with cheese age due to increased proteolysis and release of free amino acids such as glutamic acid during cheese ripening. In addition, propionic acid content as well as organic acids and nucleotides found in cheese may contribute to umami taste. Moreover, results of model cheese systems suggested that succinate has a higher impact on the perceived umami taste than glutamate (Drake et al. 2007). Recently, it was proposed that γ-glutamyl peptides found in Emmental cheese contribute to the kokumi impression of Emmental cheese (Hartmann et al. 2015). The concept of kokumi is connected to the perception of the complexity of mouthfeel, heartiness and thickness. Kokumi is supposed to contribute to the sensation of mouth-coating elicited by butter, fats and emulsions coating the tongue and mouth (Yamaguchi and Ninomiya 1998).

11.4 Relationship between sensory data and analytical measurements

In the last few decades, cheese manufacturers started to recognise the importance of understanding interactions between cheese microflora, different types of manufacturing processes, or cheese components and the perceived sensory character of a cheese. Knowledge on these relationships is necessary for the optimisation and control of the sensory profile of the final cheese product.

There are also quite a few studies which attempted to relate sensory and instrumental data for Swiss-type cheeses in order to better understand the influence of the product composition and individual volatile compounds on perceived sensory characteristics of this cheese variety.

11.4.1 *Relationship between microflora and perceived flavour*

It is well accepted that the composition and the activity of the cheese microflora is a key factor in the development of the characteristic sensory profile of a cheese variety. In general, Swiss-type cheese made from raw milk typically exhibits a more intense flavour compared to cheeses produced from pasteurised milk due to higher concentrations of free amino acids, FFA and volatile compounds formed by the indigenous milk microflora and enzymes (Beuvier et al. 1997; Bisig et al. 2010).

Observed differences in the flavour profile of mini Swiss cheeses made from raw milk of different origins were attributed mainly to differences the activity of PBA considered to be one of the major factors contributing to the characteristic flavour of Swiss-type cheeses as well as facultatively heterofermentative lactobacilli (FHL) during the secondary proteolysis (Demarigny et al. 1997). In addition, the activity and composition of the indigenous microflora was also influenced by seasonal aspects resulting in a noticeable difference of the sensory profile of cheeses made from autumns and winter milk, respectively (Demarigny et al. 1997).

Nevertheless, information available on the effect of the interaction between PBA and starter cultures used in the production of Swiss-type cheese is still scarce and the data did not show clear and conclusive results.

Depending on the adjunct strains, a significant increase in aroma or no effect was reported by Weinrichter et al. (2004), whereas Fröhlich-Wyder et al. (2002) observed significantly lower sensory ratings for texture and flavour for pilot-scale Emmental cheeses manufactured with the addition of different *Lactobacillus casei* strains.

The addition of different strains of *Lactobacillus* species to Swiss cheese made from pasteurised milk showed a strain-dependent influence on the sensory profile of the cheese samples. Significant differences were observed for the salty, brothy, sweet, buttery and burned flavour notes when grown in reconstituted milk, whereas in cheese the intensity of the flavour qualities 'sweet', 'buttery', 'burned', 'lipolysed' and 'acid' did not significantly differ. Differences in proteolytic capacity of the tested strains influenced mainly textural properties but had little impact on perceived flavour (di Palma et al. 1987). Similarly, a strain-dependent influence on the flavour of Swiss-type cheese made with *Lactobacillus* strains differing in their citrate metabolism was also found by Kocaoglu-Vurma et al. (2008). After 90 days of ripening the adjunct strains did not have a significant influence on the number of PAB and propionic acid levels. However, sensory evaluation by eight trained panellists encompassing 13 attributes of eight-month-old cheese samples showed that the use of adjunct lactobacilli strains resulted in a lower nutty flavour intensity and a slightly reduced sulphur/cabbage/brothy note of the cheeses compared to the control cheese produced with no adjunct cultures. In contrast, cheeses did not differ in the perceived intensity of the umami, sulphur/eggy and FFA/butyric acid note (Kocaoglu-Vurma et al. 2008).

The addition of mesophilic lactobacilli and enterococci to 12 pilot-scale Swiss-type cheese made from microfiltered and pasteurised milk showed a significant influence due to FHL on the formation of aroma active volatiles, whereby especially ester production derived from ethanol and propanol was positively affected by the addition of FHL. However, the adjunct cultures tested had only a minor influence on the sensory profile

(Bouton et al. 2009). No significant differences in the texture and taste attributes evaluated by a trained panel (n = 7) were found. Most of the aroma descriptors did not show a significant difference either for the flavour or for the texture attributes between cheeses produced with and without adjunct cultures with one exception: Cheeses produced with FHL and enterococci showed a significantly higher intensity for the attributes 'propionic' and 'cow shed' compared to cheeses made without adjuncts. However, no conclusion could be drawn regarding whether the observed influence of FHL on cheese characteristics was a result of the combined interaction between FHL and PAB or by FHL activity alone (Bouton et al. 2009).

11.4.2 *Relationship between volatile and non-volatile compounds and perceived flavour*

The majority of the studies concerning the relationship between instrumental and sensory data were focusing on the flavour aspect (Biede and Hammond 1979; Vangtal and Hammond 1986; Lawlor et al. 2003; Liggett et al. 2008; Castada et al. 2019), while others looked also at textural aspects (Bachmann et al. 1999).

One of the early studies concerning the relationship between sensory and chemical data focused on fractions of water-soluble volatiles and non-volatiles as well as fat-soluble compounds (Biede and Hammond 1979). Among the water-soluble volatile fractions of the seven Emmental cheeses tested, significant differences were found for the attributes 'acid', 'sweet', 'nutty', 'lipolysed', 'volatile' and 'buttery'. The perceived acidity showed a fairly high correlation (r = 0.6) with the total amount of acids whereas it could not be correlated to any of the individual FFA. Sweetness perception was also related to the fraction of the water-soluble tastants containing peptides and amino acids, lactic acid, salt, as well as the concentration of diacetyl, usually associated with a buttery note. Furthermore, this fraction was also characterised by burned, nutty and bitter notes. The sample containing the highest amount of bitter tasting amino acids such as leucine, lysine, valine and phenylalanine was also perceived as the most bitter sample. Of the fat-soluble fraction, the perceived 'nutty' and 'volatile' notes were the most prominent. However, the panel was not able to differentiate the sweet, nutty, acid and bitter intensities among samples. Cheese samples showed significant intensity differences of the buttery, nutty, burned, volatile, fermented and bitter note.

By combining the sensory intensity data (16 attributes) and analytical data determined for six European and nine US Swiss-type cheeses, an association between the sweet as well as nutty notes, typical of Swiss-type cheeses, and the amount of propionic acid, acetic acid and lipolysis in general was found (Vangtal and Hammond 1986). The nutty note was also related to low pH value. High levels of propionic and acetic acid in the oil fraction were related to a metallic note and in the case of propionic acid also with a bitter note. A positive correlation was observed for caprylic and capric acids and perceived lipolytic flavour. Cheeses described as buttery were generally perceived as low in overall flavour intensity and were associated with a low proteolysis. A fairly good prediction (range of R^2 0.63–0.85) of the flavour notes 'acid', 'metallic', 'burned', 'lipolytic flavour' and 'bitter' was achieved for selected chemical parameters using a linear multiple regression, whereas with a calculated R^2 between 0.17 and 0.46 the characteristic nutty flavour could be only poorly predicted (Vangtal and Hammond 1986).

The study by Bachmann et al. (1999) investigated the possibility to predict flavour and texture development during ripening of Emmentaler cheese based on parameters of proteolysis. A trained panel evaluated sensory characteristics such as maturity, aroma intensity, sweetness, sourness, saltiness and bitterness using a JAR scale. Perceived aroma intensity and saltiness were best predicted by the amount of free amino acids, whereas sweetness and bitterness showed the highest correlation with the lysine and proline aminopeptidase.

Studying the relationship between perceived flavour and composition of volatile and non-volatile compounds of Swiss- and other sweet-type hard cheeses including Emmentaler, Lawlor et al. (2003) observed a positive correlation between the characteristic nutty flavour and propionic acid, ethyl acetate, 2-pentanone and 2-hexanone, whereas the nutty flavour note was found to be negatively correlated with salt in moisture and ash content as well as with salt concentration confirming the results of Vangtal and Hammond (1986). In contrast, salt (NaCl) concentration was shown to positively influence the perception of the characteristic sweet note attributed to Swiss-type cheeses such as Emmentaler (Figure 11.3).

Unexpectedly, in a study by Lawlor et al. (2003) Emmentaler was rated as relatively low in sweetness – an attribute generally considered a pronounced typical flavour note of Swiss-type cheeses compared to the other seven cheeses tested. In general, perceived sweetness of cheeses seemed to be also influenced by a set of free amino acids such as threonine, serine and proline, all of which are considered to have a sweet taste (Kawai et al. 2012) as well as 2-pentanol, 2-hexanol and dimethyl sulphide. Sweetness was negatively correlated with pH value, indicating the influence of ripening duration on sweetness, as well as the concentrations of 2-octanone and ethyl hexanoate, respectively (Lawlor et al. 2002).

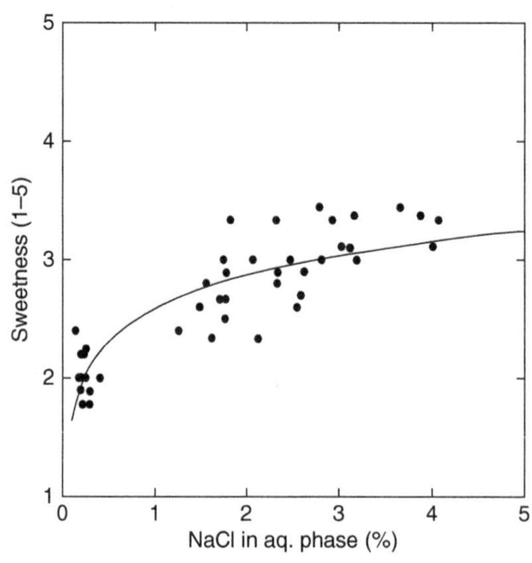

Figure 11.3 Relationship between perceived sweetness intensity and NaCl concentration in Emmental cheese. *Source:* Adapted from Fröhlich-Wyder et al. (2017).

Several studies looked at the relationship between the sensory profile of Swiss-type cheeses derived by descriptive analysis and infrared spectroscopy methods.

A study with 20 Emmental cheeses originating from six different European regions showed positive correlation coefficients for attributes such as friability and firmness, firmness and odour intensity, odour and aroma intensity with acidity, between aroma intensity and acidity as well as between aroma intensity and perceived saltiness, saltiness and acidity, saltiness and bitterness. A negative association was found between firmness and elasticity and adhesivity as well as between saltiness and adhesivity. Prediction of the perception of specific sensory attributes of Emmentaler by NIR data was best for firmness and adhesiveness (Karoui et al. 2007). The analysis of possible regional differences in the sensory characteristics revealed significant differences in perceived saltiness, bitterness and acidity, whereas no significant differences in texture and aroma notes were found (Pillonel et al. 2002).

The flavour variability of Swiss cheese manufactured in five different US production sites and its influence in relation to aroma volatiles, objective sensory analysis and consumer preference was studied by Castada et al. (2019). Swiss cheese samples from five different production plants using a set of 21 attributes following the method of Spectrum™ descriptive analysis. In addition, an untrained consumer panel (n = 100) judged visual aspects, overall liking and liking of nutty flavour of the same cheese samples. Correlation of the sensory data and volatile compounds showed a strong positive or negative correlation only for a limited number of compounds. As expected, a strong positive correlation was found between 2,3 butanedione and the buttery note. A strong positive association was also found between ethyl hexanoate and the sensory attributes dried fruit, sweet and bitter, as well as for the two aldehydes 2-methylpropanal and 3-methylbutanal and the sensory characteristics 'nutty', 'malty' and 'salty'. Furthermore, the flavour attributes 'cooked milk', 'fresh milk' and 'buttery' were positively related to γ-decalactone.

The correlation between the hedonic liking data and objective sensory data from the trained panel showed a strong positive association between nutty-malty, milkfat lactone, salty, umami and sweet taste (Castada et al. 2019).

A recent study by Agroscope (Agroscope, internal report 2020) characterised the sensory profile of eight different commercially available Swiss-type cheeses varying in origin of production, milk treatment as well as ripening time. Quantitative descriptive analysis was applied to describe the sensory profile of the cheeses using a set of 26 attributes representing odour, texture and flavour characteristics (Figure 11.4). In addition, different chemical and microbiological parameters such as water content, fat, NaCl, as well as counts of PAB were measured.

As expected, a broad variation in the sensory profile of the evaluated cheeses was observed. Not surprisingly, ripening time of the cheeses had a major impact on the overall flavour intensity with cheeses matured for longer periods showing higher mean intensity values for overall flavour. In addition, these cheeses were perceived as nuttier and more acidic and caramel like whereas the perceived buttery, milky and sweet notes decreased with increased ripening time. Moreover, the mean intensity of the taste attributes 'salty', 'umami' and 'sour' increased with increasing cheese age, sweetness intensity decreased, whereas perceived bitterness was not influenced by ripening period.

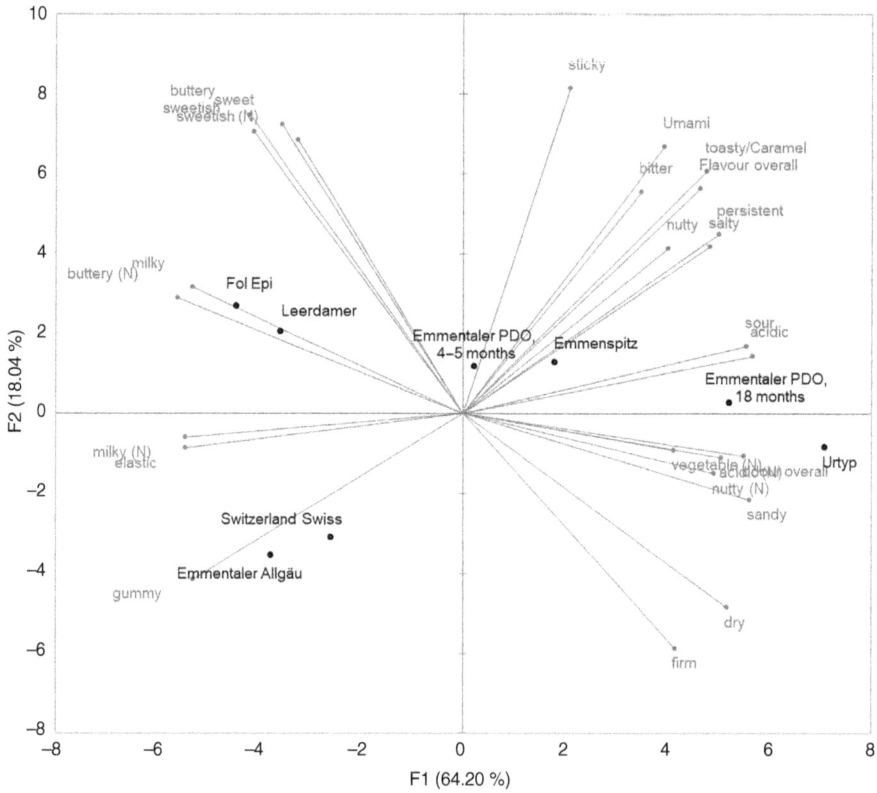

Figure 11.4 Biplot for the mean intensity values of eight Swiss-type cheeses and sensory attributes of the descriptive analysis.

In contrast, younger cheeses such as Emmentaler Allgäu, Leerdammer, Fol Epi and Switzerland Swiss showed significantly higher milky and buttery notes compared to longer ripened cheeses. These flavour notes seemed to be also influenced by milk treatment since Emmentaler Allgäu, Leerdammer, Fol Epi and Switzerland Swiss were all made from pasteurised milk.

The ripening time had also a significant influence on selected texture attributes. Young cheeses were characterised by an elastic and gummy texture, whereas older cheeses were perceived as firmer and drier. It has to be mentioned that these texture changes are observed in the majority of ripened cheeses and are not unique to Swiss-type cheeses.

Of the eight Swiss-type cheeses analysed, Spearman correlation coefficients between mean sensory intensities and mean values of the volatiles families as well as chemical parameters were calculated to get preliminary information about the potential contribution of the flavour families and composition on the perceived flavour quality of the cheeses (Table 11.3). For data interpretation, only raw data of volatiles were taken into account, i.e. no OAV were determined for the individual compounds. Results revealed a significant positive correlation between total amounts of measured volatiles and

Table 11.3 Spearman correlation coefficients for taste intensity values and selected flavour families (semi-quantitative) as well as selected chemical parameters.

Sensory descriptor	Volatile families (semi-quantitative, area under curve)									Selected chemical parameters (quantitative)											
	Volatiles overall	Sulphur compounds	Aldehydes	Esters	Indoles & derivatives	Ketones	Alcohols	Lactones	Acids overall	Formic l acid (mmol/kg)	Acetic acid (mmol/kg)	Propanoic acid (mmol/kg)	Total VCA (mmol/kg)	NaCl (g/kg)	Water (g/kg)	Fat (g/kg)	pH	D-LA (g/kg)	L-LA (g/kg)	Lactic acid (g/kg)	Proteolysis (mmol/kg)
Salty	**0.76**	–0.48	0.48	0.71	0.45	–0.38	0.31	**–0.83**	**0.74**	0.60	0.57	0.52	0.60	0.61	–0.57	0.64	0.62	0.48	–0.10	0.05	0.64
Sour	**0.89**	–0.51	0.46	**0.80**	**0.78**	–0.65	0.73	**–0.80**	**0.83**	**0.89**	0.67	0.35	0.62	0.13	**–0.83**	**0.78**	0.68	**0.85**	0.44	0.51	0.65
Umami	0.69	–0.60	0.50	0.57	0.29	–0.14	0.12	**–0.79**	0.62	0.38	0.36	0.50	0.45	0.73	–0.48	0.60	0.52	0.25	–0.29	–0.17	0.62
Sweet	–0.52	0.33	–0.29	–0.64	–0.43	**0.83**	**–0.88**	0.14	–0.67	–0.69	–0.24	0.40	0.05	–0.01	0.67	–0.43	–0.12	**–0.80**	**–0.81**	**–0.83**	–0.21
Bitter	**0.79**	–0.26	0.36	0.64	**0.74**	–0.24	0.24	**–0.81**	0.62	0.55	**0.74**	**0.83**	**0.88**	–0.07	–0.55	**0.79**	**0.98**	0.39	–0.12	–0.05	0.69
Flavour overall	0.43	–0.38	0.19	0.40	0.12	–0.24	0.17	–0.64	0.43	0.50	0.43	0.29	0.38	**0.83**	–0.31	0.24	0.21	0.39	–0.07	0.07	0.31
Milky	**–0.86**	0.33	–0.71	**–0.95**	**–0.81**	**0.83**	**–0.79**	0.64	**–0.98**	–0.71	–0.52	–0.26	–0.48	0.00	**0.83**	**–0.83**	–0.67	**–0.79**	–0.40	–0.50	**–0.76**
Buttery	–0.52	0.48	–0.43	–0.60	–0.48	0.71	**–0.90**	0.12	–0.64	–0.55	0.00	0.48	0.19	0.13	**0.74**	–0.50	–0.07	**–0.72**	**–0.83**	**–0.81**	–0.31
Sweetish	–0.57	0.45	–0.48	–0.71	–0.57	**0.76**	**–0.95**	0.26	**–0.74**	–0.62	–0.10	0.36	0.07	0.07	**0.81**	–0.55	–0.14	**–0.80**	**–0.81**	**–0.83**	–0.43
Acidic	**0.88**	–0.43	0.40	**0.79**	0.69	–0.64	0.60	**–0.81**	**0.81**	**0.88**	**0.79**	0.45	0.71	0.26	–0.71	**0.74**	**0.74**	**0.79**	0.29	0.38	0.60
Nutty	0.19	–0.26	0.24	0.29	0.43	–0.10	0.38	–0.62	0.26	0.36	0.14	0.21	0.29	0.18	–0.43	0.17	0.05	0.47	0.24	0.33	0.45
Toasty/Caramel	**0.76**	–0.45	0.62	**0.76**	0.62	–0.43	0.43	**–0.90**	**0.79**	0.60	0.52	0.55	0.62	0.44	–0.67	0.69	0.62	0.57	0.00	0.14	**0.79**
Persistent	0.57	–0.24	0.48	0.62	0.55	–0.52	0.45	**–0.79**	0.64	0.64	0.57	0.43	0.60	0.40	–0.52	0.45	0.43	0.66	0.14	0.29	0.60

Persistent: Time until the perceived flavour has vanished.
VCA: Volatile carboxylic acids.
LA: Lactic acid.
Numbers in bold indicate the significant correlation coefficient ($p \leq 0.05$).

perceived overall odour (orthonasal) and overall flavour. The positive correlation coefficient between the levels of esters, alcohols and acids and the perceived intensity of the attribute 'nutty' suggests that all these aroma families add to the nutty notes characteristic of Swiss-type cheeses. The group of aldehydes showed a significant correlation only with the overall perceived odour. However, since it was mainly Emmentaler Urtyp, which was characterised by a very high amount of aldehydes, no conclusion on the influence of aldehydes on the observed sensory data can be made. A highly significant negative Spearman correlation coefficient was determined for the acids and the milky and buttery notes indicating that the milky and buttery characteristics are reduced with an increase in acid concentration, which itself is positively related to cheese age. There was no significant correlation observed between the group of sulphur compounds and any of the evaluated sensory attributes. A reason might be the fact that correlations were based on detected (low) amounts and not on OAVs which are supposed to be high for sulphur containing compounds. A strong association was found between the NaCl content and the overall flavour intensity supporting the results observed by Karoui et al. (2007). A positive correlation coefficient was also found between sourness intensity and the concentrations of formic and acetic acid. However, no significant correlation was found between any of the sensory attributes and propionic acid content.

11.5 Relationship between consumer data and descriptive panel data

Only a few studies have looked at the relationship between volatile flavour compounds and consumer perception of Swiss-type cheeses.

Liggett et al. (2008) studied the flavour impact of different Swiss-type cheeses on the liking of American consumers. The sample set consisted of 11 American Swiss cheeses differing in various processing variables and one non-American Emmental cheese. Liking scores were collected on the classic 9-point hedonic scale (1: dislike extremely, 9: like extremely). The buttery note (diacetyl) and umami were the most important drivers of liking, whereas the cabbage as well as brothy notes were strongly linked to the dislike of the tested cheese samples (Liggett et al. 2008).

Correlation of hedonic consumer data (n = 100) for appearance, overall liking and liking of the nutty flavour of five Swiss cheeses and the results of the objective sensory data collected by a trained panel of (n = 8) showed a strong positive association between liking and the attributes 'nutty-malty', 'milkfat lactone', 'salty', 'umami' and 'sweet taste' (Castada et al. 2019). Thus, these five attributes were considered as most relevant for US consumer acceptance of Swiss cheeses.

The evaluation of four Emmentalers differing in brining duration, salting procedure during ripening and type of adjunct cultures by a naïve panel (n = 307) of consumers of Swiss-type cheeses showed a significant effect of NaCl on the overall liking (9-point hedonic scale) of the cheeses (Agroscope, Switzerland 2021). The cheeses with a longer brining duration were significantly more liked and were judged as having a more balanced and intense flavour compared to the cheese samples with a shorter brining duration. Cheeses with a shorter brining time were also often described as too soft, whereas the cheese samples with a longer brining period were judged by almost half of the participants as ideal in consistency. The perceived subjective sweetness attributed to Swiss-type cheeses was not affected by either the duration of brining or by the adjunct culture. Furthermore, test results suggested that the addition of FHL an adjunct culture consisting of *L casei* contributed to the aroma, since the two cheeses produced with the addition of adjunct cultures were judged significantly more often as JAR in aroma compared to the two cheeses without FHL.

11.6 Perception of defects of Swiss-type cheese varieties

Defects related to eye formation are often classified according to number, size, shape and distribution. Examples together with descriptors also for flavour and textural defects are given by Cakir and Clark (2009).

Castada et al. (2014) studied the volatile compounds of Swiss cheeses produced in the United States with and without visual defects (splits). The profiles of cheese volatiles revealed differences between and within production plants. Fermentation differences due to variation in milk source, microflora and production parameters were thought to be responsible for the observed differences in volatiles. Defects such as splits and cracks, normally occurring during later stages of cheese maturation, are assumed to be the result of an overproduction of CO_2 formed as by-product along with the production of flavour active volatiles during butyric acid fermentation or the decarboxylation of amino acids to biogenic amines (Castada et al. 2014).

Among the most dreaded defects is the so-called 'late blowing' mainly caused by clostridia which results in deformation of eye shape accompanied by a strong putrid flavour note (Fröhlich-Wyder et al. 2017).

Especially Swiss-type cheeses made from raw cow's milk and ripened for a longer period can exhibit a burning note. This sensation is associated with histamine concentrations higher than 200 mg/kg formed during cheese ripening by the decarboxylation of histidine in the presence of *Lactobacillus parabuchneri* (Wechsler et al. 2021). Moreover, depending on the bacteria endogenous to the milk used for cheese production, the composition of the starter culture and the type of coagulant, a bitter taste can be developed as a result of the formation of short chain peptides especially during the first three months of cheese maturation (Fröhlich-Wyder et al. 2017).

11.7 Conclusions

The two most important properties responsible for the unique sensory profile of Swiss-type cheeses are a characteristic eye pattern and the typical, often described as sweet and nutty, flavour notes.

It is widely accepted that the propionic acid fermentation common to all Swiss-type cheeses strongly contributes to the characteristic flavour of this cheese variety. However, up to date no odour active compounds uniquely attributed to the activity of *P. freudenreichii* and other species of propionic bacteria could be determined in Swiss-type cheeses.

As for any cheese variety, it is supposed that the perceived flavour characteristics of Swiss-type cheeses result from a balanced mix of a vast number of volatiles and their interactions with the composition and structure of the cheese. In addition, the flavour seems to highly depend on the strain composition of both the lactic acid bacteria and PAB used for the production of Swiss-type cheeses.

Sensory evaluation is always applied in the context of defined specific research questions such as impact of technological adaptations or influence of microbial flora. Very often sensory data is combined with chemical flavour compounds in order to get information on the impact of specific aroma active volatiles or non-volatiles on flavour perception or to build models to predict sensory characteristics from instrumental data. In the last few decades, novel types of physico-chemical analyses as well as advanced techniques for data analysis relating sensory and instrumental data have been developed to support a better understanding and prediction of flavour and texture characteristics.

However, the observed association between sensory and instrumental measurements is often weak. Moreover, the majority of studies relating to sensory and instrumental flavour data found high positive or negative correlations for only a few volatiles, illustrating the complex interactions between composition and the type of cheese matrix with human perception.

The two main reasons for the often-poor relationship are supposed to be the complex nature of a flavour as well as the inherent 'linearity' of many statistical models used to relate the two data sets.

In-depth knowledge on the influence of key factors contributing to the sensory character such as choice of bacterial cultures, production parameters or ageing strategies allows for better control and optimisation of the quality of Swiss-type cheeses. Thereby, an accurate and reproducible analysis of the sensory properties desired by producers and by consumers is of great importance to guarantee a high and consistent product quality.

References

Agroscope, Switzerland (2021). Internal report (confidential).

Bachmann, H.P., Bütikofer, U., and Meyer, J. (1999). Prediction of flavour and texture development in Swiss-type cheeses. *LWT – Food Science and Technology* 32 (5): 284–289.

Bachmann, H.P., Bütikofer, U., Fröhlich-Wyder, M.T. et al. (2011). Swiss-type cheeses. In: *Encyclopedia of Dairy Sciences*, 2e, vol. 1 (ed. P.L.H. McSweeney and J.P. McNamara), 712–720. Academic Press.

Bertuzzi, A.S., McSweeney, P.L.H., Rea, M.C., and Kilcawley, K.N. (2018). Detection of volatile compounds of cheese and their contribution to the flavour profile of surface-ripened cheese. *Comprehensive Reviews in Food Science and Food Safety* 17 (2): 371–390.

Beuvier, E., Berthaud, K., Cegarra, S. et al. (1997). Ripening and quality of Swiss-type cheese made from raw, pasteurized or microfiltered milk. *International Dairy Journal* 7 (5): 311–323.

Biede, S.L. and Hammond, E.G. (1979). Swiss cheese flavor: II. Organoleptic analysis. *Journal of Dairy Science* 62 (2): 238–248.

Bisig, W., Fröhlich-Wyder, M.T., Jakob, E., and Wechsler, D. (2010). Comparison between Emmentaler PDO and generic Emmental cheese production in Europe. *Australian Journal of Dairy Technology* 65 (3): 206–213.

Bisig, W., Guggisberg, D., Jakob, E. et al. (2019). The effect of NaCl and metabolic profile of propionibacteria on eye formation in experimental Swiss-type cheese. *International Dairy Journal* 89: 86–95.

Bouton, Y., Buchin, S., Duboz, G. et al. (2009). Effect of mesophilic lactobacilli and enterococci adjunct cultures on the final characteristics of a microfiltered milk Swiss-type cheese. *Food Microbiology* 26 (2): 183–191.

Cakir, E. and Clark, S. (2009). Swiss cheese and related products. In: *The Sensory Evaluation of Dairy Products*, 427–457.

Castada, H.Z., Wick, C., Taylor, K., and Harper, W.J. (2014). Analysis of selected volatile organic compounds in Split and nonsplit Swiss cheese samples using selected-ion flow tube mass spectrometry (SIFT-MS). *Journal of Food Science* 79 (4): C489–C498.

Castada, H.Z., Hanas, K., and Barringer, S.A. (2019). Swiss cheese flavor variability based on correlations of volatile flavor compounds, descriptive sensory attributes, and consumer preference. *Foods* 8 (2): 1–16.

Chamba, J.F. and Perreard, É. (2002). Contribution of propionic acid bacteria to lipolysis of Emmental cheese. *Le Lait* 82 (1): 33–44.

Daly, D.F.M., McSweeney, P.L.H., and Sheehan, J.J. (2010). Split defect and secondary fermentation in Swiss-type cheeses – a review. *Dairy Science and Technology* 90 (1): 3–26.

Delarue, J., Lawlor, J. B., and Rogeaux, M. (2016). *Rapid Sensory Profiling Techniques and Related Methods: Applications in New Product Development and Consumer Research.*

Demarigny, Y., Beuvier, E., Buchin, S. et al. (1997). Influence of raw milk microflora on the characteristics of Swiss-type cheeses: II. Biochemical and sensory characteristics. *Lait* 77 (1): 151–167.

Drake, M.A. and Civille, G.V. (2003). Flavor lexicons. *Comprehensive Reviews in Food Science and Food Safety* 2 (1): 33–40.

Drake, M.A. and Delahunty, C.M. (2017). Sensory character of cheese and its evaluation. In: *Cheese: Chemistry, Physics and Microbiology: Fourth Edition*, vol. 1, 517–545.

Drake, S.L., Carunchia Whetstine, M.E., Drake, M.A. et al. (2007). Sources of umami taste in Cheddar and Swiss cheeses. *Journal of Food Science* 72 (6): S360–S366.

Federal Office for Agriculture (2018). *Functional Specification Document Emmentaler - Pflichtenheft Emmentaler Switzerland Englisch*, 7. BLW. Federal Office for Agriculture. http://www.blw.admin.ch.

Foegeding, E.A. and Drake, M.A. (2007). Invited review: sensory and mechanical properties of cheese texture. *Journal of Dairy Science* 90 (4): 1611–1624.

Fox, P.F., Guinee, T.P., Cogan, T.M., and McSweeney, P.L.H. (2017). *Fundamentals of Cheese Science*, 2e. Springer.

Fröhlich-Wyder, M.T., Bachmann, H.P., and Casey, M.G. (2002). Interaction between propionibacteria and starter/non-starter lactic acid bacteria in Swiss-type cheeses. *Le Lait* 82 (1): 1–15.

Fröhlich-Wyder, M.T., Bisig, W., Guggisberg, D. et al. (2017). Cheeses with propionic acid fermentation. In: *Cheese Chemistry, Physics and Microbiology, Vol. 2, Cheese Technology and Major Cheese Groups*, 4e, vol. 2 (ed. P.L.H. McSweeney, P.F. Fox, P.D. Cotter and D.W. Everett), 889–910. London, United Kingdom: Elsevier, Academic Press.

González, M., Pérez, N., Budelli, E., and Lema, P. (2019). Acoustic monitoring of eye formation in Swiss type cheese. *Proceedings of Meetings on Acoustics* 38 (1).

González, M., Budelli, E., Pérez, N., and Lema, P. (2020). Estimation of eye formation in Swiss-type cheese using geometrical measurements. *Journal of Food Engineering* 275.

Guggisberg, D., Schuetz, P., Winkler, H. et al. (2015). Mechanism and control of the eye formation in cheese. *International Dairy Journal* 47: 118–127.

Harper, W.J., Kocaoglu-Vurma, N.A., Wick, C. et al. (2011). Analysis of volatile sulfur compounds in Swiss cheese using selected ion flow tube mass spectrometry (SIFT-MS). In: *Vol. 1068. ACS Symposium Series*, 153–181.

Hartmann, K.I., Dunkel, A., Hillmann, H. et al. (2015). Identification of physical properties and volatile and non-volatile compounds for discrimination between different Emmental-type cheeses: a preliminary study. *Dairy Science and Technology* 95 (5): 701–717.

Hartmann, K., Eugster-Meier, E., Fröhlich-Wyder, M.-T. et al. (2017). Swiss-type cheeses (propionic acid cheeses). In: *Global Cheesemaking Technology: Cheese Quality and Characteristics*, vol. 1 (ed. P. Papademas and T. Bintsis), 336–348. Wiley.

Imhof, P. (2013). Das kulinarische Erbe der Schweiz, Band 2

Karoui, R., Pillonel, L., Schaller, E. et al. (2007). Prediction of sensory attributes of European Emmental cheese using near-infrared spectroscopy: a feasibility study. *Food Chemistry* 101 (3): 1121–1129.

Kawai, M., Sekine-Hayakawa, Y., Okiyama, A., and Ninomiya, Y. (2012). Gustatory sensation of L-and D-amino acids in humans. *Amino Acids* 43 (6): 2349–2358.

Kocaoglu-Vurma, N.A., Harper, W.J., Drake, M.A., and Courtney, P.D. (2008). Microbiological, chemical, and sensory characteristics of Swiss cheese manufactured with adjunct Lactobacillus strains using a low cooking temperature. *Journal of Dairy Science* 91 (8): 2947–2959.

Lamichhane, P., Kelly, A.L., and Sheehan, J.J. (2018). Symposium review: structure-function relationships in cheese. *Journal of Dairy Science* 101 (3): 2692–2709.

Langler, J.E., Libbey, L.M., and Day, E.A. (1967). Identification and evaluation of selected compounds in Swiss cheese flavor. *Journal of Agricultural and Food Chemistry* 15 (3): 386–391.

Langsrud, T. and Reinbold, G.W. (1973). Flavor development and microbiology of Swiss cheese: a review. III. Ripening and flavor production. *Journal of Milk and Food Technology* 36 (12): 593–609.

Lawlor, J.B., Delahunty, C.M., Wilkinson, M.G., and Sheehan, J. (2002). Relationships between the gross, non-volatile and volatile compositions and the sensory attributes of eight hard-type cheeses. *International Dairy Journal* 12 (6): 493–509.

Lawlor, J.B., Delahunty, C.M., Wilkinson, M.G., and Sheehan, J. (2003). Swiss-type and Swiss-Cheddar hybrid-type cheeses: effects of manufacture on sensory character and relationships between the sensory attributes and volatile compounds and gross compositional constituents. *International Journal of Dairy Technology* 56 (1): 39–51.

Liggett, R.E., Drake, M.A., and Delwiche, J.F. (2008). Impact of flavor attributes on consumer liking of Swiss cheese. *Journal of Dairy Science* 91 (2): 466–476.

Mikš-Krajnik, M., Babuchowski, A., and Bialobrzewski, I. (2013). Impact of physiological state of starter culture on ripening and flavour development of Swiss-Dutch-type cheese. *International Journal of Dairy Technology* 66 (4): 562–569.

Nielsen, R.G., Zannoni, M., Bérodier, F. et al. (1998). Progress in developing an international protocol for sensory profiling of hard cheese. *International Journal of Dairy Technology* 51 (2): 57–64.

di Palma, V.L., Hammond, E.G., and Glatz, B.A. (1987). Effect of Lactobacilli on the properties of Swiss cheese. *Journal of Dairy Science* 70 (4): 733–737.

Pillonel, L., Badertscher, R., Bütikofer, U. et al. (2002). Analytical methods for the determination of the geographic origin of Emmentaler cheese. Main framework of the project; chemical, biochemical, microbiological, colour and sensory analyses. *European Food Research and Technology* 215 (3): 260–267.

Preininger, M., Warmke, R., and Grosch, W. (1996). Identification of the character impact flavour compounds of Swiss cheese by sensory studies of models. *European Food Research and Technology* 202 (1): 30–34.

Qian, M. and Reineccius, G.A. (2003). Quantification of aroma compounds in Parmigiano Reggiano cheese by a dynamic headspace gas chromatography-mass spectrometry technique and calculation of odor activity value. *Journal of Dairy Science* 86 (3): 770–776.

Rehn, U., Vogensen, F.K., Persson, S.E. et al. (2011). Influence of microflora on texture and contents of amino acids, organic acids, and volatiles in semi-hard cheese made with DL-starter and propionibacteria. *Journal of Dairy Science* 94 (3): 1098–1111.

Rétiveau, A., Chambers, D.H., and Esteve, E. (2005). Developing a lexicon for the flavor description of French cheeses. *Food Quality and Preference* 16 (6): 517–527.

Rodriguez, L.M., Ritvanen, T., Joutsjoki, V. et al. (2011). The role of copper in the manufacture of Finnish Emmental cheese. *Journal of Dairy Science* 94 (10): 4831–4842.

Rogers, N.R., Drake, M.A., Daubert, C.R. et al. (2009). The effect of aging on low-fat, reduced-fat, and full-fat Cheddar cheese texture. *Journal of Dairy Science* 92 (10): 4756–4772.

Roth, H.A., Radle, L.J., Gifford, S.R., and Clydesdale, F.M. (1988). Psychophysical relationships between perceived sweetness and color in lemon- and lime-flavored drinks. *Journal of Food Science* 53 (4): 1116–1119.

Saint-Eve, A., Panouillé, M., Capitaine, C. et al. (2015). Dynamic aspects of texture perception during cheese consumption and relationship with bolus properties. *Food Hydrocolloids* 46: 144–152.

Talavera-Bianchi, M. and Chambers, D.H. (2008). Simplified lexicon to describe flavor characteristics of western European cheeses. *Journal of Sensory Studies* 23 (4): 468–484.

Tarrega, A., Yven, C., Sémon, E., and Salles, C. (2008). Aroma release and chewing activity during eating different model cheeses. *International Dairy Journal* 18 (8): 849–857.

Taylor, K., Wick, C., Castada, H. et al. (2013). Discrimination of Swiss cheese from 5 different factories by high impact volatile organic compound profiles determined by odor activity value using selected ion flow tube mass spectrometry and odor threshold. *Journal of Food Science* 78 (10): C1509–C1515.

Thierry, A. and Maillard, M.B. (2002). Production of cheese flavour compounds derived from amino acid catabolism by *Propionibacterium freudenreichii*. *Le Lait* 82 (1): 17–32.

Thierry, A., Richoux, R., and Kerjean, J.R. (2004). Isovaleric acid is mainly produced by *Propionibacterium freudenreichii* in Swiss cheese. *International Dairy Journal* 14 (9): 801–807.

Thierry, A., Collins, Y.F., Abeijón Mukdsi, M.C. et al. (2017). Lipolysis and metabolism of fatty acids in cheese. In: *Cheese: Chemistry, Physics and Microbiology*, 4e, vol. 1 (ed. P. McSweeney, P. Cotter and D. Everett), 423–444. Elsevier Inc.

Thuillier, B., Valentin, D., Marchal, R., and Dacremont, C. (2015). Pivot© profile: a new descriptive method based on free description. *Food Quality and Preference* 42: 66–77.

Turgay, M., Irmler, S., Isolini, D. et al. (2011). Biodiversity, dynamics, and characteristics of Propionibacterium freudenreichii in Swiss Emmentaler PDO cheese. *Dairy Science and Technology* 91 (4): 471–489.

Turgay, M., Bachmann, H.P., Irmler, S. et al. (2021). Propionibacterium spp. and Acidipropionibacterium spp. In: *Encyclopedia of Dairy Sciences* (3) (ed. P. McSweeney and J. McNamara). Elsevier Ltd.

Vangtal, A. and Hammond, E.G. (1986). Correlation of the flavor characteristics of Swiss-type cheeses with chemical Parameters. *Journal of Dairy Science* 69 (12): 2982–2993.

Vara-Ubol, S., Chambers Iv, E., and Chambers, D.H. (2004). Sensory characteristics of chemical compounds potentially associated with beany aroma in foods. *Journal of Sensory Studies* 19 (1): 15–26.

Warmke, R., Belitz, H.D., and Grosch, W. (1996). Evaluation of taste compounds of Swiss cheese (Emmentaler). *European Food Research and Technology* 203 (3): 230–235.

Wechsler, D., Irmler, S., Berthoud, H. et al. (2021). Influence of the inoculum level of *Lactobacillus parabuchneri* in vat milk and of the cheese-making conditions on histamine formation during ripening. *International Dairy Journal* 113: https://doi.org/10.1016/j.idairyj.2020.104883.

Weinrichter, B., Sollberger, H., Ginzinger, W. et al. (2004). Adjunct starter properties affect characteristic features of Swiss-type cheeses. *Nahrung – Food* 48 (1): 73–79.

Wyder, M.T., Bosset, J.O., Casey, M.G. et al. (2001). Influence of two different propionibacterial cultures on the characteristics of Swiss-type cheese with regard to aspartate metabolism. *Milk Science International* 56 (2): 78–81.

Yamaguchi, S. and Ninomiya, K. (1998). What is umami? *Food Reviews International* 14 (2–3): 123–138.

Yee, A.L., Maillard, M.B., Roland, N. et al. (2014). Great interspecies and intraspecies diversity of dairy propionibacteria in the production of cheese aroma compounds. *International Journal of Food Microbiology* 191: 60–68.

12 Sensory Profiles of Very Hard Italian Cheeses and Related Varieties

Antonella Santillo and Marzia Albenzio

12.1 Introduction

Italy has a great range of cheese varieties and among them, some hard cheeses such as Parmigiano Reggiano, Grana Padano and Pecorino Romano are recognised worldwide for dairy excellence (Figure 12.1). According to firmness, cheeses are defined as extra-hard and hard when they have a moisture content, on a fat-free basis, of lower than 51% and ranging from 49% to 56%, respectively (Codex Alimentarius 2018).

Cheese flavour development is a dynamic biochemical process that is impacted firstly: by cheese type, composition and the thermal treatment given to the milk used, secondly: by cheesemaking conditions and thirdly: by microorganisms and enzymes, whether exogenous or indigenous to the cheese matrix. Also, within the cheesemaking conditions of hard and very hard cheeses, the ripening time and conditions greatly influence the final characteristics of the cheese. The main characteristics of the most representative Italian hard cheeses are presented in Table 12.1, drawn up by extrapolating some distinctive elements described in the cheese production specifications. The main milk sources are bovine and ovine breeds, whereas Canestrato di Moliterno cheese is found as the only hard cheese with a European designation made from a mixture of ovine (70–90%) and caprine (10–30%) milk. Furthermore, milk fat content, milk heat treatment, bacterial inoculum, rennet type, ripening time and possible heat treatment of the curd have been detailed.

12.2 Grana-type cheeses

Grana-type cheeses, characterised by a grainy structure, have been produced in the Po Valley of Italy, since the thirteenth century. Nowadays, the area of production is limited by official regulations and dairy farms and cheesemakers are organised into Consortia. Dairy cow feeding is based on the use of fresh pasture, straw and hay, and feed resources grown in the area of production.

Raw cow's milk is used in the manufacture of Grana cheeses and the production protocol provides for partial milk skimming, fermentation with natural whey starter

Sensory Profiling of Dairy Products, First Edition. Edited by John J. Tuohy.
© 2023 John Wiley & Sons Ltd. Published 2023 by John Wiley & Sons Ltd.

(a) Grana Padano

(b) Pecorino Romana

(c) Asiago

Figure 12.1 Examples of cheese varieties: (a) Grana-type, (b) Pecorino-type, and (c) Asiagos. *Source:* (a) Felicia Perretti; (b) Consorzio per la tutela de Pecorino Romano; (c) Marco Cattaneo/ Wikimedia Commons/CC BY-SA 4.0.

Table 12.1 Major characteristics of the most representative Italian hard cheese varieties as reported in the PDO and PGI specifications.

Cheese denomination	Production area	Designation[a]	Milk species[b]	Milk fat content/ Heat treatment	Bacterial inoculum	Rennet type	Heat treatment of curd	Ripening time
Grana Padano	Piedmont regions, some territories of Lombardy, Veneto and Emilia–Romagna, Trento and some municipalities in the province of Bolzano	PDO (12 June 1996, modified on 17 June 2011)	C	Skimmed (ca. 2.1–2.2%)/Raw	Natural whey culture	Calf rennet (powdered preparation)	Cooked at 53–56 °C for 10 min	9 to more than 20 months
Parmigiano Reggiano	Some provinces of Emilia–Romagna region and Lombardy	PDO (12 June 1996, modified on 5 September 2003 and 8 August 2011)	C	Skimmed (ca. 2.4–2.5%)/Raw	Natural starter culture	Calf rennet (powder preparation)	Cooked at 53–56 °C 5–15 min	12 to 20–24 months
Canestrato Pugliese	Apulia	PDO (12 June 1996)	S	Whole/Raw, thermised or pasteurised	Natural whey culture	Calf liquid	Heated at 45 °C for 5–10 min or dipping of the moulds in whey at 80 °C for 30 s, when using raw milk	3–10 months
Fiore Sardo	Sardinia	PDO (01 July 1996)	S	Whole raw	–	Lamb or kid rennet paste	Uncooked	At least 3.5 mo
Canestrato di Moliterno	Basilicata	PGI (21 May 2010)	S, G	Whole raw or thermised	Natural milk or whey cultures or commercial starters	Lamb or kid rennet paste	Uncooked	6–12 mo

(Continued)

Table 12.1 (Continued)

Cheese denomination	Production area	Designation[a]	Milk species[b]	Milk fat content/ Heat treatment	Bacterial inoculum	Rennet type	Heat treatment of curd	Ripening time
Pecorino Romano	Lazio, Sardinia and Tuscany	PDO (12 June 1996, modified on 29 October 2009)	S	Whole raw or thermised	Natural culture, *scotta fermento*, which is produced by acidifying the *scotta*, the deproteinised whey obtained from the manufacture of Ricotta	Lamb rennet paste	Cooked at 45–48 °C for ca. 10 min	Five to eight months
Pecorino Siciliano	Sicily	PDO (12 June 1996)	S	Whole/Raw	Natural milk or Whey cultures	Lamb rennet paste	Cooked at 40 °C for ca. 10 min	At least four months
Pecorino di Filiano	Basilicata	PDO (14 December 2007)	S	Whole/Raw	Natural culture	Lamb or kid rennet paste	Uncooked	At least six months
Asiago d'Allevo	Province of Vicenza, Padova, Treviso and Trento	PDO (12 June 1996, modified on 15 October 2007)	C	Partially skimmed/Raw or thermised	Natural milk or whey cultures or commercial/ selected starters	Liquid or powdered rennet	Cooked at 46–48 °C for ca. 10–15 min	At least four months
Asiago Pressato	Province of Vicenza, Padova, Treviso and Trento	PDO (12 June 1996, modified on 15 October 2007)	C	Whole/Raw or pasteurised	Natural milk or thermophilic commercial selected starters	Liquid or powdered rennet	Cooked at 42–46 °C for ca. 10–15 min	At least 20 days

[a] PDO, protected designation of origin; PGI, protected geographical indication.
[b] Milk species: C = cow; S = sheep; G = goat.
Source: Adapted from Ministry of Agricultural, Food and Forestry Policies (Italy). https: //http://www.politicheagricole.it/flex/cm/pages/ServeBLOB.php/L/IT/IDPagina/3340

culture, and slow and long ripening (Neviani and Gatti 2013; Gatti et al. 2014). In most hard and very hard cheese varieties, proteolysis is classified as strong and slow especially regarding the extensive secondary proteolysis which leads to the liberation of elevated concentrations of small peptides and amino acids which accumulate in the cheese matrix (Gobbetti and Di Cagno 2017). Also, lipids play a role in the definition of the cheese rheology, texture and flavour by acting as a source of fatty acids, which are in turn catabolised to flavour compounds or act as a solvent for flavour compounds (Collins et al. 2004). The natural whey cultures typically used as starters for Grana-type raw milk cheeses ensure rapid and complete lactose fermentation and then, together with raw milk microbiota, contribute to casein proteolysis and to aroma and flavour development through amino acid catabolism (Gatti et al. 2014). The heating temperature of the curd, the temperature during whey fermentation and the increase in acidity lead to the selection of a characteristic microbiota mainly consisting of thermophilic, aciduric and moderately heat-resistant LAB (Rossetti et al. 2009; Gatti et al. 2011). The composition of these natural whey cultures is complex and different from those present in cultures used for Parmesan cheeses produced in the United States, which are comprised of selected strains of *Streptococcus thermophilus*, and *Lactobacillus helveticus* (Govindasamy-Lucey et al. 2004). Brine salting of Italian Grana-type cheeses is performed for different lengths of time (18–25 days), depending on the cheese wheel size and allows an overall NaCl uptake of about 1.5 g/100 g of cheese, following a centripetal diffusion process (Malacarne et al. 2009; Gatti et al. 2014). Altogether, these factors play a fundamental role in the achievement of the typical sensory characteristics of these cheeses and distinctive traits in cheese biochemistry.

12.2.1 *Grana Padano*

According to the consortium, vertical tasting of Grana Padano involves tasting cheese from individual producers at different stages of ripening while horizontal tasting refers to cheeses of the same age produced by different manufacturers. The first is conducted in order to evaluate the correct evolution of the cheese during the maturation process, while the purpose of the second type of tasting assessment is to evaluate the homogeneity of the organoleptic features across the different producers. The sensory characteristics of Grano Padano are summarised in Table 12.2. The visual analysis of Grana Padano protected designation of origin (PDO) focuses on colour, which ranges from white to straw-yellow, the intensity of which increases as the ripening time progresses. The peculiar grainy texture is associated with the presence of calcium crystals in the form of tiny white spots, and tyrosine crystals, round whitish spots with a slightly bitter taste. Both crystalline inclusions are typically absent in cheese ripened up to 16 months or are found under the rind or uniformly in cheese aged over 16 and over 20 months. Recently, D'Incecco et al. (2016) studied the crystalline structures in samples of long-ripened Grana Padano and Parmigiano Reggiano cheese by using light and electron microscopy and confocal micro-Raman. It was hypothesised that spot development originated from an incomplete aggregation of curd granules and a consequent local whey stagnation; during ripening, the growth and lysis of bacterial cells entrapped in these microcavities

Table 12.2 Sensory characteristics of Grana-type cheeses as reported in the PDO specifications.

Cheese	Shape and weight	Rind	Colour	Texture	Taste, odour and flavour
Grana Padano	Cylindrical shape; diameter of 35–45 cm and height of 18–25 cm, weight 24–40 kg	Hard and smooth; natural straw colour. Thickness ca. 4–8 mm	White or straw-coloured	Typical compact texture, with or without many very small eyes; minute grain structure	Aroma and flavour: fragrant and delicate (not spicy), due to very slow ripening, during which proteolysis is the main biochemical event
Parmigiano Reggiano	Cylindrical shape diameter 35–45 cm, height 20–26 cm. Minimum weight 30 kg	Hard and smooth; natural straw colour. Thickness ca. 6 mm	Slightly straw-coloured	Compact texture, with or without many very small eyes; minute grain structure, which has a radial flake fracture when cut	Aroma and flavour: fragrant, delicate, tasty but not spicy, due to very slow ripening during which proteolysis is the main biochemical event

Four types of Grana Padano are classified according to the ripening periods and the production zones:
Grana Padano DOP (minimum ripening period of nine months);
Grana Padano DOP over 16 months (matured for more than 16 months);
Grana Padano DOP Reserve beyond 20 months (matured for at least 20 months.);
Grana Padano DOP Trentingrana (produced in the autonomous province of Trent).

influence both metabolite availability and micro-environmental conditions, which in turn regulate diffusion and crystallisation of solutes locally. Finally, among the visual parameters, the ease of breaking and the overall grainy texture are evaluated, which becomes more pronounced in the cheeses subjected to longer ageing times.

The olfactory evaluation of Grana Padano cheese does not refer to a specific vocabulary of descriptive terms, but to the name of specific aromas perceived as: milk and butter in the cheese ripened for up to 16 months; milk, cream, butter, beef stock in the cheese ripened for over 16 months; milk, cream, butter and beef stock, grass and nuts in the cheese ripened for over 20 months. The taste of Grana Padano PDO refers to taste sensations arising from lactic acid, minerals, simple and complex amino acids, and all water-soluble molecules. The parameters for the taste analysis of the cheese are sweetness and savouriness/pungency, which are inversely proportional to each another, and persistence, which increases from short to medium and long with the advancement of the ripening time. The texture features of these cheese varieties are measured through parameters such as elasticity, adhesiveness, crumbliness and solubility. Elasticity and adhesiveness are classified as strong, medium and low for cheese matured for up to

16 months, over 16 months and over 20 months, respectively, while crumbliness and solubility show an opposite trend in relation to the ripening time of the cheese.

Grana Padano cheese aroma was characterised by high-resolution gas chromatography coupled with mass spectrometry and olfactometry by Moio and Addeo (1998). Neutral and acid extracts of the volatiles concentrate contained 67 and 16 components respectively. The compounds were mainly esters, alcohols, ketones, aldehydes, nitrogen-containing compounds and lactones. Most of the neutral fraction components were recognised as potent odour compounds and the most abundant were ethyl butanoate and ethyl hexanoate. In the acid fraction, compounds with a high-olfactometric index were also identified such as *n*-butanoic and *n*-hexanoic. The backbone of Grana Padano cheese aroma seemed to consist of these acids and 14 potent neutral odour compounds imparting fruity, green, nutty and coconut notes. Ethyl esters play an important role in the formation of the fruity character (Fenster et al. 2003) and were found to be inversely proportional to the length of ripening, whereas the concentration of volatile agents with spicy, nutty and earthy notes tend to increase during maturation (Moio and Addeo 1998; Boscaini et al. 2003).

As reported previously, the sensory and appearance characteristics of Grana Padano cheese are rather generic and thus inadequate for assessing the conformity of cheeses marketed under this designation, particularly in the case of grated cheese. A survey was conducted on samples of grated Grana Padano marketed in or outside Italy and grated 'Grana-type' hard cheeses (Cattaneo et al. 2008). The study reported that (i) measurement of alkaline phosphatase activity, as a marker for milk heat treatment, (ii) identification of specific peptides in the rind linked to the very slow progress of proteolysis in the rind during Grana Padano cheese ripening and (iii) free amino acid composition were parameters which could distinguish grated Grana Padano from similar, non-PDO, grated hard cheeses. In particular, cheese authenticity was linked to the presence of threshold levels of glutamic acid ($\geq$16.0% expressed as percentage of total free amino acids), lysine ($\geq$10.8%), serine ($\geq$3.4%) and glutamine ($\leq$2.3%) thereby distinguishing Grana Padano from similar 'Grana-type' cheeses.

12.2.2 *Trentingrana*

Trentingrana is produced in small dairy cooperatives linked in a Consortium of Dairy Factories in Trento province within the area of production (province of Trento, eastern Italian Alps). The characteristics of the cheese are guaranteed by restrictions on cow feeding (silages are forbidden), traditional manufacturing protocols (lysozyme is not allowed), by the use of partially skimmed raw cow's milk and of natural starter culture obtained from the spontaneous fermentation of part of the previous day's whey (Bittante et al. 2011; Franciosi et al. 2011). The quality assessment of the cheese provides for the evaluation of the cheese ripened up to 9 and up to 18 months by official consortium inspectors according to external aspects (colour, integrity, presence of mould, swelling) and the sound obtained by beating the wheel with a special hammer to recognise inner holes or imperfections (Bittante et al. 2011). Based on the results of this evaluation, wheels are classified as 'first quality', 'second quality' or 'discarded'. Then, a second type of quality assessment is carried out, namely the evaluation of some sensory attributes, which is performed by a panel of experts on first-quality wheels sampled from each dairy.

A list of sensory attributes for Trentingrana are used for quality assessment that includes visual, odour and taste evaluations. A weighting of attributes is used in the calculation of a quality index. The heterogeneity of cooperative dairy factories and the seasonality of milk production were found to be important sources of variation of the visual aspect and flavour of Trentingrana (Bittante et al. 2011). The authors concluded that the major limitation of the evaluation system adopted by the consortium was the lack of a precise description of sensory attributes according to sensory profiling methods and they suggested the development of an objective and rigorous system based on a trained panel to monitor cheese production in cooperative dairies.

12.2.3 *Parmigiano Reggiano*

Parmigiano Reggiano is the oldest traditional cheese in Europe, and it has been produced since the thirteenth century in the Benedictine and Cistercian abbeys; it underwent a complex evolution bound up to the social, economic, cultural and technological environment and, consequently, its sensory characteristics changed over time (Zannoni 2010). The ancient Parmigiano Reggiano had rather variable characteristics due to the cheesemaking process being passed down by imitation and the lack of controls and standardisation in the fermentation process. In the traditional assessment, visual, texture and odour examinations were the most important while taste and aroma appeared to be less pivotal. Texture assessment was performed by tapping the cheese wheel with a special hammer and by using a special 'spike', odour was described only by the terms 'aromatic' and 'sharp'. Important sensory changes happened in the twentieth century when the production protocol was improved thanks to technical progress, but the ripened cheese was still rather heterogeneous and classified in three different quality classes (Fascetti 1935). Zannoni (2010) presented an interesting study on sensory information gathered about the cheese starting from fifteenth century to current days. Considering visual, olfactory, taste and texture characteristics, the main change related to the yellow colour, which became much paler, turning from yellow to straw-yellow, in the second half of the last century because saffron was no longer added. The meltable texture, visually characterised by many small eyes with uniform distribution and stringy liquid, changed to firm, friable, grainy with crystals and no holes. The olfactive, taste and aroma information became completer and more complex, and today sensory analysis is based on a quantitative descriptive scorecard for Parmigiano Reggiano developed by an experienced panel and adopted for the evaluation of sensory quality by the control body of the designation of origin (Zannoni et al. 2006). The aim of this sensory quality score is to ensure that the typical sensory characteristics are present in products that receive the PDO marks. The sensory analysis involves perception through each of the five senses. The external appearance of the product is evaluated considering the colour of the cheese, which ranges through different intensities of straw-yellow and can be more or less uniform. The appearance of both the exterior rind and the inside of the cheese is evaluated in relation to the temperature, humidity, roughness, elasticity, fattiness, the presence and dimension of the eyes, which should be less than 2 mm in diameter. Tasting of the cheese involves detecting the presence of trigeminal sensations, which describe the spicy or astringent flavour. The typical aromas are referred to as 'milky, floral, toasted', while other olfactive notes include shades of pepper, nutmeg and spices. Other descriptors present in the scorecard for the sensory analysis of Parmigiano – Reggiano for smell and aroma are butter, yoghurt, rind, boiled vegetables, nuts, citrus fruits, meat broth, cowshed and nutmeg. The ripening time is strictly linked to the intensity of the aromas. As in the past, an important and peculiar evaluation relates to the sound of the cheese wheel, the traditional assessment is done by tapping the wheel with a special hammer that reveals the degree of crunchiness or crumbliness of the cheese. Finally, the tactile analysis of Parmigiano Reggiano indicates its level of maturity and concerns elasticity, hardness, granularity, crumbliness and solubility.

12.2.4 *Reggianito*

One of the most studied Grana-type cheeses related to the Parmigiano Reggiano is named Reggianito, an Argentinean cheese developed by Italian immigrants in late nineteenth and early twentieth centuries (Wolf et al. 2010). The manufacturing process for Reggianito cheese is provided in the 'Protocolo de Calidad para Queso Reggianito' (Resolución SAGPyA No. 16/2008) and specifies for this cheese a moisture content of 27–35% w/w, a cylindrical shape of 5–10 kg, a ripening period of at least six months, and a compact, firm and brittle texture. Reggianito Argentino cheese is traditionally manufactured with whey starter cultures that provide typical and intense flavour but can cause poor quality standardisation (Hynes et al. 2003). Wolf et al. (2010) analysed the chemical composition and the volatile profile of commercial Reggianito cheeses from different producers and found that the main differences between the Argentinean cheese and Italian Grana-type cheeses related to the level of NaCl and of degree of lipolysis and proteolysis, which were found to be higher and lower, respectively in Argentinean and Italian Grana-type cheeses. Reggianito Argentino cheese manufactured using *L. helveticus SF209*, cultured in sterile whey as starter culture resulted in a significant improvement in the secondary proteolysis over six months of ripening (Hynes et al. 2003) without impairment of the sensory attributes evaluated by descriptive quantitative analysis (Perotti et al. 2008). Over 50 compounds were identified in commercial cheeses, the majority belonging to the ketone, alcohol, acid, ester and aldehyde groups, which have been previously described in Italian Grana-type cheeses (Wolf et al. 2010).

12.3 Pecorino-type cheeses

Ovine milk is widely used in Italy for Pecorino cheesemaking, especially in the Southern and Central regions, deriving from the ancient pastoral culture. The traditional agro-pastoral systems are still important activities in terms of employment and income, but also for ecosystem functioning and landscape management (Farinella et al. 2017). Recently, transhumance was inserted into the UNESCO list of the intangible cultural heritage of humanity, in recognition of traditional practices in parts of Italy, Austria and Greece.

The traditional farming practice of seasonal migration of livestock along herbal tracks connecting flatlands to uplands is practiced to provide better climate and feeding conditions for the herds throughout the year. Traditional Pecorino cheeses stem from the agro-pastoral culture and have evolved into several well-recognised and appreciated cheeses.

The sensory characteristics of the most representative Italian Pecorino-type cheeses are summarised in Table 12.3.

The organoleptic properties of Pecorino Romano, Pecorino Sardo, Fiore Sardo, Canestrato Pugliese and Canestrato di Moliterno cheeses have been characterised by trained assessor panels using quantitative descriptive sensory analysis (Di Cagno et al. 2003; Piombino et al. 2008; Trani et al. 2016), and the complex pattern of attributes describing odour, flavour, after-flavour and texture of these cheeses is summarised in Table 12.4.

Table 12.3 Sensory characteristics of the most representative Italian Pecorino-type cheese as reported in PDO and PGI specifications.

Cheese	Shape and weight	Rind	Colour[a]	Texture	Taste, odour and flavour
Canestrato Pugliese	Cylindrical shape, 10–14 cm high and 25–34 cm in diameter; weight 7–14 kg	Brown to pale yellow; wrinkled due the basket moulds manually treated with extra virgin olive oil also mixed with vinegar	Straw-yellow	Compact friable structure with small eyes	Flavour: pronounced and tends to be moderately piquant
Fiore Sardo	Cylindrical or wheel shape with curved sides, 12–18 cm high, 18–20 cm in diameter; weight 1.5–4.0 kg	From deep gold to dark brown	White and straw-yellow	Compact, hard and friable structure. The cheese body has no holes	Flavour and taste: pronounced, aromatic, moderately spicy, with a sour smell
Canestrato di Moliterno	Cylindrical shape, 10–15 cm high, 15–20 cm in diameter; weight 2–5.5 kg	Pale yellow colour, tending to brown or black, if the cheese is treated with oil or a mixture of water and ash	White to intense yellow	Hard, sometimes with nonuniformly distributed small holes	Flavour and taste: sweet taste, while spicy notes characterise the long-ripened types

Table 12.3 (Continued)

Cheese	Shape and weight	Rind	Colour[a]	Texture	Taste, odour and flavour
Pecorino Romano	Cylindrical shape, 25–40 cm high and 25–35 cm in diameter; weight 20–35 kg	Ivory white or light straw-coloured and may be coloured black when a protective film is used or if traditionally the cheese is treated with grape seed oil	White or slightly straw-yellow	Compact, hard and friable structure and may contain little holes	Flavour and taste: distinctive taste with an aromatic slightly spicy flavour. Long-ripened Pecorino Romano cheeses are more or less spicy with an intense flavour
Pecorino Siciliano	Cylindrical shape, 10–18 cm high; 35 cm in diameter; weight 4–12 kg	Yellowish-white and is much wrinkled due the basket moulds used	White or pale yellow	Compact, with little holes	Flavour and taste: ripened cheese develops a moderate piquant flavour, which is distinctive and the aroma is intense. Pepato (peppery) is a variant of Pecorino Siciliano cheese, which differs by the addition of black pepper to the curd during moulding
Pecorino di Filiano	Cylindrical shape 8–18 cm high, 15–30 cm in diameter; weight 2,5–5 kg	Golden yellow to dark brown when manually treated with extra virgin olive oil (Basilicata) and vinegar; wrinkled due the basket moulds used	White to straw-yellow	Compact, with nonuniformly distributed very small holes	Flavour and taste: ripened cheese develops a piquant taste from moderate to intense along with the advancement of ripening

[a] Colour refers to the mass of the cheese inside the rind.

Table 12.4 Descriptive sensory analysis attributes developed to characterise Pecorino-type cheeses.

Attributes			
Odour	Flavour	After-flavour	Texture
Pungent	Pungent	Cold cooked fat	Solubility
Sweety	Buttery	Sensation fast	Hardness
Pineapple	Sweet	Sensation slow	Adhesiveness
Fruity	Salty		Friability
Caramel	Mushroom		Greasiness
Dairy sweet	Caramel		Crystals
Creamy	Rancid		
Mouldy	Soapy/chemical		
Rancid	Processed		
Mushroom	Sour		
Silage	Bitter		
Cooked potatoes	Astringent		
Nuts	Creamy		
Animal fat	Fruity		
Butter	Mouldy		
Rennet	Stable		
Stable	Ewe's milk		
Ewe's milk	Cheese rind		
Cheese rind	Umami		
Smoky	Smoky		
	Fermented		

12.3.1 *Canestrato Pugliese*

For Canestrato Pugliese cheese, a specific consensual vocabulary was developed by a selected panel using quantitative descriptive analysis (Piombino et al. 2008). Measuring sessions allowed the panel to judge Canestrato Pugliese made from pasteurised and raw milk according to a set of attributes including odour, flavour and taste. Attributes for odour were: butter, ewes' milk, stable, mushroom, cream and rennet. Attributes for flavour were: butter, ewes' milk, stable, mushroom and cream and finally for taste the attributes were salty and pungent. Milk pasteurisation was a fundamental source of variation. Canestrato Pugliese made from raw milk was characterised by higher notes of ewes' milk, stable, and mushroom odour and aroma, and higher pungent and salty taste intensities. Interestingly, in Cheddar cheese pasteurisation has been reported to be probably the single factor exerting the greatest influence during cheese ripening (McSweeney et al. 1993).

The application of GC coupled with mass spectrometry and olfactometry permitted quantification and identification of the volatile compounds in the aroma extracts of Canestrato Pugliese cheese, with the most abundant class of compounds represented by FFA liberated upon hydrolysis by the action of both endogenous and microbial lipases. The FFA fraction was present at a three times higher level in Canestrato Pugliese made from raw milk than in cheese made from pasteurised milk and in particular the contribution of short-chain Fas was notable. Other compound classes such as alcohols, lactones, ketones, esters, volatile phenols, pyrazines, sulphur compounds, aromatic compounds and aldehydes have been detected in Canestrato Pugliese cheese (Piombino et al. 2008).

12.3.2 *Fiore Sardo and Pecorino Romana*

Pirisi et al. (2007) studied proteolysis and lipolysis in PDO Fiore Sardo cheese made using traditional rennet paste (from lambs fed only milk and slaughtered after suckling) and industrial lamb rennet pastes (from lambs weaned and slaughtered on an empty stomach). The proteolytic enzymes contained in the traditional rennet preparation, especially chymosin, contributed to a higher rate of proteolytic activity. The pregastric and gastric lipases were responsible for greater amounts of free fatty acids, in particular butyric acid. The use of traditional lamb rennet usually induces a higher degree of primary proteolysis, whereas bovine rennet produces a faster secondary proteolysis resulting in high concentrations of small peptides and amino acids (Santoro and Faccia 1998), due to the presence of both chymosin and pepsin. The balance of different lipases in rennet paste influences the lipolytic pattern in cheese, in fact, the high-flavour intensity comes from short- and medium-chain FFA, in particular butyric acid which contributes to the cheesy lipolysed aroma (Pinho et al. 2003). The major portion of the short-chain fatty acids are attached at the Sn-3 position of glycerol, which is the preferential site of hydrolysis of the pregastric esterase in the rennet (Woo and Lindsay 1984).

The sensory characteristics of Pecorino Romano, Fiore Sardo and Canestrato Pugliese cheeses were compared in the study of Di Cagno et al. (2003), and correlations between sensory evaluation and volatile and chemical analysis were also found. Canestrato Pugliese is described by the odour attribute pungent, the flavour terms pungent, soapy/chemical, bitter and acidic, and the after-flavour terms sensation fast and sensation slow which are described as a tingling sensation experienced while the cheese is still in the mouth or after the cheese had been swallowed, respectively. Collectively, those attributes are ascribed to the highest concentration of carboxylic acids, especially of butanoic acids found in cheese. Canestrato Pugliese samples were also characterised by a high-fruity and mushroom flavour coming from esters, that were the main components of steam distillates prepared from Canestrato Pugliese cheese. Fiore Sardo was distinguished by its caramel and dairy sweet odour, a sweet, creamy and buttery flavour and a cold cooked meat after-flavour. It was differentiated from the other two cheeses by the high concentrations of ketones and the concentrations and proportions of esters. Pecorino Romano was characterised by a very high intensity of salty flavour corresponding to a concentration of $8.7 \pm 0.1\%$, w/w NaCl, whereas Fiore Sardo and Canestrato Pugliese display a medium level of salt flavour intensity. Overall, in the production of Pecorino-type cheeses a critical consideration is the salt concentration, which depends on the production technology. Excess salt could reduce the perception of the cheese flavour complexity. Moreover, because of its adverse health effects, there is a demand to reduce the salt levels in the cheese (Rulikowska et al. 2013). However, an ideal production procedure should define the right content of salt because its concentration influences proteolytic development (Pripp et al. 2006) and indirectly affects the aroma, taste and texture properties of cheese products (Sousa et al. 2001). Recently, the Consortium of Pecorino Romano approved a new category of the cheese named 'extra quality' Pecorino Romano which is low in salt (max 3.5%, w/w NaCl) (https://www.pecorinoromano.com/application/files/7315/7851/9906/Comunicato_Stampa_-_Modifiche_al_Disciplinare_080120.pdf) and modifications to the PDO have to be submitted for the opinion of the Italian Ministry of Agriculture and the European Union.

12.3.3 *Canestrato di Moliterno*

Canestrato di Moliterno was recently acknowledged by the European Union as a protected geographical indication product; it is manufactured under artisanal conditions from December to May. The cheese is classed as 'Primitivo', ripened up to six months; 'Stagionato', ripened 6–12 months; and 'Extra', ripened more than one year (Pirisi et al. 2011). Trani et al. (2016) studied the chemical and biochemical traits, the volatile fraction and the sensory characteristics of Canestrato di Moliterno from four different certified producers; all cheeses were made in the warm season and analysed after 30 weeks of ripening. These four cheeses had broadly similar sensory characteristics, apart from salty, crystals, mouldy and sour perceptions.

The perception of crystals in the mouth is given by the presence of small hard grains in the cheese matrix, normally observed in ripened hard cheeses, such as Canestrato di Moliterno and Parmigiano. It is caused by crystallisation of poorly soluble compounds such as calcium lactate, tyrosine, cysteine and calcium phosphate (Tansman et al. 2015). In Parmigiano Reggiano, crystals are found after at least 12 months of ripening (Noël et al. 1996), whereas in Canestrato di Moliterno, they can be detected at 30 weeks. This is probably related to the weight of cheese in the mould (about 3 kg for Canestrato di Moliterno vs. about 40 kg for Parmigiano Reggiano).

Among volatile components identified in Canestrato di Moliterno cheese were a group of terpenoids containing D-limonene and p-menth-2-ene as major compounds, which are responsible for herbaceous and flower odours, and have been proposed as molecular markers of milk and cheese from grazing animals.

The enzymatic content of rennet used in cheesemaking plays a role on the biochemical events in cheese maturing. Rennet paste, in particular, is very heterogeneous and includes proteolytic, lipolytic and milk-clotting enzymes (Santillo et al. 2007) that can ultimately impact on the development of the sensory features of cheese. Indeed, the typical characteristics of Pecorino-type cheese are likely to develop from the use of traditional lamb rennet paste containing a balanced amount of milk-clotting and milk-fat-hydrolysing enzymes.

12.3.4 *Idiazábal, Manchego, Roncal and Castellano*

Addis et al. (2008) reported that many other traditional Pecorino-type ewe milk cheeses i.e. PDO Idiazábal, PDO Roncal are produced in certain Southern European countries using lamb rennet paste. In Idiazábal cheese, Vicente et al. (2001) found that the rennet type (artisanal lamb rennet vs commercial calf rennet) influenced the free amino acid content, being higher when the cheeses were made with commercial calf rennet.

Production protocols for most of the hard ovine cheese varieties produced in the Mediterranean area do not contain sensory profiles of the cheese or specify sensory standards or methodology to measure the evolution of sensory characteristics. The definition of a descriptive lexicon for the sensory profile of a cheese is important for the promotion and commercialisation of the product and to make it distinguishable and attractive on the market. This protocol should take into account the characteristics of the milk and the process variables, on one hand, and the methods of sample preparation and tasting, on the other hand. Bárcenas et al. (2001) reported scores for the odour, flavour and texture attributes of Idiazábal PDO cheeses, Manchego PDO cheese, Roncal PDO cheese and Castellano cheese evaluated by trained assessors and identified two separate groups of attributes that were negatively correlated according to principal component analysis. One group consisted of the attributes: odour intensity, sharp odour, brine odour, rennet flavour and butyric acid flavour and the other group consisted of milky odour, toasty odour, buttery odour and flavour, nutty flavour and sweet flavour. The authors concluded that the main sensory differences were basically attributable to different manufacturing technologies as well as different ripening periods. Lexicons that have been developed to describe cheese flavour attributes are complex and specific; Talavera-Bianchi and Chambers (2008) highlighted the need to share methodology and terminology for the sensory description of European cheeses and proposed a simplified lexicon for descriptive sensory analysis. The authors identified 25 basic attributes and 19 additional attributes to be used occasionally to describe targeted specific flavour characteristics.

12.4 Asiago and Montasio cheeses

Asiago and Montasio PDO cheeses, produced in the North of Italy from bovine milk, hold an important commercial position among the very hard Italian cheeses. The principal sensory characteristics of the cheese are reported in Table 12.5.

Table 12.5 The principal sensory characteristics of the Asiago and Montasio cheese as reported in PDO and PGI specifications.

Cheese	Shape and weight	Rind	Colour[a]	Texture	Taste, odour and flavour
Asiago d'Allevo	Cylindrical shape, 9–12 cm high, 30–35 cm in diameter; weight 8–12 kg	Smooth and regular, not edible	Slightly straw-yellow to amber	Compact structure with small and medium-sized holes, and becomes granular in cheese ripened for over a year	The flavour is sweet for young cheeses, but it can become pleasantly spicy after long ripening
Montasio	Cylindrical shape, 8 cm or less high, 30–35 cm in diameter; weight 6–8 kg	Brown, smooth and elastic	White or slightly straw-yellow to deeper straw-yellow and eventually amber	Granular texture with very small holes	Flavour is sweet for young cheeses and may become pleasantly spicy and moderately piquant after long ripening

[a] Colour refers to the mass of the cheese inside the rind.

12.4.1 *Asiago*

Asiago has been classified as a rennet coagulated, internal bacterially ripened, extra-hard cheese according to the classification scheme proposed by Fox et al. (2017). There are variants of the cheese named Asiago d'Allevo, which is the hard variant, Asiago Pressato (pressed Asiago), which is a semi-hard, washed-curd cheese, and Asiago d'Allevo Vecchio or Stravecchio (aged or extra aged Asiago d'allevo) with ripening times over 10 months and 15 months, respectively. The presence of small eyes is characteristic of Asiago d'Allevo Veccho and Stravecchio. Marangon et al. (2006) characterised the sensory and organoleptic profile of Asiago d'Allevo Vecchio cheese produced in several different dairy plants and in alpine small-scale on-farm plants named 'malga' after 12 months of ripening. The cheeses were analysed according to ISO 3299:2003 and high variability in the sensory profile was found among cheeses made on different farms. The most distinguishing parameters were flavour attributes (in particular acid, salty, bitter and spicy) and rheological parameters (hardness, humidity, adhesiveness). The latter parameters are closely related to the intensity of the ripening process, in terms of both proteolysis and lipolysis, and the cheese structure. The rheological parameters were determined by the application of a fixed stress or strain and the behaviour of cheese subjected to fixed stress can be described by hardness, firmness, springiness, adhesiveness, cohesiveness, gumminess and chewiness. In Asiago d'Allevo, it has been reported that the major variation in the textural properties of the cheese produced in traditional dairies was due to the lack of standardised processing and ripening conditions. Furthermore, the ripening process may affect the colour traits of cheese, as demonstrated by Marchesini et al. (2009) for Asiago d'Allevo. Cheeses produced on a single farm located in the Altopiano dei Sette Comuni (above 1000 m.a.s.l.) were sampled after 6, 12, 18 and 36 months of ripening and samples were analysed for quality traits, in order to evaluate the effect of ripening on colour and gross composition. Cheese colour is influenced by many factors such as the content of pigments, like β-carotene, in the feed and the botanical composition of pasture grazed by cattle (Marchesini et al. 2009). Cheese colour perception differs from person to person and depends especially on lighting. Combining visual and instrumental methods for colour assessment is useful to standardise personal variation and evaluation conditions. An instrumental system for colour measurement revealed the impact of ripening on Asiago d'Allevo colour; with the lightness parameter (L^*) decreasing linearly with ageing (up to 36 months), yellowness (b^*) decreasing during the first year and after 18 mo. and finally redness (a^*) significantly reducing only in the first 12 months of ripening (Marchesini et al. 2009). A corresponding decrease in lightness during cheese ripening has also been described in Emmental cheese (Pillonel et al. 2002).

The production protocol for PDO Asiago cheese provides for pasture feeding with limited concentrate supplementation during the grazing season; during the indoor season, cattle are fed stored forages and higher amounts of concentrates. The impact of the feeding regimen on the chemical characteristics and sensory features of the cheese is notable. Fatty acids and terpenoids can be considered a chemical fingerprint for the characterisation of the dairy cow feeding regimens and, therefore, are suitable biomarkers to be used as tracers in the classification of Asiago d'Allevo PDO cheeses according to the season of production (Renna et al. 2011). In particular, fatty acid and terpenoid composition are useful in distinguishing summer from autumn/winter produced cheeses, while terpenoids, which are secondary plant metabolites, are also biomarkers for the traceability of cheeses originating from PDO. In fact, total sesquiterpenoids in the cheese are about 3- and 17-fold higher in the early and late grazing seasons compared to the indoor season, respectively, and some compounds (α-thujene, δ-3-carene, dihydrocarveol isomer 1 and βionone) were detected only in early and/or late summer cheeses.

12.4.2 *Montasio*

Montasio cheese derives its name from a mountain plateau, but it is produced largely in the lowlands. Recently, the production of PDO Montasio was combined with two additional labels: 'Prodotto della Montagna' (product of mountain) and 'Solo di Pezzata Rossa Italiana' (produced exclusively with milk from Italian Simmental cows) with the specific aim to valorise the production of cheese produced in the mountain areas and from the Italian Simmental breed respectively.

Innocente et al. (2013) isolated and identified the volatile compounds of Montasio cheese at different ripening stages (60, 90, 170, 300 and 365 days), by HS-SPME coupled with GC-MS GC. Alcohols, FFA, ketones and esters were reported to be the cheese volatiles, which changed most during ripening but alcohols and FFA were the most significant contributors to the volatile profile of the Montasio cheese. Both alcohols and FFA were observed to increase up to 170 days and remain stable at longer ripening times, while ketones increased up to 300 days and esters progressively increased until the end of the ripening period. Specifically, in regard to the alcohol class, ethanol, 2,3-butanediol and 3-methyl-1-butanol were detected in the headspace of the Montasio samples. The volatile profile of the Montasio cheese samples was characterised by short-chain and medium-chain, even-numbered FFA from C2 to C10: acetic acid showed an increasing trend until the later ripening stages, butanoic and hexanoic acids increased up to 170 days, and remained stable thereafter (Innocente et al. 2013). Ethyl hexanoate was the most significant ester compound determined in the headspace of Montasio cheese samples over the ripening period (Innocente et al. 2013). Stefanon and Procida (2004) reported an even higher number of compounds i.e. 12 aliphatic aldehydes (C2–C9) and nine ketones (C3–C9) in Montasio cheese.

Bovolenta et al. (2014) analysed the volatile compounds, physicochemical characteristics and sensory properties of 60-day-old Montasio cheese produced only using milk from dual-purpose Italian Simmental cows grazing on alpine pastures. Several studies have reported the effect of the floral diversity of grazed herbage on the sensory characteristics of raw milk cheese, particularly the direct or indirect effect of volatile compounds present in the different grass species (Barham 2003). The floral composition of pastures is, in turn, strongly related to the environmental conditions and is part of the 'terroir', a basic concept of the PDO labels (Bovolenta et al. 2014). A recent study on the sensory profile of Montasio produced with milk of Italian Simmental (Bovolenta et al. 2014) used a 10-member panel trained according to specific procedures for the sensory evaluation of hard and semi-hard cheese by quantitative descriptive analysis, to develop a specific pattern of attributes related to visual appearance, odour, texture, taste and flavour. Sensory profiling was performed in a laboratory built according to the UNI EN ISO (UNI 2010) standard. In general, the panellists described the experimental cheeses as having an intense colour; small and evenly distributed eyes; an intense odour and flavour of sour milk, milk and cow; and a tender and creamy texture. However, the same panellists were unable to detect differences linked to experimental factors among two-month-old cheeses (Bovolenta et al. 2014).

A total of 61 volatiles were identified including alcohols (11), aldehydes (6), ketones (10), lactones (2), esters (5), hydrocarbons (3), carboxylic acids (7), phenolic compounds (4), monoterpenes (7), sesquiterpenes (1), sulphur compounds (4) and amines (1). The largest chemical families were represented by carboxylic acids and esters in terms of relative quantity and by alcohols and ketones in terms of the number of detected volatile compounds (Bovolenta et al. 2014). Among the carboxylic acids, acetic, butyric, hexanoic and octanoic acids were the most abundant volatiles detected in Montasio cheese. Acetic acid has a typical vinegary odour, which is sour and sharp, whereas butyric acid has a rancid cheese-like odour and hexanoic acid has a goat-like odour (Curioni and Bosset 2002). Carboxylic acids are also precursors of other aroma compounds, such as methyl ketones, alcohols, lactones, aldehydes and esters (Collins et al. 2003). Ketones have a low-perception threshold, and diacetyl and acetoin in particular confer buttery notes to cheeses (Molimard and Spinnler 1996; Moio and Addeo 1998). Esters were the second most abundant chemical group detected and these volatile compounds, derived from enzymatic esterification of free FA (Carpino et al. 2004), contribute to the sweet, floral and fruity cheese flavour notes, although the perception threshold of these compounds is generally low (Molimard and Spinnler 1996). Alcohols can provide alcoholic and floral notes and as confirmed by Innocente et al. (2013), alcohols were the most significant contributors to the volatile profile (Bovolenta et al. 2014).

12.5 Conclusions

Very hard Italian cheeses represent an important part of Italian dairy production, having a prominent share of the domestic and export cheese markets and in the field of scientific research. The main families of very hard cheeses derived from bovine, ovine and caprine milk have been discussed, gathering scientific knowledge on their sensory profile with a focus on descriptive analysis and on the molecular profiles. In most cases, the production of such cheeses is linked to ancient pastoral traditions which later evolved into processing protocols associated with the European quality marks PDO and PGI.

There is a great diversity in the sensory evaluation of very hard cheeses. For Grana-type cheeses and Asiago and Montasio cheese, both the evaluation procedures and sensory terminology have been set up as well as a quality grading system for cheese commercialisation. Further effort is needed for very hard cheeses made from the milk of minor species. This would enable processors to optimise product flavour, texture and organoleptic properties and to guarantee better monitoring of product quality for market exploitation.

References

Addis, M., Piredda, G., and Pirisi, A. (2008). The use of lamb rennet paste in traditional sheep milk cheese production. *Small Ruminant Research* 79: 2–10.

Bárcenas, P., Pérez Elortondo, F.J., Salmerón, J., and Albisu, M. (2001). Sensory profile of Ewe's milk cheeses. *Food Science and Technology International* 7 (4): 347–353.

Barham, E. (2003). Translating terroir: the global challenge of France AOC labeling. *Journal of Rural Studies* 19 (1): 127–138.

Bittante, G., Cecchinato, A., Cologna, N. et al. (2011). Factors affecting the incidence of first-quality wheels of Trentingrana cheese. *Journal of Dairy Science* 94: 3700–3707.

Boscaini, E., Van Ruth, S., Biasioli, F.G., and F., Mark, T.D. (2003). Gas chromatography–Olfactometry (GC-O) and proton transfer reaction–mass spectrometry (PTR-MS) analysis of the flavor profile of Grana Padano, Parmigiano Reggiano, and Grana Trentino cheeses. *Journal of Agricultural and Food Chemistry* 51: 1782–1790.

Bovolenta, S., Romanzin, A., Corazzin, M. et al. (2014). Volatile compounds and sensory properties of Montasio cheese made from the milk of Simmental cows grazing on alpine pastures. *Journal of Dairy Science* 97: 7373–7385.

Carpino, S., Mallia, S., La Terra, C. et al. (2004). Composition and aroma compounds of Ragusano cheese: native pasture and total mixed rations. *Journal of Dairy Science* 87: 816–830.

Cattaneo, S., Hogenboom, J.A., Masotti, F. et al. (2008). Grated grana Padano cheese: new hints on how to control quality and recognize imitations. *Dairy Science and Technology* 88: 595–605.

Codex Alimentarius (2018). WHO. http://www.fao.org/fao-who-codexalimentarius (accessed 8 May 2020).

Collins, Y.F., McSweeney, P.L.H., and Wilkinson, M.G. (2003). Lipolysis and free fatty acid catabolism in cheese: a review of current knowledge. *International Dairy Journal* 13: 841–866.

Collins, Y.F., Mc Sweeney, P.L.H., and Wilkinson, M.G. (2004). Lipolysis and catabolism of fatty acids in cheese. In: *Cheese: Chemistry, Physics and Microbiology*, volume 1 (ed. P.F. Fox, P.L.H. McSweeney, T.M. Cogan and T.P. Guinee). Elsevier Academic Press.

Curioni, P.M.G. and Bosset, J.O. (2002). Key odorants in various cheese types as determined by gas chromatography-olfactometry. *International Dairy Journal* 12: 959–984.

Di Cagno, R., Banks, J., Sheehan, L. et al. (2003). Comparison of the microbiological, compositional, biochemical, volatile profile and sensory characteristics of three Italian PDO ewes' milk cheeses. *International Dairy Journal* 13: 961–972.

D'Incecco, P., Limbo, S., Faoro, F. et al. (2016). New insight on crystal and spot development in hard and extra-hard cheeses: association of spots with incomplete aggregation of curd granules. *Journal of Dairy Science* 99: 6144–6156.

Farinella D., Nori, M., Ragkos, A. (2017). Changes in Euro-Mediterranean pastoralism: which opportunities for rural development and generational renewal? (ed. C. Porqueddu A., Franca G., Lombardi G. et al.). *Proceedings of the 19th Symposium of the European Grassland Federation*, Alghero, Italy (7–10 May 2017).

Fascetti, G. (1935). *Butter and Cheese (Burro e Formaggio)*. Milano: Hoepli.

Fenster, K., Rankin, S., and Steele, J. (2003). Accumulation of short n-chain ethyl esters by esterases of lactic acid bacteria under conditions simulating ripening parmesan cheese. *Journal of Dairy Science* 86: 2818–2825.

Fox, P.F., Guinee, T.P., Cogan, T.M., and McSweeney, P.L.H. (2017). Principal families of cheese. In: *Fundamentals of Cheese Science* (ed. P.F. Fox, T.P. Guinee, T.M. Cogan and P.L.H. McSweeney), 27–69. Boston, MA: Springer.

Franciosi, E., Settanni, L., Cologna, N. et al. (2011). Microbial analysis of raw cows' milk used for cheese-making: influence of storage treatments on microbial composition and other technological traits. *World Journal of Microbiology and Biotechnology* 27: 171–180.

Gatti, M., Bottari, B., Santarelli, M., and Neviani, E. (2011). Comparison of natural whey starters for Grana Padano cheese using sunray plots. *Annals of Microbiology* 61: 475–481.

Gatti, M., Bottari, B., Lazzi, C. et al. (2014). Invited review: microbial evolution in raw-milk, long-ripened cheeses produced using undefined natural whey starters. *Journal of Dairy Science* 97: 573–591.

Gobbetti, M. and Di Cagno, R. (2017). Extra-hard varieties, Chap 32. In: *Cheese Chemistry, Physics and Microbiology* Vol. 2 (ed. P.L.H. McSweeney, P.F. Fox, P.D. Cotter and D.W. Everett), 809–828. Elsevier Academic Press.

Govindasamy-Lucey, S., Jaeggi, J.J., Bostley, A.L. et al. (2004). Standardization of milk using cold ultrafiltration retentates for the manufacture of Parmesan cheese. *Journal of Dairy Science* 87: 2789–2799.

Hynes, E.R., Bergamini, C.V., Suarez, V.B., and Zalazar, C.A. (2003). Proteolysis on Reggianito Argentino cheeses manufactured with natural whey cultures and selected strains of *Lactobacillus helveticus*. *Journal of Dairy Science* 86: 3831–3840.

Innocente, N., Munari, M., and Biasutti, M. (2013). Characterization by solid-phase microextraction-gas chromatography of the volatile profile of protected designation of origin Montasio cheese during ripening. *Journal of Dairy Science* 96: 26–32.

Malacarne, M., Summer, A., Franceschi, P. et al. (2009). Free fatty acid profile of Parmigiano-Reggiano cheese throughout ripening: comparison between the inner and outer regions of the wheel. *International Dairy Journal* 19: 637–641.

Marangon, A., Cappellari, M., Moretti, B. et al. (2006). Definition of the sensory profile of Asiago d'Allevo cheese and the effect of the different production chains (Definizione del profilo sensoriale del formaggio Asiago d'Allevo ed effetto delle diverse filiere di produzione). *Quaderno SOZOOALP* 3: 116–127.

Marchesini, G., Balzan, S., Segato, S. et al. (2009). Colour traits in the evaluation of the ripening period of asiago cheese. *Italian Journal of Animal Science* 8 (2): 412–413.

McSweeney, P.L.H., Fox, P.F., Lucey, J.A. et al. (1993). Contribution of the indigenous microflora to the maturation of Cheddar cheeses. *International Dairy Journal* 3: 613–634.

Moio, L. and Addeo, F. (1998). Grana Padano cheese aroma. *Journal of Dairy Research* 65: 317–333.

Molimard, P. and Spinnler, H.E. (1996). Review: compounds involved in the flavor of surface mold-ripened cheeses: origins and properties. *Journal of Dairy Science* 79: 169–184.

Neviani, E. and Gatti, M. (2013). Microbial evolution in raw-milk, long ripened cheeses: Grana Padano and Parmigiano Reggiano case. In: *Cheese Ripening: Quality, Safety and Health Aspects* (ed. C.L. Randazzo, C. Caggia and E. Neviani), 133–148. New York, NY: Nova Science Publishers Inc.

Noël, Y., Zannoni, M., and Hunter, E.A. (1996). Texture of Parmigiano Reggiano cheese: statistical relationship between rheological and sensory variates. *Lait* 76: 243–254.

Perotti, M., Bernal, S., Wolf, V., and Zalazar, C. (2008). Free fatty acid profile and sensory characteristics of Reggianito cheeses made with different starter bacteria (Perfil de ácidos grasos libres y características sensoriales de quesos Reggianito elaborados con diferentes fermentos). *Grasas y Aceites* 59 (2): 152–159.

Pillonel, L., Badertscher, R., Butikofer, U. et al. (2002). Analytical methods for the determination of the geographic origin of Emmental cheese. Main framework of the project: chemical, biochemical, microbiological, color and sensory analyses. *European Food Research and Technology* 215: 260–267.

Pinho, O., Ferreira, I.M.P.L.V.O., and Ferreira, M.A. (2003). Quantification of short-chain free fatty acids in "Terrincho" ewe cheese: intravarietal comparison. *Journal of Dairy Science* 86: 3102–3109.

Piombino, P., Pessina, R., Genovese, A. et al. (2008). Sensory profiling, volatiles and odor active compounds of Canestrato Pugliese PDO cheese made from raw and pasteurized ewe's milk. *Italian Journal of Food Science* 2 (20): 225–237.

Pirisi, A., Pinna, G., Addis, M. et al. (2007). Relationship between the enzymatic composition of lamb rennet paste and proteolytic, lipolytic pattern and texture of PDO Fiore Sardo ovine cheese. *International Dairy Journal* 17: 143–156.

Pirisi, A., Comunian, R., Urgeghe, P.P., and Scintu, M.F. (2011). Sheep's and goat's dairy products in Italy: technological, chemical, microbiological, and sensory aspects. *Small Ruminant Research* 101: 102–112.

Pripp, H., Skeie, S., Isaksson, T. et al. (2006). Multivariate modelling of relationships between proteolysis and sensory quality of Präst cheese. *International Dairy Journal* 16 (3): 225–235.

Quality protocol for Reggianito cheese (Protocolo de Calidad para Queso Reggianito). (2008). Resolución SAGPyA No. 16/2008. Available in: https://www.argentina.gob.ar/normativa/nacional/resolución-16-2008-136938/texto.

Renna, M., Cornale, P., Lussiana, C. et al. (2011). Efficacy of fatty acids and terpenoids and weakness of electronic nose response as tracers of asiago d'Allevo PDO cheese produced in different seasons. *Dairy Science and Technology* 92: 203–218.

Rossetti, L., Carminati, D., Zago, M., and Giraffa, G. (2009). A qualified presumption of safety approach for the safety assessment of grana Padano whey starters. *International Journal of Food Microbiology* 130: 70–73.

Rulikowska, A., Kilcawley, K.N., Doolan, I.A. et al. (2013). The impact of reduced sodium chloride content on Cheddar cheese quality. *International Dairy Journal* 28: 45–55.

Santillo, A., Quinto, M., Dentico, M. et al. (2007). Rennet paste from lambs fed a milk substitute supplemented with *lactobacillus acidophilus*: effects on lipolysis in ovine cheese. *Journal of Dairy Science* 90: 3134–3142.

Santoro, M. and Faccia, M. (1998). Influence of mould size and rennet on proteolysis and composition of Canestrato Pugliese cheese. *Italian Journal of Food Science* 3: 217–228.

Sousa, M.J., Ardö, Y., and McSweeney, P.L.H. (2001). Advances in the study of proteolysis during cheese ripening. *International Dairy Journal* 11: 327–345.

Stefanon, B. and Procida, G. (2004). Effects of including silage in the diet on volatile compound profiles in Montasio cheese and their modification during ripening. *Journal of Dairy Research* 71: 58–65.

Talavera-Bianchi, M. and Chambers, D.H. (2008). Simplified lexicon to describe flavor characteristics of Western European cheeses. *Journal of Sensory Studies* 23: 468–484.

Tansman, G.F., Kindstedt, P.S., and Hughes, J.M. (2015). Crystal fingerprinting: elucidating the crystals of Cheddar, Parmigiano-Reggiano, gouda, and soft washed-rind cheeses using powder X-ray diffractometry. *Dairy Science and Technology* 95: 651–664.

Trani, A., Gambacorta, G., Loizzo, P. et al. (2016). Short communication: chemical and sensory characteristics of Canestrato di Moliterno cheese manufactured in spring. *Journal of Dairy Science* 99: 6080–6085.

UNI (Ente Nazionale Italiano di Unificazione) (2010). *UNI EN ISO 8589: Sensory Analysis—General Guidance for the Design of Test Rooms*. Milan, Italy: UNI.

Vicente, M.S., Ibáñez, F.C., Barcina, Y., and Barron, L.J.R. (2001). Changes in the free amino acid content during ripening of Idiazabal cheese: influence of starter and rennet type. *Food Chemistry* 72 (3): 309–317.

Wolf, I.V., Perotti, M.C., Bernal, S.M., and Zalazar, C.A. (2010). Study of the chemical composition, proteolysis, lipolysis and volatile compounds profile of commercial Reggianito Argentino cheese: characterization of Reggianito Argentino cheese. *Food Research International* 43: 1204–1211.

Woo, A.H. and Lindsay, R.C. (1984). Concentration of major free fatty acids and flavour development in Italian cheese varieties. *Journal of Dairy Science* 67: 960–968.

Zannoni, M. (2010). Evolution of the sensory characteristics of Parmigiano–Reggiano cheese to the present day. *Food Quality and Preference* 21: 901–905.

Zannoni, M., Pigoni, S., and Hunter, E.A. (2006). Validazione di una scheda per l'analisi sensoriale del Parmigiano-reggiano (Validation of a sensory scorecard for the evaluation of Parmigiano-Reggiano cheese). *Scienza e tecnica Lattiero casearia* 57 (6): 545.

13 Sensory Profiles of Iberian and Related Cheese Varieties

Elena Molina, Lourdes Amigo and Daniel Lozano-Ojalvo

13.1 Introduction

Spain and Portugal possess a large diversity of traditional cheese varieties that are generally named after the locality of origin (McSweeney et al. 2017). Cheese manufacture in Spain and Portugal, as in other Mediterranean countries in general, has often been linked to specific regions, indigenous breeds, pasture grazing, and other natural feed resources, and traditional production methods, which have resulted in many high-quality cheeses (Zabaleta et al. 2017).

The most representative cheese varieties of Spain and Portugal are manufactured from ewe's milk (Medina and Núñez 2017). They are Manchego cheese, coagulated with animal rennet, originally made in the central plain of La Mancha in Spain and Serra cheese produced by curdling milk with extracts from thistle flowers (*Cynara* spp.) in the Serra da Estrela mountains in Portugal (Freitas et al. 2000). The use of vegetal rennet has been successfully maintained in some areas of Spain and Portugal, as part of a strong cultural heritage, for the production of many traditional varieties, leading to the production of characteristic high-quality and particular cheeses, with a creamy semi-soft texture and exquisite flavour (Reis and Malcata 2011). These characteristics are in part due to the type of milk and manufacturing technology used, but mainly to the vegetable coagulant, which is very proteolytic (Roseiro et al. 2003a, b, c), although it presents lower specificity than chymosin (Ben et al. 2017; Conceição et al. 2018). However, there are many other high-quality cheeses very popular in the domestic market, but less known in the international market.

In addition to the two aforementioned cheeses made from ewe's milk, important goats' milk cheeses are produced in the South and Southeast of the Peninsula and in the Canary Islands, while the production of cow's milk cheeses is largely concentrated in the North of the Peninsula and in the Balearic & Azores archipelagos, (Marcos and Esteban 1993).

Regarding their sensory characteristics, cheeses made from goat's milk are the strongest and most aromatic. The typical goat flavour is considered a quality component in some goat cheeses (Ribeiro and Ribeiro 2010). The fat composition is well differentiated from sheep's milk, mainly with regard to the content of medium-chain fatty acids

Sensory Profiling of Dairy Products, First Edition. Edited by John J. Tuohy.
© 2023 John Wiley & Sons Ltd. Published 2023 by John Wiley & Sons Ltd.

(C6–C14), which are higher in goat's milk (Sanz Ceballos et al. 2009). Thus, ripened cheeses made from goat's milk are characterised by a sharp, sharp flavour due to the presence of a large proportion of short-chain fatty acids, such as C6 (caproic acid), C8 (caprylic acid), C10 (capric acid), C12 (lauric acid). In addition, some researchers propose that the flavour of goat cheeses may be strongly related to the presence of branched-chain fatty acids such as 3-methylbutanoic, 4-methyloctanoic and 4-ethyl-octanoic acids, which themselves contribute to the 'goaty' flavour of the cheese (Poveda et al. 2008; Buffa et al. 2001). The characteristic goat flavour in cheese is intensified when the product has pH values of six or higher. Goat cheeses are better digested because their milk contains less lactose, casein and the fat globules are smaller in diameter (less than three microns) than those in cow's milk (Park et al. 2007). However, cheeses made from goat's milk are harder, in general, compared to cheeses made from sheep's milk. This difference may be due to casein structures or their concentrations in the milk. On the other hand, goat milk cheeses have a much whiter colour due to the absence of beta-carotene in the animal's milk (Amigo and Fontecha 2011).

Cheeses made from sheep's milk have a special flavour and aroma, very different from those made from cow's milk, and a strong brand image of quality and authenticity (Bárcenas et al. 2002). In general, they have a low elasticity, firm to very firm texture, depending on the dairy breed, medium friability, medium-low adherence and very little fluidity (Miguel et al. 2002). When compared to cow's milk cheese, they have a more intense flavour due to the high percentage of acids, are soft on the palate and have a flavour and aroma that identify them (Suárez and Busetti 2005). The odour and flavour could be generalised as strong and the terms used include: butyric, brine, spicy, sharp, intensity, rennet, salty and smoky, while the terms related to lactic and sweet notes are on the negative side. Due to the seasonal nature of sheep milk production, the time of year is another factor that exerts a great influence on the characteristics of sheep milk cheeses. The lactation period affects both the composition of the milk and its microbial load, which is higher in summer. Mendía et al. (2000) reported that the sensory quality of sheep cheeses made in summer was lower compared to those made in winter and spring.

Most of the cheese produced internationally is made from cow's milk due in part to the fact that cows produce more milk than other animals such as sheep and goats. Therefore, cow's milk cheeses are more affordable. Cow's milk cheese is characterised by a creamy, mild flavour. Its fat content is lower than that of goat's or sheep's milk cheese. Many of the Iberian cheeses can be made with raw cow's milk. Their sensory characterisation has revealed that they have more diverse and intense flavour notes, while in the same cheeses made with pasteurised milk their sensory characteristics are less intense (Poveda et al. 2015).

Although there are over 100 autochthonous Iberian cheese varieties (that includes cheeses from the Iberian Peninsula and the Canary, Balearic and Azores archipelagos), most of which are well established and duly described (Spanish Cheese Catalog 2013), only a few possess a protected designation of origin (PDO) and protected geographical indication (PGI) status (Table 13.1). PDO and PGI identify the designation of cheeses where production, processing and preparation are in a specific geographical area (Barrón et al. 2017). These protection logos were established by the European Community

Table 13.1 List of the most representative cheeses from Spain and Portugal.

Cheeses	Cow's milk (c)	Goat's milk (g)	Ewe's milk (e)	Binary milk mixtures	Tertiary milk mixtures
SPAIN					
Peninsula	Afuega'l pitu[a]	Alhama de Granada	Gaztarra	De Letur (C and/or G)	Cabrales[a]
	Arzúa-Ulloa[a]	Camerano[a]	Idiazábal[a]	Del Valle del Narcea (C and G)	Gamonedo[a]
	Casín[a]	De Acehuche	Ansó-Hecho	Ahumado de Pría (C and E)	De Peñamellera
	Cebreiro[a]	De Andalucía de cabra	Arribes de Salamanca	De Burgos (C and E)	De Valdeón[b]
	De Abredo	De Aracena	Castellano	De Pastor (G and C)	Genestoso
	De Benasque	De Babia y de Lanciana	De Barros	De Grazalema (G and E)	Los Beyos[b]
	De la Montaña de León	De Buelles	De Cáceres	De la Nucia (G and E)	Mató
	De la Peral	De Cádiz	De Campo Real	De Miraflores (G and E)	Picón Bejes-Tresviso[a]
	De Oscos	De Colmenar Viejo	De Guriezo	De Servilleta (G and E)	Pido
	De Trapo	De Fresnedilla de la Oliva	De la Bureba	De Tupí (S and C)	Porrúa
	L'Alt Urgell y la Cerdanya[a]	De Gata-Hurdes	De la Calahorra	De Villalón (S and C)	Quesucos de Liébana[a]
	Nata de Cantabria[a]	De la Adrada	De La Serena[a]	De Teruel (S and G)	
	Pasiego	De la Siberia	De la Sierra de Espadán	De Torremocha del Jarama (E and G)	
	Requeixo	De la Vera	De los Pedroches	Tronchón (S and G)	
	San Simón da Costa[a]	De las Alpujarras	De Oropesa		
	Tetilla[a]	De los Montes de Toledo	Manchego[a]		
	Urbiés	De Málaga	Pata de mulo		
	Vidiago	De Murcia[a]	Roncal[a]		
		De Murcia al vino[a]	Serrat		
		De Ossera	Torta del Casar[a]		
		De Sierra Morena	Valle de Alcudía		
		Del Tiétar	Zamorano[a]		
		Garrotxa			
		Ibores[a]			
		Montsec			
		Quesaílla			

Table 13.1 (Continued)

Cheeses	Cow's milk (c)	Goat's milk (g)	Ewe's milk (e)	Binary milk mixtures	Tertiary milk mixtures
Balearic Islands				Mahón-Menorca[a] (C and S)	Mallorquín
Canary Islands		Palmero[a]		De la Gomera (G and S)	De Flor de Guía[a]
				De Lanzarote (G and S)	De Guía[a]
				Majorero[a] (G and S)	De Media Flor de Guía[a]
					De Tenerife
					Herreño
PORTUGAL					
Peninsula		Cabra Transmontano[a]	Azeitão[a]	Amarelo da Beira Baixa[a] (S and G)	Bica de Queijo
		Ribafria	Castelo Branco[a]	Mestiço de Tolosa[b] (S and G)	Queimoso
			Curado a ovelheira	Picante da Beira Baixa[a] (S and G)	
			Évora[a]	Rabaçal[a] (S and G)	
			Nisa[a]		
			Mondegueiro		
			Ovelha Amanteigado		
			Salaoio		
			Pasta Mole		
			Raiano		
			Serpa[a]		
			Serra da Estrela[a]		
			Terrincho[a]		
			Tomar		
Archipelago Azores	Ilha Graciosa				
	Ilha Terceira o Vaquinha				
	Pico[a]				
	São Jorge[a]				
	São Miguel				

[a] PDO cheeses.
[b] PGI cheeses.

(Council Regulations 2006) to recognise the quality of the products and to offer a guarantee of provenance to the consumer. Actually, 26 cheeses in Spain and 13 in Portugal have PDO status which is controlled by their respective Regulatory Councils, and two cheeses in Spain and one in Portugal are protected by PGI status.

This chapter describes the sensory characteristics of the main Iberian cheeses (understood as those produced in Spain and Portugal, both on the Peninsula and on its islands and archipelagos) with PDO and PGI including fresh and ripened cheeses made from sheep's, goat's and cow's milk and their blends. Their main characteristics and sensory properties are described below and summarised in Tables 13.2 and 13.3.

13.2 Fresh Iberian cheese varieties

There are four cheese varieties manufactured in Spain with PDO status that can be consumed fresh (Tables 13.2 and 13.3): Afuega'l Pitu, Camerano, Cebreiro and De Murcia cheeses (Figure 13.1). All of them have sensory characteristics reminiscent of the milk they are made from and typically have a mild aroma and flavour.

13.2.1 *Afuega'l Pitu*

This is an artisanal, lactic acid coagulated cheese manufactured from cow's milk to which a very small amount of rennet is added (Piñeiro-Lago et al. 2019). It has a truncated cone shape (*Atroncau*), or courgette shape (*Trapu*), and is without rind when fresh but becomes somewhat yellowish in colour as it matures (*Blancu*). Paprika can be added to achieve a reddish or orange appearance (*Roxu*). The body is continuous, closed, without eyes, soft and spreadable as a fresh cheese, while short and friable when matured. As a fresh cheese, it is soft but as it matures it becomes firmer, crumbly, tender, pasty, somewhat sticky, floury and moist. Its aroma is of medium intensity, reminiscent of yoghurt, when fresh, more lactic when matured and of paprika in the red type. The flavour is acidic, not at all salty and somewhat metallic when fresh with some refreshing notes and develops a somewhat spicy, somewhat astringent and somewhat burning flavour when matured.

13.2.2 *Camerano*

Camerano is made from raw or pasteurised goat's milk by a combination of acid and enzymatic coagulation. It has a flattened cylindrical shape, without rind but with rind marks (*cillas)*. The curd is without eyes, has a gelled lamellar texture, is very moist and soft and has bright white appearance when fresh. The cheese has a milky odour, of low intensity and the texture is not very firm. The odour is of low intensity: weakly acidified curd, to cut grass, to goat, slightly to rennet. A slight goat's milk aroma is detectable: faintly sweet and salty.

Camerano cheese is available as soft (min. 15 days of ripening), semi-hard (min. 30 days of ripening) and hard (min. 75 days of ripening) types (see respective categories in Tables 13.2 and 13.3). As the ripening period increases, the rind becomes more defined and greenish brown moulds appear. In the same way, the body changes from a gelled, very humid and soft lamellar texture of bright white colour, to a semi-hard curd of white colour and finally, in the last stages of ripening, it becomes a firm, hard and

Table 13.2 Main characteristics of the PDO and PGI Iberian cheeses.

Iberian cheeses	Milk type/Mixtures	Rennet type	Weight (Kg)	Ripening (Days)	Production area	Reference
FRESH						
Afuega'l pitu	Cow	Lactic Acid	0.2–0.6	10	Asturias	BOE (2004)
Camerano	Goat	Animal	0.20–0.80	–	La Rioja	BOE (2009)
Cebreiro	Cow	Animal	0.3–2	1	Galicia	eAmbrosia-UE (electronic)
De Murcia	Goat	Animal	0.25–3.00	–	Murcia	BOE (2001b)
SOFT OR SEMI-SOFT						
Arzúa-Ulloa	Cow	Animal	0.5–3.5	6	Galicia	eAmbrosia-UE
Azeitão	Ewe	Vegetal	0.20–0.30	21–28	Palmela, Sesimbra and Setúbal	DR (2005a)
Cabrales	Cow, ewe, goat or their mixtures	Lactic Acid/ Animal	0.75–4.00	60–120	Picos de Europa, Asturias	BOE (1990)
Camerano	Goat	Animal	0.20–1.20	15	La Rioja	BOE (2009)
De Flor de Guía	Ewe and cow or/ and goat	Vegetal	0.50–5.00	15–60	Gran Canaria Island	BOE (2008)
De Guía	Ewe and cow or/ and goat	Animal or vegetal	0.50–5.00	15–60	Gran Canaria Island	BOE (2008)
De La Serena	Ewe	Vegetal	1.00–3.00[en] 0.75–2.00	40–60	Badajoz, Extremadura	Martínez et al. (2011)
De Media Flor de Guía	Ewe and cow or/ and goat	Vegetal/ Animal	0.50–5.00	15–60	Gran Canaria Island	BOE (2008)
De Valdeón	Cow or cow, ewe and goat	Lactic Acid/ Animal	2.00–2.50	60	León, Castile-León	BOCYL (1998)
L'Alt Urgell y la Cerdanya	Cow	Animal	2.5	45	Lérida y Girona	DOCE (2000)
Los Beyos	Cow or ewe or goat	Lactic Acid	0.25–0.50	20	Asturias and North Castile-León	BOE (2010)
Mahón-Menorca	Cow	Animal	0.60–4.00	21–60	Minorca Island	BOE (1985a)
Majorero	Goat	Animal	1.00–6.00	8–20	Fuerteventura Island	BOE (1996)
Mestiço de Tolosa	Ewe and goat	Animal	0.15–0.0.40	21–28	Nisa and Portalegre	DR (2005b)
Nata de Cantabria	Cow	Animal	0.40–2.80	7	Cantabria	BOE (1985b)
Palmero (tender)	Goat	Animal	0.75–15.00	8–20	La Palma Island	BOE (2001a)

(*Continued*)

Table 13.2 (Continued)

Iberian cheeses	Milk type/Mixtures	Rennet type	Weight (Kg)	Ripening (Days)	Production area	Reference
Pico	Cow	Animal	0.50	21–28	Pico Island	Freitas and Malcata (2000)
Quesucos de Liébana	Cow, ewe and goat	Animal	02.0–0.50	30–60	Liébana, Cantabria	BOE (1994b)
Torta del Casar	Ewe	Vegetal	0.50–1.10	60	Cáceres, Extremadura	Martínez et al. (2011)
SEMI-HARD						
Amarelo da Beira Baixa	Binary	Animal	0.20–0.50	40	Castelo Branco	DR (1988, 2007)
Castelo Branco	Ewe	Vegetal	080–1.30	21–28	Castelo Branco	DR (1988, 2007)
De Murcia al Vino	Goat	Animal	0.40–2.00	30–45	Murcia	BOE (2001b)
Gamonedo	Ewe and goat	Animal	0.50–7.00	90–120	Asturias	BOPA (2003)
Ibores	Goat	Animal	0.65–1.20	60	Cáceres	DOE (2003)
Picón Bejes-Tresviso	Cow or goat	Animal	1.00–5.00	60	Liébana, Cantabria	BOE (1994a)
San Simón da Costa	Cow	Animal	0.40–1.5	30	Galicia	BOE (2005)
Serpa	Ewe	Vegetal	1.70–2.00	30–40	Serpa and Beja	Freitas and Malcata (2000)
Serra da Estrela	Ewe	Vegetal	1.00–1.70	28–60	Serra da Estrela	eAmbrosia-UE
Terrincho	Ewe	Animal	0.6–1.2	30–90	Bragança, Viseu and Guarda	eAmbrosia-UE
Tetilla	Cow	Animal	1	10	Galicia	Freitas and Malcata (2000)
SEMI-HARD OR HARD						
Camerano (semicured)	Goat	Animal	0.20–1.20	30	La Rioja	BOE (2009)
Casín	Cow	Animal	0.25–1	60	Asturias	eAmbrosia-UE
Évora	Ewe	Vegetal	0.10	30–90	Évora	eAmbrosia-UE
Idiazábal	Ewe	Animal	0.50–3.50	60	Navarre and Basque Country	BOE (1993b)
Mahón-Menorca	Cow	Animal	0.60–4.00	60->150	Minorca Island	BOE (1985a)
Majorero	Goat	Animal	1.00–6.00	20->60	Fuerteventura	BOE (1996)

Manchego	Ewe	Animal	1.00–3.50	60	Castile La Mancha	Martínez et al. (2011)
Nisa	Ewe	Vegetal	0.2–1.3	45	Nisa y Portalegre	DR (2005b)
Picante da Beira Baixa	Ewe or ewe/goat	Animal	0.40–1.00	120–150	Castelo Branco	DR (1988, 2007)
Rabaçal	Ewe and goat	Animal	0.30–0.50	20	Coimbra	DR (2009)
Roncal	Ewe	Animal	1.00–3.00	120	Roncal valley, Navarre	BOE (1991)
São Jorge	Cow	Animal	3.50–7.00	42–84	São Jorge Island	Freitas and Malcata (2000)
Zamorano	Ewe	Animal	4.00	100	Zamora, Castile-León	BOE (1993a)
HARD OR EXTRA-HARD						
Cabra Transmontano	Goat	Animal	0.60–0.90	60–120	Bragança and Vila Real	DR (2005a)
Camerano (cured)	Goat	Animal	0.20–1.20	75	La Rioja	BOE (2009)
De Murcia (cured)	Goat	Animal	0.25–2.60	120	Murcia	BOE (2001b)
Palmero (cured)	Goat	Animal	0.75–15.00	90–180	San Miguel de la Palma Island	BOE (2001a)

BOCYL: Official Bulletin of Castile and León (Boletín Oficial de Castilla y León); BOE: State Official Bulletin (Boletín Oficial del Estado); BOPA: Official Bulletin of the Principality of Asturias (Boletín Oficial del Principado de Asturias); DOCE: Official Journal of the European Communities (Diário Oficial de las Comunidades Europeas); DOE: Official Journal of Extremadura (Diário Oficial de Extremadura); DR: Republic Diary (Diário da República).

Table 13.3 Summary of the main sensory characteristics of the PDO and PGI Iberian cheeses.

Iberian Cheeses	Shape	Rind	Paste colour	Paste texture	Taste, odour and flavour
FRESH					
Afuega'l Pitu	Truncated cone or courgette	Cloth wrapped	White, yellowish or orange	Soft, no holes	Fresh, medium intensity, acidic
Camerano	Flattened cylindrical	No rind	White and bright	Laminate but gelled, very moist and soft. No holes	Low intensity. Aroma to goat milk
Cebreiro	Chef's hat	No rind	White	Soft, granular, creamy	Fresh and slightly salty
De Murcia	Cylindrical (Diameter: 20 cm) (Height: 12 cm)	No rind	Bright white	Creamy. No holes or irregularly distributed	Fresh goat milk, something sweet and refreshing.
SOFT OR SEMI-SOFT					
Arzúa-Ulloa	Lenticular or cylindrical	Thin and pliant. Medium to dark yellow	Ivory-white and pale yellow	Soft, creamy with small holes	Low-to-medium intensity Slightly acidic and salty
Azeitão	Low cylinder (Diameter: 5–11 cm) (Height: 3–5 cm)	Extremely thin, soft, smooth, straw yellow	Pale straw yellow	Creamy with few or no holes	Delicate, slightly acidic and fresh
Cabrales	Cylindrical (Diameter: 10–25 cm) (Height: 7–15 cm)	Natural, soft, thin, grey with orange-brown patches	Ivory-white with blue-green veins	Compact, but spreadable, without openings	Medium intensity, penetrating and spicy taste
Camerano (tender)	Flattened cylindrical	Clearly defined with greenish brown moulds	White to white ivory	Soft to semi-hard without holes	Intense and defined. Aroma to goat milk, natural cave and mould
De Flor de Guía (semicured)	Cylindrical (Diameter: 15–30 cm) (Height: 4–6 cm)	Thin and soft, white ivory, cloth marks	Off-white ivory	Soft, compact, creamy, buttery	Highly aromatic, lactic acid, bitter
De Flor de Guía (cured)		Thick and yellow (all ranges) to dark brown	Yellow to brownish yellow	Compact, closed. Fermentation eyes	Something spicy, slightly astringent and slightly burning

De Guía (semicured)	Cylindrical (Diameter: 15–30 cm) (Height: 4–8 cm)	Thin, ivory with well-defined cloth marks	Ivory	Compact, closed. Fermentation eyes	Acid, nothing salty and something bitter
De Guía (cured)		Thick, yellow to dark brown in colour	Yellow to brownish yellow		Something spicy, slightly astringent and slightly burning
De La Serena	Disc shaped (Diameter: 18–24 cm) (Height: 4–8 cm)	Yellow to ochre with regular surface	Ivory-white	Soft, creamy, with few small holes	Strong and (slightly) bitter. Very buttery.
De Media Flor de Guía (semicured)	Cylindrical (Diameter: 15–30 cm) (Height: 4–8 cm)	Thin, white-ivory, well-defined cloth marks	Off-white ivory	Compact, closed with the presence of fermentation eyes	Highly aromatic, lactic acid, bitter
De Media Flor de Guía (cured)		Thick, yellow to dark brown in colour	Yellow to brownish yellow		Something spicy, slightly astringent and slightly burning
De Valdeón	Cylindrical (Diameter: 25 cm) (Height: 15 cm)	Thin, soft, whitish, yellowish and greyish	Pale yellow, green/blue veins	Soft, small cavities with white and greenish blue moulds	Medium intensity. Burning sensation
L'Alt Urgell y la Cerdanya	Cylindrical (Diameter: 20 cm)	Natural, washed and orange	Cream	Soft and creamy. Small, evenly distributed, irregular-shaped eye	Low-intensity, ammoniacal background. Balanced flavour, soft and pleasant bitterness
Los Beyos	Cylindrical (Diameter: 9–10 cm) (Height: 6–9 cm)	Thin, rough. Brownish (sheep), pale yellow (goat and cow)	Ivory (sheep), white (goat), pale yellow (cow)	Firm, crumbly. No or small mechanical holes	Intense, pleasant, balanced slightly acid and slightly goaty
Mahón-Menorca (tender)	Parallepipedic (High: 5–9 cm)	Poorly developed and white	Bone white	Weak roughness, somewhat moist, a little elastic	Mild intensity, remember fresh milk. Butter, curd and may appear the grass cut
Mahón-Menorca (semicured)		Poorly developed and orange or brown	Ivory-yellow		Moderate intensity, fresh milk, butter, toasted nuts

(*Continued*)

Table 13.3 (Continued)

Iberian Cheeses	Shape	Rind	Paste colour	Paste texture	Taste, odour and flavour
Majorero (tender)	Cylindrical (Diameter: 15–35 cm) (Heigh: 6–9 cm)	Practically non-existent and white	Snow white	Compact closed, sometimes with some small eyes regularly distributed	Weak to medium intensity Slightly acid and bitter
Majorero (semicured)		Yellowish-white	Ivory-white		Medium intensity. Citric acidity
Mestiço de Tolosa	Cylindrical (Diameter: 7–10 cm) (Heigh: 3–4 cm)	Thin, toasted yellow or orange	Yellowish to orange	Well linked, presenting small eyes	Strong, clean and slightly spicy
Nata de Cantabria	Cylindrical or parallepipedic	Soft-rind, yellowish	Ivory-white to yellowish-white	Compact, small eyes distributed regularly	Low intensity
Palmero (tender)	Cylindrical (Diameter: 12–60 cm) (Height: 6–15 cm)	Smooth, yellow or brown when smoked.	Bright white	Medium firmness, small eyes	Weak intensity, goat's milk. Slightly acid, moderate salty and sometimes weak bitter taste
Pico	Flat cylinder	Smooth and yellow	Yellowish-white	Soft, pasty with holes	Intense, fairly piquant
Quesucos de Liébana	Disc or cylinder (Diameter: 8–12 cm) (Height: 3–10 cm)	Thin, wrinkled, white to yellowish or ochre to light brown	White to straw yellow	Compact, few irregularly distributed holes	Medium intensity. Delicate, slightly spicy, sharp and buttery
Torta del Casar	Cylindrical	Thin, soft, slightly cracked on the outside	Ivory	Delightfully soft, creamy	Strong, full, intense
SEMI-HARD					
Amarelo da Beira Baixa	Cylindrical (Diameter: 7–10 cm) (Heigh: 3–5 cm)	Fine, yellow or tan-yellow	Slightly yellow uniform	Closed texture, with some irregular holes	Intense but pleasant aroma, clean flavour. Slightly acidic and sea salt
Castelo Branco	Low cylinder (Diameter: 2–16 cm) (Height: 5–8 cm)	Smooth, intense yellow to brown	Yellowish-white	Uniform, slightly crumbly, unctuous	Strong aroma and flavour. Slightly piquant (in prolonged maturation)
De Murcia al Vino	Cylindrical (Diameter: 19 cm) (Height: 10 cm)	Smooth, light and characteristic red-violet colour	Matte white, ivory and yellowish	Compact, creamy and elastic. Few small eyes	Medium intensity. Fresh lactic, goat's milk, cream/butter, fresh hazelnut

Gamonedo	Cylindrical (Diameter: 10–30 cm) (Height: 6–15 cm)	Thin, sienna, reddish, greenish and bluish tones	White or yellowish with green/blue veins	Firm with irregularly distributed eyes and small size	Medium intensity. Slightly acidic, salty, smoky and spicy
Ibores	Cylindrical (Diameter: 11–15 cm) (Height: 5–9 cm)	Smooth. Waxy yellow to dark ochre.	Ivory-white	Semi-hard. Small holes, little abundant and irregularly distributed	Mild to moderate. Vegetal, floral and fruity hints. Slightly spicy
Picón Bejes-Tresviso	Cylindrical (Diameter: 15–20 cm) (Height: 7–15 cm)	Thin, soft, greasy, grey with yellow-greenish areas	White ivory with green/blue veins	Compact somewhat soft, slightly unctuous	Intense, slightly acidic, salty, bitter and spicy
San Simón da Costa	Spinning top or bale	Smoked, hard and non-elastic. Yellow-ochre	Firm, closed White to yellow	Rounded or irregular-shaped and small eyes	Medium intensity. Slightly salty, a little sour and sweet. Smoky
Serpa	Low cylinder (Diameter: 10–30 cm) (Height: 3–8 cm)	Thin, pale straw yellow	Yellowish-white or straw yellow	Closed, buttery, unctuous with few or no eyes	Strong with a dominant spicy taste
Serra da Estrela	Low cylinder (Diameter: 15–20 cm) (Height: 4–6 cm)	Yellow and smooth with regular surface	White or pale yellow	Crumbly, unctuous, few small, few or no eyes	Smooth, clean and slightly acidic bouquet
Terrincho	Cylindrical (Diameter: 13–20 cm) (Height: 3–6 cm)	Well formed, smooth and straw-coloured	White	Uniform, with some holes	High intensity of odour and aroma
Tetilla	Nipple	Thin, smooth, straw yellow	Ivory-white to soft yellow	Closed, few or no holes, evenly distributed	Slightly acidic not spicy, stringent or burning
SEMI-HARD OR HARD					
Camerano (semicured)	Flattened cylindrical	Clearly defined with greenish brown moulds	White to white ivory	Semi-hard to hard with very few holes or none	Intense, defined. Roma to goat milk, natural cave, mould
Casín	Thick disc shape (Diameter: 10–20 cm) (Height: 4–7 cm)	Thin, smooth	Dark creamy yellow	Firm. Small eyes evenly distributed	High intensity. Strong dairy flavour. Intense spiciness. Long-lasting in the mouth

(*Continued*)

Table 13.3 (Continued)

Iberian Cheeses	Shape	Rind	Paste colour	Paste texture	Taste, odour and flavour
Évora	Disc shaped (Diameter: 6–15 cm) (Height: 1.5–4 cm)	Dark yellow, smooth and slightly wrinkled	Pale yellow	Well-formed. Few small holes	Sharp, strong, acid and slightly piquant
Idiazábal	Cylindrical (Diameter: 10–30 cm) (Height: 8–10 cm)	Hard-rind, smooth, shiny, pale yellow, dark brown when smoked	White ivory-yellowish	Firm consistency, compact, few small eyes, slight elasticity	*Sui generis* odour, delicate smoky flavour, intense
Mahón-Menorca	Parallepipedic (High: 5–9 cm)	Compact, smooth, tight, greasy, brown	Yellow to brown- yellow	Rough, dry and not at all elastic. Small eyes in small numbers	Moderate-high intensity. Cream, melted butter hints. Toasted hazelnut, leather beef broth.
Majorero	Cylindrical (Diameter: 15–35 cm) (Heigh: 6–9 cm)	Yellowish	Dark ivory	Weak to medium firm	Medium-high intensity. Fruit family, nuts
Manchego	Cylindrical (Diameter: 18–22 cm) (Heigh: 8–10 cm)	Hard-rind, pale yellow or dark green by surface moulds	White to ivory-yellow	Firm, compact, no holes or many small ones	Characteristic. Pronounced but no strong, slightly buttery and oily, slightly spicy
Nisa	Low cylinder (Diameter: 10–16 cm) (Height: 3–8 cm)	White or pale yellow	White and yellow	Firm, closed with few eyes	Intense, slightly acid and reminiscent of Brazil nuts
Palmero	Cylindrical (Diameter: 12–60 cm) (Height: 6–15 cm)	Smooth, yellow or brown when smoked.	Bright white	Firm	Medium intensity, notes of butter, dry grass, rancid nuances
Picante da Beira Baixa	Cylindrical (Diameter: 10–15 cm) (Heigh: 3–5 cm)	No rind	White-dirty to white-grey	Closed texture, without eyes or with small eyes irregular	Very intense and characteristic. Strong flavour and markedly spicy
Rabaçal	Cylindrical (Diameter: 10–15 cm) (Heigh: 3–5 cm)	Yellowish-white and straw yellow, uniform	White matte or slightly yellow uniform	Firm, closed with few eyes irregularly distributed	Flavour characteristic thyme

Roncal	Cylindrical (Diameter: variable) (Height: 8–10 cm)	Natural rind, thin, rough, brown or straw yellow	White or pale yellow	Hard, pores without eyes	Mild-intense, well developed, structured, straw, nuts and mushrooms. Slightly piquant
São Jorge	Tall cylinder	Hard, dark yellow with reddish brown spots	Yellowish	Crumbly, many small and irregular eyes	Strong bouquet, clean and slightly spicy
Zamorano	Cylindrical	Hard, flat, pale yellow or dark grey	Pale yellow	Firm, compact, semi-hard, without holes	Intense but not strong, slightly piquant, buttery, oily
HARD OR EXTRA-HARD					
Cabra Transmontano	Low cylindrical (Diameter: 12–19 cm)	Well-formed and white	White	Closed. Some holes	Intense and pleasant aroma Clean with a slight spicy
Camerano (cured)	Flattened cylindrical	Clearly defined with greenish brown moulds	White to white Ivory	Hard firm with very few holes or none	Intense and slightly bitter. Goat milk, natural cave, mould
De Murcia (cured)	Cylindrical (Diameter: 19 cm) (Height: 11 cm)	Yellow waxy to ochre in the very aged	Dull white. Yellowish in the more mature	Crumbly. Few and small holes of variable size irregularly distributed	Medium-high intensity. Butter, goat, hay, nuts aromas. Roasted, caramel in the most cured
Palmero (cured)	Cylindrical (Diameter: 12–60 cm) (Height: 6–15 cm)	Smooth, yellow or brown when smoked. Small reliefs in grid	Ivory and matte tones as it matures	Small holes spread randomly and mostly irregularly	High intensity, lactic nuances, nuts, hay. Slightly bitter

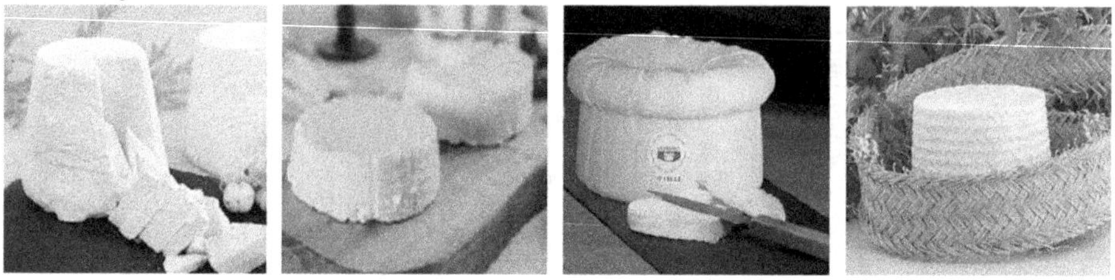

Figure 13.1 PDO fresh cheeses from Spain.

ivory-white in appearance. In all cases, there are no or very few eyes. The texture is between tender and chewy, between floury and pasty, not very adhesive, somewhat sandy, somewhat moist but not at all greasy. It can be described as malleable, soluble and creamy.

The packing of Camerano cheese under modified atmospheres has proved effective for extending the shelf life while retaining good sensory characteristics. MAP cheeses present lower moisture, greater hardness, higher acidity and higher salt content that unripened cheeses (Olarte et al. 2001).

13.2.3 *Cebreiro*

This is a fresh cheese made from pasteurised cow's milk using animal rennet. It has a cylindrical shape reminiscent of a mushroom or chef's hat, with a cylindrical base of variable diameter and a slightly wider cap with a height of no more than 12 cm. It has no rind and has a continuous white, soft and grainy body, without cavities, of laminar, granular appearance.

The odour and aroma are of low intensity and described as sour milk, curd and yoghurt. It is sour, slightly salty, somewhat metallic and gives refreshing sensations. The most abundant volatile compounds in Cebreiro cheese are ethanol, isopropanol and diacetyl, which account for the normal aroma variation among cheeses (Centeno et al. 2015).

13.2.4 *De Murcia*

This cheese is enzymatically coagulated, lightly drained and made from pasteurised Murciana-Granadina goat's milk. It has a cylindrical shape with straight or slightly curved sides and rounded edges. The rind is practically non-existent in the fresh cheeses and somewhat more noticeable and harder in the cured cheeses. It has a characteristic body appearance with the engravings of the pleats on the lateral faces, and flower or smooth on the horizontal ones. The fresh cheese has an intense white colour while the cured cheese is waxy yellow to ochre.

The body is between soft and bright white (fresh) or firm with a slight yellowish tinge (hard cheeses), practically without eyes but sometimes with unevenly distributed eyes. When the cheese is fresh, the smell is of low intensity: fresh lactic or goat's milk. In the mouth, it has a weak firmness, it is soft, fairly crumbly, tender, and between pasty and somewhat rubbery, it is not adhesive (does not stick to the surface of the mouth). The aroma is mildly reminiscent of fresh goat's milk and cream and somewhat refreshing. As it matures, the characteristics described for the fresh cheese become slightly intensified with the appearance of lactic (buttery), animal (goaty), vegetable (hay, nuts) aromas. In the more mature cheeses flavour and aroma notes of the roasted family (toasted, caramel) and other flavour sensations such as burning and somewhat spicy can be found in very mature hard cheeses.

13.3 Soft and semi-soft Iberian cheese

The soft and semi-soft cheeses produced in Spain and Portugal protected by PDO or PGI are shown in Figure 13.2 and described below.

13.3.1 *Arzúa-Ulloa*

This is a cheese made from raw or pasteurised cow's milk, with a lenticular or cylindrical shape with rounded edges. The rind is thin, elastic, medium to dark yellow, shiny, clean and smooth. It may be coated with a transparent and colourless anti-mildew plastic emulsion.

The body is soft, very creamy, of uniform colour, between ivory-white and pale yellow, shiny, without cracks and may have small 'eyes', angular or rounded, irregularly distributed. The smell is of low intensity and buttery with notes of vanilla, cream and walnut. The aroma is of medium intensity, reminiscent of cream and yoghurt. The taste is slightly acidic and slightly salty. The aroma differs markedly in 'good quality' artisanal raw milk Arzúa-Ulloa cheeses, where free fatty acids, a series of esters and a significantly higher content of aldehydes and branched-chain alcohols along with

Figure 13.2 PDO and PGI soft and semi-soft cheeses from Spain and Portugal. *Source:* (b) Raquel/Adobe Stock; (e) Unknown Source/Wikimedia Commons/Public Domain; (f) QUESERÍAS PICOS DE EUROPA, S.L.; (n) Uncyclopedia; (o) isaphoto2016/Adobe Stock.

some secondary alcohols and ketones are detected (Rodríguez-Alonso et al. 2009). The formation of these aromatic compounds must be related (in addition to the possible action of native lipoprotein lipase) to the metabolic activities of the indigenous microbiota, other than LAB, present in this cheese.

13.3.2 *Azeitão*

Azeitão cheese is an artisanal, matured cheese, obtained by slow draining of the curd after coagulation of raw sheep's milk with an infusion of thistle. It is cylindrical in shape, with slightly convex sides and a slightly rounded top, with no edges. The rind is soft, well-formed, thin, smooth, whole, straw yellow in colour and uniform. The body is semi-soft and buttery, closed or with few eyes, and white or slightly yellowish. The aroma is characteristic and clean, and the taste is slightly acidic, salty and spicy.

The Azeitão, Nisa and Évora cheeses contain considerably high values of trans fatty acids, mainly elaidic and vaccenic acid (trans-9 18:1 and trans-11 18:1). This is a consequence of the grazing system (Partidário et al. 2008). Some differences in the fatty acid profiles, especially during ripening, are probably responsible for the natural variation in flavour observed among these cheeses.

13.3.3 *Cabrales*

This cheese is by far the most famous of the Spanish blue-veined cheese. It has PDO status since 1981. It is of mixed coagulation, acid-enzymatic, and is made with raw cow's, sheep's or goat's milk, or a mixture of two or all three types of milk from herds that graze in mountain pastures and is mould ripened. Traditionally, this cheese used to be wrapped in leaves and ripened in natural caves with high humidity. Nowadays, it is wrapped in aluminium foil, which must have special characteristics: bottle green in colour and with the letters DOP Cabrales on the entire surface.

Its shape is cylindrical with flat faces that in some cases allow a slight concavity. The vertical faces of the cheese are straight or slightly convex. The rind is natural, soft and thin with an orange-brown to greyish colour, and the appearance to the touch is slimy due to significant microbial growth. The microbial growth sometimes gives a red or yellow colouration at certain points. The colour of the body is ivory-white, which may vary depending on the type of milk used for its preparation. This colour should be homogeneous throughout the body, although there can be areas with streaks of bluish-green colour due to the development of the Penicillium genus that grows in the body. The effect of inoculation with commercial *Penicillium roqueforti* spores on microbiological, chemical and sensory properties has been studied (Flórez et al. 2006). It can present fermentative eyes in the mass, but these must be few in number. It has a smell and aroma of medium intensity: penetrating, clean, pleasant, slightly reminiscent of something lactic, natural cave or mould. In the more mature cheeses, there are smells and aromas of spices, nuts (hazelnuts, almonds. . .), also spicy notes, although the musty smell is maintained. Occasionally, a slight smell of ammonia is detected in the very ripe ones, because of extensive maturation. The flavour is of low intensity: balanced, and slightly acid, correct saltiness, and not bitter at all.

13.3.4 *De Flor de Guía, De Guía and De Media Flor de Guía*

These cheeses are mixed, low-acid and primarily enzymatically coagulated cheeses. They are produced on the island of Gran Canaria with milk from Canary Island sheep or a mixture of this (min. 60%) with that of other species, cow (max. 40%), or goat (max. 10%), from any of the Canary Island breeds.

De Flor de Guía cheese is coagulated only with vegetable rennet, *Cynara cardunculus L. var. Fierce* or *Cynara escolymus*, while in Media Flor de Guía animal rennet and 50% of flower thistle are used and in De Guía cheese rennet or another authorised coagulating enzyme is used, even of vegetable origin and authorised ferments.

They have a cylindrical shape with a rind whose thickness and colour vary according to the ripening time. In De Flor de Guía and De Media Flor de Guía cheeses of short ripening duration, the outer rind is quite thin, elastic and soft, off-white ivory in colour and cake-like in shape. In the cheeses of De Guía, the outer rind is quite thin and ivory-coloured for the short-ripened cheeses, and the longer-ripened cheeses have a thick, dark brown rind that runs through the whole range of cheeses. The interior appearance is compact, closed with the presence of fermentation eyes, off-white body in the fresh ones and yellow to brownish yellow in the cured ones. They have a medium intensity odour, reminiscent of the lactic family (milk), vegetable family (pasture, hay), fruity family (nuts) and animal family (sheep's milk). The aroma is mild, with the same attributes as the smell, but the vegetal notes of the thistle and the pastures on which the cattle feed are more intense. The taste is acidic, slightly salty and pleasantly bitter, mainly due to the intense action of the vegetable coagulant used during the maturation of the cheese and slightly spicy, astringent and burning.

13.3.5 *De La Serena*

De La Serena cheese is enzymatically coagulated with extracts prepared from dried wild cardoon flowers of *C. cardunculus L.* It has a discoidal shape, with remarkably flat horizontal faces and convex lateral surfaces. The rind is semi-hard in the immature cheeses but hard in the aged ones and is waxy yellow to ochre in colour. The body is soft to semi-hard, ivory-white in the immature ones and waxy yellow in the more mature ones and may have small irregularly distributed eyes. It has a smooth and very unctuous texture.

The smell is of medium intensity and reminiscent of: the lactic family (yoghurt, sweet cream), sweet, animal family (sheep's milk), vegetable family (hay, wet grass), depending on the time of year and stage of maturity of the cheese (Arqués et al. 2007). The aroma is perceived as being from soft to more intense, depending on the maturation, similar to the characteristics already described for the smell and buttery. In particular, ethyl esters and some alcohols seem to play a key role in the aroma of La Serena cheese (Carbonell et al. 2002a). The flavour is somewhat acidic, not at all salty and from somewhat bitter to very bitter, depending on the age of the cheese. It also has other sensations such as slight astringency but not burning or spicy.

A seasonal variation of volatile components has been recorded in the La Serena cheeses. Those made in winter showed a lower abundance of most of them, except dimethyl sulphide and dimethyl disulphide (Carbonell et al. 2002b). Ethanol and ethyl esters were more abundant in spring and autumn cheeses, while branched-chain alcohols and branched-chain alkyl esters were more abundant in summer cheeses (Fernández-García et al. 2008).

13.3.6 *De Valdeón*

De Valdeón is a blue cheese of dual acid-enzymatic coagulation, made with cow's milk or a mixture of cow's, sheep's and/or goat's milk. It has a cylindrical shape, with slightly concave upper and lower sides. Its rind is natural, rough, soft, thin and whitish or yellowish to shades of grey in colour. Internally, it is compact, somewhat soft, slightly unctuous with areas of varying degrees of cohesion, ivory-white that evolves to cream in colour, bright intensity, with the presence of a faintly marked halo, depending on the degree of maturation. It has galleries and irregular cavities of variable size, filled with bluish-green mould.

De Valdeón has a smell of medium intensity, reminiscent of: slightly acidic, lactic, cow, natural cave, mould odours. As maturation progresses, spicy, fruity and even aromatic aromas and scents appear, although the musty smell is maintained. The aroma is reminiscent of type milk from which it made, especially if it is goat's milk. The flavour is of medium intensity, slightly acidic and slightly salty. It has an intense, slightly burning or itching sensations on the tongue, that are more pronounced when the ripening time is longer, especially when made from goat's or sheep's milk.

13.3.7 *L'Alt Urgell y la Cerdanya*

This cheese is a low-acid and predominantly enzymatically coagulated cheese, made with pasteurised whole milk from Friesian cows. It is cylindrical in shape, with flat horizontal and slightly convex vertical surfaces. The rind is natural, with a characteristic odour developed by the specific organisms that grow on the surface. It is slightly moist and orange in colour. There are abundant eyes: small, irregularly shaped and evenly distributed throughout the cream-coloured body. The smell is of low intensity with a mild ammonia background, typical of washed rind cheeses. The aroma is reminiscent of: dried mushrooms, moss, toasted nuts and the flavour is balanced: slightly acidic and salty. A soft and pleasant bitterness can be detected.

13.3.8 *Los Beyos*

Los Beyos cheese is a dual coagulated cheese: very acidic and not very enzymatic, made with pure raw or pasteurised cow's, sheep's or goat's milk, so its sensory characteristics are different depending on the milk of origin. It has a cylindrical shape, with a diameter almost identical to its height and a thin, rough rind of creamy yellow (cow), light brown (sheep), pale yellow (goat) colour. Its interior aspect is firm, closed, without eyes and scarce presence of openings. The colour is pale yellow (cow), ivory (sheep), white (goat) and the cheese is crumbly when cut. The body can have medium to high firmness with medium (cow) to high friability (sheep and goat types).

The odour is of low intensity: lactic, yoghurt, acid whey, cow, goat or sheep (clean) and wet, musty cave. The aroma is of light intensity, with notes of fresh lactic, buttery, cow's milk, grains or nuts (walnut) and in the very young and in spring, grass and cut grass. The taste is somewhat acidic and salty. It is also slightly spicy and in the most mature: goaty, somewhat astringent and somewhat burning.

13.3.9 *Mahón-Menorca*

Mahón-Menorca cheeses are enzymatically coagulated and are produced with cow's and sheep's milk with a maximum proportion of 5% sheep's milk. Artisan Mahón-Menorca cheeses are made with raw milk. Among Spanish cheeses, the production of Mahón-Menorca cheese is the second only to Manchego.

It has a rectangular shape with rounded edges and rind that is not very developed with a yellowish-white colour as a soft cheese, and orange to brown in the semi-hard and hard types. The body is of weak roughness, somewhat moist, a little elastic at the immature stage. As the ripening process progresses, these characteristics change and in the semi-hard or hard cheeses the texture will be rough, dry and not at all elastic. Variations in the textural parameters of Mahón-Menorca cheese can be considered as an indicator of cheese maturity (Benedito et al. 2000). It has small number (~10) of small eyes: more or less round in shape and pea distributed irregularly in the body.

The odour is of mild intensity, reminiscent of fresh milk or butter. In semicured cheeses, it is of moderate intensity, reminiscent of: fresh milk, butter and toasted nuts, and if they are artisan-produced, also of leather, curd and caves with moderate-high intensity. When the cheese is aged, the odour intensity becomes moderate-high with notes of: melted butter, leather and toasted hazelnut. In artisan-cured cheeses, meat broth, olive oil and rind odours are also detectable and the intensity is high. The aroma is of mild intensity, with notes of milk and cream, butter, curds and cut grass. In semi-hard cheeses, the intensity is moderate or moderately high (if they are artisan), reminiscent of milk, butter, curd and cut grass together with toasted nuts (hazelnuts). Leather and meat stock aromas are also detectable in artisan-produced semi-hard Mahón-Menorca. In the hard Mahón-Menorca, the aroma is of moderate-high intensity and high in the artisan-produced cheeses. Notes of cream, butter, cooked potato, toasted hazelnut, leather and meat broth appear, and in the artisan cheeses notes of olive oil, rind, rind, walnuts, nuts and spices can also be detected. The flavour is salty and acidic, soft and balanced in soft Mahón-Menorca cheeses, while in the semi-hard cheeses the flavour is moderately salty and acidic but balanced. In hard Mahón-Menorca cheeses, the flavour is moderate – highly salty and acidic but balanced, although a slightly pungent and astringent sensation may develop. In artisan-produced Mahón-Menorca cheeses, the flavours are similar but more intense. Changes in the volatile fraction during the ripening of Mahón cheese have been studied by Mulet et al. (1999) who identified 21 main aromatic components. According to the statistical analysis by principal components, the variables were grouped by their chemical characteristics (fatty acids, methylketones, ethyl esters) and their ripening time (0/18 days, 35 days, 53/67/82 days).

13.3.10 *Majorero*

This cheese which is enzymatically coagulated is made from goat's milk of the Majorero breed on the island of Fuerteventura (Spain). It is a pressed cheese, with a compact cut curd, creamy texture and developed lactic flavour. The moulds have a characteristic plaited palm outline.

Majorero cheese is cylindrical in shape, with a rind that is practically non-existent, somewhat unctuous to the touch and white in colour in the case of soft cheeses, yellowish-white for semicured cheeses and yellowish brown for the most mature cheeses. During maturing, its surface may be evenly coated with paprika, gofio and oil, giving it a characteristic colour and appearance, red for the former and beige for the latter. It has the engraved impressions of the moulds or 'pleitas' on the side surface (diamond pattern of plaiting for the palm 'pleita') and the flat sides are marked with the typical 'pinta' or 'quesera', parallel grooves from the moulding bases.

The body is weakly firm, closed, with a few small, evenly distributed eyes. The colour is snow white for soft cheeses, ivory-white for semicured cheeses and darker ivory for mature cheeses. There is an area near the rind or transition border of different colour and texture, which does not exist in soft cheeses, but is present in semicured and cured cheeses and increases in thickness with ripening. In semi-hard cheeses, the body is weak to medium in firmness, of less than medium friability, tender to chewy, slightly pasty, somewhat sticky, fine-grained, somewhat moist, somewhat creamy, somewhat meltable and soluble. Cured cheeses have the same but slightly more pronounced characteristics.

Pasteurised milk cheeses have a higher moisture content and are creamier and less hard than raw milk cheeses (Fontecha et al. 1990). The smell is of faint intensity, of: goat's milk, some whey and a little yoghurt in the soft cheeses. If the goats are grazing, there are hints of butter and nuts. In semi-soft cheeses, the smell is of weak intensity, with predominantly lactic aromas of goat's milk, whey and yoghurt in order of decreasing importance. Notes of butter and nuts may also appear in cheeses from herds grazing in arid areas. In mature cheeses, the aroma is similar to the above, but with a greater intensity of more evolved notes of butter, nuts, fermented fruits and hay. The aroma of soft Majorero cheeses is of weak to medium intensity, surpassing the smell, but no other descriptors other than those already described for the smell are detected, although the following predominate: the fruit aromas, citric notes, followed by the lactic notes. In semicured cheeses, the aroma is of medium intensity. No descriptors other than those already described for the smell are detected, although the citric acidity of the fruity aromas predominates, followed by the lactic, buttermilk and yoghurt. Occasionally, vegetable notes, such as hay, can be detected. In semi-hard Majorero cheeses, the aroma is of medium-high intensity, and in addition to the descriptors already given for the smell, the fruit notes: nuts, oil and citrus fruits diminish, followed by the lactic notes, butter, hay and hints of mould. The flavour of Majorero is slightly acidic, salty and slightly bitter in the young cheeses. In the semicured cheeses, the sensation of acidity is muted, it has a desirable salinity and is slightly bitter. In the cured Majorero cheese, it is slightly acidic, salty and slightly bitter. There is also a slightly spicy sensation, somewhat astringent, though not at all burning, in matured Majorero. Most of the sensory parameters of Majorero cheese change significantly during the ripening period, in addition to an intensification of the smell and aroma (Fresno and Alvarez 2012). On the other hand, a specific starter culture has been developed to impart some sensory characteristics peculiar to artisan Majorero cheese to industrially produced cheese (Requena et al. 1992). The texture properties and other sensory characteristics imparted by this starter have been studied in Calvo et al. (2007) and Castillo et al. (2007). In addition, the effect of an alternative goat diet on the quality of Majorero cheese has also been studied (Álvarez et al. 2007).

13.3.11 *Mestiço Tolosa*

This is a semi-soft, mature cheese, made from a mixture of ewe's and goat's milk in the proportions of 20/80%, 40/60% or 60/40%. The curds are drained after coagulation, by the action of either animal rennet or *C. cardunculus* L., and treated and matured within the geographical processing area set out in the specification. It has a cylindrical shape. The rind is slightly rough, thin, whole, sometimes greasy but resistant, toasted yellow or orange colour and uniform. The texture is well bound, presenting few small eyes, with a colour ranging from yellowish to orange. The flavour is strong, clean, slightly spicy and pleasant.

During the maturation process, the Queijo Mestiço de Tolosa is immersed in a mixture of bell pepper powder and water, which gives it a characteristic colour.

13.3.12 *Nata de Cantabria*

Nata de Cantabria cheese is a low-acid, predominantly enzymatically coagulated cheese. It has a cylindrical or rectangular shape, with a thin, soft and waxy rind and yellowish bone colour.

The body is compact, closed, sometimes with some small eyes distributed regularly. Ivory-white to yellowish-white coloured in the most cured cheeses. Young cheeses are soft and have little firmness but older cheeses are firm, though not crumbly. In younger cheese, the texture is somewhat pasty comprising of particles of fine granularity with a tendency to adhere to the palate and fatty, creamy, melting, soluble mouthfeel.

The smell is of low intensity; lactic acid and slightly buttery and the aroma is slightly reminiscent of butter. The flavour is slightly acidic, a little salty and refreshing.

13.3.13 *Palmero*

This cheese is made with enzymatic coagulation made from raw goat's milk from the Palmero breed and animal rennet, sometimes natural kid's rennet. The cheeses may be smoked according to a traditional practice, using almond (*Prunus dulcis*), dried prickly pear (*Opuntia ficus indica*) and Canary Island pine (*Pinus canariensis*) shells, which darken the rind of the cheeses to varying degrees (Fresno et al. 2013). It is cylindrical in shape. The rind has a smooth lateral surface. The smoked ones may have different shades of dark brown, and bands of different intensity may appear. When the surface has been treated (smeared) with oil, gofio or flour, the rind acquires the characteristic appearance of these cheeses.

Inside, Palmero is bright white in colour, acquiring ivory and matt shades as it matures. The surface and body are free of cavities and cracks, and may have small eyes distributed randomly and mostly irregularly. As a soft cheese, it has a medium firmness. The hard cheeses (six months of maturation) are firmer, almost hard, quite crumbly, stick to the palate, are very slightly moist but feel dry in the mouth and have a granular but soluble consistency. Palmero cheeses made with natural rennet are harder, more elastic, breakable and darker than those made with commercial rennet.

The smell of Palmero is of weak intensity in the younger cheeses and a little more intense in the older matured ones. In both cases, the descriptor goat's milk and whey are used. If the ewes graze, there are hints of butter and nuts, peeled almonds and herbaceous notes.

As for the aroma, when the cheese is soft, the same characteristics can be detected as in the smell and with the same intensity, as well as an aroma of dry grass. When the cheese matures, notes of butter, dry grass, rancid nuances and rind can be detected without being aggressive. When the cheese is smoked, in addition to the typical cheese components, typical smoke components are also detected, the main ones being acids and phenolic derivatives. The latter compounds play an important role in the taste of this cheese, and their relative proportions together with the presence of specific smoke components derived from pine leaves serve to distinguish Palmero cheese from other smoked cheese varieties (Guillén et al. 2004). As a soft cheese, the flavour is slightly acidic, with just the right amount of salt, sometimes with a faint bitter taste. It is the same in mature Palmero cheeses, although with a little more intensity.

13.3.14 *Pico*

Pico cheese is made in small artisan dairies on the island of Pico (Azores) with raw milk coagulated with animal rennet. It is a flat cylindrical cheese, rounded and regular, with convex edges. The rind is yellow. The consistency of the body is soft, pasty, yellowish-white in colour, with holes, friable and very creamy. It has a strong and characteristic smell. As a mature cheese, it has a salty and intensely piquant flavour. Pico undergoes proteolysis, similar to that of other raw milk Iberian cheeses, but the free fatty acid content is higher, which could contribute to its sensory characteristics (Camara et al. 2017).

13.3.15 *Quesucos de Liébana*

These are low-acid cheeses and coagulation is primarily enzymatic. Made with pasteurised cow's/sheep's/goat's milk, and if they have a maturation period of 60 days, the milk can be raw. It can be smoked. They are from the Liébana valley, in Cantabria (Spain).

The cheeses have a cylindrical or disc shape. The rind is defined but thin, somewhat rough, off-white to yellowish-white in matured cheeses and ochre or light brown in smoked cheeses. The interior is compact, continuous, closed, sometimes with some mechanical eyes (few) irregularly distributed, white in the younger ones to straw yellow in the cured ones. The colour is more or less intense depending on the season and the type of milk used. The body has low firmness, it is soft in the younger cheeses and firmer in the cured ones, somewhat crumbly, chewy, floury, somewhat adhesive in the case of cured ones, of medium granularity, somewhat dry in the case of very cured ones, not very fatty, the younger cheeses melt in the mouth to a degree but the cured ones not at all. The smell and aroma are of medium intensity of butter and smoke in the smoked ones. The flavour is not very intense, somewhat lactic, not at all salty. The smoked cheeses are described as somewhat astringent and slightly spicy, sensations that will be more pronounced the more mature the cheese is or if they are made from goat's or sheep's milk.

13.3.16 *Torta del Casar*

Torta del Casar is made from raw Merino sheep's milk and coagulated with vegetable rennet from the dried flowers of the thistle *C. cardunculus* L. Traditionally, this cheese was wrapped in banana leaves, but nowadays aluminium foil is also used. It has a cylindrical shape with remarkably flat faces and a convex lateral surface. The rind is semi-hard, with a uniform colour of ochre tones without the addition of colourants, and may have small cracks on its surface. The body is soft to very soft, spreadable, sometimes capable of diffusing through cracks if present in the rind and white to yellowish in colour. Torte del Casar may have rounded eyes which are characteristic of maturation and distributed in the body. It is soft, even liquid but more or less firm in the most mature (Añejos) cheeses, not friable, very chewy, tender, the less soft ones may be pasty, somewhat adhesive, have little or no graininess and are not very moist. Moderate or high creaminess is a feature of very mature cheeses as well as meltability and a degree of solubility.

The odour of Torta del Casar is of low-to-medium intensity with lactic notes (yoghurt, sweet cream, depending on the time of year manufactured and age of ripening of the cheese); vegetable notes (hay, wet grass), and animal notes (sheep's milk). The aroma is from mild to more intense, depending on the stage of ripening, reflecting the characteristics of smell already described and butter. Delgado et al. (2010) described the great importance of carboxylic acids and protein degradation as a unique characteristic associated with the typical aroma of Torta del Casar cheese. Acids were the most abundant volatile compounds, followed by alcohols and carbonyls. The taste is characteristic, intense, slightly salty, barely acidic and with a slight bitterness caused by the use of vegetable rennet (Ordiales et al. 2013).

13.4 Semi-hard Iberian cheeses varieties

Among the Spanish and Portuguese cheeses, there are 11 semi-hard cheeses with PDO (Table 13.1, Figure 13.3). Their sensory characteristics are described below and summarised in Tables 13.2 and 13.3).

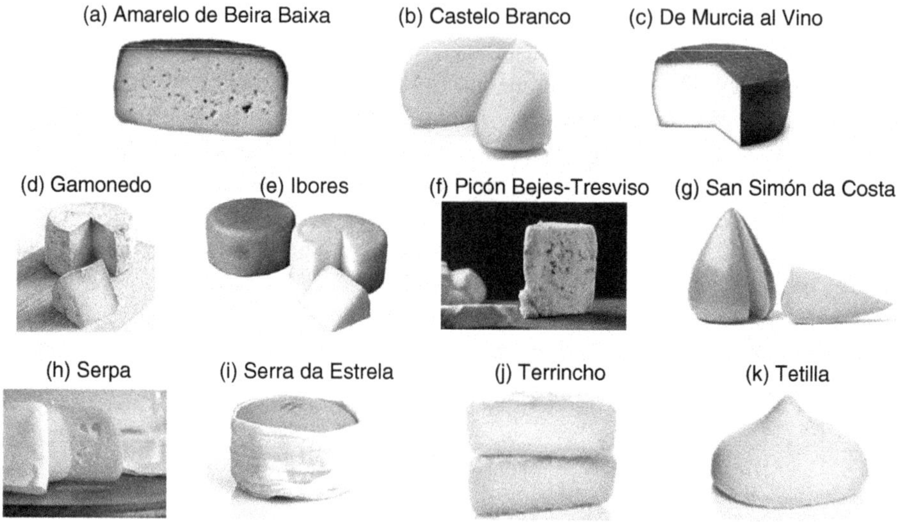

Figure 13.3 PDO semi-hard cheeses from Spain and Portugal. *Source:* (d) Hansgeel/Adobe Stock; (f) Rojo/Adobe Stock; (g) Picture Partners/Adobe Stock.

13.4.1 *Amarelo da Beira Baixa*

Amarelo da Beira Baixa cheese is made from raw sheep's milk or a mixture of sheep and goat milk which is coagulated with animal rennet. It is cylindrical in shape, low, smooth-sided and with a well-formed, fine and yellow or roasted yellow rind. It has a semi-hard yellowish body, with some irregular holes. It has a very marked aroma and is consumed as a mature cheese, with a rustic and rough taste, with a hint of sea salt.

13.4.2 *Castelo Branco*

This cheese is made from raw sheep's milk by slow draining of the curd after coagulation by the action of an infusion of cardoon (*C. cardunculus L.*). It is round, convex on the sides and slightly convex on the top, and may have slight edges. The rind is smooth, well formed, fine, uniform, of an intense straw yellow to brown colour. The body is uniform, slightly crumbly, unctuous, of a uniform yellowish-white colour. The cheese has an intense flavour and aroma described as acidic, sheepy-like and pungent (Ferreira et al. 2009).

13.4.3 *De Murcia al Vino*

This is a dual coagulated cheese, with low-acidity and high-enzymatic addition rate, made with pasteurised Murcian-Granadine goat's milk. The cheeses are soaked in red wine from the Region of Murcia for a variable length of time, depending on the degree of maturity. De Murcia al Vino cheese is cylindrical in shape with straight but not sharp edges and a slightly rounded side. The rind is smooth (no engravings), very light, washed twice with red wine from the region, which gives it a characteristic garnet-violet colour and a wine smell. The texture is compact, closed, sometimes with a few small eyes. It has a matt white colour with shades ranging from ivory to yellowish in the most cured cheeses, with a creamy and elastic texture. The smell is of medium-low intensity. When assessing the aroma, fresh lactic, goat's milk, cream/butter and fresh hazelnut aromas may be detected. The flavour is of medium intensity, slightly acidic and not very salty in the case of young cheeses. In more mature cheeses, the acidity will decrease and the salty flavour will increase. It is slightly astringent.

13.4.4 *Gamonedo*

Gamonedo is a blue cheese smoked with ash or beech wood for two to three weeks prior to ripening in caves. It is traditionally ripened in natural limestones caves and is typically marketed wrapped in leaves of the maple tree or fern. It is a cylindrical cheese, with noticeably flat faces. The rind, acquired during the smoking phase, is thin of a toasted sienna colour, which later, during maturation in caves, acquires reddish, greenish and bluish tones. Later, during its storage period in caves, when populated with fungi, it acquires reddish, greenish and bluish tones. The body is firm, crumbly, tender when chewed, between floury and doughy, depending on the degree of maturation, somewhat sticky or slightly sticky. At maturity, it is somewhat adhesive, granular, sandy, like some grainy pears, slightly moist, somewhat creamy, somewhat meltable. Eyes are irregularly distributed and of small size. Gamonedo has a white or yellowish-white colour, with slight blue-green outcrops on the edges. The smell is of medium intensity and described as natural cave, musty, soft notes of smoke. The aroma is slightly reminiscent of goat's or sheep's milk, depending on the percentage of milk of each species the cheese is made from. The taste is lightly acidic, slightly salty, sometimes a touch bitter and in the case very mature cheeses, slightly spicy, somewhat astringent and somewhat burning. Its physical, chemical and sensory characterisation during ripening has been reported by González de Llano et al. (1992) as well as volatile flavours compounds by González de Llano et al. (1990).

13.4.5 *Ibores*

Ibores cheese is made with raw milk from Serrana, Verata and Retinta goats and their crossbreeds. The milk is coagulated by the dual action of low-acid and high-enzyme addition. The cheese is cylindrical in shape, with remarkably flat horizontal faces and straight or slightly convex sides. The rind is smooth and semi-hard in the case of long cured cheeses, whose natural colour is waxy yellow to dark ochre. It is traditionally dusted with paprika, as well as smeared with oil, and may have various colourings due to the different moulds which colonise the rind; from greyish in the natural rind, to orangey red in the 'pimentonados', to yellow-ochre in the cheeses smeared (Martínez et al. 2011). The firmness of the cheese, in the case of cheeses that are not very mature, is semi-hard, not crumbly, with small holes, not very abundant and unevenly distributed. The colour is ivory-white. The smell is of mild-to-moderate-intensity raw goat's milk.

The aroma has vegetal, floral and fruity hints, which are imparted by the extensive rearing of livestock on natural pastures, heather, thyme, rockrose, rosemary and acorns. Delgado et al. (2011) have reported that the most characteristic compounds in Ibores cheese aroma were butanoic, hexanoic and octanoic acids, some alcohols (2-butanol and 2-heptanol), ethyl esters of hexanoic and butanoic acids, some methyl ketones (2-butanone, 2-pentanone and 2-heptanone) and δ-decalactone which were produced by lipolysis. The taste is moderately acidic at the correct salt content. It is also slightly spicy, growing in intensity as it matures, but without being excessive.

13.4.6 *Picón Bejes-Tresviso*

This cheese is a blue cheese made from raw cow's, sheep's or goat's milk, or mixtures of two or all three types of milk from animals grazed on mountain pastures and undergoes ripening in natural limestone caves. It is a cylindrical cheese with flat faces that in certain cases are slightly concave. The rind is natural, soft and thin. The colour it acquires in the caves is orange-brown to greyish, and to the touch is slimy due to the microbial development on the outside of the cheese during the ripening process.

The body is compact, somewhat soft, slightly unctuous with areas of varying degrees of cohesion and the presence of a weakly marked halo depending on the degree of ripening. The mouthfeel is somewhat oily, creamy, meltable and very soluble. The colour of the body is ivory-white but may vary depending on the type of milk used for production. There are bluish-green zones and streaks due to the development of the fungi of the *Penicillium* genus, which should grow in all the cheese mass equally. The colour of the mould is bluish-green. It should not show marked signs of puncture of the cheese mass. The smell is intense: of cow/sheep/goat, natural cave, mould and in very mature cheeses slightly of ammonia. The aroma is reminiscent of the species of origin of the milk. It is very rich in free fatty acids, methyl ketones and 2-alkanols according to Prieto et al. (2000). The taste is slightly acidic, slightly salty and slightly bitter, with other sensations such as slightly spicy, fiery and somewhat astringent which become more pronounced the more mature it becomes or due to production from goat's or sheep's milk.

13.4.7 *San Simón da Costa*

San Simón da Costa cheese is made from pasteurised or raw cow's milk which is enzymatically coagulated. If it is made from raw milk, it must be matured for 60 days. The cheese is smoked with birch wood just before delivery to the market. It has an intermediate shape between a spinning top and a bale, with the upper part ending in a beak. The rind is hard and inelastic, shiny, smooth, waxy, smoked, yellowish-ochre in colour and somewhat oily. The cheese body is firm, closed and if it has eyes, they are not very numerous, rounded or irregular in shape and small in size. It is white to yellow in colour. The smell is of medium intensity, of lactic (buttery) and roasted (smoky) nature. The aroma has the same characteristics as those already described for the smell, but with increased intensity. The taste is slightly salty, a little sour and a little sweet. In very mature cheeses, other sensations are found, such as a little spicy, a little astringent but not at all burning.

13.4.8 *Serpa*

Serpa cheese, another of the most popular cheeses from Portugal, is made from the milk of Merino sheep breeds (Araújo-Rodrigues et al. 2020) and has a delicate subtly spicey flavour, due the vegetal coagulant used for coagulation. It is a buttery semi-hard cheese with few or no eyes, obtained by slow draining of the curds after coagulation of raw sheep's milk by the action of a thistle infusion. It is shaped like a short cylinder (disc), regular with bulging sides and a little at the top, without defined edges. The rind is malleable, allowing some fluctuation, well formed, slightly rough and thin with a pale straw yellow uniform colour. The body is closed, buttery, with an easily deformable cut area. It is very unctuous, with few or no eyes. It has a yellowish-white or straw yellow colour, which darkens in contact with air. The aroma and flavour are generally strong with a dominant spicy taste.

13.4.9 *Serra da Estrela*

This is the most notable cheese variety in Portugal, made with raw milk from the Bordaleira sheep breed and is considered one of the seven wonders of Portuguese gastronomy (Inácio et al. 2020). It is a semi-hard to extra-hard textured cheese, slightly crumbly, unctuous, orange/brown in colour, with few or no eyes, obtained by prolonged maturation.

Throughout ripening and storage, intense proteolytic activity occurs, due to coagulation with extracts prepared from dried wild thistle flowers of *C. cardunculus* L. Volatile compounds are associated with its creamy, unctuous texture, with few or no eyes and has a smooth, clean, slightly acidic bouquet. Free fatty acids are quantitatively the dominant flavour compounds present; Butyric and isovaleric acids levels dictate the overall aroma of Serra da Estrela cheese. Acetic, isobutyric and isovaleric acids increase in concentration during the ripening process up to 90 days, and remained constant thereafter (Tavaria et al. 2004). Thus, acids are the most abundant volatile compounds, followed by alcohols and carbonyls (Dahl et al. 2000).

Serpa and Serra da Estrela cheeses may seem similar, but the origin of the milk and the production areas is enough to obtain two very different cheeses in terms of flavour and texture (Roseiro et al. 2003c).

13.4.10 *Terrincho*

Terrincho cheese is a semi-soft cheese made from sheep's milk of the Churra da terra Quente or Terrincha breeds. It is a cured cheese, cylindrical in shape and has a well-defined rind which is smooth and straw yellow in colour. The body has a uniform white colour with some holes. It is slightly oily. It stands out for its high intensity of animal odour and aroma and for its high acidity and astringency (Ferreira et al. 2006).

13.4.11 *Tetilla*

Tetilla cheese is a dual coagulation cheese: low-acid and highly enzyme, made with pasteurised cow's milk. It has a conical, concave-convex shape in the form of a teat or pear. The rind is appreciable, thin, smooth, clean, mould-free and elastic, straw yellow in colour. The interior of the cheese is soft, closed and uniform, with no or few holes, evenly distributed, ivory-white to soft yellow in colour. The smell, of lactic acid and of the milk from which it originates, is of low intensity. The slightly intense aroma is milky, reminiscent of butter and a hint of vanilla in the more mature cheeses. The flavour is slightly acidic, with just the right amount of salt. It is not spicy, astringent or burning.

Recently, selected strains of *Kocuria varians* and Yarrowia lipolytica have been studied for use as adjunct cultures in industrially produced Tetilla to obtain cheeses with a similar sensory profile to traditional 'high-quality' cheeses, which would contribute to the quality and uniqueness of Tetilla cheese (Centeno et al. 2017).

13.5 Semi-hard or hard Iberian cheese varieties

There are 13 semi-hard or hard Iberian cheeses, all of them with PDO (Table 13.2, Figure 13.4). Four of them have been described in previous sections (Camerano, Mahón-Menorca, Majorero and Palmero cheeses) and the other 10 are described below. Tables 13.2 and 13.3 also summarise the characteristics of the 14 semi-hard or hard cheeses.

Figure 13.4 PDO semi-hard or hard cheeses from Spain and Portugal. *Source:* (b) Sebastiana Raw/Adobe Stock; (d) Michael Marquand/Adobe Stock.

13.5.1 *Casín*

Casín cheese is made from raw cow's milk of ancestral origin using enzymatic coagulation. It has an irregular cylindrical-disc shape, with a smooth, thin, poorly defined crust, since, due to the successive kneading, the outside and inside form a compact and bound, clean, dry or slightly unctuous curd mass. It has a dark creamy yellow colour with whitish tones. Casín cheeses have floral or geometric motifs, symbols or names, or each producer's own seal on its upper surface.

Its interior is shiny, without cracks, may have some small eyes (angular or rounded and irregularly distributed), has a uniform colour of between ivory-white and pale yellow in young cheeses. Mature cheeses have very compact body, may have few eyes, and are an intense yellow colour but are paler in the centre. The smell is of high intensity, reminiscent of lactic acid, and aged butter, as is the aroma, although with greater intensity. The flavour is slightly acidic, with the right amount of salt and somewhat bitter. It also gives other sensations, such as intense spiciness.

A full description of the microbial populations of Casín cheese has been done (Alegría et al. 2009), although the direct relationship with the cheese flavour is not available yet.

13.5.2 *Évora*

This cheese is made from raw sheep's milk of the regional White Merina breed and coagulated with vegetable rennet from cardoon (*C. cardunculus* L.). It is a small, hard, disc-shaped cheese with a bulging side and a little convex on the upper side, without defined edges. The rind is hard and semi-hard, well-formed, smooth or slightly rough and yellowish in colour, darkening when exposed to air. The body is serrated and well bound, unctuous, with a few small eyes. Évora cheese has a characteristic aroma and flavour; slightly sharp and more pronounced in the case of hard cheeses.

13.5.3 *Idiazábal*

Idiazábal cheese is made from raw milk from Latxa and/or Carranzana sheep using enzymatic coagulation. The cheeses are cylindrical in shape, with flat horizontal faces and slightly convex vertical faces and are uniform with rounded or sharp edges. The rind is hard and smooth, free of foreign matter, with slight impressions of the cloths used during moulding. Slight marks on the faces from the box or tray they ripen in are sometimes visible. Idiazábal cheese is uniform in colour, ranging from pale yellow or whitish grey to dark brown in the case of smoked cheeses. The body is compact and closed, without cracks, without eyes or with some randomly distributed ones, mostly irregular in shape and smaller than a 'short' grain of rice. The colour of the interior of the cheese is homogeneous, varying from ivory-white to straw yellow.

The smell is of medium intensity: lactic in nature with animal (natural rennet) and with roasted notes in smoked products. For the aroma, no sensations other than those already described for the smell are experienced, except for sweet cereal notes. In raw milk Idiazábal cheeses, the characteristic sensory attributes for the aroma are present at three months, whereas in pasteurised milk cheeses it takes six months (Barrón et al. 2005, 2007). The flavour is sweet (weak in intensity), sour (weak to medium in intensity) and salty (medium in intensity). Other sensations may appear such as spicy (weak to medium intensity), barely astringent and not burning.

It is also notable that significant seasonal differences can be found in Idiazábal cheeses; cheeses made from spring milk have more intense fruity and sweet overtones than winter-produced cheeses (Abilleira et al. 2010).

13.5.4 *Manchego*

Manchego cheese is made exclusively with sheep's milk from the Manchego breed using enzymatic coagulation. It is cylindrical in shape, with noticeably flat sides. The rind is hard, closed and free of parasites. The rind has a pale yellow colour but may be greenish-blackish in the old cheeses.

The lateral surface is typically engraved with the design of the plaiting of the 'pleita' of esparto grass, and the flat faces are marked with the typical 'flower' striations of the bases of the moulds. The body is firm, compact, somewhat hard when cured, with some small, irregularly distributed eyes but sometimes closed without eyes. The interior of the cheese is white to ivory-white to yellowish-white in the more mature cheeses. The smell is of medium intensity, predominantly lactic in young cheeses but evolves into different nuances with ageing. Hints of nuts, hazelnut and spicy appear during ripening and increase in intensity. The aroma is pleasant, of sheep's milk and more intense in the artisanally produced cheeses. The flavour is slightly acidic, slightly salty, sometimes a little sweet in very matured cheeses. A slight spicy sensation may appear in semicured cheeses, which is accentuated in cured cheeses.

Raw milk Manchego cheeses ripen faster than pasteurised milk cheeses, developing a more intense flavour and aroma, and higher scores for most of the gustative attributes (Ballesteros et al. 2006; Poveda et al. 2015). In general, as maturation increases and due to the effect of proteolysis, there is an increase in free amino acids and other related compounds (Poveda et al. 2004). At the end of the ripening period (240 days), raw milk cheese develops a more bitter taste than pasteurised milk cheeses, due to the release of bitter peptides, as a consequence of intense proteolytic activity (Poveda et al. 2015). In artisanal Manchego cheese, differences in the olfactory characteristics compared to corresponding industrially produced cheeses were detected after six months of ripening (Gómez-Ruiz et al. 2002).

13.5.5 *Nisa*

This is a cured cheese made from ewe's milk by infusion of cardoon (*C. cardunculus* L.). The body is of semi-hard consistency, compact, with some small eyes, obtained by slow draining of the whey. Its colour is yellowish-white. It has small holes. The sheep from whose milk this cheese is made graze in areas with typical Mediterranean flora, which gives the milk, and, therefore, the cheese, a particular flavour. The flavour of Nisa cheese is very pronounced and reminiscent of Brazil nuts. Nisa is best consumed between three and five months of age, when the flavour is at its best and the texture is quite soft and buttery.

13.5.6 *Picante de Beira Baixa*

Picante de Beira Baixa cheese is produced in the area of Castelo Branco (Portugal) and made from raw sheep's milk or a mixture of sheep and goat milk, which is coagulated with animal rennet. It is characterised by a high ripening depth index (Freitas et al. 1997), which is a measure of starter peptidase activity.

The appearance of the cheese is that of a low cylinder (disc), with straight, smooth faces and defined edges and without a rind. The body is closed, semi-hard or hard, without eyes or with small irregular eyes, of a dirty white to greyish white colour. It has a strong and characteristic aroma and sharply spicy flavour. The proportion of ovine and caprine in the cheese milk contributes considerably to the characteristic piquant flavour (Freitas and Malcata 1998).

13.5.7 *Rabaçal*

This is an artisanal cured cheese made from a mixture of sheep and goat milk, in a ratio of approximately two parts sheep's milk to one-part goat's milk, by the action of animal rennet. It is cylindrical in shape. The rind can be white to straw yellow in colour, smooth, dry, slightly oily and uncoated. The body is of medium consistency with no or few holes (small and irregular) and matt white in colour. The characteristic flavour of Rabaçal cheese is due to a plant called 'Santa Maria', a spontaneous thyme, which is an abundant forage for the sheep and goats in the area where this cheese is made.

13.5.8 *Roncal*

Roncal cheese is made from the raw milk of sheep of the Latxa breed using enzymatic coagulation. It is cylindrical in shape with noticeably flat sides and sharp edges. The rind is natural, smooth, hard, thick, rough to the touch, oily, with or without surface mould growth, and brown or straw-coloured. The body is off-white to yellowish-white in colour, depending on the stage of maturity. Roncal has no eyes but has numerous irregular, pinhead-sized cavities. It is very firm, being semi-hard as a young cheese and hard, crumbly with advancing age. The smell is of high intensity, mainly of: sheep's milk with notes of cream, meat broth, olive, toffee, liquid caramel, toasted, acetic acid, sweat, acidified milk, ammonia, leather, stable straw or wool (Larráyoz et al. 2007). The aroma, which is detected in a mild form, has notes of: nuts, pineapple, cider, roasted touches, the prevalence and intensity depending on the stage of ripening. The flavour is slightly acidic, not at all or very slightly bitter, very slightly sweet and medium salty. Sixty-eight volatile flavour compounds have been reported, including hydrocarbons, fatty acids, esters, sulphur and carbonyl compounds and (especially) alcohols. All of them appeared in all seasons, but in different concentrations (Izco and Torre 2000). Other sensations detectable are somewhat astringent and somewhat spicy, perceptions that are more pronounced the more mature the cheese is.

13.5.9 *São Jorge*

This cheese is obtained by draining and pressing after coagulation of raw whole cows' milk produced on the São Jorge Island of Azores. It is a hard or semi-hard cylindrical cheese, regular in shape. The rind is of a hard consistency, sometimes dark yellow in colour sometimes with reddish brown spots, smooth in appearance, well-formed and sometimes coated with paraffin or other suitable colourless plastic coatings.

The body is firm textured, sometimes brittle, yellowish in colour, with many small and irregular eyes, unevenly distributed. As regards aroma and flavour, São Jorge has a strong bouquet, clean and slightly spicy features which are accentuated with ageing, probably due to extensive primary proteolysis and the high concentration of peptides found in this cheese (Kongo et al. 2009).

13.5.10 *Zamorano*

Zamorano cheese is made from raw or pasteurised sheep's milk from the Castellana and Churra breeds using animal rennet. It is cylindrical in appearance with flat horizontal faces and straight sides bearing the 'pleita' mark (plaiting of the esparto grass). The rind is hard, pale yellow or dark grey in colour and well defined. It is medium firm when young and very firm when very ripe, crumbly, tender and doughy. The body of the cheese is slightly sticky, floury, moist in a young cheese and dry in a very ripe one, not very oily, not creamy, non-melting and somewhat soluble. The smell is of medium intensity, reminiscent of sheep's milk, cooked butter and hay in young cheeses and evolving to notes of dried fruits, nuts and a touch spicy in the matured cheese. The aroma is mild and reminiscent of matured (sheep's milk) curd, volatile fatty acids and nutty. Zamorano cheese has a volatile profile similar to Manchego cheese, with aldehydes, ketones, ethanol and propanol, and ethyl and propyl esters, as the most abundant compounds, although their proportions differ (Fernández-García et al. 2004). This profile is affected by seasonal variations (Fernández-García et al. 2004), and colour intensity and hardness are significantly affected by breed (Lurueña-Martínez et al. 2010). The taste is somewhat acidic, and can be somewhat astringent depending on maturity and somewhat spicy in very mature cheeses.

13.6 Hard or extra-hard Iberian cheese varieties

Finally, four cheeses can be found in the section of (extra) hard cheeses (Tables 13.2 and 13.3, and Figure 13.5). However, three of them, Camerano, De Murcia and Palmero cheeses, have been previously described in other sections, so only the characteristics of Cabra Transmontano cheese are shown below.

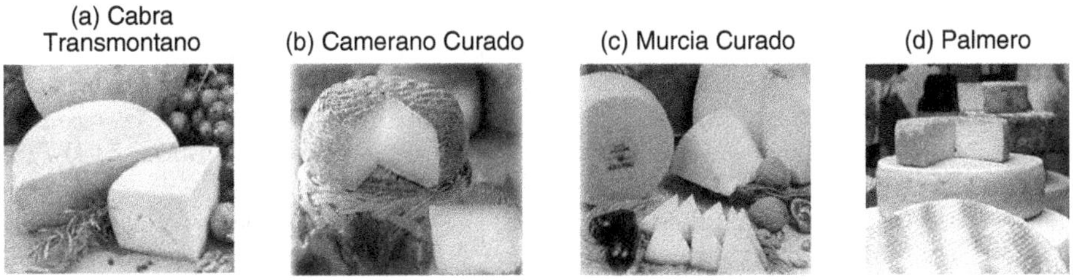

Figure 13.5 PDO hard cheeses from Spain and Portugal. *Source:* (a) Artfood/Adobe Stock.

13.6.1 *Cabra Transmontano*

Cabra Transmontano cheese is a matured, extra-hard cheese, obtained by slow-drying of the curd after the coagulation of raw Serrana goat's milk by the action of the animal rennet.

It has a regular short cylindrical (plate) shape with defined edges. The rind is semi-hard, intact, well-formed, smooth and white. The body is closed, cast and not buttery, not very smooth, with a few holes and with a uniform white colour. The aroma and taste are intense to pleasant, clean, generally with a slightly spicy aftertaste (Mendonça et al. 2017).

References

Abilleira, E., Schlichtherle-Cerny, H., Virto, M. et al. (2010). Volatile composition and aroma-active compounds of farmhouse Idiazabal cheese made in winter and spring. *International Dairy Journal* 20: 537–544.

Alegría, A., Álvarez-Martín, P., Sacristán, N. et al. (2009). Diversity and evolution of the microbial populations during manufacture and ripening of Casín, a traditional Spanish, starter-free cheese made from cow's milk. *International Journal of Food Microbiology* 136: 44–51.

Álvarez, S., Fresno, M., Méndez, P. et al. (2007). Alternatives for improving physical, chemical, and sensory characteristics of goat cheeses: the use of arid-land for-ages in the diet. *Journal of Dairy Science* 90: 2181–2188.

Amigo, L. and Fontecha, J. (2011). Goat milk. In: *Encyclopedia of Dairy Sciences*, 2e, vol. 3 (ed. J.W. Fuquay, P.F. Fox and P.L.H. McSweeney), 484–493. San Diego: Academic Press.

Araújo-Rodrigues, H., Tavaria, F.K., dos Santos, M.T.P.G. et al. (2020). A review on microbiological and technological aspects of Serpa PDO cheese: an ovine raw milk cheese. *International Dairy Journal* 100: 104561.

Arqués, J.L., Garde, S., Fernández-García, E. et al. (2007). Volatile compounds, odor, and aroma of La Serena cheese high-pressure treated at two different stages of ripening. *Journal of Dairy Science* 90: 3627–3639.

Ballesteros, C., Poveda, J.M., González-Viñas, M.A., and Cabezas, L. (2006). Microbiological, biochemical and sensory characteristics of artisanal and industrial Manchego cheeses. *Food Control* 17: 249–255.

Bárcenas, P., Pérez-Elortondo, F.J., Salmerón, J., and Albisu, M. (2002). Sensory characterisation of ewe's milk cheeses using direct and indirect similarity measures: a comparison. *Science of Food and Agriculture* 82: 435–442.

Barrón, L.J.R., Redondo, Y., Aramburu, M. et al. (2005). Variations in volatile compounds and flavour in Idiazabal cheese manufactured from ewe's milk in farmhouse and factory. *Journal of the Science of Food and Agriculture* 85: 1660–1671.

Barrón, L.J.R., Redondo, Y., Aramburu, M. et al. (2007). Volatile composition and sensory properties of industrially produced Idiazabal cheese. *International Dairy Journal* 17: 1401–1414.

Barrón, L.J.R., Aldai, N., Virto, M., and de Renobales, M. (2017). Cheeses with protected land- and tradition-related labels: traceability and authentication. In: *Global Cheesemaking Technology: Cheese Quality and Characteristics* (ed. P. Papademas and T. Bintsis), 100–119. New York: Wiley.

Ben, A., Besbes, S., Attia, H., and Blecker, C. (2017). Milk-clotting properties of plant rennets and their enzymatic, rheological, and sensory role in cheese making: a review. *International Journal of Food Properties* 20: 1–18.

Benedito, J., Gonzalez, R., Rosselló, C., and Mulet, A. (2000). Instrumental and expert assessment of Mahón cheese texture. *Journal of Food Science* 65: 1170–1174.

BOCYL (2021). Regulation of Valdeón cheese and its Regulatory Council. Order of 27 April 2021 of the Agriculture and Livestock Counseling. *Boletín Oficial de Castilla y León (Official Bulletin of Castile and Leon)* 85: 22837–22854. 9 July 1998.

BOE (1985a). Regulation of the PDO Mahón-Menorca cheese and its Regulatory Council. Order of 24 June 1985 of the Ministry of Agriculture, Fisheries and Food. *Boletín Oficial del Estado (State Official Bulletin)* 160: 21162–21168. 5 July 1985.

BOE (1985b). Regulation of the PDO Nata de Cantabria cheese and its Regulatory Council. Order of 29 October 1985 of the Ministry of Agriculture, Fisheries and Food. *BoletínOficialdel Estado (State Official Bulletin)* 272: 35816–35821. 13 November 1985.

BOE (1990). Regulation of the PDO Cabrales cheese and its Regulatory Council. Order of 29 June 1990 of the Ministry of Agriculture, Fisheries and Food. *BoletínOficial del Estado (State Official Bulletin)* 166: 20077–20083. 12 July 1990.

BOE (1991). Regulation of the PDO Roncal cheese and its Regulatory Council. Order of 29 June 1991 of the Ministry of Agriculture, Fisheries and Food. *Boletín Oficial del Estado (State Official Bulletin)* 63: 8489. 12 July 1991.

BOE (1993a). Regulation of the PDO Zamorano cheese and its Regulatory Council. Order of 10 October 1993 of the Ministry of Agriculture, Fisheries and Food. *Boletín Oficial del Estado (State Official Bulletin)* 120: 115311. 25 October 1993.

BOE (1993b). Regulation of the PDO Idiazábal cheese and its Regulatory Council. Order of 29 June 1993 of the Ministry of Agriculture, Fisheries and Food. *Boletín Oficial del Estado (State Official Bulletin)* 289: 34591–34596. 12 July 1993.

BOE (1994a). Regulation of the PDO Picón Bejes-Tresviso cheese and its Regulatory Council. Order of 1 March 1994 of the Ministry of Agriculture, Fisheries and Food. *Boletín Oficial del Estado (State Official Bulletin)* 57: 7585–7589. 8 March 1994.

BOE (1994b). Regulation of the PDO de Liébana cheese and its Regulatory Council. Order of 7 March 1994 of the Ministry of Agriculture, Fisheries and Food. *Boletín Oficial del Estado (State Official Bulletin)* 68: 9272–9278. 21 March 1994.

BOE (1996). Ratification of the PDO Majorero cheese and its regulatory council. Order of 6 September 1996 the Ministry of Agriculture, fisheries and food. *Boletín Oficial del Estado (State Official Bulletin)* 223: 27883–37890. 14 September 1996.

BOE (2001a). Ratification of the PDO Palmero cheese and its Regulatory Council. Order of 31 August 2001 of the Ministry of Agriculture, Fisheries and Food. *Boletín Oficial del Estado (State Official Bulletin)* 218: 34210–34217. 11 September 2001.

BOE (2001b). Ratification of the PDO Murcia cheese and Murcia al Vino cheese and its Regulatory Council. Order of 10 October 2001 of the Ministry of Agriculture, Fisheries and Food. *Boletín Oficial del Estado (State Official Bulletin)* 256: 39201–99209. 25 October 2001.

BOE (2004). ORDEN APA/1893/2004, de 7 de junio, por la que se ratifica el Reglamento de la Denominación de Origen Protegida «Afuega'l Pitu». *Boletín Oficial del Estado (State Official Bulletin)* 247: 22436–22443. 18 Junio 2004.

BOE (2005). Ratification of the PDO San Simón da Costa and its Regulatory Council. Order of 17 May 2005 of the Ministry of Agriculture, Fisheries and Food. *Boletín Oficial del Estado (State Official Bulletin)* 128: 18147–18153.

BOE (2008). Regulation of PDO Flor de Guía, media Flor de Guía and Guía cheese and its Regulatory Board. Order of 15 April 2008 of the Ministry of the Environment and Rural and Marine Environment. *Boletín Oficial del Estado (State Official Bulletin)* 137: 26260–26263. 6 June 2008.

BOE (2009). Granting transitory national protection to the PDO Camerano cheese. Resolution of May 18, 2009, of the General Directorate of Industry and Food Markets. *Boletín Oficial del Estado (State Official Bulletin)* 162: 56343–56362. 6 July 2009.

BOE (2010). Resolution of the IPG Los Beyos cheese. Order 4 February, 2010 of the Ministry of the Environment and Rural and Marine Environment. *Boletín Oficial del Estado (State Official Bulletin)* 48: 18128–18138. 24 February 2010.

BOPA (2003). Regulation of the PDO Gamoneu or Gamonedo cheese and its Regulatory Board. Resolution of 6 August 200 of the Rural and Fisheries Counseling. *Boletín Oficial del Principado de Asturias (Official Bulletin of the Principality of Asturias)* 10930–10940. 22 August 2003.

Buffa, M.N., Trujillo, A.J., Pavia, M., and Guamis, B. (2001). Changes in textural, microstructural and colour characteristics during ripening of cheeses made from raw, pasteurized or high-pressure treated goat's milk. *International Dairy Journal* 11: 927–934.

Calvo, M.V., Castillo, I., Díaz-Barcos, V. et al. (2007). Effect of hygienized rennet paste and a defined strain starter on proteolysis, texture and sensory properties of semi-hard goat cheese. *Food Chemistry* 102: 917–924.

Camara, S.P.A., Dapkevicius, A., Rosa, H.J.D. et al. (2017). Physicochemical, biochemical, microbiological and safety aspects of Pico cheese: assessment throughout maturation and on the final product. *International journal of dairy technology* 70: 542–555.

Carbonell, M., Núñez, M., and Fernández-García, E. (2002a). Evolution of the volatile components of ewe raw milk La Serena cheese during ripening. Correlation with flavour characteristics. *Lait* 82: 683–698.

Carbonell, M., Núñez, M., and Fernández-García, E. (2002b). Seasonal variation of volatile components of ewe raw milk La Serena cheese. *Lait* 82: 699–711.

Castillo, I., Calvo, M.V., Alonso, L. et al. (2007). Changes in lipolysis and volatile fraction of a goat cheese manufactured employing a hygienized rennet paste and a defined strain starter. *Food Chemistry* 100: 590–598.

Centeno, J.A., Rodríguez-Alonso, P., Carballo, J., and Garabal, J.I. (2015). A comparative biochemical study of two industrially produced short-ripened cow's milk cheeses with PDO status: rennet-curd Tetilla cheese and acid-curd Cebreiro cheese. *International Journal of Dairy Technology* 68: 291–298.

Centeno, J.A., Garabal, J.I., Docampo, F. et al. (2017). Recovering traditional raw-milk Tetilla cheese flavour and sensory attributes by using Kocuria Varians and Yarrowia lipolytica adjunct cultures. *International Journal of Food Microbiology* 251: 33–40.

Conceição, C., Martins, P., Alvarenga, N. et al. (2018). *Cynara cardunculus*: use in cheesemaking and pharmaceutical applications. In: *Technological Approaches for Novel Applications in Dairy Processing* (ed. N. Koca), 73–107. London, UK: Intech Open.

Council Regulation (2006). On the protection of geographical indications and designation of origin for agricultural products and foodstuffs. N° 510/2006 of 20 March 2006. *Official Journal of the European Union* 93: 12–25. 31 March 2006.

Dahl, S., Tavaria, F.K., and Malcata, F.X. (2000). Relationships between flavour and microbiological profiles in Serra da Estrela cheese throughout ripening. *International Dairy Journal* 10: 255–262.

Delgado, F.J., González-Crespo, J., Cava, R. et al. (2010). Characterisation by SPME-GC-MS of the volatile profile of a Spanish soft cheese (P.D.O. Torta del Casar) during ripening. *Food Chemistry* 118: 182–189.

Delgado, F.J., González-Crespo, J., Cava, R., and Ramírez, R. (2011). Formation of the aroma of a raw goat milk cheese during maturation analysed by SPME-GC-MS. *Food Chemistry* 129: 1156–1163.

DOCE (2000). Reglamento (CE) n° 2446/2000 de la Comisión Europea (Regulation [CE] n° 2446/2000 of the European Commission). *Diario Oficial de las Comunidades Europeas (Official Journal of the European Communities)* 281: 12–13.

DOE (2003). Regulation of PDO Ibores cheese and its Regulatory Council. Decree 69/2003 of 20 May 2003. *Diario Oficial de Extremadura (Official Journal of Extremadura)* 61: 7653–7667. 27 May 2003.

DR (1988). Creation of the Beira Baixa Cheeses Demarcated Region. *Diário da República (Republic Diary)* 121, Decreto Regulamentar n° 22/88: 25 May 1988.

DR (2005a). Recognition of private control and certification of PDO Cabra Transmontano and Azeitão cheeses. *Diário da República (Republic Diary)* 78: 6479. Aviso n°4313/2005 (2ª série), 30 June 2005.

DR (2005b). Recognition of private control and certification of PDO Nisa and PIG Mestiço de Tolosa cheeses. *Diário da República (Republic Diary)* 78: 6479–6480. Aviso n°4314/2005 (2ª série), 30 June 2005.

DR (2007). It establishes that only producers authorized by the Cheese Producers Association of the Castelo Branco District can benefit from the use of Castelo Branco, Amarelo da Beira Baixa and Picante da Beira Baixa cheeses. *Diário da República (Republic Diary)* 107: 15090–15093. Despacho 10518/07 (2ª série) 4 June 2007.

DR (2009). CODIMACO is recognized as a Control and Certification Organism for PDO Rabaçal cheese. *Diário da República (Republic Diary)* 161: 33952. Aviso n° 14787/2009 (2ª série), 20 August 2009.

eAmbrosia-UE (n.d.) (electronic) https://ec.europa.eu/info/food-farming-fisheries/food-safety-and-quality/certification/quality-labels/geographical-indications-register (accessed 6 December 2022).

Fernández-García, E., Carbonell, M., Gaya, P., and Núñez, M. (2004). Evolution of the volatile components of ewe's raw milk Zamorano cheese. Seasonal variation. *International Dairy Journal* 14: 701–711.

Fernández-García, E., Imhof, M., Schlichtherle-Cerny, H. et al. (2008). Terpenoids and benzenoids in La Serena cheese made at different seasons of the year with a *Cynara cardunculus* extract as coagulant. *International Dairy Journal* 18: 147–157.

Ferreira, I.M.P.L.V.O., Veiros, C., Pinho, O. et al. (2006). Casein breakdown in Terrincho ovine cheese: comparison with bovine cheese and with bovine/ovine cheeses. *Journal of Dairy Science* 89: 2397–2407.

Ferreira, I.M.P.L.V.O., Pinho, O., and Sampaio, P. (2009). Volatile fraction of DOP "Castelo Branco" cheese: influence of breed. *Food Chemistry* 112: 1053–1059.

Flórez, A.B., Ruas-Madiedo, P., Alonso, L., and Mayo, B. (2006). Microbial, chemical and sensorial variables of the Spanish traditional blue-veined Cabrales cheese, as affected by inoculation with commercial *Penicillium roqueforti* spores. *European Food Research Technology* 222: 250–257.

Fontecha, F.J., Peláez, C., Juárez, M. et al. (1990). *Journal of Dairy Science* 73: 1150–1157.

Freitas, A.C. and Malcata, F.X. (1998). Lipolysis in Picante cheese: influence of milk type and ripening time on free fatty acid profile. *Lait* 78: 251–258.

Freitas, C. and Malcata, F.X. (2000). Microbiology and biochemistry of cheeses with Appellation d'Origine Protegée and manufactured in the Iberian Peninsula from ovine and caprine Milks. *Journal of Dairy Science* 83: 584–602.

Freitas, A.C., Fresno, J.M., Prieto, B. et al. (1997). Effects of ripening time and combination of ovine and caprine milks on proteolysis of Picante cheese. *Food Chemistry* 60: 219–229.

Freitas, A.C., Macedo, A.C., and Malcata, F.X. (2000). Review: technological and organoleptic issues pertaining to cheeses with denomination of origin manufactured in the Iberian Peninsula from ovine and caprine milks. *Food Science and Technology International* 6: 351–370.

Fresno, M. and Alvarez, S. (2012). Chemical, textural and sensorial changes during the ripening of Majorero goat cheese. *International Journal of Dairy Technology* 65: 393–400.

Fresno, M., Álvarez, S., and Briggs. (2013). Traditional cheesemaking practices, old tools for healthy and desirable cheeses. In: *Handbook of Cheese in Health: Production, Nutrition and Medical Sciences* (ed. V.R. Preedy, R.R. Watson and V.B. Patel), 751–764. The Netherlands: Wageningen Academic Publishers.

Gómez-Ruiz, J.A., Ballesteros, C., González-Viñas, M.A. et al. (2002). Relationships between volatile compounds and odour in Manchego cheese: comparison between artisanal and industrial cheeses at different ripening times. *Lait* 82: 613–628.

González de Llano, D., Ramos, M., Polo, C. et al. (1990). Evolution of the volatile components of an artisanal blue cheese during ripening. *Journal Dairy Science* 73: 1676–1683.

González de Llano, D., Ramos, M., Rodríguez, A. et al. (1992). Microbiological and physicochemical characteristics of Gamonedo blue cheese during ripening. *International Dairy Journal* 2: 121–135.

Guillén, M.D., Ibargoitia, M.L., Sopelana, P. et al. (2004). Components detected by means of solid-phase microextraction and gas chromatography/mass spectrometry in the headspace of artisan fresh goat cheese smoked by traditional methods. *Journal of Dairy Science* 87: 284–299.

Inácio, R.S., Gomes, A.M.P., and Saraiva, J.A. (2020). Serra da Estrela cheese: A review. *Journal of Food Processing and Preservation* 44: e14412.

Izco, J.M. and Torre, P. (2000). Characterisation of volatile flavour compounds in Roncal cheese extracted by the "purge and trap" method and analysed by GC-MS. *Food Chemistry* 70: 409–417.

Kongo, M., Gomes, A.M., Malcata, F.X., and McSweeney, P.L.H. (2009). Microbiological, biochemical and compositional changes during ripening of São Jorge – a raw milk cheese from the Azores (Portugal). *Food Chemistry* 112: 131–138.

Larráyoz, P., Mendía, C., Torre, P. et al. (2007). Sensory profile of flavour and odor characteristics in Roncal cheese made from raw ewe's milk. *Journal of Sensory Studies* 17: 415–427.

Lurueña-Martínez, M.A., Revilla, I., Severiano-Pérez, P., and Vivar-Quintana, A.R. (2010). The influence of breed on the organoleptic characteristics of Zamorano sheep's raw milk cheese and its assessment by instrumental analysis. *International Journal of Dairy Technology* 63 (2): 216–223.

Marcos, A. and Esteban, M.A. (1993). Iberian Cheeses. In: *Cheese: Chemistry, Physics and Microbiology. Volume 2, Major Cheese Groups*, 2e (ed. P.F. Fox), 173–219. London: Chapman & Hall.

Martínez, S., Franco, I., and Carballo, F.J. (2011). Spanish goat and sheep milk cheeses. *Small Ruminant Research* 101: 41–54.

McSweeney, P.L.H., Ottogalli, G., and Fox, P.F. (2017). Diversity and classification of cheese varieties: an overview. In: *Cheese*, Chemistry, Physics and Microbiology, 4e (ed. P. McSweeney, P. Cotter and D. Everett), 781–808. Academic Press.

Medina, M. and Núñez, M. (2017). Cheese from ewe and Goat milk. In: *Cheese*, *Chemistry, Physics and Microbiology*, 4e (ed. P. McSweeney, P. Cotter and D. Everett), 781–808. Academic Press.

Mendía, C., Ibañez, F.C., Torre, P., and Barcina, Y. (2000). Influence of the season on proteolysis and sensory characteristics of Idiazábal cheese. *Journal of Dairy Science* 8 (3): 1899–1904.

Mendonça, A., Sousa, F., Fernandes, A. et al. (2017). Consumers' perception about sensorial characteristics of the extra-long maturation Transmontano goat cheese. *Revista de CiênciasAgrárias* 40: 309–320.

Miguel, E., Blázquez, B., Onega, E., and Ruiz de Huidoro, F. (2002). Los quesos de oveja de razas autóctonas de la comunidad de Madrid (Sheep cheeses from autochthonous breeds of the community of Madrid). *Evaluación de los parámetros de calidad físico-química, microbiológica y sensorial de los quesos de oveja Rubia de El Moler y Colmenareña (Evaluation of physico-chemical, microbiological and sensory quality parameters of Rubia de El Moler and Colmenareña sheep cheeses). SEOC* 335: 229–334.

Mulet, A., Escriche, I., Rossello, C., and Tarrazó, J. (1999). Changes in the volatile fraction during ripening of Mahón cheese. *Food Chemistry* 65: 219–225.

Olarte, C., González-Fandos, E., and Sanz, S. (2001). A proposed methodology to determine the sensory quality of a fresh goat's cheese (Cameranos cheese): application to cheeses packaged under modified atmospheres. *Food Quality Preference* 12: 163–170.

Ordiales, E., Martín, A., Benito, M.J. et al. (2013). Role of the microbial population on the flavour of the soft-bodied cheese Torta del Casar. *Journal of Dairy Science* 96: 5477–5486.

Park, Y.W., Juárez, M., Ramos, M., and Haenlein, G.F.W. (2007). Physico-chemical characteristics of goat and sheep milk. *Small Ruminant Research* 68: 88–113.

Partidário, A.M., Ribeiro, J.C.S., and Prates, J.A.M. (2008). Fatty acid composition and nutritional value of fat in three PDO ewe's milk Portuguese cheeses. *Dairy Science Technology* 88: 683–694.

Piñeiro-Lago, L., Franco, I., Campo-Deaño, L., and Tovar, C.A. (2019). Effect of storage time on microstructure, viscoelastic and biochemical parameters of Afuega'l Pitu cheese (PDO): *Atroncau Blancu* and *Roxu* varieties. *LWT-Food Science and Technology* 116: 1085671.

Poveda, J.M., Cabezas, L., and McSweeney, P.L.H. (2004). Free amino acid content of Manchego cheese manufactured with different starter cultures and changes throughout ripening. *Food Chemistry* 84 (2): 213–218.

Poveda, J.M., Sánchez-Palomo, E., Pérez-Coello, M.S., and Cabezas, L. (2008). Volatile composition, olfactometry profile and sensory evaluation of semi-hard Spanish goat cheeses. *Dairy Science and Technology* 88: 355–367.

Poveda, J.M., Chicón, R., and Cabezas, L. (2015). Biogenic amine content and proteolysis in Manchego cheese manufactured with Lactobacillus paracasei subsp paracasei as adjunct and other autochthonous strains as starters. *International Dairy Journal* 47: 94–101.

Prieto, B., Franco, I., Fresno, J.M. et al. (2000). Picón Bejes-Tresviso blue cheese: an overall biochemical survey throughout the ripening process. *International Dairy Journal* 10: 159–167.

Reis, P.J.M. and Malcata, F.X. (2011). Current state of Portuguese dairy products from ovine and caprine milks. *Small Ruminant Research* 101 (1–3, 133): 122.

Requena, T., de la Fuente, M.A., Fernández de Palencia, P. et al. (1992). Evaluation of a specific starter for the production of semi-hard goat's milk cheese. *Lait* 72: 437–448.

Ribeiro, A.C. and Ribeiro, S.D.A. (2010). Specialty products made from goat milk. *Small Ruminant Research* 89: 225–233.

Rodríguez-Alonso, P., Centeno, J.A., and Garabal, J.I. (2009). Comparison of the volatile profiles of Arzúa-Ulloa and Tetilla cheeses manufactured from raw and pasteurized milk. *LWT – Food Science and Technology* 42: 1722–1728.

Roseiro, L.B., García-Risco, M.R., Barbosa, M., and Ames, J. (2003a). Evaluation of Serpa cheese proteolysis by nitrogen content and capillary zone electrophoresis. *International Journal of Dairy Technology* 56 (2): 99–104.

Roseiro, L.B., Gómez-Ruiz, J.A., García-Risco, M.R., and Molina, E. (2003b). Vegetable coagulant (*Cynara cardunculus*) use evidenced by capillary electrophoresis permits PDO Serpa cheese authentication. *Lait* 83: 343–350.

Roseiro, L.B., Wilbey, R.A., and Barbosa, M. (2003c). Serpa cheese: technological, biochemical and microbiological characterisation of a PDO Ewe's milk cheese coagulated with *Cynara cardunculus* L. *Lait* 83: 469–481.

Sanz Ceballos, L.S., Ramos Morales, E., de la Torre Adarve, G. et al. (2009). Composition of goat and cow milk produced under similar conditions and analysed by identical methodology. *Journal of Food Composition and Analysis* 22: 322–329.

Spanish Cheese Catalog (2013). Ministry of Agriculture, Fisheries and Environment. https://www.alimentosdespana.es/images/es/Fichas%20Cat%C3%A1logo%20Electr%C3%B3nico%20de%20Quesos%20de%20Espa%C3%B1a_tcm69-567354.pdf (accessed 6 December 2022).

Suárez, V. and Busetti, M. (2005). Lechería ovina en Argentina (Sheep dairy in Argentina). *Boletín de Divulgación técnica (Bulletin of Technical Dissemination). INTA Anguil* 195–204.

Tavaria, F.K., Silva-Ferreira, A.C., and Malcata, F.X. (2004). Volatile free fatty acids as quality indicators for consumption of Serra da Estrela cheese. *Journal of Dairy Science* 87: 4064–4072.

Zabaleta, L., Albisu, M., and Barrón, L.J.R. (2017). Volatile compounds associated with desirable flavour and off-flavour generation in ewe's raw milk commercial cheeses. *European Food Research and Technology* 243: 1405–1414.

14 Sensory Evaluation in Processed Cheese Innovation

Silvani Verruck, Saionara Sartor, Mônica Q. Freitas, Tatiana C. Pimentel, Erick A. Esmerino, Adriano G. da Cruz and Marice N. Oliveira

14.1 Introduction

Milk and milk products have a long history of use for human nutritional purposes (Balthazar et al. 2017; Rozenberg et al. 2016). In several countries worldwide, these products are part of the official nutritional recommendations and are considered excellent sources of nutritional components as part of a well-balanced diet (Verruck et al. 2019a). The content of protein, lipids, minerals and vitamins in dairy products makes these products very versatile for both direct consumption and use as ingredients (Tunick and Van Hekken 2015). It is known that generally three portions of dairy products are recommended daily in order to have a balanced diet in nutrients (Rozenberg et al. 2016). These portions would be sufficient to supply an adult's calcium demand (1000 mg) in the day, as recommended by the World Health Organization/Food and Agriculture Organization of the United Nations (WHO/FAO) (Rippin et al. 2017).

One of the most widespread and highly accepted dairy products worldwide is cheese (Ferrão et al. 2016). Cheese is defined as a fresh or mature product obtained from milk coagulation, easily digestible and rich in nutritional components. There is an estimative that more than 1000 varieties of cheese are produced from the same raw materials, usually bovine, ovine, caprine or buffalo milk, lactic acid bacteria (LAB), coagulant and NaCl (McSweeney et al. 2017). Cheese can be classified based on the type of milk used, manufacturing process, fat content, type of fermentation and microbiota (Santiago-López et al. 2018). One of the types of cheese that is gaining a large market and consumer interest in recent years is processed cheese.

Processed cheese is a stable oil-in-water emulsion that is formulated from natural cheeses/dairy proteins and edible oil/fat ingredients (Ferrão et al. 2016). This cheese-based food is produced by comminuting, blending and melting one or more natural cheeses and optional ingredients into a smooth homogeneous blend with the aid of heat, mechanical shear and emulsifying salts (Guinee 2016). About two million tonnes of processed cheese is produced annually, which accounts for approximately

Sensory Profiling of Dairy Products, First Edition. Edited by John J. Tuohy.
© 2023 John Wiley & Sons Ltd. Published 2023 by John Wiley & Sons Ltd.

14% of natural cheese utilisation (McSweeney et al. 2017). Processed cheese can be produced in a wide range of flavours, consistencies and shapes; thus, it is particularly popular for ingredient applications (McSweeney et al. 2017). This popularity could be attributed to several factors including the diversity they offer in flavour, textural-based attributes and cooking properties, easy customisation to cheese ingredient applications and adaptability to the fast-food trade, attractive packaging in convenient formats and shapes, popularity with young children and lower cost relative to natural cheese

A positive sensory perception is one of the most important characteristics of the flavourful image that dairy products, including processed cheeses, continue to enjoy with consumers (Torres et al. 2017). Processed cheeses can be produced with a wide spectrum of consistencies ranging from firm, elastic and sliceable to creamy, smooth and spreadable (Guinee 2017). These kinds of cheeses can differ in flavour characteristic, mainly dependent on the cheese-base used. The texture-based attributes can also be varied, e.g. sliceability, shredability, spreadability and consistency are dependent on the structure chosen for the product (Guinee 2017). Processed cheese is available on the market under different technological forms, such as foil-covered portions (blocks, triangles), slices (stacked or individually wrapped in plastic film), fingers, tubes and tubs (Guinee 2016). Another important sensory characteristic of processed cheese is its cooking properties that include different levels of meltability, viscosity and flowability (Ferrão et al. 2016; Guinee 2017).

There is increased pressure from consumers to increase the health benefits associated with processed cheeses (Talbot-Walsh et al. 2018). In general, processed cheeses are known to have high-salt and fat contents and if consumed in excess, can bring some harm to the consumer (Ferrão et al. 2016). Thus, the perception of improving health benefits of processed cheese consumption has led to multiple developments in the healthy cheese market sector, mainly related to the reduction of fat and sodium (Coyle et al. 2020; Haron et al. 2020; Johnson et al. 2009; Ni Mhurchu et al. 2011; Pravst et al. 2017). Reformulating processed cheese is a challenge for the food industry because different technological protocols must be observed, including data collection, finding standards and food types to replace (Barr 2010; Ferrão et al. 2016; Pérez-Cueto et al. 2012). However, food reformulation is a realistic option that food companies can use to create healthier products (Golan and Unnevehr 2008; Uauy et al. 2009). Modifications to the formulation of processed cheeses can include limiting the quantities of ingredients, such as sodium salts, trans-fatty acids (TFA) and saturated fatty acids (SFA) but also the addition of compounds with beneficial health proprieties (Ferrão et al. 2016; Ferrão et al. 2018; Nogueira et al. 2018; Santiago-López et al. 2018; Solhi et al. 2020b). All these technological modifications must pass the sensory analysis of the final product, which will tell if the structural and textural modifications are acceptable or not when compared to regular processed cheese (Talbot-Walsh et al. 2018).

In the light of these observations, the objective of this chapter is to demonstrate the state of the art and the recent advances in nutritional and functional components of processed cheeses and their sensory perception by consumers.

14.2 Processed cheese products

Processed cheese, also known as melted cheese or pasteurised processed cheese, is a product made from natural cheeses and other dairy ingredients to which emulsifiers, salt and colouring are added (McSweeney et al. 2017). It is produced by comminuting, blending and melting one or more natural cheeses and optional ingredients into a smooth homogeneous blend with the aid of heat, mechanical shear and emulsifying salts (Guinee 2016). Processed cheese is generally classified according to three major categories: processed cheese, processed cheese foods and processed cheese spreads, based on the requirements for moisture content, fat content and the amount and number of allowed optional ingredients (Johnson et al. 2009). The ingredients and composition specifications of these processed cheese categories are shown in Table 14.1. According to the U.S. Food and Drug Administration standards, the product can be categorised as processed cheese only when prepared using natural cheese and salts (≤3%), and has ≤43% moisture and fat content of ≥47% (Ferrão et al. 2016). Products that do not meet these requirements should be classified in another way, such as processed cheese foods or processed cheese spreads. Processed cheese products not included in this categorisation can be classified as processed cheese foods, processed cheese spreads or processed cheese analogues (Hanaei et al. 2015). Processed cheese are available on the market in different technological formats, such as foil-covered

Table 14.1 Categories of processed cheese with ingredients and compositional specification.

Category	Ingredients	Moisture (%)	Fat (%)	Fat in dry matter (%)	pH
Pasteurised processed cheese	Cheese: cream, anhydrous milk fat, dehydrated cream (≤5%), water; emulsifying salts (≤3%); salt; organic acids; food-grade colours, spices and flavourings; mould inhibitors (≤0.3%) when product is in the form of slices or cuts in consumer packs.	≤43	–	≥47	≥5.3
Pasteurised processed cheese foods	As for pasteurised processed cheese, but with the following extra optional ingredients: dairy ingredients, such as milk, cream, skim milk, buttermilk, cheese whey, whey proteins.	≤44	≥23	–	≥5.0
Pasteurised processed cheese spread	As for pasteurised processed cheese, but with the following extra optional food-grade hydrocolloids, such as guar gum, xanthan gum, gelatin, carboxymethylcellulose and/or carrageenan at levels <0.8%, and food-grade sweetening agents such as sugar, dextrose, corn syrup, glucose syrup and hydrolysed lactose.	40–60	≥20	–	≥4.0

Source: Adapted from Guinee (2017).

portions (blocks, triangles), slices (stacked or individually wrapped in plastic film), fingers, tubes and tubs (Guinee 2016).

The principles of manufacturing processed cheeses involve the controlled destabilisation of the para-casein calcium phosphate network of natural cheese through the addition of emulsifying salts, such as sodium phosphate and citrate salts, which improve the emulsifying properties of casein (Figure 14.1) (Guinee 2017). The addition of emulsifying salts aims to partially solubilise the protein, allowing it to bind to free water and emulsify the free fat released during processing (heating and shearing). This process disrupts the calcium phosphate protein network that results in the exposure of hydrophilic and hydrophobic regions of the individual casein proteins. Subsequently, when shear and heat are applied, the hydrophilic sections interact with the aqueous phase, while the hydrophobic sections interact with the fat phase to form an emulsion (Johnson et al. 2009). The process represents a structural transformation of the cheese from its natural form to a fine oil-in-water (O/W) emulsion in a concentrated dispersion of para-casein(ate) (Guinee 2017). In natural cheese, native fat globules are surrounded by a fat globule membrane (proteins and phospholipids), while in processed cheeses the fat globules are surrounded by an additional layer of rehydrated para-caseinate (McSweeney et al. 2017). The main factors that provide the final processed cheese with its unique microstructure are the use of emulsifying salts and other ingredients (dairy ingredients, water, condiments, flavours, colours and preservatives) during formulation in combination with processing conditions as summarised in Table 14.1 (Talbot-Walsh et al. 2018). The structural change that occurs in this process is central to the formation of a stable processed cheese with different functionalities (for example, rheological and cooking properties) (Guinee 2016). The unique microstructure of the processed cheese means that it can be used for different functions (non-melted and melted texture) and in

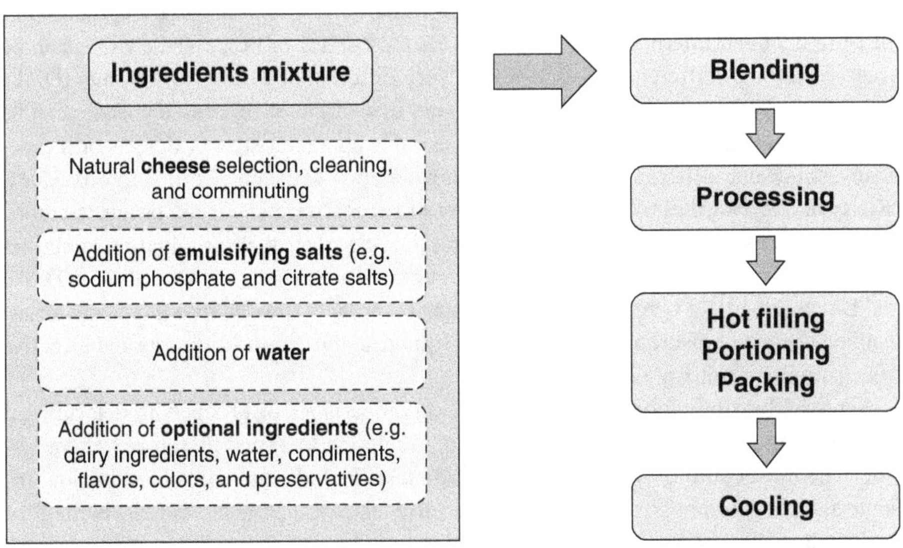

Figure 14.1 Overview of processed cheeses manufacture. *Source*: Adapted from Guinee (2017).

different shapes (blocks, block slices, individually wrapped slices, strips and sauces). The process results in a homogeneous cheese with an extended shelf life (Johnson et al. 2009).

14.3 Sensory characteristics of processed cheese

Consumer acceptance and willingness to repeat the purchase of a product depends on sensory properties such as aroma (ortho-nasal), flavour (aromatics in the mouth, basic flavours and taste), texture and appearance attributes (Drake and Delahunty 2011). Thus, sensory quality is recognised as one of the main aspects related to the marketing of cheeses and processed cheeses (Gulzar et al. 2020). Traditional methods for analysing dairy products are commonly used to assess the quality of these products. However, in a society with constant changes and the demand for unique sensory characteristics in products, more and more innovative sensory methods are being tested (Belsito et al. 2017; Esmerino et al. 2017; Gulzar et al. 2020).

Most sensory properties are complex by definition and cannot be properly detected or represented solely by analytical techniques as they are stimulated by the integrated involvement of many different compositional and structural properties of the product (Drake and Delahunty 2011). However, due to the sophisticated functioning of human sensory systems, even a slight change in product formulation can be detected as a change in sensory character. Thus, sensory evaluation has routinely been applied in quality assurance by the dairy industry and increasingly also as a powerful research and marketing tool (Kussy and Aylward 2009).

Traditionally processed cheeses were evaluated for flavour (acidic, rancid, bitter, salty, sweet, creamy and mouldy), texture (viscosity, elasticity, firmness, tooth pull) and appearance (melting, oiliness, browning of the edge and surface rupture) (Gulzar et al. 2020). However, due to the large number of shapes and textures that processed cheeses can have these attributes can vary considerably. The nomenclature, method of measurement and interpretation of sensory characteristics of processed cheeses can be accessed through different techniques with very different objectives and results (Drake and Delahunty 2011). For example, the sensory evaluation of processed cheeses can be performed to determine the presence of undesirable characteristics or defects or if there is any noticeable difference in sensory characteristics between two or more products (Kussy and Aylward 2009). The distinctions and characteristics of the products can be verified using methodologies for scoring and judging quality, discrimination, quantitative descriptive assessment and consumer acceptability assessment. According to Drake and Delahunty (2011), sensory assessment can be broadly divided into three categories: analytical sensory discrimination tests, descriptive analytical sensory tests and affective or consumption sensory tests.

Sensory discrimination tests involve direct comparisons of products to determine if there is a noticeable difference between them in relation to a specific sensory characteristic. The most commonly used discrimination tests include duo-trio, classification, triangle and paired comparison tests. In the duo-trio test, evaluators are asked which of the two products is most similar to a third reference product, allowing a common reference to be used repeatedly as a benchmark control. In the classification test, several products are compared with each other for a single designated attribute and the evaluator is asked to classify them in the order of increasing intensity of that attribute. In the triangle test, evaluators are asked to choose which two of three products are similar, or which product is most different from the other two. In the paired comparison test, two products are compared to each other, usually one feature at a time, and evaluators are asked to indicate which product has more of the designated feature (Drake and Delahunty 2011).

Descriptive sensory assessment refers to a set of techniques that seek to differentiate between a range of products based on all their sensory characteristics and to determine a quantitative description of all sensory attributes that can be identified (Kussy and Aylward 2009). The sensory characteristics of processed cheese that can be quantified include all the aroma, appearance, flavour, texture, residual taste and even sound properties of the product that distinguish it from other products (Drake and Delahunty 2011). There are several different descriptive analysis methods described in the literature, including the texture profile method, flavour profile method, quantitative descriptive analysis TM, spectrum TM method, free choice profile and quantitative flavour profile (Gulzar et al. 2020). Although they differ in approach, each descriptive method has three steps: the selection of a panel of evaluators, the development and definition of terminology or vocabulary, and the quantification of these sensory attributes (Drake and Delahunty 2011). Common terms used to describe sensory attributes of processed cheese are summarised in Table 14.2 and Figure 14.2. Descriptive sensory analysis is considered one of the most powerful sensory tools in the research of processed cheeses. Descriptive sensory analysis differs from other sensory testing methods, as it aims to profile all the perceived sensory characteristics of a product (Gulzar et al. 2020).

Consumer sensory acceptance tests use rating scales that measure relative aversion and taste, as well as discrimination tests that are based on preferences and scales that ask the consumer how they feel about a particular sensory characteristic (Drake and Delahunty 2011). When consumers are used the evaluators must be subjective or untrained consumers of the product class being evaluated. Sensory tests that include home use, focus groups or identification of attractive product concepts (joint analysis) are also used to determine consumer experiences with products (Drake and Delahunty 2011). From these tests, consumer demand for diversification of processed cheese through the development of innovative products is noticeable. The healthier eating consumer trend and promotion by government agencies are leading towards healthier eating habits (Sánchez-Macas et al. 2012). Thus, consumers have a preference for processed cheeses with less fat and salt content, in addition to being able to offer some extra functional benefit (Ferrão et al. 2016; Ferrão et al. 2018; Talbot-Walsh et al. 2018). However, modification in the formulation of traditional processed cheeses in order to reduce sodium, fat or add some functional compound may result in sensory changes that will be discussed in the next section.

14.3.1 *Sensory impact of sodium replacement*

Salt consists of sodium and chloride ions and has been widely used as a natural preservative to prolong the shelf life of food products, as well as to prevent the growth of microorganisms (Haron et al. 2020). Therefore, sodium appears in high concentrations

Table 14.2 Terms commonly used to describe the sensory attributes of processed cheese.

Attribute	Description
Aged cheese; sulphur	Sulphurous aromatic typically associated with aged Cheddar cheese, hard-boiled eggs and struck matches
Browned/toasted	Sweet, browned aromatics associated with slight overheating
Burnt	Harsh aromatic associated with overheating
Diacetyl (buttery)	Aromatic associated with the chemical compound diacetyl
Cardboardy (oxidised)	Aromatics reminiscent of wet cardboard also described as flat or stale
Milkfat (creamy)	Sweet aromatics associated with milkfat and lactone compounds found in milkfat
Sour aromatic	Sour aromatics reminiscent of the fermentation of milk products
Unclean (dirty socks)	Sharp, sweaty, aromatics that also generally provide a lingering unpleasant aftertaste
Fruity	Non-specific fruity aromatics reminiscent of ripe fruit
Free fatty acid (rancid)	Aromatic associated with short-chain free fatty acids primarily butyric acid. At low concentrations, this is often perceived as a delayed response after swallowing or expectorating
Methyl ketone	Sweet aromatic or flavour suggestive of methyl ketones, typical of blue cheese
Musty	Earthy aromatics reminiscent of a poorly ventilated cellar, musty hay or mould
Old oil (rancid like)	Flat, stale, rancid oil flavour and aroma
Soapy	Phosphate flavour characteristic of oxidised fats and long-chain fatty acids
Yeasty	Aroma and flavour of fermented bread dough
Acetaldehyde	Pungent, penetrating aroma and taste of green apple
Brothy	Aromatics associated with boiled meat or vegetable soup stock
Cooked	Aromatics of cooked milk
Nutty	Aromatics typically associated with nuts
Whey (whey taint)	Aroma and flavour of cheese whey
Sweet taste	Like sugar; a basic taste which triggers taste receptors
Sour taste (acidic)	A basic taste that triggers taste receptors. Sharply acidic vinegary or tart
Salty taste	Salty taste at an intensity beyond expectation
Bitter	A basic taste sensation (somewhat delayed) similar to strong black coffee, quinine or tonic water
Umami	The basic taste associated with nucleotides and monosodium glutamate
Astringent	Mouthfeel sensation elicits a drying, puckering sensation of the oral tissues

Source: Adapted from Kussy and Aylward (2009).

among processed foods, including processed cheese (Ni Mhurchu et al. 2011). However, excessive sodium intake in the diet is a well-recognised risk factor as one of the main causes of chronic non-communicable diseases, mainly with effects on the increase in systolic arterial hypertension (Pravst et al. 2017; WHO World Health Organization 2020). In most countries, the average daily salt intake ranges from 9–12 g per person, which is well above the World Health Organization's recommendation of 5 g/day (WHO World Health Organization 2020). It is estimated that 2.5 million deaths could be prevented each year if global salt consumption were reduced to the recommended level of 5 g/day (Pravst et al. 2017; WHO World Health Organization 2020). Considering the importance of the issue, several countries are establishing salt reduction strategy programmes

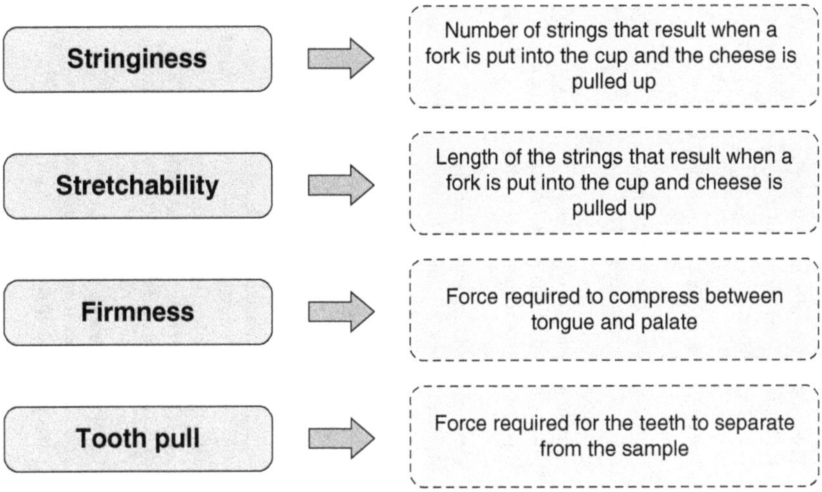

Figure 14.2 Common descriptors and definitions of processed cheese texture attributes. *Source:* Adapted from Gulzar et al. (2020).

intending to reduce the global population's sodium intake by 30% by 2025 (Ni Mhurchu et al. 2011; Ferrão et al. 2016; Pravst et al. 2017; Coyle et al. 2020; Haron et al. 2020).

However, the sodium salts added to processed cheese formulations have an important technological function and are difficult to replace. In addition to being used to enhance flavour, decrease water activity and as a natural preservative, they are added largely due to their emulsifying properties (Ferrão et al. 2018). The addition of sodium phosphate salts as emulsifying agents is an additional source of sodium that contributes to the increase in the overall sodium content of the final product (Johnson et al. 2009). Sodium-based emulsifying salts (responsible for 44–48% of total cheese sodium), natural cheese and sodium chloride are the three main ingredients that contribute to the high-sodium content present in processed cheeses (Nogueira et al. 2018). Therefore, the reduction of sodium is a major challenge during the processing of processed cheeses since it performs important functions in the cheese matrix, affecting the flavour, rheology and shelf life of the processed product by controlling the activity of enzymes and microorganisms (Johnson et al. 2009). When there is a decrease in the levels of sodium salts in processed cheeses, other food additives need to be added to supply the lack of these ingredients, in addition to guaranteeing the safety, flavour, general quality and texture of the processed cheeses (Ferrão et al. 2016). In addition, aspects related to physical–chemical properties, such as firmness, melting rate and oil loss, represent additional factors to be investigated during the manufacture of processed cheeses (Ferrão et al. 2018). However, considering that even a slight change in product formulation can be detected as a change in sensory character due to the sophisticated functioning of human sensory systems, these tests must be carried out (Drake and Delahunty 2011). Any formulation or process modifications must be verified by the sensory analysis of the final product to confirm if any structural and textural changes are detectable compared to regular processed cheese (Talbot-Walsh et al. 2018). Some studies on the substitution of sodium salts by other ingredients on the sensory characteristics of processed cheeses are summarised in Table 14.3.

Table 14.3 Studies on sodium replacement in processed cheeses and verification by sensory analysis.

Product	Reduced-sodium formulation	Type of sensory test used	Sensory test details	Main results	Reference
Block-type processed cheese	Partial substitution of sodium by potassium in emulsifying salts (sodium citrates; potassium citrates; potassium phosphates; polyphosphates)	Discrimination testing: triangle test	Nineteen assessors from a trained panel. Five most-experienced assessors evaluated the three best formulations	The three best recipes had contents of 0.39–0.55% sodium and 0.47–0.72% potassium. The three samples were well accepted. The typical bitter taste of potassium-containing salts was barely perceptible	Hoffmann et al. (2012)
Processed Mozzarella cheese	Partial replacement of sodium chloride (NaCl) with potassium chloride (KCl) at 1%. The level ranged from 25 to 75% (w/w). Replacement of sodium-containing salts by potassium-based emulsifying salts: potassium citrate and di-potassium phosphate (1 : 1). The level ranged from 1 to 2% (w/w).	Descriptive analysis	Eight trained judges. A descriptive sensory scorecard with a 90-point scale was used. Maximum scores for flavour (50); body and texture (30) and colour and appearance (10) were used.	Sodium was successfully reduced using a combination of KCl and potassium-based emulsifying salts. The product with maximum sensory scores was 30.96% of KCl and 2% of emulsifying salt. A 71.27% reduction in overall sodium content was obtained, without jeopardising the sensory quality.	Chavhan et al. (2015)
Processed mozzarella cheese	Partial substitution of sodium chloride (NaCl) with potassium chloride (KCl) at 1%. Total replacement of sodium-based emulsifying salts with potassium-based emulsifying salts, a blend of potassium citrate and dipotassium phosphate (1 : 1). Products were evaluated after regular intervals of 7 days up to 56 days.	Descriptive analysis	Eight trained judges. A descriptive sensory scorecard with a 90-point scale was used. Maximum scores for flavour (50); body and texture (30) and colour and appearance (10) were used.	The flavour of the samples up to 28 days of storage did not change. After 28 days, a low-sodium sample obtained lower scores for flavour (development of bitterness). Samples with low-sodium content showed higher scores for body and texture, being firmer. Colour and appearance attribute decreased during storage; however, no difference in both treatments.	Khetra et al. (2015)

Processed cheese	Replacement of traditional citrate- and phosphate-based emulsifying salts by selected hydrocolloids: modified starch (with bound sodium octenyl succinate), low-methoxy pectin (alone or combined with lecithin), locust bean gum, k-carrageenan and I-carrageenan, at concentrations in the final product of 1.0%, w/w.	Descriptive analysis	Five expert assessors trained. The criteria of the homogeneity and hardness were evaluated.	The product with 1.0% (w/w) lecithin was sensorially homogeneous without the obvious release of fat. k- and i-carrageenan were potential substitutes for traditional emulsifying salts, mainly at 1.0%. These products were very hard and impossible to spread. Modified starch, locust bean gum and low-methoxy pectin could not be recommended as replacements for traditional emulsifying salts.	Černíková et al. (2010)
Processed cheeses and cheese	Replacement of traditional emulsifying salts (phosphate- and citrate-based emulsifying salts) with 1 g/100 g κ-carrageenan. The samples were studied in dependence on the maturity level of cheese (1–16 weeks' maturity).	Descriptive analysis	Twenty selected assessors trained. Twelve assessors were between 21 and 30 years old and five assessors were between 31 and 40 years old. Homogeneity, rigidity and flavour were evaluated on a 7-point scale.	Both samples (with the addition of traditional emulsifying salts and carrageenan) were homogeneous. The texture of cheeses with the addition of 1 g/100 g κ-carrageenan (KC) was very hard and impossible to spread. With the increasing degree of ripening of cheese, the rigidity of traditional emulsifying salt (ES) samples was decreasing. No significant off-flavours were noticed in either ES or KC samples. The flavour of ES was slightly better than KC.	Hladká et al. (2014)

(*Continued*)

Table 14.3 (Continued)

Product	Reduced-sodium formulation	Type of sensory test used	Sensory test details	Main results	Reference
Spreadable processed cheese (SPPC)	Replacing sodium-based commercial emulsifying salts (blend of sodium neutral tetrapyrophosphate, sodium hexametaphosphate and sodium tripolyphosphate) with monopotassium phosphate anhydrous The total amount of emulsifying salts in each formulation represents 1.2% (w/w) of the cheese-base	Consumer acceptability testing	A total of 119 panellists (88 female and 31 males aged 18–62). The participants rated the appearance, odour, flavour, texture and overall liking, and also purchase intention	The replacement of sodium-based at 25 and 50% by potassium phosphate kept an acceptable flavour. Appearance and texture attributes have been improved and samples received better scores regarding purchase intention and overall liking. Partial substitution of up to 50% of Na-based emulsifying salt with K-based salts reduced sodium content in SPCC, improvement of sensory quality	Nogueira et al. (2018)

To reduce sodium, sodium salts can be replaced by potassium salts, since these salts help maintain the salty taste (Gomes et al. 2011; Felicio et al. 2013). In this sense, Nogueira et al. (2018) replaced sodium-based commercial emulsifying salts (blend of sodium neutral tetra pyrophosphate, sodium hexametaphosphate and sodium tripolyphosphate) with monopotassium phosphate anhydrous in a spreadable processed cheese (Brazilian requeijão). The authors reported that the replacement of 25 and 50% of sodium using potassium phosphate was enough to maintain an acceptable flavour, by suppressing some degree of sourness. One of the biggest limitations of the use of potassium salts is the residual bitter taste left when they are used at high concentrations, thus, studies have used masking agents both to minimise the strange metallic taste caused by the addition of potassium salts at high concentrations and to obtain a greater sodium reduction (Ferrão et al. 2018).

Nogueira et al. (2018) also verified that the appearance and texture attributes had been improved by the use of potassium salts. Furthermore, samples containing potassium salts received better scores for purchase intention and overall liking. The same was not observed in the work of Ferrão et al. (2018) where their results indicate that the use of potassium chloride increased the bitter taste of the reduced-sodium samples. The authors reported that sodium chloride allows the perception of other flavours, which was related to the lower scores for the attributes of salty taste and acid taste of the reduced-salt formulations.

Another strategy for the replacement of traditional citrate- and phosphate-based emulsifying salts was the use of selected hydrocolloids: modified starch (with bound sodium octenyl succinate), low-methoxy pectin (alone or combined with lecithin), locust bean gum, k-carrageenan and i-carrageenan (Černíková et al. 2010). These authors reported that fat release and flavour perception in processed cheese containing 1.0% (w/w) lecithin was affected. Furthermore, modified starch, locust bean gum and low-methoxy pectin could not be recommended as replacements for traditional emulsifying salts. Also, k- and i-carrageenan appear to be potential substitutes for traditional emulsifying salts, mainly at a concentration of 1.0%. However, the products were considered very hard and impossible to spread due to the hardness of the protein matrix in the processed cheese (Černíková et al. 2010). It is important to highlight that the application of emulsifying salts also results in chain peptisation, dispersion, hydration and swelling of proteins, in addition to the emulsification and stabilisation of fat (Guinee 2017). Thus, low-methoxy pectin, locust bean gum and modified starch probably do not have sufficient emulsifying capacity to stabilise fat during the production of processed cheeses, while k- and i-carrageenan have the ability to bind to hydrophobic parts of proteins (especially the casein fractions) in the presence of calcium ions and thus stabilise both water and fat (Černíková et al. 2010). The same behaviour was observed when traditional emulsifying salts (phosphate- and citrate-based emulsifying salts) was replaced by with 1 g/100 g κ-carrageenan (Hladká et al. 2014). The texture of processed cheeses was very hard and impossible to spread during the entire 16-week experiment. No significant off-flavours were noticed, but the flavour of the sample produced with emulsifying salts was slightly better than the sample with κ-carrageenan (Hladká et al. 2014).

14.3.2 *Sensory impact of fat replacement*

Fat is one of the main constituents in processed cheeses and affects both the texture and palatability of these cheese. Consumers are increasingly concerned about their health and the risk factors associated with consumption of high-fat foods. The industry has an important public health role because it must offer these consumers alternatives to low-fat content products (Talbot-Walsh et al. 2018). Low-fat products have been available to consumers for decades, to enable the daily intake of fats to be kept within a healthy limit. However, whenever changes are made in the formulation of processed cheese, it is necessary to take into account the mechanical properties, texture and sensory acceptance by consumers (Ferrão et al. 2016).

For a product to be labelled as 'reduced fat', it must have a reduction of at least 25% compared to the conventional product (Symbols et al. 2010). To be called 'low fat', it must have a maximum of 3 g of fat per serving. For processed cheese this means a fat requirement of <3 g of fat in 50 g (Symbols et al. 2010), which equates to a reduction of about 81% in the total fat. For processed cheese foods and processed cheese spreads the reductions are about 76 and 72% of total fat, respectively (Johnson et al. 2009). Considering that approximately 90% of the total fat in processed cheese comes from the natural cheese used in its manufacture, it is essential that a low-fat natural cheese must be used for the manufacture of low-fat processed cheese (Johnson et al. 2009). This is one of the simplest alternatives from an industrial point of view; however, the use of a low-fat natural cheese can lead to several sensory issues compared with use of the standard cheese. For example, the taste and texture of natural cheese is also be affected by fat reduction and directly affects the sensory characteristics of processed cheese made from it (Ferrão et al. 2016).

Processed cheese with reduced or low-fat content can be classified into two main categories; those with or without fat substitutes (Roller and Jones 2010). Fat substitutes are categorised as fat substitutes or fat mimetics (Roller and Jones 2010). Fat substitutes are represented by sucrose fatty acid esters and polyesters, carbohydrate fatty acid esters, various emulsifiers (such as mono- and diglycerides, lecithin) and structured lipids, such as medium-chain triglycerides (Ferrão et al. 2016). These lipid-based macromolecules are physically and chemically similar to oils and fats (Ferrão et al. 2016). On the other hand, fat mimetics are usually protein macromolecules or carbohydrate-based, such as modified starches and hydrocolloids. Due to their chemical bond with water and their organoleptic and physical properties, these molecules are called fat imitators (Gaonkar and McPherson 2016). Fat mimetics can create a perception of creaminess and plasticity similar to products with fat due to their polar and water-soluble properties; however, they are not capable of carrying aromas and flavours as fat does (Kumar 2012). Several workers have tried to develop processed cheeses with different fat substitutes and mimetics, however, due to the difficulty of recreating the ideal texture, only a few processed cheeses with low-fat content or with reduced fat content are commercially available (Figure 14.3) (Ferrão et al. 2016). Some sensory studies on fat substitution in processed cheeses are summarised in Table 14.4.

Table 14.4 Studies on sensory aspects of fat reduction in processed cheeses.

Product	Fat-reduced Formulation	Sensory test used	Sensory test details	Main results	Reference
Process Cheese	Addition of exopolysaccharide-producing cultures in base cheese (cheddar). Three treatments were used: (i) RFC: reduced-fat control cheese made with a non-EPS-producing commercial Cheddar starter culture (DVS 850, 0.02% w/w), (ii) EPSA: reduced-fat cheese made with the ropey strain *Lactococcus lactis ssp. cremoris* JFR1 (2% vol/wt), (iii) EPSB: reduced-fat cheese made with a capsule-forming non-ropey *Streptococcus thermophilus* 3855S (0.7% vol/wt).	Descriptive analysis	Six experienced panellists. Hand firmness, hand springiness, first-bite firmness, first-bite fracturability, chew-down degree of breakdown, chew-down cohesiveness, chew-down adhesiveness, chew-down smoothness and residual smoothness of mouth-coating were evaluated on a 9-point scale. Overall cheese flavour scores were also used.	The addition of EPS improved the texture. EPS reduced-fat process cheese was softer, less chewy and gummy. The flavour and chew-down cohesiveness were not affected. There were sensory attribute differences between treatments of young (2 d) and cheeses stored for 1 month.	Hassan et al. (2007)
Processed cheese analogue	Addition of pectin gel in processed semi-solid cheese	Descriptive analysis	Ten members in the sensory panel. The cheeses were evaluated for aroma (creamy, milky, acid), colour (neutral, white), texture (presence of holes and eyes, elasticity, stickiness) and mouthfeel (creamy, hard, crumbly).	Low-fat samples had lower hardness, gumminess, chewiness and adhesiveness. Low-fat samples with pectin gel addition were similar to the full-fat cheese analogues. A low-fat cheese analogue made with pectin gel addition was well received by the sensory panel.	Liu et al. (2008)

(Continued)

Table 14.4 (Continued)

Product	Fat-reduced Formulation	Sensory test used	Sensory test details	Main results	Reference
Processed cheese	Addition of pectin (0.2, 0.4, 0.6 and 0.8% w/w) in processed cheeses with 40% w/w dry matter and 50% w/w fat in dry matter.	Descriptive analysis	20–24 selected assessors trained. Seven-point hedonic scale was used for assessment of appearance, rigidity, spreadability and flavour.	All samples with the pectin addition were more rigid and less spreadable. The rigidity increased and spreadability decreased with the increasing levels of pectin (>0.6% w/w). Appearance and flavour were not changed by pectin addition. Samples with 0.2% w/w of pectin were rated highest.	MacKů et al. (2008)
Processed cheese	Addition of different concentrations of selected 1-monoglycerides (1-monocaprylin, 1-monocaprin, 1-monolaurin, 1-monopalmitin and 1-monostearin) in a concentration of 0.25% w/w in processed cheeses with 45 and 50% (w/w) fat in dry matter.	Descriptive analysis	Sixteen selected assessors trained. Appearance, rigidity, spreadability, flavour and off-flavour intensity were assessed by way of 7 point's ordinal scale.	The addition of 1-monoglycerides did not influence the visual appearance. Changes in 'rigidity' and 'spreadability' was only observed for cheeses with 50% (w/w) fat in dry matter containing added 1-monopalmitin and 1-monostearin). In the group of processed cheeses with 45% (w/w) fat in dry matter, no changes in 'rigidity' and 'spreadability' for any of the 1-monoacylglycerols was observed. Additions of 1-monocaprylin, 1-monocaprin and 1-monolaurin resulted in the detection of unacceptable off-flavours.	Bunka et al. (2007)

Processed cheeses	Addition of boiled and mashed potatoes at a level of 20, 30 and 40%. A blend of young and old Cheddar cheese was used in the preparation.	Descriptive analysis	Eight expert judges. The judged parameters were: appearance, body and texture, and flavour.	The colour and appearance of the cheese were not affected, up to 30% potato addition. Body and texture and flavour scores were lower than the control values. The 40% level resulted in a decrease in the body and texture scores as well as reduced flavour intensity, which had a negative effect on the overall acceptability. Samples with 40% potato were found to be very soft and sticky but those with 20 and 30% were found to have a smooth firm body and a better mouthfeel. Body and texture of samples with 20% potato were similar to that of the control. Potato incorporation up to 30% resulted in a smooth body with a better mouthfeel. An acceptable quality cheese was obtained with up to 30% potato incorporation.	Rafiq and Ghosh (2017)
Spreadable cheese-like	Addition of soy protein concentrate (5, 10 and 15 g/l) in a spreadable cheese-like product, developed with skimmed cow's milk	Consumer acceptability testing	45-member panel selected. The attributes evaluated were colour, flavour, spreadability and overall acceptance	The sample with 5 g/l of soy protein concentrate had an acceptable flavour. For spreadability, the order of acceptance was: 10 > 15 > 5 (g/L). The sensory attributes of the formulations with and the control were similar and were rated acceptable	Rinaldoni et al. (2014)

(*Continued*)

Table 14.4 (Continued)

Product	Fat-reduced Formulation	Sensory test used	Sensory test details	Main results	Reference
Spreadable cheese analogues – requeijão cremoso	Replacement of milk fat by vegetable fats: butter oil, partially hydrogenated soybean fat and soybean oil.	Acceptance test	Hundred non-trained consumers. The attributes of appearance, colour, flavour, firmness, creaminess and overall impression were measured using a 9-point-structured hedonic scale. The attribute spreadability was also evaluated.	The cheese manufactured with butter oil was rated better. The spreadable cheese analogue manufactured with partially hydrogenated soybean fat also showed good sensory acceptability.	Cunha et al. (2013)
Spreadable processed cheese	Replacement of 25 and 50% of the dairy fat with vegetable fat.	Acceptance test	Thirty non-trained panellists. Appearance, colour, flavour, creaminess, firmness, spreadability and overall impression were evaluated a 9-point hedonic scale The purchasing intention was evaluated using a 5-point hedonic scale.	Traditional cheese and the analogue with 50% vegetable fat were rated better than the analogue with 25% vegetable fat. Cheese with 50% of vegetable fat had optimal acceptance for the attributes of appearance, colour, creaminess and firmness. Flavour was the only parameter for which the traditional processed cheese received higher mean scores. There were no differences between the samples concerning spreadability. The overall impression was no difference between the traditional processed cheese and the analogue with 50% vegetable fat.	Cunha et al. (2010)

Processed cheese	Addition of 5% modified and non-modified starch in processed cheeses manufactured in the pH range 4.5–6. Non-modified potato and corn and pre-modified waxy maize starch containing 19.9 ± 0.95, 17.8 ± 0.23 and 2.7 ± 0.49% amylose, respectively, were used.	Consumer acceptability testing	An untrained panel of 32 members (with age ranging from 23 to 55, comprised of 21 females and 11 males). Texture, firmness, flavour, aftertaste and overall liking were evaluated, on a 9-point hedonic scale.	Decreasing the cheese pH to 4.8 led to high acceptability in terms of texture and firmness. The low pH negatively affected the flavour with increased 'acidic' and 'sour' tastes. Cheeses prepared with corn starch had a higher pH (5.8) compared to the other cheeses (5.3), and the sensory characteristic were acceptability.	Talbot-Walsh et al. (2019)
Processed cheese	Fat reduction by addition of inulin (10%). Partial fat replacement: chia oil (2.09%) and microencapsulated chia oil (2.45%). Three variations of the standard formulation were tested. Variations in the fat percentage of the milk cream (40% fat) were also tested	Consumer acceptability testing	Fifty panellists (30 female, 20 male; age range 18–59) were selected. Overall liking was evaluated using a 9-point structured hedonic scale (acceptability index ≥70%). CATA questions were also applied based on descriptive terms for the attributes of appearance, texture, colour, odour and taste	Inulin can be used as a partial substitute for fat (36% reduction), without detectible sensory differences. Unpleasant odour, oil flavour, rancid residual flavour, unpleasant flavour were characteristic of chia oil, reducing the acceptability of that sample. Microencapsulation was able to mask the oil taste. The use of inulin and encapsulated chia oil gave a product with reduced fat content and good acceptability	Cardoso et al. (2020)

Figure 14.3 Commercial processed cheese products. *Source:* Wageningen University.

An ideal fat substitute should have melting, flowing, smoothing and lubricating properties that resemble fat. In addition, food adsorption and absorption by taste receptors on the tongue should increase (Schädle et al. 2020). However, there are still no commercially available molecules that have all of these properties combined, so it is common to use combinations of ingredients to reproduce the properties of fat. Some studies have tested fat substitutes based on fat, protein or carbohydrate, each with specific properties and modes of action (O'Connor and O'Brien 2011; Chavan et al. 2015).

Starches from different dietary sources are often used to increase the viscosity and firmness of processed cheeses (Ferrão et al. 2016). However, the texture of processed cheeses can be affected by many factors, including the shelf life of the natural cheese, the pH of the dough to be melted, the processing and storage conditions, the composition of the natural cheese, presence and concentration of ions, type and concentration of sugars and emulsifiers, among other factors (Silva et al. 2012). Among these other factors, the incorporation of a hydrocolloid can also cause changes in the product structure and texture (MacKů et al. 2008). Considering that texture is the most important characteristic for consumers of processed cheeses (Silva et al. 2012), the use of hydrocolloids could improve the texture and consistency of processed cheeses with reduced fat contents. These hydrocolloids immobilise water and form a gelled protein or polysaccharide network that leads to increased product viscosity. Examples of hydrocolloids that can be used are carrageenan, locust bean gum, xanthan gum, guar gum, polydextrose, modified starch or pectin (Liu et al. 2008; MacKů et al. 2008; Silva et al. 2012).

The addition of exopolysaccharide-producing cultures in base cheese (cheddar) for the production of reduced-fat process cheese was evaluated by Hassan et al. (2007). The sensory proprieties: hand firmness, hand springiness, first-bite firmness, first-bit fracturability, chew-down degree of breakdown, chew-down cohesiveness, chew-down adhesiveness, chew-down smoothness and residual smoothness of mouth-coating were evaluated. The results showed that the addition of exopolysaccharide-producing cultures improved the texture proprieties, that is, the reduced-fat process cheese was softer, less chewy and gummy (Hassan et al. 2007).

The addition of pectin gel in processed semi-solid cheese analogue and processed cheese with low fat was evaluated (Liu et al. 2008; MacKů et al. 2008). The low-fat cheese analogue made with pectin gel addition was well received by the sensory panel and showed lower hardness, gumminess, chewiness and adhesiveness and low-fat samples with pectin gel addition were more similar to the full-fat cheese analogues (Liu et al. 2008). On the other hand, all processed cheese samples with added pectin were more rigid and less spreadable. The rigidity increased and spreadability decreased with an increasing amount of pectin (MacKů et al. 2008). One of the main problems resulting from fat reduction in processed cheese is the development of a firmer, less unctuous texture, compromising the plasticity of the product (Ferrão et al. 2016).

Another strategy for reducing the fat content of processed cheese is the addition of different concentrations of selected 1-monoglycerides including: 1-monocaprylin, 1-monocaprin, 1-monolaurin, 1-monopalmitin and 1-monostearin (Bunka et al. 2007). In practice, the addition of 1-monoglycerides did not influence the visual appearance of the processed cheeses. Changes in 'rigidity' and 'spreadability' was observed only for processed cheeses with 50% (w/w) fat in dry matter. However, the additions of 1-monocaprylin, 1-monocaprin and 1-monolaurin resulted in unacceptable off-flavours (Bunka et al. 2007). To date, no raw material has been able to replace all the properties of fat in processed cheeses, as its presence is closely related to both palatability and sensory characteristics (Ferrão et al. 2016). Fat reduction by addition of inulin (10%) and partial fat replacement with chia oil (2.09%) and microencapsulated chia oil (2.45%) has also been tested (Cardoso et al. 2020). The author reported that inulin can be used as a partial substitute for fat (36% reduction), without sensory differences from the standard formulation. on the other hand, terms such as unpleasant odour, oil flavour, rancid residual flavour, unpleasant flavour were characteristic of chia oil, reducing the acceptability of the sample. Microencapsulation was able to mask the oil taste and the use of inulin, and encapsulated chia oil gave a reduced-fat cheese with high acceptability (Cardoso et al. 2020). The use of inulin as a functional compound (prebiotic) will be explored later in this chapter.

Fat reduction is also of economic interest, as the use of milk fat contributes significantly to manufacturing costs. Therefore, an alternative to reduce the costs associated with the high-fat content in processed cheeses is to replace dairy fats with fats of vegetable origin (Cunha et al. 2010, 2013; Rinaldoni et al. 2014). Although the total fat in the product is not reduced through this substitution, some consumers prefer this type of product due to the lower cholesterol content compared to milk fats (Rinaldoni et al. 2014). However, it is necessary to evaluate the presence of trans-fat in this substitution.

There is evidence that the consumption of this type of fat, when it exceeds 1% of the total energy value of the diet, can cause the elevation of bad cholesterol (LDL and VLDL) and the reduction of good cholesterol (HDL), generating risk factors for developing cardiovascular diseases and metabolic syndromes. In this sense, the use of oleogel has been proposed as an alternative, which in addition to easy and versatile processing, reduces the level of cholesterol and saturated fat, suppresses appetite while still being able to transport bioactive compounds of absorption in the digestive tract (Huang et al. 2018). Oleogel or organogel is a mixture of a common oil, which can be corn, soy, sunflower, peanut or other according to the required properties, together with a small amount of structuring or gelling agents, such as carnauba wax, rice bran, sunflower, monoglycerides, diglycerides or free fatty acids, which will provide a more rigid, gel-like appearance (Hwang 2020). However, to date, the sensory implications of using oleogels in processed cheeses has not yet been evaluated.

Recent new findings on the nutritional significance of dairy product constituents, such as the fat, associate their consumption with several health benefits (Verruck et al. 2019a). Thus, it is also possible that dairy fats with a high content of bioactive substances, such as conjugated linoleic acids (CLA), could be used to produce processed cheese with some extra consumer health benefits. CLA, sphingomyelin, butyric acid, ether lipids, β-carotene and vitamins A and D present in dairy fat have been attributed with anti-carcinogenic, and cardiovascular diseases prevention potential and immunomodulatory activity (Verruck et al. 2019b). The fortification of processed cheeses with other substances that could bring some functional benefit will be discussed later.

14.3.3 *Sensory impact of processed cheese fortification*

Improve the functionality of food is one of the ways to attract new consumers for a given product. A functional food can be considered one containing some added functional element that provides some additional health benefit to the consumer (Karimi et al. 2015). For processed cheeses, the addition of probiotics, prebiotics, vitamins, algae and vegetable by-products are the main functional elements of interest (Talbot-Walsh et al. 2018). The sensory characteristics of fortified processed cheeses from some functional studies is summarised in Table 14.5

The addition of probiotic microorganisms to processed cheeses has received little attention, mainly because of the high-processing temperatures and technological limitations (Talbot-Walsh et al. 2018). Probiotic microorganisms are those that when regularly consumed in a certain amount provide potential health benefits to the consumer, such as: decreased constipation, increased immunity and aid the digestion of lactose (Hill et al. 2014). Ehsannia and Sanjabi (2016) were pioneers in the use of probiotics in processed cheese during manufacture (included before heat treatment). These authors used spores of *Bacillus coagulans* (10^7 and 10^8 colony-forming unit [CFU]/g) that are recognised for their resistance to high temperatures. At the end of the 60-day storage period, probiotic cheeses contained more than 10^6CFU/g of viable probiotic organisms, which is higher than the minimum viable probiotic concentration needed to be legally classified as a probiotic cheese. The sensory analysis showed that despite not showing differences in colour, odour and texture scores, the inclusion of probiotic spores in cheese had a negative effect on the scores for flavour and general acceptability. Thus, the sensory profile processed cheese appears to be affected by probiotic incorporation (Ehsannia and Sanjabi 2016). To overcome the problem of heat treatment, long storage times and the high concentrations of salt frequently found in processed cheese, an alternative would be the use of paraprobiotics. Scientific research has reported that non-viable (dead) probiotic microorganisms, therefore, non-cultivable, known as paraprobiotics, can also provide health benefits (Taverniti and Guglielmetti 2011; de Almada et al. 2016; Guimarães et al. 2019). These findings imply that bacterial viability is not an essential requirement to exert health benefits, which presents a potential opportunity in the area of functional processed cheese.

Table 14.5 Sensory characteristics of some fortified processed cheese products.

Product	Ingredient fortification	Sensory test used	Sensory test details	Main results	Reference
Pasteurised process cheese	Fortification of vitamin D_3 (360 IU/100 g)	Discrimination testing – triangle test	The experienced panel consisted of four judges	Fortification did not have any effect on the flavour No effect on final product sensory acceptability	Upreti et al. (2002)
Processed cheese	Addition of probiotic spores (107 and 108 colony-forming unit [cfu]/g) of *Bacillus coagulans*	Acceptance test	Sixty non-trained panellists. The acceptability of appearance, texture, taste, flavour, colour and general perception was evaluated using a 5-point hedonic scale.	No difference in colour, odour and texture scores was observed. The inclusion of probiotic spores in cheese had a negative effect on the scores for flavour and general acceptability. Sensory attributes may be affected by probiotic incorporation into processed cheese.	Ehsannia and Sanjabi (2016)
Processed cheese	Addition of nisin (12.5 ppm).	Descriptive analysis	Seven-member trained panel. Colour, taste and odour were evaluated in an intensity scale test of 9 points.	The addition of nisin had no influence on the sensory attributes of samples, as nisin does not cause intensive alteration in sensory properties.	Smigic et al. (2018)
Sweet processed cheese spread	Addition of pumpkin paste (0, 5, 10 and 15% w/w). Production of a salt-free healthy processed cheese spread containing different ratios of pumpkin, as a source of potassium and fibre.	Descriptive analysis	Twenty trained members. Softness, spreading ability, stickiness, smoothness, colour and all acceptability were evaluated.	The softness of treated samples was better and acceptable. Control samples had higher scores for spreadability, stickiness and smoothness. Colour was greatly affected by adding pumpkin paste. The sample with 10% (w/w) pumpkin paste was preferable and acceptable	Kassem et al. (2017)

(Continued)

Table 14.5 (Continued)

Product	Ingredient fortification	Sensory test used	Sensory test details	Main results	Reference
Functional spreadable processed cheese	Fortification with *Chlorella vulgaris* alga. Enrichment with 2, 4 and 6% dried *C. vulgaris* powder and 4% *C. vulgaris* slurry.	Descriptive analysis	Fifteen members. The surface appearance, firmness of body, spreading, stickiness, smoothness of texture, breakdown properties, oil separation, flavour, saltiness and overall preference were evaluated.	All treatments were acceptable for surface appearance, flavour, spreading and oil separation. Treatments made with dried algae had low scores for smoothness and breakdown properties. Treatment with 2% algae was the best, followed by 4%, but the 6% level was unacceptable. The best treatments were: addition of 2% *C. vulgaris* algae in powder form and 4% *C. vulgaris* algae in the slurry form at the end of the processing.	Tohamy et al. (2018)
Processed cheese (PC)	Fortification with asparagus powder (0, 0.5, 1 and 1.5% w/w)	Descriptive analysis	Fifteen selected assessors. Colour, spreadability, rigidity, flavour and total acceptance were evaluated using a 5-point hedonic scale.	The fortified samples were more rigid compared with the control. The highest scores for flavour, colour and total acceptance were for the control PC sample. The fortified samples also had relatively high scores for flavour, colour and total acceptance.	Solhi et al. (2020a)

Processed cheese (PC)	Fortification with tomato powder (0, 1, 2 and 4% w/w). Evaluation of the PC samples containing tomato powder (PCTs) during 90 days of storage.	Descriptive analysis	Fifteen selected assessors. The appearance (colour), spreadability, rigidity, flavour and total acceptance were evaluated using the 5-point hedonic scale.	Maximum scores for flavour, colour and total acceptance were given to the PC with 2% w/w of tomato powder. As tomato powder increased, the rigidity decreased and the spreadability increased.	Solhi et al. (2020b)
Requeijão cremoso processed cheese	Reduction of fat and sodium contents and addition of xylooligosaccharides (XOS) and flavour enhancers (arginine and yeast extract). Use of 3.3 w/w % XOS as a fat substitute and prebiotic agent and 1 w/w % yeast extract as a flavour enhancer.	Descriptive analysis	A trained panel (11 assessors, 7 women and 4 men aged 22 to 28 years), with experience. Nineteen sensory descriptors were obtained and divided into 7 groups, as follows: appearance (brightness); apparent texture (consistency, adhesiveness, gelatinous appearance and homogeneity); aroma (cheese, buttery, sweet, acid); flavour (buttery, cheese); taste (salty, sweet, bitter, acid); oral texture (adhesiveness, consistency); and residual sensation (oil film).	Reduction of fat and salt contents and the addition of XOS, arginine or yeast extract resulted in products with different sensory characteristics. The effect of XOS as a fat replacer was observed and sensory properties of the low-fat and full-fat products were similar. Arginine and yeast extract enhanced the flavour, reduced the bitter taste and improved the salty taste of the products. It was possible to obtain a functional formulation, with 50% sodium reduction, 80% fat reduction and prebiotic status, using 3.3 w/w % XOS as a fat substitute and prebiotic agent and 1 w/w % yeast extract as a flavour enhancer.	Ferrão et al. (2018)

(*Continued*)

Table 14.5 (Continued)

Product	Ingredient fortification	Sensory test used	Sensory test details	Main results	Reference
Processed cheese spread	Addition of inulin (0, 4, 6 and 8%).	Descriptive analysis	Ten trained panellists. Attributes evaluated were flavour, body and texture, colour, appearance, spreadability and total sensory score.	The control cheese had a shiny yellowish colour, the addition of inulin resulted in dull colour and appearance. Spreadability was similar to the control, up to 6% inulin addition. Cheese spread with a 6% inulin level was observed to be optimum. At 8% inulin addition, there was a decrease in flavour, body, texture, spreadability and total sensory score.	Giri et al. (2017)
Requeijão cremoso processed cheese	Addition of prebiotic, using (GOS) (0, 1.5, 3 or 4 g/100 g) as prebiotic food components	Consumer acceptability testing	The affective test was performed with 80 consumers (50 women, 30 men, age 19–48 years) using a 9-point hybrid hedonic scale. Appearance, aroma, taste, texture and overall impression were evaluated	The addition of the 3 and 4 g/100 g of GOS enhanced the aroma and taste of queijo cremoso samples, with satisfactory mean values for all the sensory attributes, evaluated and were rated acceptable by the selected consumers	Belsito et al. (2017)

Processed cheese spread	Incorporation of black pepper was from 0, 5%, 4%, 3% and 2%, corresponding to treatments 1, 2, 3, 4 and 5	Consumer acceptability testing	Thirty untrained panellists. Among 5 treatments prepared, the last 3 treatments were used for sensory analysis. Colour, flavour, texture, spreadability and overall acceptability were evaluated using a 9-point hedonic scale	There were no significant differences in colour and appearance between the three treatments. However, the values for flavour, spreadability and overall appearance were significant different. Treatment 4, containing 3% black pepper was selected as the best treatment because of its better sensory characters except for the colour	Rifky et al. (2018)
Functional processed cheese	Addition of tomato juice (10, 20 and 30%)	Consumer acceptability testing	Firmness, spreading, stickiness, crumbliness, acceptability was evaluated	The samples were acceptable and had good firmness and crumbliness. The treatment with 30% tomato juice and the control sample had better spreadability, stickiness and acceptability	Mehanna et al. (2017)

A more attractive alternative for food companies is the use of prebiotic ingredients in the manufacture of processed cheese. The most recent definition of a prebiotic considers it a substrate selectively used by host microorganisms that confer health benefits (Gibson et al. 2017). Most known prebiotics are carbohydrates, but other substances, such as polyphenols and polyunsaturated fatty acids, may be included in the current definition (Gibson et al. 2017). The addition of prebiotics may result in changes in the quality attributes of processed cheese due to interactions between the functional ingredient and the components of the food matrix (Verruck et al. 2019). Prebiotics used in processed cheeses include inulin, galactooligosaccharide (GOS) and xylooligosaccharides (XOS) (Belsito et al. 2017; Giri et al. 2017; Ferrão et al. 2018; Cardoso et al. 2020). Inulin-type fructans are used as ingredients in the development of new products not only due to their functional properties but also for technological reasons (Verruck et al. 2019c). Inulin can also be used as a fat substitute in dairy products since it can promote a similar sensation on the tongue when ingested. Its higher molecular weight, compared to oligofructose, makes it less soluble and able to form microcrystals when mixed with water or milk. These microcrystals are not perceived during consumption, but interact to form a smooth creamy texture (Mensink et al. 2015). Giri et al. (2017) added 0, 4, 6 and 8% of inulin in processed cheese spread and the sensory evaluation showed that spreadability was similar to the control up to the 6% level of addition but at 8% inulin level, there was a decrease in flavour, body, texture, spreadability and total sensory score was found. Thus, the cheese spread with a 6% inulin level was observed to be optimum (Giri et al. 2017). The effectiveness of XOS as a fat replacer was observed for requeijão cremoso processed cheese, resulting in low-fat and the full-fat cheeses with similar sensory characteristic. The authors developed a functional formulation, with 50% sodium reduction, 80% fat reduction and prebiotic potential using 3.3 w/w % XOS and 1 w/w % yeast extract to enhance the flavour (Ferrão et al. 2018). In requeijão cremoso, a similar type of processed cheese, the incorporation of GOS enhanced the aroma and taste of experimental samples, with satisfactory mean values achieved for all sensory attributes evaluated (Belsito et al. 2017).

Consumer sensory research has been carried out on processed cheese fortified with vitamin D. The fortification of pasteurised process cheese with vitamin D_3 (360 IU/100 g) did not have any effect on the flavour of the product and no effect on the sensory acceptability of the final product was observed (Upreti et al. 2002). Vitamin D has good heat stability and does not degrade over time, which makes it very promising for addition to processed cheeses (Talbot-Walsh et al. 2018). It is estimated that 50–70% of Americans do not ingest the recommended daily requirement of vitamin D (Ganesan et al. 2011), so incorporating this ingredient into processed cheeses could supply the daily requirement for vitamin D supplementation.

Recently, the addition of macronutrients derived from natural sources, for example, carotenoids, antioxidants, vitamins and amino acids are being investigated to increase functional appeal of processed cheeses. Extracts of tomatoes and mushrooms, pumpkin paste, *Chlorella vulgaris* seaweed, asparagus, tomato powder and black pepper have been incorporated in the manufacture of processed cheeses to address consumer interest in foods with potential health benefits (Kassem et al. 2017; Rifky et al. 2018; Tohamy et al. 2018; Solhi et al. 2020a; Solhi et al. 2020b). The sensory implications of these incorporations varied and are detailed in Table 14.5.

14.4 Conclusions

The growing demand from consumers for healthier food products has stimulated processed cheese innovations. Studies on the reduction of sodium and fat in processed cheeses are at the forefront in the development of products to meet new nutritional labelling categories. However, formulation changes can lead to major changes in the sensory characteristics of processed cheese products, which are not always positive in the assessment of consumers. Therefore, sensory profiling and consumer analysis are critical disciplines in the quest for processed cheese innovations and the development of new products to meet market opportunities and drive growth in consumption.

References

de Almada, C.N., Almada, C.N., Martinez, R.C.R., and Sant'Ana, A.S. (2016). Paraprobiotics: evidences on their ability to modify biological responses, inactivation methods and perspectives on their application in foods. *Trends in Food Science and Technology* 58: 96–114. https://doi.org/10.1016/j.tifs.2016.09.011.

Balthazar, C.F., Pimentel, T.C., Ferrão, L.L. et al. (2017). Sheep milk: physicochemical characteristics and relevance for functional food development. *Comprehensive Reviews in Food Science and Food Safety* 16 (2): 247–262. https://doi.org/10.1111/1541-4337.12250.

Barr, S.I. (2010). Reducing dietary sodium intake: the Canadian context. *Applied Physiology, Nutrition and Metabolism* 35: 1–8. https://doi.org/10.1139/H09-126.

Belsito, P.C., Ferreira, M.V.S., Cappato, L.P. et al. (2017). Manufacture of Requeijão cremoso processed cheese with galactooligosaccharide. *Carbohydrate Polymers* 174: 869–875. https://doi.org/10.1016/j.carbpol.2017.07.021.

Bunka, F., Pavlínek, V., Hrabě, J. et al. (2007). Effect of 1-monoglycerides on viscoelastic properties of processed cheese. *International Journal of Food Properties* 10 (4): 819–828. https://doi.org/10.1080/10942910601113756.

Cardoso, L.G., Bordignon Junior, I.J., Vieira da Silva, R. et al. (2020). Processed cheese with inulin and microencapsulated chia oil (*Salvia hispanica*). *Food Bioscience* 37 (February): https://doi.org/10.1016/j.fbio.2020.100731.

Černíková, M., Buňka, F., Pospiech, M. et al. (2010). Replacement of traditional emulsifying salts by selected hydrocolloids in processed cheese production. *International Dairy Journal* 20 (5): 336–343. https://doi.org/10.1016/j.idairyj.2009.12.012.

Chavan, R.S., Khedkar, C.D., and Bhatt, S. (2015). Fat replacer. In: *Encyclopedia of Food and Health* (ed. B. Caballero, P.M. Finglas and F. Toldrá), 589–595. Academic Press http://dx.doi.org/10.1016/B978-0-12-384947-2.00271-3.

Chavhan, G.B., Kanawjia, S.K., Khetra, Y., and Puri, R. (2015). Effect of potassium-based emulsifying salts on sensory, textural, and functional attributes of low-sodium processed Mozzarella cheese. *Dairy Science and Technology* 95 (3): 265–278. https://doi.org/10.1007/s13594-014-0207-0.

Coyle, D.H., Shahid, M., Dunford, E.K. et al. (2020). Contribution of major food companies and their products to household dietary sodium purchases in Australia. *International Journal of Behavioral Nutrition and Physical Activity* 17 (1): https://doi.org/10.1186/s12966-020-00982-z.

Cunha, C.R., Dias, A.I., and Viotto, W.H. (2010). Microstructure, texture, colour and sensory evaluation of a spreadable processed cheese analogue made with vegetable fat. *Food Research International* 43 (3): 723–729. https://doi.org/10.1016/j.foodres.2009.11.009.

Cunha, C.R., Grimaldi, R., Alcântara, M.R., and Viotto, W.H. (2013). Effect of the type of fat on rheology, functional properties and sensory acceptance of spreadable cheese analogue. *International Journal of Dairy Technology* 66 (1): 54–62. https://doi.org/10.1111/j.1471-0307.2012.00876.x.

Drake, M.A. and Delahunty, C.M. (2011). Sensory evaluation. In: *Encyclopedia of Dairy Sciences*, 2e (ed. J.W. Fuquay, P.F. Fox and P.L.H. McSweeney), 105–110. London, UK: Elsevier.

Ehsannia, S. and Sanjabi, M.R. (2016). Quality characterization of processed cheese inoculated by *Bacillus coagulans* during cold storage: compositional and sensorial attributes and probiotic microorganism viability. *Journal of Food Processing and Preservation* 40 (4): 667–674. https://doi.org/10.1111/jfpp.12646.

Esmerino, E.A., Tavares Filho, E.R., Thomas Carr, B. et al. (2017). Consumer-based product characterization using pivot profile, projective mapping and check-all-that-apply (CATA): a comparative case with Greek yogurt samples. *Food Research International* 99: 375–384. https://doi.org/10.1016/J.FOODRES.2017.06.001.

Felicio, T.L., Esmerino, E.A., Vidal, V.A.S. et al. (2016). Physico-chemical changes during storage and sensory acceptance of low sodium probiotic Minas cheese added with arginine. *Food Chemistry 196*: 628–637.

Ferrão, L.L., Silva, E.B., Silva, H.L.A. et al. (2016). Strategies to develop healthier processed cheeses: reduction of sodium and fat contents and use of prebiotics. *Food Research International* 86: 93–102. https://doi.org/10.1016/j.foodres.2016.04.034.

Ferrão, L.L., Ferreira, M.V.S., Cavalcanti, R.N. et al. (2018). The xylooligosaccharide addition and sodium reduction in requeijão cremoso processed cheese. *Food Research International* 107 (December 2017): 137–147. https://doi.org/10.1016/j.foodres.2018.02.018.

Ganesan, B., Brothersen, C., and McMahon, D.J. (2011). Fortification of Cheddar cheese with vitamin D does not alter cheese flavor perception. *Journal of Dairy Science* 94 (7): 3708–3714. https://doi.org/10.3168/jds.2010-4020.

Gaonkar, A. and McPherson, A. (2016). Ingredient interactions: effects on food quality. https://books.google.com/books?hl=pt-BR&lr=&id=0HPLBQAAQBAJ&oi=fnd&pg=PP1&ots=WcWcQSdb6s&sig=DEC438tU39_h8U7VXAyoXraksno (accessed 6 December 2022).

Gibson, G.R., Hutkins, R., Sanders, M.E. et al. (2017). Expert consensus document: the International Scientific Association for Probiotics and Prebiotics (ISAPP) consensus statement on the definition and scope of prebiotics. *Nature Reviews Gastroenterology & Hepatology* 14 (8): 491–502. https://doi.org/10.1038/nrgastro.2017.75.

Giri, A., Kanawjia, S.K., and Singh, M.P. (2017). Effect of inulin on physico-chemical, sensory, fatty acid profile and microstructure of processed cheese spread. *Journal of Food Science and Technology* 54 (8): 2443–2451. https://doi.org/10.1007/s13197-017-2686-5.

Golan, E. and Unnevehr, L. (2008). Food product composition, consumer health, and public policy: introduction and overview of special section. *Food Policy* 33 (6): 465–469. https://doi.org/10.1016/j.foodpol.2008.05.004.

Gomes, A.A., Braga, S.P., Cruz, A.G. et al. (2011). Effect of the inoculation level of *Lactobacillus acidophilus* in probiotic cheese on the physicochemical features and sensory performance compared with commercial cheeses. *Journal of Dairy Science 94* (10): 4777–4786.

Guimarães, J.T., Balthazar, C.F., Scudino, H. et al. (2019). High-intensity ultrasound: a novel technology for the development of probiotic and prebiotic dairy products. *Ultrasonics Sonochemistry* 57: 12–21. https://doi.org/10.1016/J.ULTSONCH.2019.05.004.

Guinee, T.P. (2016). Cheese: pasteurized processed cheese products ☆. *Reference Module in Food Science* https://doi.org/10.1016/b978-0-08-100596-5.00680-6.

Guinee, T.P. (2017). Pasteurized processed and imitation cheese products. In: *Cheese: Chemistry, Physics and Microbiology: Fourth Edition*, vol. 1, 1133–1184. https://doi.org/10.1016/B978-0-12-417012-4.00046-6.

Gulzar, N., Sameen, A., Muhammad Aadil, R. et al. (2020). Descriptive sensory analysis of pizza cheese made from mozzarella and semi-ripened Cheddar cheese under microwave and conventional cooking. *Foods* 9 (2): 214. https://doi.org/10.3390/foods9020214.

Hanaei, F., Cuvelier, G., and Sieffermann, J.M. (2015). Consumer texture descriptions of a set of processed cheese. *Food Quality and Preference* 40 (PB): 316–325. https://doi.org/10.1016/j.foodqual.2014.05.018.

Haron, H., Hiew, I., Shahar, S. et al. (2020). A survey on salt content labeling of the processed food available in Malaysia. *International Journal of Environmental Research and Public Health* 17 (7): https://doi.org/10.3390/ijerph17072469.

Hassan, A.N., Awad, S., and Mistry, V.V. (2007). Reduced fat process cheese made from young reduced fat cheddar cheese manufactured with exopolysaccharide-producing cultures. *Journal of Dairy Science* 90 (8): 3604–3612. https://doi.org/10.3168/jds.2007-0141.

Hill, C., Guarner, F., Reid, G. et al. (2014). Expert consensus document: the international scientific association for probiotics and prebiotics consensus statement on the scope and appropriate use of the term probiotic. *Nature Reviews Gastroenterology & Hepatology* 11 (8): 506–514. https://doi.org/10.1038/nrgastro.2014.66.

Hladká, K., Randulová, Z., Tremlová, B. et al. (2014). The effect of cheese maturity on selected properties of processed cheese without traditional emulsifying agents. *LWT – Food Science and Technology* 55 (2): 650–656. https://doi.org/10.1016/j.lwt.2013.10.023.

Hoffmann, W., Gärtner, J., Lück, K. et al. (2012). Effect of emulsifying salts containing potassium on the quality of block-type processed cheese. *International Dairy Journal* 25 (1): 66–72. https://doi.org/10.1016/j.idairyj.2011.11.010.

Huang, H., Hallinan, R., and Maleky, F. (2018). Comparison of different oleogels in processed cheese products formulation. *International Journal of Food Science & Technology* 53 (11): 2525–2534. https://doi.org/10.1111/ijfs.13846.

Hwang, H.S. (2020). A critical review on structures, health effects, oxidative stability, and sensory properties of oleogels. *Biocatalysis and Agricultural Biotechnology* 26: 101657. https://doi.org/10.1016/j.bcab.2020.101657.

Institute of Medicine (2010). *Front-of-Package Nutrition Rating Systems and Symbols: Phase I Report* (ed. E.A. Wartella, A.H. Lichtenstein and C.S. Boon). Washington, DC. Available in: https://www.ncbi.nlm.nih.gov/books/NBK209851/ (accessed 6 December 2022).

Johnson, M.E., Kapoor, R., McMahon, D.J. et al. (2009). Reduction of sodium and fat levels in natural and processed cheeses: scientific and technological aspects. *Comprehensive Reviews in Food Science and Food Safety* 8 (3): 252–268. https://doi.org/10.1111/j.1541-4337.2009.00080.x.

Karimi, R., Azizi, M.H., Ghasemlou, M., and Vaziri, M. (2015). Application of inulin in cheese as prebiotic, fat replacer and texturizer: a review. *Carbohydrate Polymers* 119: 85–100. https://doi.org/10.1016/j.carbpol.2014.11.029.

Kassem, J.M., Abbas, H.M., Mohamed, A.G. et al. (2017). Sweet processed cheese spread analogue as a novel healthy dairy product. *International Journal of Dairy Science* 12 (5): 331–338. https://doi.org/10.3923/ijds.2017.331.338.

Khetra, Y., Chavhan, G.B., Kanawjia, S.K., and Puri, R. (2015). Storage changes in low sodium-processed Mozzarella cheese prepared using potassium-based emulsifying salts. *Dairy Science and Technology* 95 (5): 639–649. https://doi.org/10.1007/s13594-015-0248-z.

Kumar, R. (2012). An investigation into improvement of low fat cheddar cheese by the addition of hydrocolloids. University of Minnesota. https://conservancy.umn.edu/handle/11299/165556 (accessed 6 December 2022).

Kussy, D. and Aylward, E. (2009). Pasteurized process cheese. In: *The Sensory Evaluation of Dairy Products* (ed. S. Clark, M. Costello, M.A. Drake and F. Bodyfelt), 387–401. Springer https://doi.org/10.1007/978-0-387-77408-4_12.

Liu, H., Xu, X.M., and Guo, S.D. (2008). Comparison of full-fat and low-fat cheese analogues with or without pectin gel through microstructure, texture, rheology, thermal and sensory analysis. *International Journal of Food Science and Technology* 43 (9): 1581–1592. https://doi.org/10.1111/j.1365-2621.2007.01616.x.

MacKů, I., Buňka, F., Pavlínek, V. et al. (2008). The effect of pectin concentration on viscoelastic and sensory properties of processed cheese. *International Journal of Food Science and Technology* 43 (9): 1663–1670. https://doi.org/10.1111/j.1365-2621.2008.01734.x.

McSweeney, P.L.H., Ottogalli, G., and Fox, P.F. (2017). Diversity and classification of cheese varieties: an overview. In: *Cheese: Chemistry, Physics and Microbiology: Fourth Edition*, vol. 1 (ed. P.L.H. McSweeney, P.F. Fox, P.D. Cotter and D.W. Everett), 781–808. Academic Press http://dx.doi.org/10.1016/B978-0-12-417012-4.00031-4.

Mehanna, N.S., Hassan, F.A.M., El-Messery, T.M., and Mohamed, A.G. (2017). Production of functional processed cheese by using tomato juice. *International Journal of Dairy Science* 12 (2): 155–160. https://doi.org/10.3923/ijds.2017.155.160.

Mensink, M.A., Frijlink, H.W., van der Voort Maarschalk, K., and Hinrichs, W.L.J. (2015). Inulin, a flexible oligosaccharide I: review of its physicochemical characteristics. *Carbohydrate Polymers* 130: 405–419. https://doi.org/10.1016/j.carbpol.2015.05.026.

Miele, N. (2013). Calorie reduction in food: Sensory performance of a new sweetener and fat replacer optimization.

Ni Mhurchu, C., Capelin, C., Dunford, E.K. et al. (2011). Sodium content of processed foods in the United Kingdom: analysis of 44,000 foods purchased by 21,000 households1-3. *American Journal of Clinical Nutrition* 93 (3): 594–600. https://doi.org/10.3945/ajcn.110.004481.

Nogueira, E.B., Costa-Lima, B.R.C., Torres, F. et al. (2018). Effect of potassium-based emulsifying salts on the sensory and physicochemical parameters of low-sodium spreadable processed cheese. *International Journal of Dairy Technology* 71 (3): 717–722. https://doi.org/10.1111/1471-0307.12519.

O'Connor, T.P. and O'Brien, N.M. (2011). *Butter and Other Milk Fat Products. Fat Replacers.* Oxford, UK: Elsevier Ltd.

Pérez-Cueto, F.J.A., Aschemann-Witzel, J., Shankar, B. et al. (2012). Assessment of evaluations made to healthy eating policies in Europe: a review within the EATWELL project. *Public Health Nutrition* 15: 1489–1496. https://doi.org/10.1017/S1368980011003107.

Pravst, I., Lavriša, Ž., Kŭar, A. et al. (2017). Changes in average sodium content of prepacked foods in Slovenia during 2011–2015. *Nutrients* 9 (9): https://doi.org/10.3390/nu9090952.

Rafiq, S.M. and Ghosh, B.C. (2017). Effect of potato incorporation on the physico-chemical, textural and sensory properties of processed cheese. *Journal of Food Measurement and Characterization* 11 (2): 776–780. https://doi.org/10.1007/s11694-016-9448-3.

Rifky, A.L., Shabry, M.H., and Mubarak, A.J. (2018). Development of black pepper incorporated processed cheese spread for the local market. *International Journal ofAcademic and Applied Research (IJAAR)* 2 (4): 6–10.

Rinaldoni, A.N., Palatnik, D.R., Zaritzky, N., and Campderrós, M.E. (2014). Soft cheese-like product development enriched with soy protein concentrates. *LWT – Food Science and Technology* 55 (1): 139–147. https://doi.org/10.1016/j.lwt.2013.09.003.

Rippin, H., Hutchinson, J., Jewell, J. et al. (2017). Adult nutrient intakes from current National Dietary Surveys of European Populations. *Nutrients* 9 (12): 1288. https://doi.org/10.3390/nu9121288.

Roller, S. and Jones, S. (2010). Handbook of fat replacers. https://books.google.com/books?hl=pt-BR&lr=&id=wP8l58oKtisC&oi=fnd&pg=PA1&ots=cZVVKu6BKJ&sig=slCtdfQWS7ILuRbcWrqjlTLT0r8 (accessed 6 December 2022).

Rozenberg, S., Body, J.J., Bruyère, O. et al. (2016). Effects of dairy products consumption on health: benefits and beliefs – a commentary from the Belgian Bone Club and the European Society for Clinical and Economic Aspects of osteoporosis, osteoarthritis and musculoskeletal diseases. *Calcified Tissue International* 98 (1): 1–17. https://doi.org/10.1007/s00223-015-0062-x.

Sánchez-Macas, D., Moreno-Indias, I., Álvarez, S. et al. (2012). Sensory analysis of full-, reduced- and low-fat cheese elaborated with raw goat milk. *Journal of Applied Animal Research* 40 (2): 124–132. https://doi.org/10.1080/09712119.2011.633880.

Santiago-López, L., Aguilar-Toalá, J.E., Hernández-Mendoza, A. et al. (2018). Invited review: bioactive compounds produced during cheese ripening and health effects associated with aged cheese consumption. *Journal of Dairy Science* 101 (5): 3742–3757. https://doi.org/10.3168/jds.2017-13465.

Schädle, C.N., Eisner, P., and Bader-Mittermaier, S. (2020). The combined effects of different fat replacers and rennet casein on the properties of reduced-fat processed cheese. *Journal of Dairy Science* 103 (5): 3980–3993. https://doi.org/10.3168/jds.2019-17694.

Silva, R.C.S.N., Minim, V.P.R., Vidigal, M.C.R.T. et al. (2012). Sensory and instrumental consistency of processed cheeses. *Journal of Food Research* 1 (3): https://doi.org/10.5539/jfr.v1n3p204.

Smigic, N., Miocinovic, J., Tomic, J. et al. (2018). The effect of nisin and storage temperature on the quality parameters of processed cheese. *Mljekarstvo* 68 (3): 182–191. https://doi.org/10.15567/mljekarstvo.2018.0303.

Solhi, P., Azadmard-Damirchi, S., Hesari, J., and Hamishehkar, H. (2020a). Effect of fortification with asparagus powder on the qualitative properties of processed cheese. *International Journal of Dairy Technology* 73 (1): 226–233. https://doi.org/10.1111/1471-0307.12635.

Solhi, P., Azadmard-Damirchi, S., Hesari, J., and Hamishehkar, H. (2020b). Production of the processed cheese containing tomato powder and evaluation of its rheological, chemical and sensory characteristics. *Journal of Food Science and Technology* 57 (6): 2198–2205. https://doi.org/10.1007/s13197-020-04256-1.

Talbot-Walsh, G., Kannar, D., and Selomulya, C. (2018). A review on technological parameters and recent advances in the fortification of processed cheese. *Trends in Food Science and Technology* 81 (August): 193–202. https://doi.org/10.1016/j.tifs.2018.09.023.

Talbot-Walsh, G., Kannar, D., and Selomulya, C. (2019). pH effect on the physico-chemical, microstructural and sensorial properties of processed cheese manufactured with various starches. *LWT* 111 (April): 414–422. https://doi.org/10.1016/j.lwt.2019.05.036.

Taverniti, V. and Guglielmetti, S. (2011). The immunomodulatory properties of probiotic microorganisms beyond their viability (ghost probiotics: proposal of paraprobiotic concept). *Genes and Nutrition* 6 (3): 261–274. https://doi.org/10.1007/s12263-011-0218-x.

Tohamy, M.M., Ali, M.A., Shaaban, H.A.G. et al. (2018). Production of functional spreadable processed cheese using *Chlorella vulgaris*. *Acta Scientiarum Polonorum. Technologia Alimentaria* 17 (4): 347–358. https://doi.org/10.17306/J.AFS.2018.0589.

Torres, F.R., Esmerino, E.A., Carr, B.T. et al. (2017). Rapid consumer-based sensory characterization of requeijão cremoso, a spreadable processed cheese: performance of new statistical approaches to evaluate check-all-that-apply data. *Journal of Dairy Science* 100 (8): 6100–6110. https://doi.org/10.3168/jds.2016-12516.

Tunick, M.H. and Van Hekken, D.L. (2015). Dairy products and health: recent insights. *Journal of Agricultural and Food Chemistry* 63 (43): 9381–9388. https://doi.org/10.1021/jf5042454.

Uauy, R., Aro, A., Clarke, R. et al. (2009). Who scientific update on trans fatty acids: summary and conclusions. *European Journal of Clinical Nutrition* 63: S68–S75. https://doi.org/10.1038/ejcn.2009.15.

Upreti, P., Mistry, V.V., and Warthesen, J.J. (2002). Estimation and fortification of vitamin D3 in pasteurized process cheese. *Journal of Dairy Science* 85 (12): 3173–3181. https://doi.org/10.3168/jds.S0022-0302(02)74405-6.

Verruck, S., Balthazar, C.F., Ramon, S.R. et al. (2019a). Dairy foods and positive impact on the consumer's health. *Advances in Food and Nutrition Research* https://doi.org/10.1016/BS.AFNR.2019.03.002.

Verruck, S., Balthazar, C.F., Ramon, S.R. et al. (2019b). Dairy foods and positive impact on the consumer's health. In: *Advances in Food and Nutrition Research*, 1e (ed. F. Toldra), 95–164. Academic Press https://doi.org/10.1016/bs.afnr.2019.03.002.

Verruck, S., Dantas, A., and Prudencio, E.S. (2019c). Functionality of the components from goat's milk, recent advances for functional dairy products development and its implications on human health. *Journal of Functional Foods* 52 (September 2018): 243–257. https://doi.org/10.1016/j.jff.2018.11.017.

WHO. World Health Organization. (2020). Salt Reduction. Geneva, Switzerland: WHO. https://www.who.int/en/news-room/fact-sheets/detail/salt-reduction (accessed 6 December 2022).

15 Sensory Attributes of Fat-Rich Dairy and Ethnic Indian Products

Bhavbhuti M. Mehta and Suneeta Pinto

15.1 Introduction

The sensory quality of fat-rich dairy products (like various types of cream, butter, ghee, anhydrous milk fat) is considered an important factor that influences the overall quality of the product. Quality is related to the acceptable levels of sensory attributes viz. appearance, aroma, flavour and texture in relation to its sensory properties. Fat-rich dairy products are considered very important as they contain obviously high fat (10–70%) and ultimately affect the economy of the dairy industry.

Cream is used in many forms and for a variety of purposes. Different types of cream are classified according to their fat content (e.g. light, coffee cream, half cream, double cream) and their function (e.g. whipping cream, coffee). One of the important factors determining the consumer acceptability of cream is visual assessment of its 'body' or 'viscosity'. Cream is still considered a luxury product and it must have excellent flavour. Since it has a high-fat content (10–48%), any off-flavour from milk lipids will be concentrated. Therefore, good-quality raw milk is required to ensure high-sensory acceptability of the resulting cream.

Butter is a water-in-oil emulsion. There can be variation in the sensory properties of butter depending on the source, season and type of butter produced. The sensory properties of the butter may be described in terms of appearance, texture and taste. Spreadability and hardness are two very important characteristics of butter spreads. They have a large impact on consumer acceptability.

Ghee has a rich nutty flavour and unique grainy texture. The different methods for ghee manufacture, like desi or traditional method, creamery butter method, direct cream method, pre-stratification method and continuous method, affect the sensory quality of ghee. Ghee is regional specific; consumers in different regions prefer different types of ghee flavour and hence, judging of ghee is most difficult. Butter oil (anhydrous milk fat) lacks a characteristic flavour, aroma and texture.

Sensory evaluation of fat-rich dairy products should be carried out by experienced and trained panellists. The sample of the dairy product, the people, the testing environment and test methods used are four variables that affect the sensory evaluation. A person with normal sensory ability, including sensitivity to odour and taste, should be selected for

Sensory Profiling of Dairy Products, First Edition. Edited by John J. Tuohy.
© 2023 John Wiley & Sons Ltd. Published 2023 by John Wiley & Sons Ltd.

a judging panel. The person should have a willingness to work as a panellist and those who do not like milk products should not be allowed on the panel. A judge who is working in the sensory panel should also be trained to distinguish and discriminate between normal (control) and defective samples.

15.2 Cream and cream products

Cream is a fluid milk product, comparatively rich in fat, in the form of an emulsion of fat-in-skimmed milk, obtained by physical separation from milk (Codex Alimentarius Commission 2003). Many types of creams can be used for consumer products such as half-and-half, whipping cream and coffee cream or sometimes cultures can be added to cream to produce soured cream and a variety of other products. Cream consists of a concentration of natural fat globules from milk. Manufacturing of cream involves separating the cream phase from milk, standardising the cream to the desired fat content and pasteurisation with a view to making it safe and increase its shelf life. Homogenisation is applied to certain cream products to improve the properties of the product (Varnam and Sutherland 2001). For HTST pasteurisation of cream, heating at 75 °C for 15 seconds for cream containing <20% fat or >80 °C for 15 seconds for cream containing >20% fat is recommended by the International Dairy Federation (IDF 1996). In UHT processing of cream, the product is heated to 135–150 °C for a few seconds.

Creams are mostly classified according to their fat content (%) such as: light coffee cream (<10), coffee cream (15–18), single or half cream (15–25), cream or full cream (30–40) and double cream (45–50) (Smiddy et al. 2009). Due to its high-fat content, the mouthfeel of these products differs markedly from that of milk. Codex Alimentarius (1976) defines cream as 'the fluid milk product comparatively rich in fat, in the form of an emulsion of fat-in-skimmed milk, obtained by physical separation from milk'. The Food Safety Standards Authority of India (FSSAI) (2011) defines cream as the fluid product comparatively rich in fat, in the form of an emulsion of fat-in-skimmed milk, obtained by physical separation from cow's milk, buffalo milk or milk of any other species as defined under FSSAI (viz. cow, buffalo, goat, sheep and camel) or a mixture thereof. The product shall contain minimum 10.0% (m/m) milk fat. Acidity of the finished products, other than fermented and acidified creams, should not be more than 0.15% lactic acid. The cream is also classified based on its use, viz. fluid cream products, whipped cream, sour cream products, plastic cream, etc. All of the cream-based products are either pasteurised or ultra-pasteurised, and may be homogenised. If cream is to be stored, it should be held at temperatures less than 5 °C. The shelf life of cream is 8–10 days at 6 to 8 °C provided post heat treatment microbial contamination is low (Muir and Banks 2003).

There are a variety of cream products available which are popular in different parts of the world. It is not possible to present a uniform international classification for cream products as various countries use different names to describe similar products. A classification of various types of cream based on their fat content is shown in Table 15.1.

Table 15.1 Classification of creams based on fat content.

Cream types	Fat content (%)[a]
Primary Classification	
• Light coffee cream	<10
• Coffee cream	15–18
• Half cream or single cream	15–25
• Cream or full cream	30–40
• Double cream	45–50
General Classification	
• Clotted cream	55–70
• Coffee cream	10–30
• Cream	10–40
• Cream liqueur	15–16
• Cultured, fermented or sour cream	10–40
• Canned or sterilised cream	20–25
• Double cream	48–55
• Dried cream	40–70
• Half-and-half cream	10–18
• Half cream	12–18
• Heavy cream	36–45
• High-fat/plastic cream	70–80
• Light cream	12–30
• Light whipping cream	30–36
• Malai	25
• Whipping cream	30–48
• Table cream	20–25

[a] Compiled from literature sources.

Cream/full cream

The generic designation cream, or full cream, generally refers to a cream containing 30–40% fat widely used as an ingredient in soups, sauces, dressings, ice cream, desserts, custards and cakes.

Half-and-half/single cream

This cream normally contains about 10–18% fat. The major use of this product is as pouring cream for use in desserts and soups, in coffee and as a cereal or fruit topping.

High-fat cream/plastic cream

A high-fat cream usually contains 70–80% milk fat. High-fat spreadable creams are popular in many indigenous products around the world. It can also be used directly for manufacture of butter oil. It is prepared by passing cream (35–40% fat) through a specially designed 'plastic cream separator' having a relatively wide disc spacing with the channels for distribution arranged about halfway down the length of the discs. Before separation, the product is pasteurised at 71–77 °C for 15 minutes and cooled to 60–66 °C before separation.

Coffee cream

Coffee cream is used for whitening of coffee as well as imparting a pleasant flavour to coffee. Such creams usually contain 10 or 12% fat, less frequently 15, 18 or even 20% (Hoffmann and Buchheim 2006). Since it is used as an additive to coffee, the coffee 'whitening power' and freedom from 'feathering' are important functional properties. Feathering is a term used to describe the formation of an unpleasant scum or curd on the surface of coffee when cream is added (Smiddy et al. 2009). Cream, standardised to the desired fat percent is heat-treated at 90–95 °C and cooled to 6 °C is then subjected to the first two-stage homogenisation step (1500 and 500 psi in first and second stage, respectively) at 70 °C, followed by sterilisation. The product can be sterilised by continuous sterilisation in a UHT plant, followed by aseptic packaging (Hoffmann and Buchheim 2006). If required, stabilising salts, viz. phosphates and citrates can be added prior to pasteurisation. In case of UHT sterilisation, a second homogenisation step follows at a pressure of 1500 psi at ~70 °C. The coffee cream is then cooled to 25 °C and aseptically filled in plastic cups or cans.

Whipping cream

Whipping cream is one of the premium food products and is consumed for its pure flavour. This type of cream contains 35–40% fat. For manufacturing whipping cream, cream standardised to the desired fat percent is pasteurised at ~80 °C. The pasteurised cream is quickly whipped resulting in the formation of a coarse foam with the average diameter of air bubbles of ~150 µm, on further whipping, air bubble size is reduced approximately threefold (Noda and Shiinoki 1986) to form a firm and homogenous product containing about 50% v/v of air. Emulsifiers like glycerol monostearate (~1%) and mono-glycerides (~0.1%) are added prior to pasteurisation to improve the whipping properties. Other additives like calcium lactate (up to 0.1%), buttermilk powder, whey protein concentrates can be added to improve its whipping properties. The quality criteria for this product are stiffness, dryness, stability (as exhibited by resistance to air cell collapse and drainage or leakage) and overrun.

Aerosol-whipped cream

Aerosol-whipped cream is a foamed cream that is produced by an aerosol can. The can contains cream under pressure. The gas used is nitrous oxide which functions as a propellant. The foam formation depends on the amount of gas dissolved in the cream. The pressure in the aerosol should not exceed 1.5–2.0 MPa (Wijnen 1997).

Sterilised cream

The fat content is standardised followed by preheating the cream at a temperature of 140 °C for two seconds and then it is homogenised at 50–75 °C using either single or two-stage processes. Then stabilising salts are added to prevent the 'grainy' texture defect. The cream is filled into cans which are seam closed. Sterilisation of the can is carried out at suitable temperature, e.g. 26 minutes at 119.5 °C for continuous retort process.

Acidified cream

It is the milk product obtained by acidifying cream, reconstituted cream or recombined cream, or any combination of these, by the action of acids or acidity regulators, or both to achieve a reduction of pH with or without coagulation (FSSAI 2011).

Fermented/cultured/sour cream

Cultured cream/sour cream is utilised as a topping in many foods such as fruits, vegetables and can also be incorporated in soups. They are manufactured in many countries and their fat content generally ranges from 10 to more than 40% (Puhan 1988). The texture of cultured cream can be improved by homogenisation or by addition of suitable hydrocolloids. The fat content of sour cream can vary from 10 to 40% (Tamime and Marshall 1997). A freshly produced cultured cream should have a pH of ~4.5, resulting in a slightly acid, mild 'cheesy' or 'buttery' flavour and the texture should be uniform, viscous and creamy (Smiddy et al. 2009). According to FSSAI (2011), fermented/cultured/sour cream obtained by fermentation of cream, reconstituted cream or recombined cream, by the action of suitable microorganisms that results in reduction of pH with or without coagulation. Where the content of (i) specific microorganism(s) is(are) indicated, directly or indirectly, in the labelling or otherwise indicated by content claims in connection with sale, these shall be present, viable, active and abundant in the product to the date of minimum durability. If the product is heat-treated after fermentation, the requirement for viable microorganisms shall not apply (FSSAI 2011). Cream which is added with stabilisers, such as caseinates or hydrocolloids is standardised to the desired fat percent and heated for 15 seconds to 30 minutes at 85–95 °C or for a few seconds at 120–130 °C and homogenised. It is then cooled to the inoculation temperature of 20–24 °C and filled in cream containers to produce set-style cultured cream, or transferred to fermentation tank and culture containing mesophilic lactic acid bacteria is added. Mesophilic cultures commonly used in the production of cultured creams include *Lactococcus lactis* subsp. *lactis*, *L. lactis* subsp. *cremoris*, *L. lactis* subsp. *lactis* var. *diacetylactis* and *Leuconostoc mesenteroides* subsp. *cremoris*, with *Lactobacillus acidophilus* also being increasingly used, owing to its probiotic status (Tamime and Marshall 1997). Fermentation takes 14–24 hours at 20–24 °C. When the desired pH is reached, the cream is cooled rapidly to a temperature of 5 °C in order to inhibit further fermentation. It can be produced by direct acidification or by subjecting cream to fermentation. *Leuconostoc citrovorum* is generally preferred to produce a desirable aroma. Cream may also be chemically acidified, using glucono-δ-lactone (GDL) or food-grade acids (Kwan et al. 1982).

Clotted cream

Clotted cream containing up to 60% fat is used for confectioneries and cake toppings. It is traditionally prepared by heating cream. The heating process induces rapid fat rise and agglomeration of fat on the surface along with protein denaturation. Superficial skimming gives the product which is thick and spreadable and having a highly cooked flavour.

Malai

FSSAI (2011) has defined '*malai*' an indigenous cream product as the product rich in milk fat prepared by boiling and cooling of cow milk, buffalo milk or milk of any other species as defined under FSSAI or a mixture thereof. It is characterised by the presence of insoluble mass, principally fat and denatured protein, formed on heating and cooling of milk. It is the symbol of richness in traditional Indian products. Malai is a form of white clotted cream that can be formed as a thick layer of fat and coagulated proteins on the surface of boiled and cooled milk. Malai resembles British 'clotted cream'. Malai contains at least 25% milk fat.

15.2.1 *Sensory attributes of creams*

The desirable quality attributes of cream products are varied depending upon the consumers preferences and some of the products are mentioned below.

Cream

It should be fresh, clean and sweet with a rich creamy flavour, good consistency and mouthfeel. It should be smooth, uniform, fairly viscous but easily pourable at refrigeration temperatures. It should be neutral flavour profile with no distinct aftertaste and should not leave the mouth with a greasy feeling on swallowing. It should have a good appearance with no discoloration or visible phase separation and be free of sediments when emptied from the packet. It can possess some degree of cooked flavour. Good-quality table cream should have a pleasantly rich, clean, sweet, nutty flavour. The body should be smooth, uniform and more viscous depending on its fat percentage. Cream, like milk, should have an almost a neutral flavour profile with no distinct aftertaste. The physical appearance should be good. It should be free of sediments, have no discoloration or visible phase separation. Since cream is pasteurised at a higher temperature than milk, it commonly possesses some degree of cooked flavour. The cooked flavour is very apparent immediately after processing; however, the intensity of the cooked flavour diminishes during storage (Boelrijk et al. 2003). The slight 'cooked' or 'sulphurous' flavour of pasteurised milk results from the formation of such volatile substances as ammonia, hydrogen sulphide, mercaptans and volatile phosphorus compounds that occur with protein breakdown (de Wit and Nieuwenhuijse 2008).

Half-and-half and coffee cream

When the package is opened after shaking, the product should possess a clean, fresh flavour. The product should be homogenous, i.e. free from lumpiness and separation of serum. When contents are emptied, there should be no sedimentation. It should possess a clean, sweet, nutty flavour and impart a pleasing flavour to coffee. Product should have a uniform smooth texture and viscous mouthfeel. It should be stable when poured into hot coffee and should be easily mixable with no flakes on the surface and exhibit neither 'feathering' nor 'oiling-off'. It should impart a pleasant colour to the coffee.

Whipped cream

It should be easy to whip and produce a fine foam with good increase in volume on whipping. The foam should be stable and should not leak fluid on storage. The air cells should not be visible. It should be firm enough to retain its shape and remain stable during subsequent handling. It should have a light, fluffy mouthfeel. Whipped cream should be firm enough to retain its shape and remain stable during deformation, i.e. in decoration, and not exhibit coarsening of air cells and show negligible leakage of liquid (Pagote 2002). Sour cream should have a clean, slightly acidic/sour and pleasing taste and rich flavour. Typical cultured cream products should be uniform (without creaming), creamy and viscous with a slightly acidic, mild 'cheesy' or 'buttery' flavour (Hoffmann and Buchheim 2006). It should have a uniform white to yellowish colour. Clotted cream has an exceedingly rich flavour with a peculiar boiled taste and rough appearance with a white flaked surface.

Sour cream

It should have a clean, rich, slightly acidic/sour and pleasing taste and rich creamy flavour. It should possess a pleasantly aromatic flavour and can have a slightly acidic, mild buttery flavour. It should be uniform, creamy appearance and be viscous in consistency. The texture should be homogenous and smooth. The appearances should be uniform with a white to yellowish colour and slightly creamy appearance.

Clotted cream

Clotted cream should have an exceedingly rich cooked, caramelised flavour with a peculiar boiled taste and rough appearance with a white flaked surface. It has a nutty flavour with a golden crust or grainy texture with slight butteriness. It should be creamier than ordinary cream with a characteristic rich aroma and flavour.

15.2.2 *Sensory evaluation of creams*

Usually, the same off-flavours that are found in skim milk, low-fat milk and whole milk are found in various types of creams. They may appear to have different characteristics, but much of that is due to the different flavour background. Flavour-producing chemical compounds that are fat soluble are more concentrated in cream. Due to their higher fat content, and the optional presence of stabilisers and emulsifiers, the mouthfeel of these products differs markedly from that of milk. In addition to sensory qualities, important functional properties such as whippability and coffee-whitening properties should also be tested by recommended or standardised procedures (Scott et al. 2003).

The sensory properties of various cream products may be assessed by applying the milk scorecard and scoring guide, with a few modifications. The score card for judging cream is same as for market milk. According to the Indian 100-point BIS score card for milk, maximum score of 40 is allotted to flavour, 30 to body, 20 to odour and 10 to colour and appearance (BIS 1975c). Additional evaluation categories may be desirable for some of these products, particularly in the case of those that have certain functional properties. A logical test for whipping cream is a determination of its whipping properties and in case of coffee cream (half-and-half) or light cream the coffee 'whitening power' and freedom from 'feathering' are important functional properties.

Techniques for evaluation

Samples should be presented in odourless glass containers with lids. A sample of 50 ml should be sufficient for the test. The number of samples in one session should not exceed 5. Coding of samples should be done using a 3-digit random number. Carefully transfer bottle/container from refrigerator/cold room (5–10 °C) to the sensory evaluation laboratory in a vertical position, avoiding undue agitation. Note if there is any serum separation or sediment at the bottom of the bottle. Ascertain the condition and external appearance of the container and record observation as clean, attractive, dirty, loose lid, leaky, unsealed, dented, etc.

Open the lid and smell the contents immediately and examine smell and appearance of cream simultaneously. Check for any off-flavour if any, clean, foamy, fat separation, etc. Examine the inner side of the lid for adhering fat or foam. Take about 20 ml cream in a 100 ml beaker. Warm it to 15–21 °C and stir the sample properly to detect the odour of the cream. Pour the cream from one vessel to another and check its consistency. The cream should show an even consistency, firm in body with a velvety surface. There should be no clots or lumps in the cream. Note the taste. Rinse the mouth with warm water at regular intervals to cleanse the mouth. As the work entails considerable strain on the palate, it is unwise to swallow even small quantities of choice cream. Defects such as oiling-off and feathering in coffee should be checked by adding the cream to hot coffee.

Training for evaluation of cream

The panellists should be familiarised with all desirable and undesirable characteristics of cream. The common undesirable characteristics may be lactic acidic, weedy, feed, barny, oxidised and burnt flavours; ropiness and other microbial defects; sediments, adulterant, neutralisers and foreign matter. A series of 20 judgements shall be obtained from each prospective panellist utilising a pair of samples with known differences. To compensate for the known variations from day to day, a qualifying test should be given for a three-day period. Those who make an acceptable (75% correct) scores should be chosen. The panellists should be checked once in three months for their consistency and acuity (Solanky 1990).

Preparation of cream samples for training

A control sample of cream (40% fat) free from objectionable flavour, appearance and colour defects should be served along with cream (40% fat) samples having following defects:

Acidic – Prepare cream samples with varying lactic acidity by incubating fresh cream at 37 °C for varying periods so as to produce 0.1, 0.12, 0.14 and 0.16% acidity. Cream samples with developed acidity should be chilled to arrest further increase in acidity.

Sunlight – Expose capped sample bottles to direct sunlight ranging from 30 minutes to 3 hours to develop varying intensities of sunlight flavour defect.

Oxidised – Prepare cream samples with this defect by bubbling oxygen gas from a cylinder for 10, 15 and 20 minutes or alternatively by blowing air into a series of bottles of milk with the help of a rubber aspirator. The bottles should be stored at 37 °C prior to training session.

Barny – Expose milk sample to barn conditions for 10, 20, 30 and 40 minutes. Cap and store at 5 °C. Bring the sample temperature to 40 °C before presenting to the panellist.

Burnt – Boil a small quantity of milk for 15 minutes. Add aliquot portions from this overheated milk ranging from 5 to 25% to cream to develop different levels of burnt flavour.

Metallic – Prepare a 0.3% solution of ferrous sulphate in distilled water. To each bottle of 500 ml cream, add 0.5, 1.0 and 2.0 ml of ferrous sulphate solution to simulate varying degree of metallic taint.

15.2.3 *Flavour defects in cream*

A range of flavour defects in cream products have been defined and characterised. The absence of these defects is a basic criterion of quality assurance and control systems used in cream processing.

Acidic: This highly objectionable defect can be detected by a sour/acidic smell as well as its sour taste with unclean odour. This defect is caused by the growth of lactic acid bacteria and other acid producing organisms in milk which convert lactose into lactic acid resulting in a characteristic sour flavour which imparts a tingling effect. This defect in cream favours feathering, a phenomenon associated with protein denaturation, when the cream is added to a hot acidic solution such as coffee (Nelson and Trout 1965).

Barny: This off-flavour is reminiscent of the odour of a poorly maintained barn and leaves a persistent, unclean aftertaste. It is a transmitted off-flavour caused by poor ventilation and foul-smelling environment. It can be perceived by sniffing and tasting. A characteristic unpleasant aftertaste is most noticeable immediately after sample expectoration. The 'barny' or 'unclean' odours are transferred to the milk through the cow's respiratory system in the same manner as feed flavours (Bassette et al. 1986; Bruhn 2002).

Bitter: This defect can be detected by the sense of taste only. Bitterness in cream is detected at the base of the tongue and this taste sensation tends to persist for a relatively long time. Bitterness is associated with rancidity and can also be caused by consumption of bitter weeds by the cows before milking. It can also be caused by growth of certain types of microorganisms, especially growth of some types of psychrotrophic bacteria. Proteolysis of milk proteins often results in bitter flavours and further degradation of amino acids produces putrid flavours in milk (Shipe et al. 1978). Proteinases act on milk proteins, particularly the caseins, to form peptides and amino acids. Certain small peptides impart a bitter or astringent taste to milk and milk products (Driessen 1989).

Cowy: Cowy is a characteristic flavour of cream that is mainly attributed to the presence of low-molecular-weight fatty acids, nitrogen heterocyclic compounds, phenolics, lactones, phytol and acetone derivatives, and volatile carbonyl products present in fresh milk. Methyl sulphide contributes significantly to the characteristic flavour of milk (Bendall 2001). Usually, a 'cowy' flavour defect implies a distinct cow's breath-like odour. It has distinct, persistent unpleasant, medicinal chemical aftertaste with acetone bodies in milk, i.e. ketosis in cows.

Feed: Feed off-flavour can be detected by the sense of smell. It imparts aromatic taints to milk which is sometimes pleasant (i.e. alfalfa). The characteristic odour/taste of feed off-flavours varies with the type of feed consumed by the animal. This off-flavour is most noticed if feeds are given to cows within 0.5–3 hours before milking (Mounchilli et al. 2005). To prevent this type of off-flavour problematic feeds should not be fed to lactating animals from three hours before milking. This off-flavour is often associated with a sudden change in the type of feed given to animals.

Fermented/Fruity: The off-flavour is detected by its odour which may resemble that of vinegar (fermented) or fruity such as pineapple, apples or other fruits (fruity). This off-flavour is commonly found in stored pasteurised cream or in raw cream which has been stored for a long time and is caused by the growth of psychrotrophic bacteria, especially certain *Pseudomonas* sp. (e.g. *Pseudomonas fragi*) (Cormier et al. 1991). Certain microorganisms produce aromatic fermentation end products that seriously taint cream; this off-flavour is variously described as 'fermented' or 'fruity' (Morgan 1976; Hayes et al. 2002). A fruity flavour in dairy products is associated with post-pasteurisation contamination by *P. fragi* producing ethyl esters (Bassette et al. 1986; Deeth 1986).

Lacks freshness (stale): This taste response indicates a loss of fine pleasing taste, slightly chalky. It could be a 'forerunner' of either oxidised or rancid off-flavour defect.

Malty: This off-flavour definite or pronounced is suggestive of malt which is caused by the growth of *S. lactis* var. *maltigenes* especially when the cream is stored at temperatures greater than >18.2 °C for two to three hours resulting in a bacterial population in millions. This off-flavour is accompanied by an acid/sour taste. A 'malty' flavour in milk, sometimes described as 'cooked', 'caramelised' and 'burnt' can be associated with the metabolism of *L. lactis* biovar *maltigenes* (Bassette et al. 1986; Deeth 1986)

Metal-induced oxidised off-flavour: This off-flavour which occurs most often in raw and unhomogenised pasteurised cream is caused by metal catalysed oxidation of milk lipids. Terms like metallic, oily, cardboardy, stale, tallowy, painty and fishy are used to describe this off-flavour. The off-flavour is quickly perceived in the mouth and has a relatively short adaptation time.

Light-induced oxidised off-flavour: Light-induced flavours can develop when cream is exposed to sunlight, fluorescent light or even diffused daylight (Jensen and Poulsen 1992). This off-flavour has been described as burnt, burnt protein, burnt feathers, cabbagey, medicinal or chemical-like, light-activated or sunlight flavour or sunshine flavour which is attributed to singlet oxygen oxidation of serum proteins and free amino acids (Min and Boff 2002) as well as oxidation of lipids when the milk is exposed to light. This off-flavour usually occurs in milk after exposure to light accompanied with the degradation of Vitamin C (Whited et al. 2002). The most damaging wavelengths are in the UV range between 440 and 490 nm, while 310–440 nm and 490–500 nm also contribute to accelerated degradation (Robertson 2005). Min and Boff (2002) reported that methyl mercaptan, dimethyl disulphide and methionine sulfoxide are by-products of light-induced methional degradation in the presence of riboflavin, protein and oxygen.

Rancid: This off-flavour is extremely unpleasant, soapy, bitter and unclean aftertaste and is caused due to volatile fatty acids formed through enzymatic hydrolysis of fat which is more intense in cream than in milk. Lipolytic rancidity is caused by endogenous and/or exogenous lipases that hydrolyse milk fat triglycerides (Shipe et al. 1978). Short-chain free fatty acids, particularly butyric acid (C_4) and caprylic acid (C_6) and lauric acid (C_{12}) in the sn1 and sn3 positions of a triacylglycerol are mostly responsible for the unpleasant rancid flavour (Deeth and Fitz-Gerald 2006). Post-pasteurisation contamination of cream with psychrotrophs can lead to the production of heat-stable extracellular enzymes, particularly lipases and proteinases. These enzymes act on fat and proteins producing 'unclean', 'bitter', 'putrid' and 'rancid' flavours (Downey 1980; Walstra et al. 2006).

15.2.4 *Common body and texture defects in cream*

Three possible defects are commonly found when half-and-half is added to hot coffee: feathering, oiling-off and off-colour (in coffee). Of these, feathering is probably the most commonly encountered and the most objectionable. Whipping cream is subject to the same general body defects as light cream, but to different degrees of intensity.

Feathering: The major defect in coffee cream is feathering, which is manifested by formation of a particulate precipitate when added to coffee (Scott et al. 2003). This results in reduced whitening effect and significant oiling-off, which is caused by breakdown of the fat emulsion. Cream with developed acidity (i.e. more than 0.12% lactic acid) is more susceptible to this defect. If the cause of the defect is developed acidity, it will be accompanied by a sour/acid off-flavour.

Oiling-off: This defect is characterised by the presence of droplets of butter oil on the surface of the coffee and the defect tends to be seasonal. A cream plug, partial churning and/or coalescence of fat globules may be seen in the cream before its addition to hot coffee. Cream that has this defect generally appears shiny and usually has a thin body with the presence of a distinct skim milk layer. It is observed more often when cows have just been placed on pasture or green grass. This defect is closely associated with the cream plug defect. In the intensified state of oiling-off, a cream plug often forms. The possible causes of this defect are freezing of the cream product or improper homogenisation.

Cream plug: This defect is characterised by a lack of uniformity in the body of the cream which results in obstruction of flow of cream from the container. This is due to a distinct, thick cream plug those forms on the surface of the container sometimes accompanied by butter particles. When a thick layer of cream plug is found in the remaining portion of the cream, it has a distinctly thin body. When such cream is poured into hot coffee, droplets of milk fat are generally noted on the surface.

Separation of a skim milk layer: This type of defect is commonly found in cream products with lower fat contents. This happened because of the rising of fat particles which is often referred as creaming-off. When the cream is stored for an extended period, the skim milk layer becomes very distinct. The defect is described as a bluish, watery like layer that may be from one-sixteenth to one-half inch in depth, at the bottom of the product container. Sometimes an old, stale or oxidised off-flavour may be noted and associated with a cream displaying this type of body defect.

Thin body: Thin appearing body is a quite common body defect of some light creams. It is evidenced by a tendency to drip as it is slowly poured from the container, and/or a tendency to definitely 'splash' (similar to milk) as the product is poured onto a flat surface, from a distance of 6 in. or more. Thin body may sometimes be associated with the cream plug defect, but it will rarely be associated with the separation of a skim milk layer. Growth of microorganisms like *Bacillus cereus, Bacillus licheniformis, Bacillus coagulans, Bacillus subtilis and Geobacillus stearothermophilus* in pasteurised cream stored at 5 °C can cause bitter off-flavours and cream thinning due to lipolysis, proteolysis and acid production.

15.3 Butter

Butter means the fatty product principally in the form of a water-in-oil-type emulsion derived from pasteurised cream. Typically, butter contains 80–84% milk fat which is characterised by a relatively high level of short-chain fatty acids (C_4–C_{10}) (Frede 2002a). Butter contains 80% milkfat (typically, 80–81%), 17% moisture, 1% carbohydrates and protein and 1.2–1.5% sodium chloride (with no salt, the milkfat increases to 82–83%). The pH of sweet-cream butter (unfermented) is about 6.4–6.5. Many countries allow sodium chloride and lactic cultures as the only nonmilk additives in butter (Milner 1995). It is rich in phospholipids as lecithin and glycoprotein, and in fat soluble vitamins as A and D (Schimidt 2005). According to FSSAI (2011), the maximum moisture, milk solids-not-fat and common salt expressed as % (m/m) permitted in table butter in India is 16% and 2% and 3%, respectively. The minimum milk fat is 80% in case of table butter and 76% in case of white butter.

The textural and rheological properties of milk fats are strongly dependent on the thermal and structural properties of the triglycerides (Lopez et al. 2001). Differences in the fatty acid composition of butters result in alterations of butter crystallisation and melting point. Spreadability and hardness are two very important characteristics of plastic fats (Prentice 1972). They have a large impact on consumer acceptability. Butter is difficult to spread at refrigeration temperatures and is too soft or oily at room temperature (deMan and Wood 1958b; Kaylegian and Lindsay 1992; Shukla et al. 1994). The manufacture of butter is illustrated in Figure 15.1.

15.3.1 *Sensory attributes of butter*

Butter should possess a pleasant, creamy, sweet, clean, nutty flavour and a delicate aroma. It should also be dense and taste fresh. It should have as slightly cooked flavour. No visible water droplets on the surface and it should look dry in appearance. When removed from

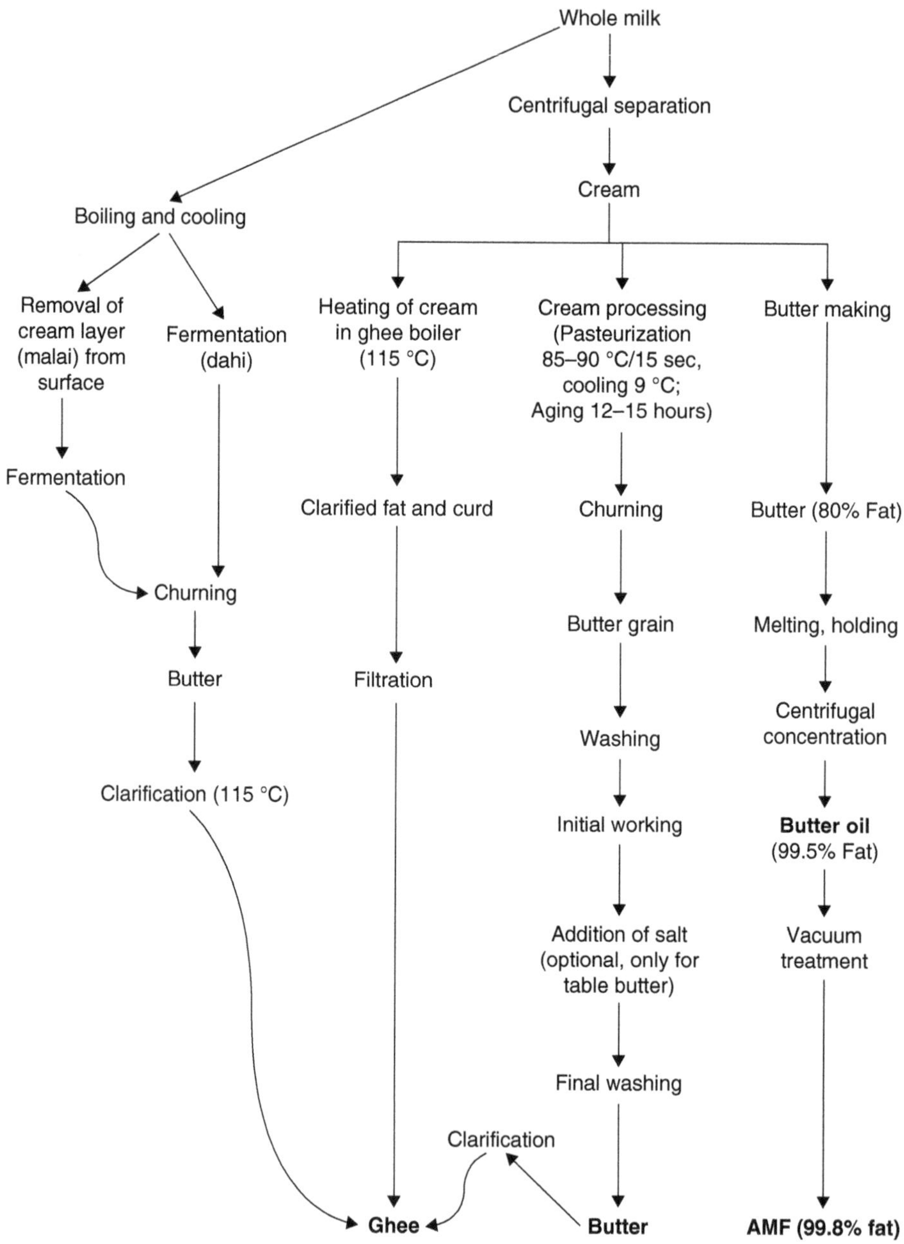

Figure 15.1 Flow diagram for butter, ghee, butter oil and anhydrous milk fat.

the refrigerator, it should be firm, smooth and exhibit a waxy, compact and close-knit texture. The consistency should be smooth, so that the butter is easy to spread and melts readily in the mouth. It should have a pale yellow colour similar to the colour of milk fat. The colour of butter may vary from light creamy white to dark creamy yellow or uniform light straw colour. Sour cream/cultured cream butter should exhibit a distinct culture flavour and an aroma with moderate levels of diacetyl and a pleasing buttery-like aroma with a slight to definite level of 'cooked' flavour. The water content should be dispersed in fine droplets so that the butter looks dry. It should have close-knit, smooth, waxy texture without any free visible moisture droplets. Sour cream butter should smell of diacetyl.

The flavour, body and texture of butter depends on the type of milk (cow, buffalo, goat, sheep, etc.); the diet, stage of lactation and dietary supplementation of the animal and the season of the year in which it is produced (Nickerson 1995; Chen et al. 2004). Additionally, processing, storage conditions, addition of starter culture and salt affect the sensory attributes of butter.

Aliphatic aldehydes are potentially major contributors to both desirable and undesirable flavours of milk fat. At very low concentrations (i.e. at or below their flavour threshold values), some aldehydes are thought to contribute to the desirable flavour of butter. Hept-4-enal is claimed to contribute a creamy flavour to butter (Urbach and Gordon 1994). When cream is heated, the trace amounts of β-keto acids present in milk fat are decarboxylated to yield methyl ketones. These volatiles are considered to contribute to the rich flavour associated with baked goods that contain butter. In addition to milkfat, butter contains moisture, curd (milk proteins, milk ash, lactose and other minor constituents) and commonly salt. The aroma of butter is derived primarily from the volatile compounds present in the fat fraction (Mallia et al. 2008). Lactones, ketones and aldehydes are the key aroma compounds in sweet-cream butter. Skatole was also found as a key odorant of sweet-cream butter Budin et al. (2001). Other aroma compounds like hydrogen sulphide, acetaldehyde, dimethyl sulphide, diacetyl, hexanal, 2-methylbutanal. 3-methylbutanal, butanoic acid, dimethyl trisulphide, hexanoic acid, delta-hexalactone, nonanal, delta-octalactone and gamma-dodecalactone have also been reported in the headspace of sweet-cream butter (Peterson and Reineccius 2003).

There is a wide variation in sensory expectations of butter among consumers. The key discriminating sensory characteristics of butters included colour intensity, diacetyl, cooked, grassy and milk fat flavours; and salty taste (Krause et al. 2007).

15.3.2 *Sensory evaluation of butter*

The sensory evaluation laboratory in which the scoring of butter is to be carried out should be clean, well-lit and well-ventilated. Ideally, the temperature of the room should be between 15 and 18°C. There should not be strong, offensive or irritating odours within the room or from nearby areas.

Preparation of butter samples for training

A control sample of fresh butter (prepared under standard condition either from unripened or ripened cream) representing desirable characteristics like flavour, texture and colour should be served along with the samples having following defects:

Flat: Convert fresh pasteurised cream into butter and wash excessively.
Stale: Store butter in a closed container at 10 °C or below for 72 hours.
Neutraliser: Prepare butter from cream having more than 0.2% lactic acid, neutralised to 0% lactic acid and pasteurised at 70 °C and above for 30 minutes.
Oily: Make butter from plastic cream and pasteurise at 70 °C for more than 30 minutes.

Tallowy: Expose butter to sunlight for three hours or more in glass beakers. This may be accelerated by incorporating 0.5–1 ppm of chlorides of iron or copper before exposure to sunlight.
Leaky: Overwork (four to five times) butter from neutralised cream at 60 °C.
Greasy: Same as leaky.

Procedure for training

The panellists should be familiarised with all the desirable characteristics of fresh butter. They should be trained to distinguish between 0.5 and 2 ppm of diacetyl in aqueous solutions. They should also be trained to distinguish samples with common defects, such as flat, oxidised, rancid, mouldy flavours; mottled colour; greasy, leaky and oily texture; and presence of foreign matter. Panellists should also be familiarised with common adulterants of butter (BIS 1975b).

Tempering of butter

The temperature of butter samples during this evaluation is an important factor in detecting body and texture. Butter should be placed in the tempering room well in advance to allow tempering to about 10 °C (the required time will depend on relative size and temperature of the butter sample). The temperature of sample should be maintained between 7 and 13 °C.

A butter trier should be used for drawing samples from butter blocks or package. Facilities for cleaning the trier (soft tissue or absorbent paper) and disposal of waste butter should also be provided. Use of hot water for cleaning the trier should be avoided. Butter samples should be presented in cubes wrapped in butter paper for evaluation. It may be cut in small slices to see any impurities and curdled portions. A sample of about 25 g should be presented. The number of samples in one session should not exceed five. Coding of the samples should be done using a 3-digit random number.

Sequence of observations to be taken

i. Observe the cleanliness and neatness of the package. Remove the cover/packaging material and observe the sample for its evenness and finish of the wrapping material.
ii. Insert the butter trier diagonally near the centre of the package and draw a sample plug of butter. Immediately after withdrawing the plug, pass the butter sample slowly under the nose, slowly inhale and note any aroma present. The amount of butter clinging to the back of the trier should be noted. Check the colour of the butter, especially for uniformity. Examine the body and texture for shortness by pressing the ball of the thumb against the sides of the butter plug (core sample) until it shows a break. Note the presence or absence of free moisture and the degree of clarity.
iii. Break approximately a ½–1 in. piece from the lower end of the butter plug and taste it. Chew the sample till it melts. Roll the melted butter within the mouth till it reaches body temperature. Check for the possible presence of undissolved salt between the teeth and/or between the tongue and the roof of the mouth. Also note the manner in which the butter melts. After swallowing the sample, note the aftertaste or for the presence of any persistent off-flavour sensation. Record your observations on the butter score card.

The maximum marks allotted to various attributes using a 100-point score card for butter as suggested by IS:7769 (BIS 1975b) is as follows – flavour: 50; body and texture: 30; colour and appearance: 15; and package: 5.

15.3.3 *Colour defects in butter*

The colour of butter may vary from light creamy white to dark creamy yellow or uniform light straw colour. The shade of butter colour is a natural shade of yellow which is commonly associated with milk fat.

Most of the colour defects arise from faulty working of butter during manufacture especially over- and under-working of butter and can be eliminated by proper working of butter. Speed of working also influence colour of butter. High-working speeds produce small water droplets, and the butter may become dry, while low-working speeds result in larger water droplets, especially at low temperatures, and the butter becomes 'wet' (Walstra et al. 2006). Colour defects can result from the size, number and distribution of water droplets and in case of whipping butter the size and distribution of air cells. Growth of microorganisms, including mould, can cause colour defects and surface taints in butter. If butter is not properly worked, there will be fewer water droplets, and the bacterial cells will have a better chance of multiplying and causing defects (Manners et al. 1987; Walstra et al. 2006). Any materials in direct contact with butter, which contains >5 mg/l copper may also result in bleached spots and/or a tallowy flavour since copper acts as a catalyst for oxidation. The presence of copper may result in green spots, while iron causes brown or black spots. Localised oxidation around particles of iron-containing additives, such as parsley, can result in white spots due to oxidative bleaching of the β-carotene (Manners et al. 1987).

Packaging materials for butter should be impermeable to light, fat, moisture and oxygen. Type of packaging material and conditions of storage also influence the colour of butter, e.g. butter which is inadequately protected against moisture evaporation tends to exhibit an intense or high-coloured surface. The colour resulting from surface moisture loss from evaporation is termed as a 'primrose' defect (Manners et al. 1987).

Wavy, mottled, streaked, colour specks, bleached, dull, pale; lack of colour uniformity, mould or yeast discoloration, uneven colour, unnatural colour, high-coloured surface compared to butter underneath, etc. are some of the colour defects seen in butter. Spoilage of butter by mould, including discoloration, typically involves lipolytic genera, such as *Aspergillus*, *Cladosporium*, *Geotrichum* and *Penicillium* (Varnam and Sutherland 2001). 'Mottled' refers to spots of lighter and deeper shades of yellow, caused by an uneven distribution of moisture due to insufficient working. A mottled colour defect is generally due to improper salt incorporation while a 'streaky' effect, that is, different shades of yellow in the butter. 'Streaks' are recognisable as an area of light colour surrounded by more highly coloured portions. 'Waviness' is an unevenness that appears as waves of different shades of yellow. Insufficient blending of two different butter sources is the usual cause. Speckled butter can arise due to the inclusion of curd particles that have precipitated during heat treatment.

15.3.4 *Flavour defects in butter*

Most of the frequently encountered butter off-flavours are derived from the cream. Off-flavours of butter are associated with the milkfat but some flavour defects may also result from the action of microorganisms on other milk constituents such as milk proteins, milkfat, lactose and other flavour defects can originate due to improper storage conditions. Some off-flavours may result from faulty cream or milk; handling, processing or churning; and certain other flavour defects may develop in the finished butter. Common butter flavour defects include: acid, barny, cheesy, coarse, feed, flat, foreign, fruity, high-salt, musty, metallic/oxidised, old cream, onion/garlic, rancid (lipase), storage, scorched, unclean, unnatural and yeasty.

Acidic: An acidic or sour off-flavour in butter usually develops from either churning high-acid cream, over-ripened cream, excessive use of lactic starter culture, use of too much lactic acid in unsalted butter or excess retention of buttermilk in the butter wherein lactose can be fermented. This defect is recognised by the acidic taste which is perceived when the butter is tasted as well as the sour smell (Bradley and Smukowski 2009).

Cheesy: This defect is reminiscent of the taste of ripened Cheddar cheese. It persists in the mouth and fails to clear out. In some extreme cases, a cheesy off-flavour in butter may somewhat resemble the odour of Limburger cheese or putrid meat. Sometimes this off-flavour is accompanied by bitter aftertaste which is due to proteolysis resulting in bitter peptides.

Coarse: This defect is associated with butter which lacks that sweet, pleasing, delicate flavour. The product lacks the fine, pleasant flavour sensation which is desirable in butter with good quality.

Cooked: A cooked flavour in butter can be described as a smooth, nutty-like, custard-like character is produced by pasteurising sweet cream at a relatively high temperature. It is rather desirable to have a definite cooked flavour in freshly churned butter. Butter exhibiting a slight to definite 'cooked' flavour 'cleans up' completely and leaves absolutely no aftertaste, other than a rather pleasant one. An intense heat treatment of the cream may give the butter a cooked flavour (Walstra et al. 2006).

Feed: The off-flavour caused by 'feed' can usually be detected by the aroma and can be confirmed on the palate when tasting. With most feed flavour defects, the mouth usually cleans up after the sample is expectorated. Most forms of dry feeds, such as hay, grain concentrates, citrus pulp, silage, green alfalfa and various grasses generally lead to 'normal' feed flavour note in butter. Even when fed in large quantities, these feeds only have a slight objectionable effect on butter flavour. Green alfalfa tends to produce a characteristic, mild, sweet flavour (with a possible bitter-sweet tinge). When cows are placed on fresh grass pasture, the butter produced may exhibit a typical 'grassy' off-flavour. This defect can be minimised or eliminated by following proper feeding routines for dairy cows. Generally, if cows are not fed between 0.5 and 3.5 hours of milking time, feed off-flavours are considerably minimised in butter.

Flat: Butter which lacks a characteristic pleasing 'buttery' flavour is generally criticised as being 'flat'. The product does not have a characteristic butter flavour. This defect can arise in case of lightly salted or unsalted butter. A flat defect is generally caused by and lack of flavouring compounds like acids, diacetyl, other carbonyls and other volatile compounds. A low-salt content in butter may result in a flat flavour. The defect may also be caused by excessive washing of butter granules during manufacture due to dilution of churning cream with water.

Fishy flavour: This type of defect may develop in table butter upon storage, especially in the one having high-salt content and made from cultured cream (i.e. ripened). Hydrolysis of phospholipid to form trimethylamine is one of the reasons attributed for the 'fishy' flavour defect in butter.

Malty: The 'malty' off-flavour which is similar to the odour of malted milk is sometimes perceived in butter. The flavour sensation can be perceived throughout the tasting period and tends to persist even after the sample has been expectorated. The malty off-flavour results from the outgrowth of *L. lactis* ssp. *maltigens* in either milk or cream that has been cooled inadequately.

Metallic: This off-flavour is the flavour sensation perceived when a copper penny is held between the teeth. This flavour defect conveys a slightly astringent and puckery sensation to the mouth interior. The metallic note may be detected as soon as the butter is placed into the mouth. This off-flavour persists after the sample has been expectorated; a somewhat bitter taste or other objectionable aftertaste may appear at the end of the tasting period. This off-flavour is a precursor to 'oxidised' off-flavour.

Neutraliser: The presence of a 'neutraliser' off-flavour in butter can be observed immediately after the sample has melted in the mouth. The aftertaste of added neutraliser in butter is persistent. This flavour note may be soda cracker-like or somewhat alkaline, suggestive of bicarbonate of soda or similar compounds. The soda neutralisers may also produce a bitter-like aftertaste, sometimes referred to as 'limey'. A neutraliser off-flavour in butter results from the adulteration of the cream with concentrated solutions of neutraliser to counter high levels of lactic acid found in poor quality cream.

Oxidised: The presence of pro-oxidant metals like copper, iron and nickel can significantly accelerate the rate of lipid oxidation development in milk and processed dairy products. This defect is more likely to be found in butter made from sour cream. Due to souring of the cream (or the milk) a considerable part (30–40%) of the 'added' copper (i.e. copper entered by contamination) moves to the fat globules. Adding salt to sour cream butter considerably accelerates the autoxidation.

The oxidation of unsaturated fatty acids to form a group of aldehydes in butter creates a series of different off-flavours, viz. oxidised. Different flavour sensations are perceived in various stages of development of oxidised butter, such as metallic, oily, tallowy, painty and/or fishy. A characteristic cardboard-like flavour and often an associated puckery mouthfeel are the usual distinguishing features. The term 'oxidised' best describes the metal-induced form of oxidised flavour. Off-flavours in fat-containing foods are also related to the presence of methyl ketones, which are responsible for the ketonic rancidity–musty, stale off-flavour of desiccated coconut (Kellard et al. 1985), caused by oxidation of medium-chain fatty acids by moulds (Kochar 1993). The exposure of butter to light has been reported to result in the oxidation of cholesterol, giving rise to 5-cholesten-3b,7a-diol and the 7b-epimer and, possibly, 6-cholesten-3b,5a-diol (Luby et al. 1986).

Rancid (lipolytic): Hydrolytic rancidity in butter is characterised by off-flavours variously described as bitter, unclean, wintery, butyric and rancid or lipolytic. The defect is more likely to develop during storage. The rancid off-flavour of butter is unmistakably objectionable, and may be soapy and/or bitter. Butter from sour cream will contain a larger amount of free fatty acids and will more readily attain a soapy-rancid flavour. The FFA can also contribute to the sensory aspects of dairy products, with elevated levels of short-chain FFA being responsible for rancidity (Deeth 2006). Rancidity of butter somewhat resembles the strong, disagreeable off-flavour of darkened, decayed nuts, baby breath, gym bags or dirty sneakers. The odour is pungent and is that of volatile short-chain fatty acids. Often this off-flavour gives the taste impressions of soapiness and frequently, definite or intense bitterness. A rancid off-flavour is the result of hydrolysis of milk fat through the enzymatic action of lipase, which liberates fatty acids. A rancid off-flavour is attributed to the free, short-chain fatty acids and the resultant salts of these fatty acids (e.g. technically a soap). A characteristic of the rancid off-flavour is a certain astringent mouthfeel, perceived at the base of the tongue and upper throat. psychrotrophic yeasts, moulds and bacteria can cause rancidity and surface spoilage in butter (Thomas and Druce 1971). Volatile lipid oxidation products, which are also an important group of off-odorants, can also contribute to rancid off-flavour. In butter, elevated levels of 1-octene-3-one are associated with rancid off-flavour, whereas 1-pentene-3-one has a characteristic fishy off-flavour (Badings 1970; Belitz et al. 2008). Yeast, particularly *Candida lipolyticum*, *Torulopsis* spp., *Cryptococcus* spp. and *Rhodotorula* spp., are capable of growth and can cause lipolysis in butter at low temperatures (Varnam and Sutherland 2001).

Storage: Butter held for considerable time (>6 months to several years) in frozen storage may gradually absorb odours from the storeroom environment. Under these circumstances the delicate flavour characteristics of butter are lost and the consequent flavour deterioration is referred to as the storage defect. After extended storage, butter made from fresh, clean flavoured, sweet cream seems to undergo this chemical change much more slowly than butter that was made from lower quality cream.

15.3.5 *Body and texture defects in butter*

Hardness and spreadability are inversely related characteristics of butters and potentially the most important aspects of texture and rheological influence on consumer perception (Wright et al. 2001). The body of good-quality butter when evaluated at 7–13 °C should be firm, smooth and exhibit a waxy, compact, close-knit texture. The texture of butter is essential to its quality as it determines spreadability and strongly affects appearance, taste, mouthfeel and its suitability for many uses. Butter should be sufficiently firm against sagging. Butter derives its firmness largely from fat crystals that are aggregated into a network. It should be readily spreadable and thereby not too 'long' (extensible) or too 'short' (crumbly, flaky). Oiling-off (i.e. separation of liquid fat) should not occur. On the other hand, the butter should be sufficiently soft as to be easily spreadable on bread. This causes great problems because the firmness and the spreadability closely depend on the composition of the fat and on the temperature. It should be easily deformable in the mouth without being greasy, should melt rather quickly and should thereby feel cool.

When cut, the appearance of quality butter should present a somewhat jagged, irregular, wrought iron-like surface. Water droplets and air cells should be uniformly distributed and closely bound. A certain degree of springiness is desirable to prevent a 'dead' appearance (Mulder and Walstra 1974). Examples of body-texture defects that have been associated with table butter include short, crumbly, leaky, gummy, mealy or grainy, weak, sticky.

Crumbly/brittle: A crumbly textured butter lacks the desired cohesion and the fat crystals do not hold together and tend to fall apart. Some of the butter usually adheres to the back of the trier and reflects a rough appearance. Crumbliness in butter maybe as a result of relatively large fat crystals and a deficiency in liquid fat. Butter that has been warmed and then cooled slowly during storage develops large crystals which can result in this defect. Under-working and incorrect churning temperatures may contribute to a hard and crumbly texture. Extremely rapid cooling may result in the formation of impure crystals, or a mixture of middle- and high-melting triacylglycerols with higher levels of solid fat, thus resulting in hard and crumbly butter (Manners et al. 1987).

Flaky: Pumping of butter may also result in an increase in air content, leading to increased oxidation, and 'flaky' or 'split' butter due to cohesion and adhesion of the butter to the walls of the pipes (Manners et al. 1987; Varnam and Sutherland 2001). The flaky butter defect can be prevented by maintaining a low-air content in the butter, generally below 0.2% (Manners et al. 1987).

Greasy: A 'greasy' butter is extremely smooth and melts immediately when the sample is placed in the mouth. The mouth is greasy after expectorating or swallowing the sample. This defect is more prevalent in the summer months. This defect may be suggested by the extreme ease with which a trier sample is removed from the product. Overworking of butter especially at higher churning temperatures results in this defect. Another cause of this defect is presence of high levels of lower melting point triglycerides in the butter.

Gummy: 'Gummy' bodied butter tends to stick to the roof of the mouth and may leave a gum-like impression. This defect is more predominant during the winter months. Gumminess is due to an abnormally high percentage of high-melting triglycerides, which increases the melting point of butter causing a firmer butter which is unspreadable. This defect can be minimised or eliminated by slow cooling of cream, use of a higher churning temperature, or working the butter for a longer period.

Leaky: When butter has 'leaky' defect, visible droplets of moisture can be observed on the back of the trier after a sample is removed. Because of the large size of water droplets, the moisture cannot be retained within the product. This defect is mostly seen in underworked butter. Traces of free moisture can sometimes be found in cold butter. The defect is also observed in butter which has been in frozen storage for an extended period. This defect can be minimised by a process called microfixing, which is a process that softens the frozen butter after thawing and restores the original water-in-oil emulsion.

Mealy/grainy: Butter having this defect lacks a smooth and waxy texture. When a sample of partially melted butter is compressed between the tongue and roof of the mouth, a distinct 'grainy' sensation is perceived. A mealy texture may be caused by the manufacture of butter from poor quality (high-acid) cream that has been neutralised, allowing milk fat to 'oil-off' at some stage in the pasteurisation process, or adding too much oiled-off rework.

Short: A 'short' body refers to butter that lacks the desirable characteristics of plasticity and waxiness. This defect is noted when a sample plug of butter has a tendency to break sharply when moderate thumb pressure is applied or even when the plug is removed from the block. A short-textured butter exhibit marked brittleness. Low-curd content in butter resulting from excessive washing of butter, the presence of high levels of high-melting point triglycerides and cooling of butter rapidly at very low temperatures are some of the possible causes of this defect.

Sticky: A 'sticky' bodied butter adheres to the trier and appears to be quite dry. Usually, it is difficult to secure a uniform, smooth-surfaced plug from such butter. The butter plug will appear 'ragged' or 'rough'. When crumbly or brittle-textured butter is overworked, the entire mass tends to become sticky. A sticky body and crumbly texture are often present concurrently in butter. A sticky body is observed most frequently in late autumn and winter when there is a greater predominance of high-melting triglycerides in churning cream. Hence, a sticky body is primarily a feed-related defect; it appears to be more prevalent in areas where alfalfa is the major roughage fed to milk cows. Various temperature treatments of cream and butter, as well as churn working conditions, seem to markedly affect the occurrence of 'sticky' defect.

In the batch production process, an over-filled churn will take longer to produce butter, and may result in an increase in temperature that will cause the butter to become sticky during working. This moist open-textured butter is often referred to as 'leaky butter' and can cause problems with packaging and printing. High temperatures resulting in sticky butter may also be attributed to churning the cream too soon after the heat treatment. Also, an under-filled churn will churn too rapidly and may produce high-moisture butter and create fat losses in the buttermilk. However, the addition of too much water should be avoided as it may require so much working that the butter becomes sticky (Manners et al. 1987).

Weak: A 'weak' body is typically indicated by a quick meltdown or an exaggerated softness of the butter, when it is exposed to room temperatures. A weak-bodied butter often produces an imperfect plug; there is a tendency for the trier to 'cut in' on the plug. When the ball of the thumb is pressed against a plug of 'weak' butter, difficulty is often encountered in defining a distinct 'breaking point' for the plug. A weak body is due to a state of incomplete milk fat crystallisation which may be caused by inadequate tempering of pasteurised cream, or due to a relatively high proportion of low-melting triglycerides in the cream. Churning at too high a temperature and incorporating too much air or nitrogen during whipping may also lead to a weak-bodied butter. However, a weak-bodied butter generally spreads well. Overworking may result in a weak body resembling thick cream (Varnam and Sutherland 2001).

Gritty: A gritty texture may result from undissolved salt, which will attract surrounding moisture causing free moisture, discolouration and possibly a mottled colour defect (Manners et al. 1987; Varnam and Sutherland 2001). The presence of 'grittiness' or 'grit' (undissolved salt) can be detected most easily by biting gently between the molars. If undissolved salt is present, a gritty effect is usually noticed at once. Soft, rounded or hard, flat particles above 80 mm in size exhibit gritty properties or characteristics. By contrast, when particles are hard, angular and within the size range of 11–22 mm, then grittiness is noted (Tyle 1993).

15.4 Dairy spreads

The demand for butter which is spreadable in the cold condition immediately after removing from refrigerator has led to the development of dairy spreads made with a reduced milk fat content, blends of butter (55–80%) and vegetable oil (20–35%), fractionated milk fat and chemically interesterified milk fat (Rousseau and Marangoni 1999; Frede 2002b). The vegetable oil, such as un-hydrogenated soya bean oil, sunflower oil and canola oil, may be used, but to be effective it must be present at levels of 15–35% (Varnam and Sutherland 2001). Other types of oils such as cotton seed oil, rapeseed oil and olive oil are also used for the manufacture of spreads. The composition of the spread blend is designed to achieve a spreadable consistency and mouthfeel in the product. According to FSSAI (2011), fat spread in India means a product in the form of water-in-oil emulsion, of an aqueous phase and a fat phase of edible oils and fats excluding animal body fats. The individual oil and fat used in the spread shall conform to the respective standards prescribed by these regulations. Fat spread shall be classified into three groups: (i) milk fat spread: containing exclusively milk fat; (ii) mixed fat spread: containing a mixture of milk fat with any one or more of hydrogenated, unhydrogenated refined edible vegetable oils or interesterified fat; (iii) vegetable fat spread: containing a mixture of any two or more of hydrogenated, unhydrogenated refined vegetable oils or interesterified fat. All the three types of spread should confirm to the following requirements: the fat content should not be more than 80% and not less than 40% by weight; moisture content not more than 56% and not less than 16% by weight; melting point of extracted fat (capillary slip method) in case of vegetable fat spread not more than 37 °C. Technology of spread manufacture involves the following steps: selection of ingredients, preparation of aqueous and fat phase separately, and their mixing, heat treatments, emulsification, cooling, working, filling and packaging.

The technology of spread manufacture involves the following steps: selection of ingredients, preparation of aqueous and fat phase separately and their mixing, heat treatments, emulsification, cooling, working, filling and packaging. Fats are usually selected from milk fat and its fractions, or vegetable fats/oils or a blend of both. Source of milk fat includes cream (40–50% fat), butter, butter oil, low-melting fat fraction (to supply about 30% of fat), chemically modified milk fat. Vegetable oils such as corn, safflower, sunflower, soybean and groundnut oils have been preferred for spreads. In dairy spreads, the fat source is milk fat to which can be added hard (Keogh 2006) or soft milk fat fractions (Verhagen and Warnaar 1984). In margarine spreads, the fat source is usually either soybean oil or sunflower oil blended with a hydrogenated vegetable oil, typically in the ratio 3 : 1. In addition to their nutritional value, milk proteins impart creamy taste to the product and contribute to the water holding capacity and viscosity to the aqueous phase and improve emulsion stability of the product. The general flow diagram for production of fat spread is mentioned in Figure 15.2.

15.4.1 *Sensory defects in dairy spreads*

When removed from the refrigerator, dairy spreads: should be soft enough to be spreadable. It should have a uniformly light of pale straw yellow colour. It should have a clean dairy flavour with a good mouthfeel. It should not have a crumbly texture. It should

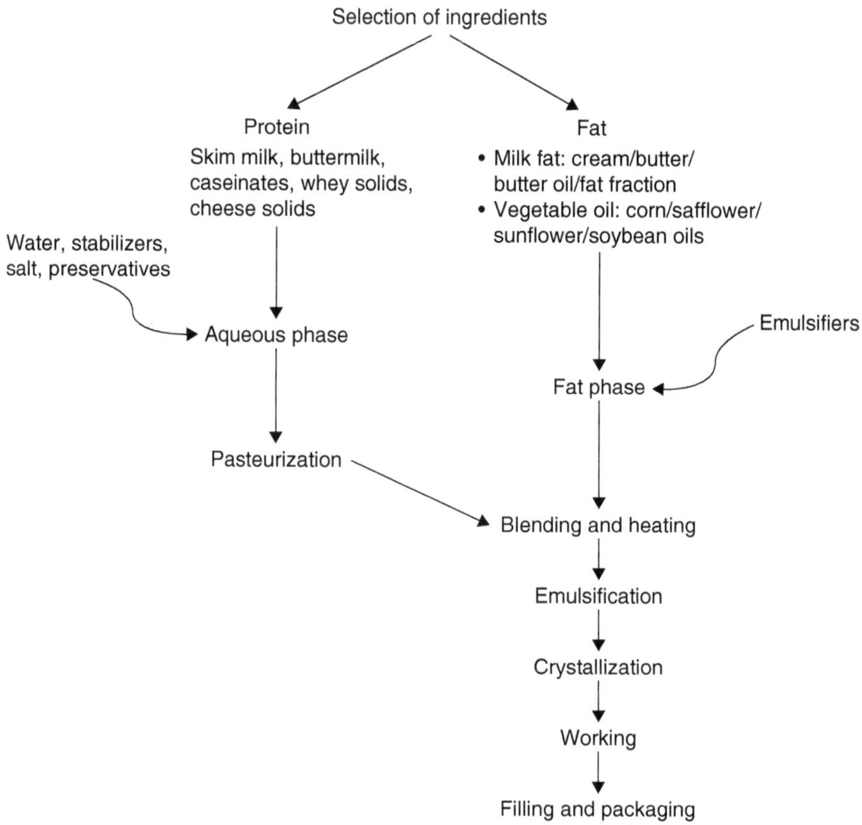

Figure 15.2 Flow diagram for dairy spread production.

possess a smooth creamy texture and melt quickly in the mouth and should not have a waxy mouthfeel. It should be firm, homogenous and should not have any leakiness. The product should have uniform light to medium yellow colour. It should have good spreadability with desirable flavour release with a buttery flavour. It should possess a fine close-knit plastic texture.

Compared to butter, dairy spreads contain greater amounts of water with increased droplet size, and as a result the microbiological stability of these products is reduced (Rousseau and Marangoni 1999; Varnam and Sutherland 2001). Apart from size, shape and arrangement of fat crystals, their amount measured as solid fat content and expressed in relation to temperature is important in controlling the physical properties of spreads (Frede 2002b).

As with butter, the temperature profile of dairy spreads must be carefully controlled during crystallisation and working to obtain metastable β crystals. 'graininess' defect may result from the formation of large coarse triclinic β-type crystals (Frede 2002b). To attain plasticity, it is necessary to have the liquid and solid fat present in the correct ratio. Sufficient crystals must be present to retain the liquid phase, but an excess leads to a very rigid structure, lack of plasticity and brittleness. In addition, some types of spreads contain globular fat, which limits the tendency of higher melting point crystals to form an over-rigid network, resulting in a brittle texture (Varnam and Sutherland 2001). Smaller water droplet size results in a paler colour and less flavour release in the mouth, and leads to a higher fat oxidation rate. Dairy spreads exhibit body defects and oiling-off at temperatures above 16 °C, particularly the spreads whose spreadability when removed from the refrigerator approaches that of soft margarines (Frede 2002b).

15.5 Ghee/Anhydrous milk fat/butter oil

The Codex Alimentarius standard for milk fat products (FAO/WHO 2006) states that anhydrous milkfat, anhydrous butteroil, butteroil and ghee are fatty products derived exclusively from milk and/or products obtained from milk by means of processes that result in almost total removal of water and non-fat solids. Ghee is a fat-rich dairy product widely used in India and the Indian subcontinent. It has a rich nutty flavour and unique grainy texture. Chemically, ghee is a complex lipid of mixed glycerides together with a small amount of free fatty acids, phospholipids, sterols and their esters, fat soluble vitamins (A, D, E and K), carotenoids, carbonyl compounds, hydrocarbons, charred casein, moisture and traces of elements like copper and iron. It occupies a unique position among edible fats because of its pleasing caramelised flavour and granular texture. It is obtained by a process of heat clarification and desiccation. Traditionally, ghee is prepared by culturing whole milk with lactic acid bacteria, prior to further processing steps. The characteristic ghee flavour is due to a complex mixture of compounds produced during the various stages of processing. During heating, the various interactions among protein, lactose and minerals as well as degradation of FFA and lactose are responsible for generation of different flavour compounds. The synonym of ghee is clarified butter, anhydrous milk fat, butterfat, butter oil. Clarified milk fat products, similar to ghee, are used in other parts of the world including the Middle East (Roghan), Africa (Samna), Uganda, Ethiopia and Sudan (De 2001). It has a shelf life of 6–8 months at ambient tropical temperatures.

Ghee has an especially developed flavour and physical structure as a result of its method of manufacturing. According to IDF (1977), ghee should have minimum 96% milk fat, 0.3% maximum moisture, 0.3% maximum free fatty acids (expressed as oleic acid) and peroxide value (as milliequivalent of oxygen/kg fat) less than 1.0. According to Indian Standards (FSSAI 2011), ghee shall conform to the following requirements: moisture – maximum 0.5% (m/m); milk fat – minimum 99.5% (m/m); butyro-refractometer reading at 40 °C : 40 to 44 (depends on feed of the animal); Reichert Meissl value, minimum 21–28 (depends on feed of the animal); FFA as Oleic Acid, maximum, %: 3.0; Baudouin test: negative.

In many countries, surplus milk fat is conserved in the form of butter oil, rather than butter, because of savings in cold storage capacity and the anticipated long shelf life at ambient temperature. The butter oil (anhydrous milk fat) lacks the characteristic flavour, aroma and texture that are otherwise required in ghee, although the chemical composition is almost identical to that of ghee. The Indian standard FSSAI (2011) has defined two types of butter oil: milk fat butter oil and anhydrous milk fat butter oil. Milk fat butter oil is a fatty product derived exclusively from milk or products obtained from milk, or both, by means of processes which result in almost total removal of water and milk solids-not-fat. It shall conform to the following requirements: moisture, maximum, 0.4% (m/m); milkfat, minimum, 99.6% (m/m); Butyro-refractometer reading at 40 °C, 40–44 (depends on feed of the animal); Reichert Meissl value, minimum 21–28 (depend on feed of the animal); FFA, maximum, 0.4% as oleic acid; peroxide value maximum 0.6 (milliequivalent of oxygen/kg fat). Butteroil should have a rich, clean, pleasing, bland smell characteristic of milk fat with a fine crystalline texture. It should have uniform deep straw yellow colour characteristic of milk fat if it is manufactured from cow milk fat and white with green tinge if it is manufactured from buffalo milk fat. Butter oil is darker in colour, less granular in appearance and a bland/flat flavour than that of ghee. A 100-point score card can be used for judging butter oil. The maximum score allotted to various attributes are: flavour 45; texture 25; acidity: 15; colour: 10 and package: 5.

15.5.1 *Sensory quality of ghee and butter oil*

A consumer judges the quality of ghee by its taste and aroma and accepts it on this basis. The main parameters used by consumers for sensory evaluation of ghee are flavour, texture, colour and freedom from suspended particles. The preference for ghee flavour varies from region to region. There is a wide variation in preferences among consumers for flavours, colour and texture of ghee. These variations are mainly due to the different methods used for manufacturing of ghee – e.g., manufacture of ghee by direct cream method has a darker colour than that ghee prepared by pre-stratification method. This is because of the longer heating period in the former case. Rapid cooling of ghee produces a product with very fine crystals resulting in a smooth texture. Whereas slow cooling of ghee produces a few crystals of large mass crystals suspended in a liquid fat with typical grainy texture. The texture of the ghee is also influenced by the fatty acid composition of the milk fat (Berg 1988). It should have large, uniform grains with very liquid fat. On melting, the ghee should be clear, transparent and free of suspended fines. The colour should be bright yellow for cow ghee and white, with or without a yellowish or greenish tinge, for buffalo ghee. Upon prolonged storage at ambient temperature, ghee undergoes oxidative changes. The reaction of oxygen with unsaturated fatty acid leads to the production of strong and disagreeable odours in ghee. The acceptability of ghee largely depends on the extent to which oxidative deterioration has occurred (Mehta et al. 2015). Other common flavour defects in ghee are smoky, burnt, acidic, rancid and tallow. The waxy texture and hard body are not liked by Indian consumers. Breakdown of granulation is responsible for a greasy body that lowers the score of ghee. Brown discoloration is considered as a defect in ghee.

Flavour of ghee

Good-quality ghee should have a rich, nutty, slightly cooked, caramelised, clean, sweet, pleasant odour and an agreeable taste. Ghee flavour is best described as a lack of oiliness or of blandness and sweetly rather than sharply acidic. The preferred ghee flavours range from 'slightly curdy' to 'pronounced curdy' and 'cooked' to 'caramelised'.

More than 100 flavour compounds have been identified in ghee (Wadhwa and Jain 1990). The lactones and carbonyls (aldehydes and ketones) are key components in the flavour of ghee (Abraham and Srinivasan 1980; Yadav and Srinivasan 1992; Mohammed et al. 1998). Short-chain fatty acids C_6–C_{10} and C_{12}, which accounting only for 5–10% of total fatty acids, contribute significantly to ghee flavour. Free fatty acids (C_4–C_{18}) contribute to the normal flavour of ghee. The C_6–C_{10} are considered to contribute significantly to the flavour of ghee. Esters compounds (ethyl esters of C_3, C_5, C_6, C_8, C_{14}, C_{16}, C_{18} and $C_{18:1}$ fatty acids and propyl ester of C_4 fatty acid) are also important in the flavour of ghee (Wadodakar et al. 1996). Carbonyls and lactones play a major role in the flavour of ghee. There is wide variation in reported values of the carbonyls content of ghee. Of flavouring compounds in ghee, about 50% are carbonyls. Different types of monocarbonyls (Alkan-2-ones, Alkanals, Alk-2-enals, Alk-2,4-dienals) and polar dicarbonyls (diacetyl, methyl glyoxal, α-keto glutaric acid, furfural and hydroxy methyl furfural) are associated with flavour of ghee (Rao and Ramamurthy 1984; Wadhwa and Jain 1990; Yadav and Srinivasan 1992). Lactones contribute a coconut-like aroma to ghee and this aroma is typical of the characteristic flavour of ghee (Wadodakar et al. 1996; Yadav and Srinivasan 1992). The temperature of clarification plays a major role in the formation of lactones and variation of it in ghee is mainly due to the temperature of clarification. The lactone content in ghee increases with increase in temperature of clarification (Wadhwa and Jain 1984). The δ-lactones, δ-octalactone, δ-decalactone, δ-dodecalctone, γ-lactone and n-aliphatic lactones (*n*-saturated δ- and γ-lactones) are reported in ghee.

Body and texture of ghee

The body and texture of ghee plays an important role in its organoleptic evaluation. Indian consumers like the granular texture of ghee. A good ghee sample should have well-developed granules dispersed fairly, thickly and uniformly over the entire mass. Ghee that is hard, greasy or has a waxy texture is not liked by consumers (Joshi 1974). The texture of ghee depends upon the source of fat, method of preparation, temperature of clarification, rate of cooling, amounts of FFAs, rate of seeding and temperature during storage.

When ghee is stored at room temperature, it crystallises into three distinct fractions or layers: (i) oily, (ii) granular semi-solid at the bottom and (iii) hard flakes portion floating on the surface and sticking to the sides of the container. Layer formation in ghee could be prevented by storing it at 20 °C or below immediately after preparation. Ghee thus solidified could subsequently be stored at higher temperature without formation of layers. The liquid portion of ghee varies with storage temperature, shape and size of container, repeated heating and agitation, ripening of cream/butter, storage and handling, external seeding, etc.

Colour of ghee

Ghee obtained from cow's milk is yellow in colour due to presence of β-carotene and constitutes a desirable criterion in areas where it is preferred. The ghee obtained from buffalo milk is whitish in colour but it may also have greenish tinge depending on the region and feeding schedule. A straw yellow colour is generally obtained when ghee is prepared using mixed milk. Ghee produced by direct cream method has a darker colour compared to that by the creamery butter process. The colour of ghee always appears deeper to the eyes when melted than when in solid form.

Butter oil/anhydrous milk fat

The flavour should be bland, clean without any objectionable taste or odour. The flavour should not be too pronounced. It should have a smooth texture with very fine grains. There should be no lingering of greasiness or waxiness in the mouth after tasting.

15.5.2 *Sensory evaluation of ghee*

The various sensory attributes like desirable flavour, texture, colour and purity have relative weight for quality assessment by consumers. For example, out of total score of 100, the relative scores of 50 for flavour (aroma and taste), 30 for texture, 10 for colour and 10 for freedom from suspended impurities (ghee residue) are recognised for judging of ghee. Ghee prepared by the desi method has a richer flavour and better textural properties. Ghee prepared from fermented cream influences the whitish colour of buffalo ghee due to conversion of biliverdin to bilirubin, imparting a yellowish greenish tinge (Chandravandana et al. 1977).

Judging and grading of ghee

Consumers from different Indian regions prefer different types of ghee flavour and hence judging of ghee tends to vary. Some consumers like acidic or mild curdy ghee whereas some like moderately cooked or slight curdy ghee. Therefore, the judge should consider the different flavour preferences of the region for which ghee is marketed. Though regional preference should be considered, the most desirable attributes of ghee must be prioritised while judging it.

The following are some of the terms associated with the sensory evaluation of ghee (BIS 1975a):

i. *Brown/burnt*: If clarification temperature of butter is kept at 130 °C or above, it leads to a brownish to dark discoloration with overheated flavour in ghee.
ii. *Curdy/kachcha*: This flavour defect of ghee is as a result of under clarification during ghee making and the flavour resembles that of curd.
iii. *Greasy*: This is associated with undesirable texture in grains of ghee.
iv. *Smoky*: Can develop when clarification is carried out in a fireplace (*Chulah*) using cow dung and the smoke comes in direct contact with milk fat.

Persons with normal sensory ability should be selected in the judging panel. Panellists should also be trained to distinguish and discriminate between normal (control) and defective ghee samples. The control fresh ghee samples should have desirable qualities of colour, flavour, texture and freedom from ghee residues while defective samples have curdy, greasy, acidic, burnt, smoky and oxidised off-flavour.

A representative sample of ghee should be drawn from the bulk or large lot in a clean and dry glass bottle. Ghee sample of 30 ml or 25 g should be sufficient for sensory evaluation; however, preferably 50 ml sample is given in a beaker. The sensory evaluation of ghee is mainly depending on temperature. Whenever possible, the samples should be presented at room temperature. However, depending on sensory attributes like body and texture as well as taste and odour, the serving temperature of the sample varies. Many judges feel that odour, taste and residue can be better detected when ghee is tempered to about 40 °C. The body and texture are best judged when the sample remains in the crystalline condition. Melted cow ghee should be stored at 25–30 °C and buffalo ghee at 30–35 °C for 12–24 hours for proper crystallisation before evaluation. The taste and aroma of the sample can be judged when ghee remains in a melted (at 45 °C) condition. Warm saline water at a temperature of 40 °C for rinsing the mouth between the samples is very effective for fatty foods like ghee. In the sensory evaluation of ghee, first start with visual observations of the sample. The colour of the melted ghee samples should be judged first and then make observations on suspended impurities. Open the cap of the glass bottle and immediately inhale the smell. Subsequently, for taste and aroma, a table spoonful of ghee is placed in the mouth. The body and texture should be evaluated at the end with the help of a spatula or glass rod.

Selection and training of panellists

Candidates should have a willingness to work as a panellist and those who do not like ghee and other similar milk products should not be allowed in the panel. They should have normal sensitivity for odour and taste. Ideally, panellist should be given extra training with various ghee samples so that they will distinguish and discriminate between the samples. A fresh control ghee sample can be prepared either from butter or cream which represents all the quality attributes of ghee like colour, flavour, texture and free from ghee residues (BIS 1975a). The control ghee sample should be served along with defective samples for training purpose.

Defective ghee samples can be prepared as follows:

Acidic: It can be prepared from aged cream or butter having acidity of 0.5–1.5% lactic acid and final ghee samples having 1.5–4.5% oleic acid.

Smoky: It can be prepared in an open pan on a smoky fire. Keep the ghee sample on a smoky fire for 30 minutes for absorption of smoky flavour in it. It can be then added at 10–30% in control ghee and serve to the trainee.

Burnt: It can be prepared by heating ghee sample at 130 °C for 20 minutes to induce burnt flavour. It can be then added at 10–30% to control ghee and served to the trainee.

Oxidised: Ghee is filled into a container, sparged with oxygen gas and kept at 79 °C for 60 minutes.

Curdy: To fresh control ghee add 5–10 g of curd and leave at 40 °C for 10–12 hours before serving to the trainee.

Greasy texture: Ghee is filled into a glass container and exposed to direct sunlight for a period of three hours to develop greasy defects.

Adulterated: The fresh control ghee sample is admixed with animal body fat or hydrogenated vegetable fats at 10–30%.

Sensory evaluation should always start with the visual observation of the sample. Colour should be judged first followed by the texture, odour, taste and aroma. Suspended impurities should be judged after melting the ghee (if already solidified).

15.6 Conclusions

The demand for dairy products depends on the consumer. Regardless of the nutrient status, if the sensory quality of the product is poor, they will be rejected. Fat-rich dairy products are considered the heart of the industry as milk pricing is largely based on the fat content. The sensory quality of the products mainly depends on the initial quality of raw milk, product manufacturing steps and care of the products in the supply chain. Before placing fat-based products on the market, sensory assessment by well trained and experienced judges is a critical step in product quality control. Sensory evaluation is a science-based technique so the preparation of samples, creating a suitable environment for judging and the selection and training of personnel should be carried out under expert supervision. Consistently meeting consumer demand for high-quality fat-rich dairy products is very important for the dairy industry and can be assured by using proper sensory evaluation techniques.

References

Abraham, M.J. and Srinivasan, R.A. (1980). Effect of ripening cream with selected lactic acid bacteria on the quality of ghee. *Journal of Dairy Research* 47: 411–415.

Badings, H.T. (1970). Cold storage defects in butter and their relation to auto-oxidation of unsaturated fatty acids. *Netherlands Milk and Dairy Journal* 24: 147–256.

Bassette, R., Fung, D.Y.C., and Mantha, V.R. (1986). Off-flavours in milk. *CRC Critical Reviews in Food Science and Nutrition* 24: 1–52.

Belitz, H.D., Grosch, W., and Schieberle, P. (2008). *Food Chemistry*. Berlin: Springer-Verlag.

Bendall, J.G. (2001). Aroma compounds of fresh milk from New Zealand cows fed diVerent diets. *Journal of Agricultural and Food Chemistry* 49: 4825–4832.

Berg, V.J.C.T. (1988). *Dairy Technology in the tropic and subtropics. Center for Agricultural Publishing and Documentation (Pudoc)*, 290. The Netherlands: Wagneningen.

BIS (1975a). IS: 7770-1975. *Method for sensory evaluation of ghee (clarified butter fat)*. Manak Bhavan, New Delhi: Bureau of Indian Standards, Indian Standards Institution.

BIS (1975b). IS: 7769-1975. *Method for sensory evaluation of Table butter*. Manak Bhavan, New Delhi: Bureau of Indian Standards, Indian Standards Institution.

BIS (1975c). IS:7768 (1975) (Reaffirmed 2002). *Method for sensory evaluation of milk*. Manak Bhavan, New Delhi: Indian Standards Institution.

Boelrijk, A.E.M., de Jong, C., and Smit, G. (2003). Flavour generation in dairy products. In: *Dairy Processing Improving Quality* (ed. G. Smit), 130–148. Cambridge, UK: CRC Press and Wood Head Publishing Limited.

Bradley, R.L. and Smukowski, M. (2009). Butter. In: *The Sensory Evaluation of Dairy Products*, 2e (ed. S. Clark, M. Costello, M.A. Drake and F. Bodyfelt), 135–166. New York: Springer.

Bruhn, J.C. (2002). *Dairy Processing; Milk Flavours. Dairy Research and Information Centre, Department of Food Science and Technology*. Davis: University of California.

Budin, J.T., Milo, C., and Reineccius, G.A. (2001). Perceivable odorants in fresh and sweet cream butter. In: *Food Favours and Chemistry: Advances of the New Millennium* (ed. A.M. Spanier, F. Shahidi, T.H. Parliment, et al.), 85–96. Cambridge: Royal Society of Chemistry.

Chandravandana, M.V., Daniel, E.V., and Dastur, N.N. (1977). Nature's watermarks in ruminant milk. *Indian Dairyman* 29: 233.

Chen, S., Bobe, G., Zimmerman, S. et al. (2004). Physical and sensory properties of dairy products from cows with various milk fatty acid compositions. *Journal of Agricultural and Food Chemistry* 52: 3422–3428.

Codex Alimentarius (1976). *Standard for cream and prepared creams*. Codex Stan 288. Rome: Food and Agriculture Organization of the United Nations.

Codex Alimentarius Commission (2003). *Codex standard for creams and prepared creams*, Codex Stan. A-9-1976, Rev. 1–2003. Rome: FAO/WHO.

Cormier, F., Raymond, Y., Champagne, C.P., and Morin, A. (1991). Analysis of odour-active volatiles from *Pseudomonas fragi* grown in milk. *Journal of Agricultural and Food Chemistry* 39: 159–161.

De, S. (2001). *Outlines of Dairy Technology*, 75–81. Delhi: Oxford University Press.

Deeth, H.C. (1986). *The appearance, texture, and flavour defects of pasteurized milk*, Bulletin no. 200. (ed. D.O. Cerf), 22–26, Brussels, Belgium: International Dairy Federation.

Deeth, H.C. (2006). Lipoprotein lipase and lipolysis in milk. *International Dairy Journal* 16 (6): 555–562.

Deeth, H.C. and Fitz-Gerald, C.H. (2006). Lipolytic enzymes and hydrolytic rancidity. In: *Advanced Dairy Chemistry*, vol. 2, Lipids (ed. P.F. Fox and P. McSweeney), 481–556. New York: Springer.

deMan, J.M. and Wood, F.W. (1958b). Oiling-off properties of butter. *Journal of Dairy Science* 41: 369–374.

Downey, W.K. (1980) Risks from pre- and post-manufacture lipolysis. *Flavour Impairment of Milk and Milk Products due to Lipolysis*. Document No. 118, 4–18. Brussels: International Dairy Federation.

Driessen, F.M.(1989). Inactivation of lipases and proteinases (indigenous and bacterial). *Heat-induced changes in milk*, (ed. P.F. Fox). Document No. 238, 71–93. Brussels: International Dairy Federation.

FAO/WHO (2006) *Codex Alimentarius standard for butter, A-1-1971*. Revised 1–1999, amended 2003 and 2006. Rome: Food and Agriculture Organization of the United Nations. https://www.fao.org/fao-who-codexalimentarius/sh-proxy/en/?lnk=1&url=https%3A%2F%2Fworkspace.fao.org%2Fsites%2Fcodex%2FStandards%2FCXS+279-1971%

Frede, E. (2002a). Butter: Properties and Analysis. In: *Encyclopedia of Dairy Sciences*, vol. 1 (ed. H. Roginski, J.W. Fuquay and P.F. Fox), 220–226. London: Academic Press.

Frede, E. (2002b). Milk-fat based spreads. In: *Encyclopedia of Dairy Sciences*, vol. 3 (ed. H. Roginski, J.W. Fuquay and P.F. Fox), 1859–1868. London: Academic Press.

FSSAI (2011). Food Safety and Standards Authority of India Regulations, New Delhi, India.

Hayes, W., White, C.J., and Drake, M.A. (2002). Sensory aroma characteristics of milk spoilage by Pseudomonas species. *Journal of Food Science* 67: 448–454.

Hoffmann, W. and Buchheim, W. (2006). Significance of milk fat in cream products. In: *Advanced Dairy Chemistry*, 3e, vol. 2, Lipids (ed. P.F. Fox and P.L.H. McSweeney), 365–375. New York: Springer.

IDF (1977). Anhydrous milkfat, anhydrous butteroil or anhydrous butterfat, butteroil or butterfat, ghee: standards of identity. Standard 68A. Brussels, Belgium: International Dairy Federation.

IDF (1996). *UHT cream*. Document No. 315, 4–34. Brussels: International Dairy Federation.

Jensen, G.K. and Poulsen, H.H. (1992). Sensory aspects. *Pasteurisation of cream*. Document No. 271, 26–31., Brussels: International Dairy Federation.

Joshi, C.H. 1974. Seasonal variation in physico-chemical composition of ghee possessing marketable attributes. Ph.D. thesis. Sardar Patel University, Anand, India.

Kaylegian, K.E. and Lindsay, R.C. (1992). Performance of selected milk fat fractions in cold spreadable butter. *Journal of Dairy Science* 75: 3307–3317.

Kellard, B., Busfield, D.M., and Kinderlerer, J.L. (1985). Volatile off-flavour compounds in desiccated coconut. *Journal of the Science of Food and Agriculture* 36: 415–420.

Keogh, M.K. (2006). Chemistry and technology of butter and milk fat spreads. In: *Advanced Dairy Chemistry, Lipids*, 2e (ed. P.F. Fox and P.L.H. McSweeney), 333–363. Boston, MA: Springer.

Kochar, S.P. (1993). Oxidative pathways to the formation of off-flavours. In: *Food Taints and off-Flavours* (ed. M.J. Saxby). Glasgow, UK: Blackie Academic & Professional.

Krause, A.J., Lopetcharat, K., and Drake, M.A. (2007). Identification of the characteristics that drive consumer liking of butter. *Journal of Dairy Science* 90: 2091–2102. https://doi.org/10.3168/jds.2006-823.

Kwan, A.J., Kilara, A., Friend, B.A., and Shahani, K.M. (1982). Comparative B-vitamin content and organoleptic qualities of cultured and acidified sour cream. *Journal of Dairy Science* 65: 697–701.

Lopez, H.W., Levrat-Verny, M.A., Coudray, C. et al. (2001). Class 2 resistant starches lower plasma and liver lipids and improve mineral retention in rats. *The Journal of Nutrition* 131: 1283–1289.

Luby, J.M., Gray, J.I., Harte, B.R., and Ryan, T.C. (1986). Photooxidation of cholesterol in butter. *Journal of Food Science* 51 (4): 904–907.

Mallia, S., Escher, F., and Schlichtherle-Cerny, H. (2008). Aroma-active compounds of butter: a review. *European Food Research and Technology* 226: 315–325.

Manners, J., Tomlinson, N., and Jones, M. (1987). *Butter and Related Products*, 60–70. Werribee, Australia: Victorian College of Agriculture and Horticulture.

Mehta, B.M., Aparnathi, K.D., and Darji, V.B. (2015). Comparison of different methods of monitoring the secondary stage of oxidation of ghee. *International Journal of Dairy Technology* 68: 589–594.

Milner, J. (1995). Butter and low-fat dairy spreads. In: *LFRA Microbiological Handbook*, B-1–B-15. Leatherhead, Surrey, UK: Leatherhead RA.

Min, D.B. and Boff, J.M. (2002). Chemistry and reactions of singlet oxygen in foods. *Comprehensive Reviews in Food Science and Food Safety* 1: 58–72.

Mohammed, L.S., Roger, K.A., and Judith, N. (1998). A review paper: current knowledge of ghee and related products. *International Dairy Journal* 8: 677–688.

Morgan, M.E. (1976). The chemistry of some microbiologically induced flavour defects in milk and dairy foods. *Biotechnology and Bioengineering* 18: 953–965.

Mounchilli, A., Wichtel, J.J., Bosset, J.O. et al. (2005). HS-SPME gas chromatographic characterization of volatile organic compounds in milk tainted with off-flavour. *International Dairy Journal* 15: 1203–1215.

Muir, D.D. and Banks, J.M. (2003). Factors affecting the shelf life of milk and milk products. In: *Dairy Processing, Improving Quality* (ed. G. Smit), 185–207. New York: Woodhead Publishing.

Mulder, H. and Walstra, P. (1974). *The Milk Fat Globule: Emulsion Science as Applied to Milk Products and Comparable Foods*. Wageningen: PUDOC.

Nelson, J.A. and Trout, G.M. (1965). Judging and Grading Cheese. In: *Judging Dairy Products*, 4e (ed. S. Clark, M. Costello, M.A. Drake and F. Bodyfelt). Milwaukee, WI: Olsen Publishing Co. 80 pp.

Nickerson, S. (1995). Milk production: factors affecting milk composition. In: *Milk Quality* (ed. F. Harding), 15–17. London, UK: Blackie Academic & Professional.

Noda, M. and Shiinoki, Y. (1986). Microstructure and rheological behaviour of whipping cream. *Journal of Texture Studies* 17: 189–204.

Pagote, C.N. 2002. Cream and consumer cream products. *Lecture compendium Advances in Fat-rich dairy products* (14th short course), 22–29. Karnal, India: National Dairy Research Institute (NDRI).

Peterson, D.G. and Reineccius, G.A. (2003). Characterization of the volatile compounds that constitute fresh sweet cream butter aroma. *Flavour and Fragrance Journal* 18: 215–220.

Prentice, J.H. (1972). Rheology and texture of dairy products. *Journal of Texture Studies* 3: 415–458.

Puhan, Z. (1988). *Results of the questionnaire 1785B "fermented milk"*. Bulletin 227, pp. 138–164. International Dairy Federation, Brussels.

Rao, D.V. and Ramamurthy, M.K. (1984). Polar carbonyls in cow and buffalo ghee. *Indian Journal of Dairy Science* 37: 98–102.

Robertson, G.L. (2005). *Food Packaging: Principles and Practice*, 2e. Boca Raton, Florida: CRC Press (Taylor & Francis Group).

Rousseau, D. and Marangoni, A.G. (1999). The effects of interesterification on physical and sensory attributes of butterfat and butterfat-canola oil spreads. *Food Research International* 31: 381–388.

Scott, L.L., Duncan, S.E., Sumner, S.S., and Waterman, K.M. (2003). Physical properties of cream reformulated with fractionated milk fat and milk-derived components. *Journal of Dairy Science* 86: 3395–3404.

Shimidt, K.F. (2005). *Elaboración artesanal de mantequilla yogurt y queso*, 116p. Zaragoza, España: Editora Acribia A.A.

Shipe, W.F., Bassette, R., Deane, D.D. et al. (1978). Off-flavour in milk. *Journal of Dairy Science* 61: 855–869.

Shukla, A., Bhaskar, A.R., Rizvi, S.S.H., and Mulvaney, S.J. (1994). Physicochemical and rheological properties of butter made from supercritically fractionated milk fat. *Journal of Dairy Science* 77: 45–54.

Smiddy, M.A., Kelly, A.L., and Huppertz, T. (2009). Cream and related products. In: *Dairy Fats and Related Products* (ed. A.Y. Tamime), 195–245. Wiley Blackwell.

Solanky, M.J. (1990). Cream products for the consumer. *A Compendium of Lectures Delivered at Refresher Course on Technology of Fat Rich Dairy Products*, 61–72. SMC College of Dairy Science, Gujarat Agricultural University, Gujarat, India (16–30 June 1990).

Tamime, A.Y. and Marshall, V.M.E. (1997). Microbiology and technology of fermented milks. In: *Microbiology and Biochemistry of Cheese and Fermented Milk*, 2e (ed. B.A. Law), 57–152. London: Blackie Academic & Professional.

Thomas, S.B. and Druce, R.G. (1971). Psychrotrophic micro-organisms in butter. A review. *Dairy Industries* 36 (75–80): 145–150.

Tyle, P. (1993). Effect of size, shape and hardness of particles in suspension on oral texture and palatability. *Acta Psychologica* 84 (1): 111–118.

Urbach, G. and Gordon, M.H. (1994). Flavours derived from fats. In: *Fats in Food Products* (ed. D.P.J. Moran and K.K. Rajah), 347–405. London: Blackie Academic and Professional.

Varnam, A.H. and Sutherland, J.P. (2001). *Milk and Milk Products; Technology, Chemistry and Microbiology*. Gaithersburg, Maryland: Aspen Publishers.

Verhagen, L.A.M. and Warnaar, L.G. (1984). Low calorie spread based on a low melting butterfat fraction. U.S. Patent. 4, 436, 760.

Wadhwa, B.K. and Jain, M.K. (1984). Studies on lactone profile of ghee. Part IV: variations due to temperature of clarification. *Indian Journal of Dairy Science* 37: 334–342.

Wadhwa, B.K. and Jain, M.K. (1990). Chemistry of ghee flavour – a review. *Indian Journal of Dairy Science* 43: 601–607.

Wadodakar, U.R., Murthi, T.N., and Punjrath, J.S. (1996). Isolation of ghee volatiles by vacuum degassing, their separation and identification using gas chromatography/mass spectrometry. *Indian Journal of Dairy Science* 49: 185–198.

Walstra, P., Wouters, J.T.M., and Guerts, T.J. (2006). Cream products. In: *Dairy Science and Technology*, 2e (ed. P. Walstra, J.T.M. Wouters and T.J. Geurts), 447–466. Boca Raton: CRC Press.

Whited, L.J., Hammond, B.H., Chapman, K.W., and Boor, K.J. (2002). Vitamin A degradation and light-oxidized flavour defects in milk. *Journal of Dairy Science* 85: 351–354.

Wijnen, M.E. (1997). Instant Foam Physics: Formation and Stability of Aerosol Whipped Cream. PhD thesis. Wageningen Agricultural University, The Netherlands.

de Wit, R. and Nieuwenhuijse, H. (2008). Kinetic modelling of the formation of sulphur-containing flavour components during heat-treatment of milk. *International Dairy Journal* 18 (5): 539–547.

Wright, A.J., Scanlon, M.G., Hartel, R.W., and Marangoni, A.G. (2001). Rheological properties of milkfat and butter. *Journal of Food Science* 66: 1056–1071.

Yadav, J.S. and Srinivasan, R.A. (1992). Advances in ghee flavour research. *Indian Journal of Dairy Science* 45: 338–334.

16 Sensory Applications in Ice Cream and Frozen Desserts

R. Andrew Wilbey

16.1 Introduction

Ice cream and similar desserts differ from most other dairy products in having the majority of the moisture content present as ice. This results in their properties being highly temperature dependent.

A typical ice cream product contains a continuous liquid phase in which the gaseous and solid phases are dispersed. The gaseous phase is a fine dispersion of air bubbles, typically making up from 10% to over 50% by volume and ideally with a mean bubble diameter of 50–100 μm. The bubble surface is initially stabilised by the more hydrophobic proteins such as β-casein but this is partially displaced by fat globules and free fat as the freezing progresses.

The solid phases are primarily ice crystals and fat globules, the latter being both undamaged and as agglomerates including damaged fat globules plus free fat that has leaked from the damaged globules during the freezing process. For smooth ice cream the ice crystal size should be below 20 μm; above 50 μm then iciness will become increasingly evident. Some crystallised sugars and undissolved polysaccharides from stabilisers may also be present. Figure 16.1 illustrates the microstructure of an ice cream.

The continuous phase is an aqueous solution of the lactose and minerals from the dairy ingredients and the added sugars (or sugar alcohols if a no-added-sugar formulation) plus a colloidal dispersion of the milk and maybe other proteins plus any polysaccharide stabilisers. The molar concentration of these solutes will determine the initial freezing point of the ice cream mix, at which point ice will begin to separate from the solution. The solution thus becomes more concentrated and more viscous, and the temperature must be lowered further in order for more ice to separate. Most of the hardness of ice cream is attributable to the proportion of water frozen as ice (Wilbey et al. 1998). However, as melting progresses then the extent of the fatty matrix and the viscoelastic properties of the stabilisers become more important.

Four factors control the sensory quality of ice cream and frozen desserts: the formulation, the ingredients, processing and storage.

Sensory Profiling of Dairy Products, First Edition. Edited by John J. Tuohy.
© 2023 John Wiley & Sons Ltd. Published 2023 by John Wiley & Sons Ltd.

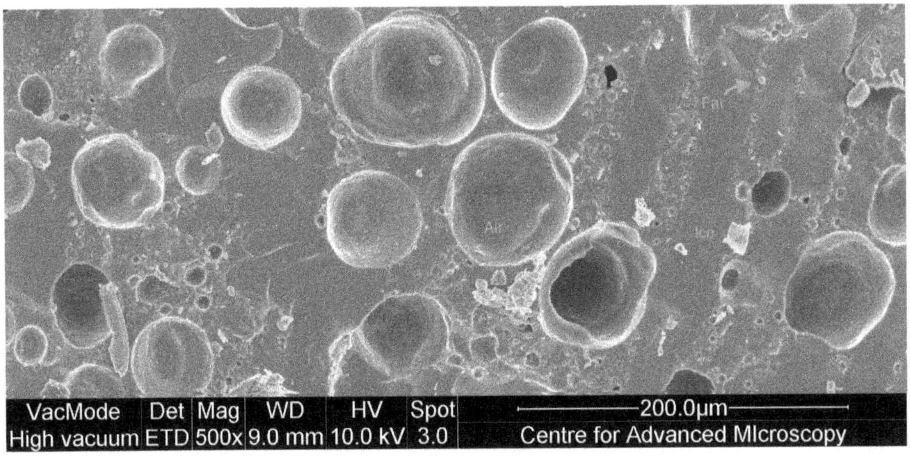

Figure 16.1 Scanning electron micrograph of ice cream.

16.1.1 *Formulation*

The fat content is usually the major contributor to the richness of a product. Typical ice creams fall within a range of 5–15% fat by weight but there is an increasing number of products with fat levels outside this range. Fat content and type may be restricted by local legislation. Some countries may only permit milk fat: where alternative fats are permitted then the more saturated fats, e.g. palm oil, palm kernel oil and coconut oil, are the most commonly used, as liquid fats tend to act as antifoams. If dairy ice cream is being produced then milk fat must be the sole fat other than fats supplied from ingredients such as emulsifier, egg, chocolate or nuts.

Sugars have several roles, providing sweetness and body to the ice cream as well as balancing with the fat and depressing the freezing point. Between 5 and 12% fat the balancing sweetness when formulating an ice cream mix can be estimated using the linear relationship:

$$\text{Sweetness} = 7 + 0.67 \times \text{Fat\%}$$

Sweetness is expressed as the sum of sweetnesses of the contributing sugars, assuming a relative sweetness of 1 for sucrose. This relationship provides a starting point for much of the ice cream in the UK market, but higher notional sweetness levels are needed to balance the bitter notes from the cocoa solids in chocolate ice cream. For fat levels above 12%, the sweetness levels flatten off. Rather than calculating the individual contributions to freezing point depression in milli degrees Celsius (m°C) then it is easier to work in terms of the relative effects, the freezing point depression factor (FPDF) being based upon the molecular weight of sucrose divided by that of the sugar. Total FPDF values between 20 and 25 are sufficient to confer soft scoop properties at −18 °C while FPDF values below 20 give ice creams with progressively harder textures. This approximation assumes that the contribution to freezing point depression from the milk-solids-not-fat (MSNF) is virtually constant and can be omitted from the calculation. Sorbets, where there is no MSNF, will have a FPDF of 30–35.

Egg yolk has been used in traditional formulations at 1% for the emulsifying role, or at higher levels to provide some stabilisation too. For most commercial formulations, mono-di-glyceride blends and complex polysaccharides are most commonly used for the emulsifier and stabiliser functions. Commercial emulsifier-stabiliser blends are typically used at 0.4–1%, the level being inversely proportional to the fat content of the mix.

The ratio between the total MSNF and the moisture content of the mix is critical, the optimum being 1 part of MSNF to 6 water. A lower proportion of MSNF will result in a watery product while higher proportions run the risk of a sticky texture as well as a greater risk of sandy defect (due to lactose crystallisation), plus an unnecessary increase in ingredient cost. Where other protein sources are being used then the optimum can be expressed as 1 part of protein to 17 parts of water, assuming that the water binding properties of the proteins are similar to that of milk proteins. Some formulations seek to reduce costs by lowering the contribution of MSNF and increasing the level of polysaccharides.

16.1.2 *Ingredients*

Milk and cream are the prime ingredients in dairy ice cream – both can be regarded as dispersions of milk fat globules in skim milk. Unsalted butter, anhydrous milkfat and anhydrous butteroil have been used as alternative milk fat sources – high-quality fats are essential if one is to avoid introducing stale and oxidised notes. The sensory properties of these dairy ingredients are covered in other chapters.

Vegetable fats have been used for non-dairy ice creams and vegan products. A high proportion of saturated fatty acids is needed to confer similar melting properties to milk fat plus the high level of solid fat at aeration. High-quality fats and care in their handling are essential to avoid off-flavours resulting from oxidation (palm oil) and/or hydrolysis (especially with lauric fats such as palm kernel oil or coconut oil).

Sucrose is the most common sugar, typically used as a highly refined and otherwise bland sweetener. Less refined variants will introduce their characteristic flavours as well as the risk of a greater microbial load. Partial substitution by dextrose, commonly available as dextrose monohydrate, provides an easy route for increasing FPDF without increasing overall sweetness. Glycerol is a more powerful alternative for increasing FPDF, as can be seen in Table 16.1.

Glucose syrups are partially hydrolysed starches and may be used for several roles. The degree of hydrolysis is normally expressed as the Dextrose Equivalent (DE). High-DE syrups primarily act as lower cost sweeteners while mid-range syrups combine some sweetening with increased body. The low-DE syrups and maltodextrins act primarily to improve mouthfeel and can play an important role in low-fat formulations, especially in sorbets to prevent crust formation. Spray dried versions of glucose syrups are more expensive but much easier to use, particularly for those working on a small scale. Maltodextrins have also been used as fat mimetics in low-fat frozen desserts.

Emulsifier-stabiliser combinations are the most commonly used additives in commercial ice creams. Monoglyceride based blends are the most commonly used

Table 16.1 Comparative data for sweetness and effect on the freezing point.

Ingredient	Molecular weight (Da)	Relative sweetness	FPDF[a]
Sucrose	342	1.0	1.0
Lactose	342	0.2	1.0
Dextrose (anhydrous)	180	0.8	1.9
Dextrose monohydrate	198	0.8	1.8
Invert sugar[b]	180	1.3	1.9
Fructose	180	1.7	1.9
Glucose syrup solids 62 DE	≈300[b]	0.7	1.1
Glucose syrup solids 42 DE	≈430[b]	0.5	0.8
Glucose syrup solids 28 DE	≈640[b]	0.4	0.5
High-fructose corn syrup	≈200[b]	1	1.8
Xylitol	152	0.5	1.9
Erythritol	122	0.6	2.8
Glycerol	46	0.8	3.7

[a] Freezing point depression factor.
[b] These products are mixtures and the values are approximate averages.

emulsifiers; any diglycerides present will have no emulsifying function. While the initial function may be to aid fat dispersion, the more important function is to destabilise the surface of the fat globules during ageing and freezing. Unsaturated monoglycerides are more effective as destabilising agents, though with greater susceptibility to oxidation. Phospholipid preparations, commonly referred to as 'lecithin', may also be used and the phospholipid and lipoprotein in egg yolk provides an emulsification/de-emulsification role for this traditional ingredient.

Stabilisers are predominantly highly hydrophilic carbohydrate polymers with viscosity enhancing and sometimes matrix forming characteristics. Part of their function is to reduce migration rates within the frozen ice cream matrix, so that ice crystal growth (and ideally lactose crystal growth) is slowed and shelf life thus increased. Some stabilisers may also inhibit accretion at the crystal surface. Viscosity enhancement can increase the perception of creaminess during meltdown on the palate: carboxymethyl cellulose is good in this respect while excessive levels of guar gum can lead to sliminess (Imeson 1997). Usually, two or more stabilisers are needed to achieve the optimum sensory properties and stability.

Chocolate, cocoa powders or other sources of cocoa solids are normally added at the blending stage before heat treatment but other flavouring agents should be blended into the completed mix immediately before freezing as heat treatment could damage the flavouring and, in the case of batch heat treatment, lead to high losses of the volatile components of the flavouring. This puts great emphasis on the hygienic handling of flavourings, whether intense flavourings or bulk additions.

16.1.3 *Processing*

The best and safest approach to making ice cream is to include a pasteurisation step in the process, as illustrated in Figure 16.2.

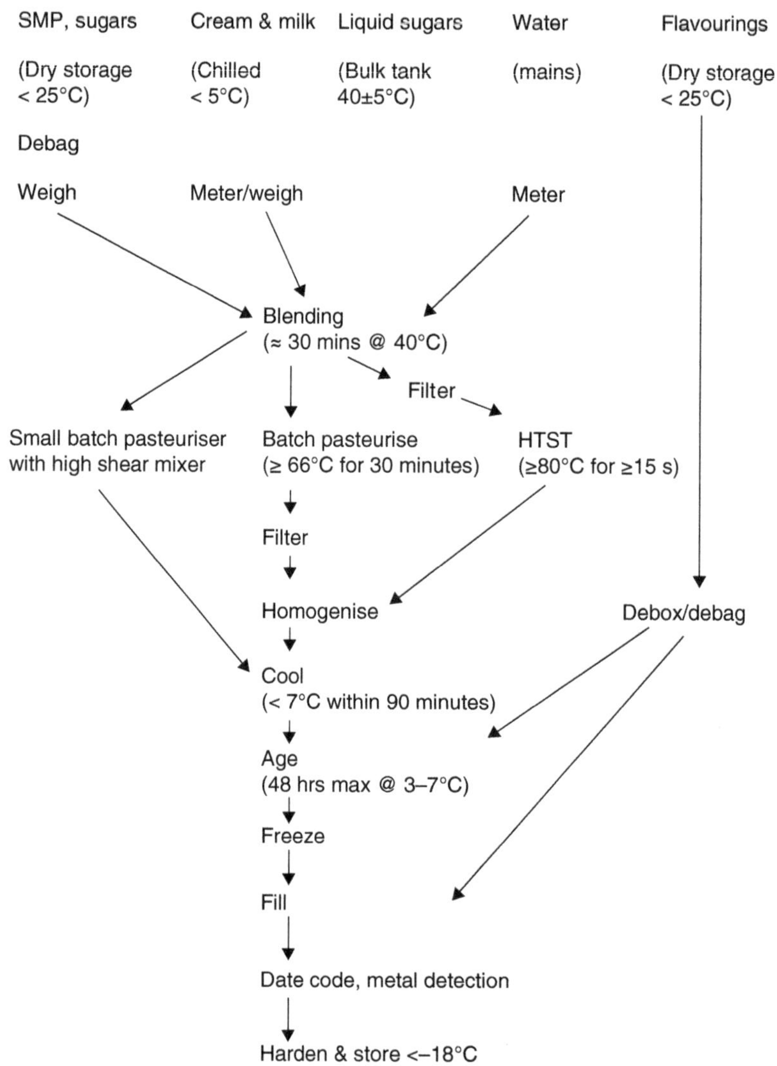

Figure 16.2 Generic process flow diagram for ice cream production.

All but the largest mix production processes use batch mixing. Milk or water is measured into the tank, the stirrer started and any milk powders added carefully. Dispersal of the milk powders is easier at low temperature, the warmer the temperature of the liquid then the better the mixing should be, in order to avoid clumping. Sugars are best added when the temperature is above 20 °C to increase the rate of dissolution. Emulsifier-stabiliser blends are best preblended with some of the sugars: stabilisers must be premixed with sugars to minimise losses when added to the mix. Cocoa powders are also better preblended with sugars to aid dispersal in the mix.

Vegetable fats, unless in pellet form, are best melted before addition to the mix at 35–40 °C and cream should be added at a similar temperature so that the fat is readily dispersible.

In small batch pasteurisers the emulsification is achieved using a high-shear mixer but in larger plants the treatment of the mix at pasteurisation temperature with a high-pressure homogeniser will achieve better results. Batch pasteurisation, for instance at 66 °C for 30 minutes, is simple but wastes energy as there is seldom any heat recovery. For the larger plants using plate heat exchangers for high-temperature-short-time treatment, when feeding mix at ≈35 °C, heat treating at 80 °C for 15 seconds and then first stage cooling using regeneration can make a useful energy saving. Pasteurisation conditions need to be more severe than for milk and cream as the lower a_w of the ice cream mix increases bacterial resistance to the heat treatment (Patel and Wilbey 1994). Using more severe heat treatments than those above can introduce cooked notes that may be acceptable for chocolate and caramel flavoured ice creams but less desirable for fruit flavours.

Cooling the mix after pasteurisation and homogenisation should be carried out as quickly as possible. Small modern batch pasteurisers have built-in refrigeration that will cool the mix from pasteurisation temperature to ≈ 4 °C in about 30 minutes and the same vessel can be used to hold the mix at that temperature till needed. Larger plants can cool the mix to 3–8 °C by pumping through a plate heat exchanger to achieve the cooling in about 30 seconds, though some of the hot mix will remain in the mix tank for up to about half an hour, prolonging the average heat treatment and the potential for the generation of cooked flavours. Further cooling can take place by circulating chilled water or glycol through the jacket of the ageing tank.

Ageing the mix has several benefits. Further hydration of the milk proteins and stabilisers will increase viscosity and hence the perception of creaminess. Further crystallisation of the fats will take place – most of the changes in the milk fat will be achieved within two hours but more time is needed if palm oil is the main fat. More subtle changes take place at the surface of the fat globule, where emulsifiers will partially displace milk proteins from the fat-water interface, making it more susceptible to shear in the subsequent freezing process. Ideally, the ageing should be for more than six hours with the optimum being an overnight hold. Thereafter there is no gain in the whipping properties but an increasing risk of microbial growth (and associated off-flavours) on extended storage.

Ice cream freezers are essentially a form of scraped-surface heat exchanger where the surface of the barrel may be approaching −20 °C. Ice crystals will form primarily on microscopic irregularities at the barrel surface, plus there may be some nucleation on motes within the mix once it has been sufficiently supercooled. The ice crystals are scraped off the barrel by the blades and blended into the mix. Initially these crystals will thaw but once the bulk of the mix is cooled to below its freezing temperature then the proportion of ice will build up. In batch freezers, the process is relatively slow, and it can take several minutes to convert the mix to ice cream, during which time air is beaten into the product. However, with continuous freezers the volume of the freezer barrel is relatively small in relation to throughput and the mix is frozen quickly so that the ice cream exiting from the freezer is slightly supercooled with a small proportion of ice still be formed.

With most ice cream freezers, the exit temperature is around −6 °C and only about 50% of the water is present as ice, depending upon the mix formulation. Soft scoop ice cream mixes have a higher molar concentration of solutes and thus a lower proportion of ice at any given temperature, giving a lower viscosity at extrusion and a relatively soft product, even at −18 °C.

Soft serve freezers are designed for dispensing ice cream into single portions for immediate consumption. Other freezers are typically used to produce ice cream that will be packed into either individual portions or bulk containers, then chilled further prior to storage and distribution.

It is critically important that the ice cream does not warm up during the packaging operation and that it is further frozen as quickly as possible, for instance using blast freezing, to avoid the creation of large ice crystals. The biggest threat to the quality of frozen desserts comes from temperature fluctuation in storage and particularly in distribution. The higher the FPDF of the mix then the greater the shift in frozen water for a given change in temperature, leading to a proportionately shorter shelf life. The simpler structure of sorbets makes them susceptible to temperature abuse and/or excessive FPDF, leading to 'bleeding' where the liquid aqueous phase drains down to the base of the container, leaving aerated ice crystals in the upper layer.

16.2 Sampling and presentation

Comparisons between soft serve ice creams can be difficult to organise, since the product changes rapidly after extrusion. If the aim is to compare different formulations in a given make of machine, then nominally identical machines should be used simultaneously to produce the fresh products for evaluation. At the next iteration of the comparison, then the machine-mix combinations should be changed, in case one of the machines is performing differently to the other(s). Where the aim is to compare two or more different freezer models with a standard mix, then the problem is just to ensure sufficient space for the freezers in the preparation area so that the samples can be served to the panellists with a similar time delay between extrusion and evaluation.

Stored ice cream presents fewer problems than soft serve ice cream. The simplest situation is where pre-packed single portions are available in plain packaging (or at least where the packaging is the same) and the packaging should be marked with a 3-digit random code. With otherwise similar stick products, the code can be written on the stick. Where the shape of the product might influence the panellists then similar sized portions of the product should be placed on a small, coded, dish. With choc ices then either serve as half portions or cut smaller portions but then discard the ends so that a similar ratio of couverture to ice cream is maintained.

With bulk packs (from 500 ml to 5 l), the pack is best tempered to −10 °C before scooping out portions for testing. Aim for a similar appearance between the samples, so stick to one design and size of ice cream scoop. Scoops with a built-in cutter may be preferred for softer products, while all-stainless construction is preferable if the scoops are to be cleaned in a dishwasher. The carrier for the sample should also be standardised, whether a small plastic or paper cup, glass or ceramic saucer/bowl. Plastic petri dishes are very convenient but not recyclable. Each sample carrier should be marked with its sample code.

Ice cream-based gateaux may need to be first viewed in their entirety before slicing into portions if the overall visual quality is important. Use a sharp knife plus a stainless cake slice to aid the transfer. A cheese trier is useful for removing a core sample from a gateau or block of ice cream where a sample is needed for analysis. Sampling tools are shown in Figure 16.3.

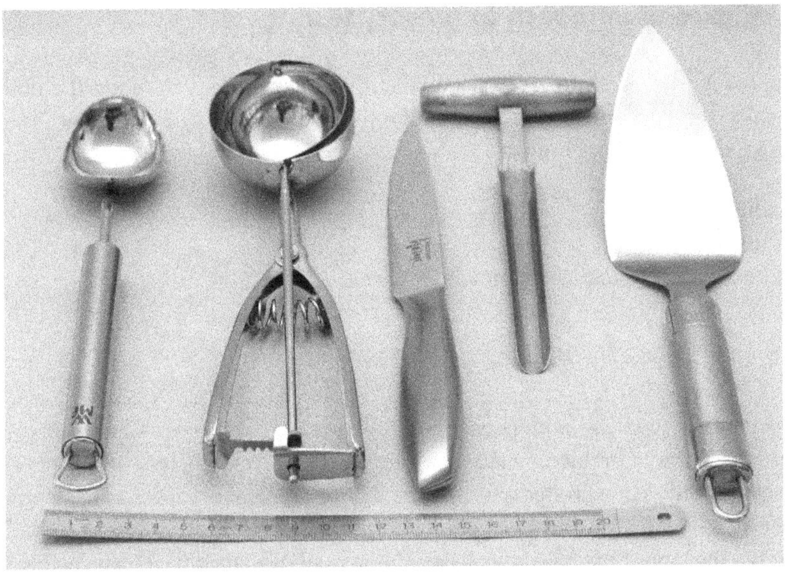

Figure 16.3 Utensils to aid sampling of frozen desserts.

Colour photographs should be taken using standardised illumination and from a standardised angle. Flash, halogen or 'cool white' LED lighting as close as possible to standard north light will give truer colour than most fluorescent lighting and minimise the need for colour correction. The background should have a neutral, matt finish, preferably off-white or light grey and be seam-free. This can be achieved using a fine textured cloth or by fixing an A2 sheet of card so that it is gently curving between the horizontal and vertical. Where dimensions may be of interest then place a ruler within the field of vision. Best results will be obtained by standardising the focal length of the camera lens, e.g. 50–60 mm for a full-frame SLR or 35 mm for a half-frame camera.

For all the stored samples, the temperature should be standardised and the easiest approach is to prepare the samples the day before, cover and store overnight at −10 °C. Samples can then be taken direct from the overnight freezer and placed on the trays for presentation to the tasters with the minimum of delay. Colder temperatures should be avoided as the colder product will make sensory evaluation both more difficult and run the risk of inducing severe pain if excessive ice cream contacts the roof of the taster's mouth.

Tempered samples may also be offered in a standardised 0.5 or 1 l container for tasting by a panel. Single-use plastic or wooden spoons/spatulas should be provided, with a fresh spoon per sample per person. Where possible the plastic spoons should be uncoloured but if coloured then it is preferable that the colour be uniform and as close to a neutral shade as possible. Wooden spoons/spatulas have a slightly rough finish that may interfere with any textural response but would be acceptable if a plastic-free regime is desired. The alternative would be to supply stainless-steel spoons and have a dishwasher available for cleaning the large number needed.

Sorbets and sherbets should be treated as for ice cream.

16.3 Choice of approach to sensory testing

The approach to sensory testing of ice cream will depend largely upon the objectives and ideally should be managed by an appropriately qualified professional. These fall into three, albeit related, groups:

i. Quality assurance
ii. Competitions
iii. Research and product development

16.3.1 *Quality assurance*

Improvements to process hygiene and control over the past 40 years have led to higher standards and fewer problems caused by ingredients. Nevertheless, it is important to make sensory checks on products to confirm that no atypical product has been produced, whether due to an off-flavour from an ingredient or to a processing error.

Few but the largest organisations have a large pool of trained panellists readily available and for most companies it is necessary to rely on staff for routine checks on the product. The primary question to be asked is whether the product tastes 'normal' and if there is any doubt then that sample should be re-tasted against a sample from an earlier batch. If the recent sample is judged inferior, then more detailed examination will be needed. This highlights the need for an accepted list of descriptors and appropriate training of the panellists so that at the very least they have a common understanding of the terms being used.

One of the best known and long-standing scoring systems is that approved by the American Dairy Science Association (ADSA) and used, inter alia, for inter-collegiate evaluation and other contests in the USA (Bodyfelt et al. 1988, Alvarez 2008; Tharp and Young 2013). Various modifications have been made to the original scheme over the succeeding years, but the principles have remained the same (Marshall et al. 2003; Alvarez 2009). Points are allocated for flavour (maximum 10), body and texture (maximum 5), plus appearance and colour (maximum 5), giving an overall maximum score of 20. Points are then deducted according to the severity of the defect.

16.3.2 *Competitions*

Competitions are essentially an exercise in ranking a series of samples against published criteria. Judges should have experience of that type of product and follow similar criteria to those set out by the ADSA. Usually, the classes are based on flavour and/or production method, the 'Open' class being the most demanding to judge.

In the UK, competitions are commonly offered as part of agricultural shows such as the Royal Highland Show and the Royal Welsh Show, as well as the National Ice Cream Competition run by the Ice Cream Alliance (ICA), which is explained in detail below.

Rules for the national competition are published by the ICA several months ahead and upon registration each entrant is supplied with standardised plain 500 ml containers. One tub is needed, other than for the open class where two are required.

The biggest risk to the samples is temperature abuse in transit. The ICA minimises this by arranging regional hubs for collecting the entries before transport at −20 °C to the competition venue. Entrants also have the option to transport their own entries. On arrival, the entries are stored in chest freezers set at −17 to −19 °C (sorbets are usually held colder) with one or two classes per freezer, depending upon the number of entries.

The entries are removed from the freezer shortly before the time of judging, 10–15 minutes at an ambient temperature below 20 °C being sufficient for the products to warm up. The containers are presented one at a time to the panel of four judges, each container being labelled with the class, an entry number and a QR code, and passed along the judges. Once all judges have tasted the entry, then the lid is replaced, and it is returned to the freezer. (The coronavirus pandemic resulted in changes to the organisation of the competition to avoid cross-infection.)

Scoring originally used cards (Figure 16.4) but since 2019 has been carried out using tablets with wireless data transfer to the hub computer. Each judge first registers on their tablet for that class and then identifies each sample by scanning the QR code on the side of the container. The judge then inspects the sample for appearance before dipping a single-use plastic spoon into it and assessing the texture and flavour. A stainless-steel knife and spoon are available to each judge, in case a sample is unduly firm. Dilute lime juice is provided, and best practice is to take a small sip between entries, so that fatigue during multi-tasting can be minimised. A typical style of layout is shown in Figure 16.5.

Sample No:

National Ice Cream Competition
Score Card

Judges No.

Dairy Ice Cream (Artisan): Vanilla				
	Max Marks	Marks Given	Comments (if any)	Judging Supervisor
Appearance	**9**			
Texture & Body	**20**			
Palatability & Flavour	**70**			
Bonus Mark	**1**			
Total Marks Awarded				

Judges Signature____

Figure 16.4 Printed score card previously used by the Ice Cream Alliance for the National Ice Cream Competition.

Figure 16.5 Typical layout of a judge's place setting.

Table 16.2 Scoring guidelines for the ICA National Ice Cream Competition.

	Range	Guideline	Expected
Appearance	0–10	5–10	
Texture	0–20	10–20	≥15
Flavour	0–70	55–70	≥65

With large entries in a class then the early products tasted should be re-judged later in the sequence. Where there is a tie in the top rankings then a taste-off between those entries may be carried out.

The scoring is summarised in Table 16.2. Entries scoring below the guideline values are expected to be rejected. Though notionally each entry is scored out of 100, the range of scores used is somewhat less in practice. Judges are also encouraged to add comments to the scoresheet; a list of suggested comments is given in Table 16.3.

The top-scoring ice cream or sorbet in each class is awarded the gold award, with other awards as listed in Table 16.4.

16.3.3 *Research and product development*

In an ideal world, research on an ingredient will lead eventually through to its application in a commercial product. In the early stages of such work, the interest may be as much in understanding how the ingredient functions as in its effect on the model product, whereas with product development the emphasis is towards the benefit to the product. While it is usually far easier to seek understanding of the functional properties of an ingredient in terms of its effect on physical properties, any benefits need to be confirmed in sensory terms.

Table 16.3 Suggested potential comments[a] on a competition entry.

Appearance		Texture and body		*Palatability and flavour*	
1	Good colour	12	Good texture	23	Good sample
2	Poor colour	13	Poor texture	24	Nice flavour
3	Acceptable colour	14	Icy texture	25	Lacks flavour
4	Inconsistent colour	15	Too high overrun (airy)	26	Strong flavour
5	Too light colour	16	Too low overrun (dense/heavy)	27	False taste
6	Too much colour	17	Needs more body	28	Granular/bitter taste
7	Good inclusions	18	Inconsistent texture	29	Has an aftertaste
8	Poor inclusions	19	Broken/flaky texture	30	Too sweet/very sweet
9	Too many inclusions	20	Not a nice combination	31	Does not taste like named flavour
10	Good/nice ripple	21	Inclusions do not freeze well	32	Chewy to taste
11	Good all-round combination	22	Poor balance with inclusions	33	Greasy aftertaste
				34	Good creamy flavour

[a] Judges have the option to use the number rather than the comment in full.
Source: Courtesy of the Ice Cream Alliance.

Table 16.4 Example of the ranking of ICA awards for each class.

Ranking	Award
1st	Gold
2nd	Silver
3rd	Bronze
4th	Special diploma of merit
5th	Special diploma of merit/Diploma of merit
6th	Diploma of merit
Others	Diploma if ≥85% of the score for the Gold award

There are several challenges in characterising the sensory properties of ice cream and similar frozen desserts; standardisation of production, storage and tempering of samples is essential. Ideally, the potential panellists should cover a range of ages and be of both sexes, with more candidates than needed in order to accommodate losses through illness or poor performance. The candidates need training sessions to either agree descriptive terms or to use already established terms as well as to understand the reporting system whether paper-based or one of the established computer programmes for sensory analysis. Each session should have a relatively small number of samples, supplied in random order and coded with a 3-digit random code. Scoring may use a numbered scale for each attribute, accompanied by appropriate descriptors for the extremes. A useful alternative that avoids a numerical judgement is to employ a 100 mm line, again with appropriate descriptors for the extremes, where the panellist simply marks the line. With paper-based systems, the score can be read off easily by using a ruler, while computer-based systems will log the judgement automatically. With the latter then the length of the line on the screen is not critical as the software will automatically adjust for the scale.

16.4 Characterising the sensory properties of ice cream and frozen deserts

Guinard et al. (1997) investigated interactions between fat (10–18%) and sucrose (8–18%) on the sensory properties of vanilla ice creams, using quantitative descriptive analysis based on Stone and Sidel (1993). A panel of 15 students first drew up a list of descriptors based on tasting commercial ice creams: 1 for colour, 7 for flavour and 7 for texture/mouthfeel and were then trained against standards for each attribute. Intensity scoring used 165 mm line scales, mainly anchored at the extremes by 'low' and 'high'. Samples were served in random order at −10 °C, four or five per session over four sessions, allowing each sample to be presented in duplicate. Data was subjected to analysis of variance (ANOVA), response surface methodology and principal components analysis (PCA). Sweetness perception increased with sucrose content and, unexpectedly, with fat content. Vanilla flavour, custard/eggy and almond notes also increased with sucrose content. Fat content had a greater effect on melting perception than sucrose content, possibly reflecting a greater effect of the fatty matrix than freezing point depression. PCA indicated that the first two components contributed 77 and 9% of the variance

In their investigation of the effect of fat level on the perception of flavour chemicals in ice cream, Liou and Grün (2007) recruited a panel of 12, aged between 22 and 35 and with an even split between the sexes. Samples were tempered at for at least 24 hours at −13 to −15 °C and served between −11 and −15 °C. Three 2-hour training sessions were run on three consecutive days, using 150 mm unstructured line scales anchored at 0 = low intensity and 150 = high intensity. The Compusense 5 computer programme (Compusense, Guelph, Canada) was also used to randomise presentation of the samples and the formal evaluation was spread over a one-month period. ANOVA was applied to the sensory data while descriptive data were analysed using SAS statistical software (SAS Institute, Cary, NC, USA). There was not a consistent response to fat levels with the flavour compounds tested. The protein-based fat mimetic was better at mimicking the flavour profile of full-fat ice cream than the carbohydrate-based mimetic.

Descriptive analysis can also help to understand differences between markets and market segments. Thompson et al. (2009) carried out a comparative study of artisanal Italian ice cream and ice creams of similar nominal flavour produced in the USA. The study used five highly trained panellists, who spent 15 hours in developing a common lexicon using USA-produced samples before commencing the analysis. Each panellist assigned a score on a 15-point scale for each of the parameters for each sample then a discussion was held to reach a consensus on the scores for that product. The 18 Italian samples were evaluated over a 5-day period in Florence, Italy while the USA samples were examined in Riverside, MO, USA. The data was analysed by PCA using Unscrambler 9.6 software. For most flavours, the Italian products were given higher scores for their characteristic notes. Overall, the Italian fruit-flavoured products were more sour and astringent but less sweet or bitter than the USA-made ice creams, while the Italian vanilla and chocolate varieties were sweeter.

Soukoulis et al. (2008) recruited nine panellists for assessment of sensory attributes of ice creams with various stabilisers. Training involved three two-hour sessions. Samples were tempered at −15 °C, as were those used for texture analysis, and supplied as two 10 g scoops. The panellists were placed in partitioned booths and invited to score using 10-point scales and the data analysed using Statistica v7 software (StatSoft Inc., Tulsa, OK, USA). Correlation was evident between the sensory and rheological results, aided by PCA and cluster analysis In subsequent work on interactions with bulk sweeteners, the same sample size, number of panellists and 10-point scale were used (Soukoulis et al. 2010b). Correlations were found between some sensory characteristics and physical properties, e.g. coarseness, wateriness and gumminess, with rheological and thermal properties, overrun, mean air cell size and melting rates.

Inclusion of yoghurt and other fermented milks into frozen dessert formulations can create new problems as the reduced pH of the mix will reduce the water binding properties of the milk proteins as well as modifying the texture. Soukoulis and Tzia (2008) reported a comparison between direct and indirect acidification of mixes using six experienced panellists in two sessions, evaluating flavour, texture and appearance, plus melting characteristics in the second session. Two 10 g scoops, tempered at −15 °C for 24 hours, were used and the data was analysed using Statistica 7 software. Use of fermented milk was more highly scored than fermented mix, while addition of xanthan gum and partial substitution of skim milk solids by whey solids also increased creaminess and overall acceptance. A subsequent investigation into probiotic ice creams with a panel of nine used an increased sample size with two 15 g scoops (Soukoulis et al. 2010a). Texture was assessed at both scooping and in the mouth. A second group of 51 regular consumers of typical dairy ice cream were recruited to evaluate series of 15 g samples without a break and express a degree of liking using a scale from 0 = unacceptable to 5 = acceptable and 10 = excellent, like extremely. Partial least squares regression was used to analyse the acceptability data.

McGee et al. (2014) provided an example of a paper-based investigation, with a panel of eight using the USDA scorecard for caprine ice creams, following four training sessions that included lists of descriptive terms. Water and crackers were provided for clearing the palate between samples.

Panellists may be requested to concentrate on characteristics at a single time point or to describe variation in intensity of a flavour or textural characteristic with time. If time dependence is being examined, then fewer samples are normally offered. Hyvönen et al. (2003) started with 15 panellists in a study of the perception of melting and flavour release from dairy and non-dairy ice creams with different fat contents, examining eight samples in duplicate over four sessions. The Compusense 5 computer programme (Compusense, Guelph, Canada) was used for training and the subsequent time-intensity study. Luke-warm water was used as a mouthwash after each sample. This study was complemented by a second study of sensory attributes using single point ratings by a panel of 35, rating 12 samples over four sessions, presented in random order with 3-digit codes. Statistical analysis included PCA and paired t-tests.

The commercial production of ice cream is largely dependent on the use of stabilisers to limit the rate of crystal formation and hence degradation of the texture on storage. These hydrocolloids can impart subtle changes to the texture and flavour profile of an ice cream or similar dessert. Varela et al. (2014) evaluated six ice creams containing different combinations of milk, cream, egg yolk, sugar and hydrocolloid (guar gum and carrageenan). The sensory evaluation used temporal dominance of sensations (TDSs) with 14 panellists who had some prior experience in sensory descriptive analysis, generating their descriptive terms in a preliminary session and focussing on in-mouth sensations. The attributes were listed with 100mm unstructured scales anchored with 'weak' and 'strong' on computer screens for the three formal assessments on different days. Each assessor was asked to click a start button on placing the ice cream in his or her mouth and then indicate the dominant attribute at each time interval, continuing after swallowing till no aftertaste remained. Fizz software v 2.45 (Biosystems, Counternon, France) was used for data collection. Hydrocolloids were shown to reduce the first impact of iciness and coldness perception in favour of creaminess.

16.5 Impact of formulation changes on the sensory profile of ice cream

16.5.1 *Fat reduction*

Many authors report attempts to modify ice creams by partial substitution with ingredients that have 'healthier' profiles and often accompanied by fat reduction. Salem et al. (2016) reported partial fat replacement by whey protein, oat, wheat germ or modified starch using 10 experienced panellists and a report sheet as recommended by Watts et al. (1989). ANOVA showed no clear significant effects of the substitutions. Maize starch has a long history as a thickener in traditional ice cream formulations; Cody et al. (2007) evaluated two rice flours in vanilla ice creams with zero, 4 and 10% fat. Fresh and stored (seven-week old) samples were assessed using descriptive analysis by eight panellists, who had generated 16 sensory attributes and definitions in their training sessions. Heat-shocked samples were also included and a 150mm line scale was used. ANOVA used SAS v8.02 software (SAS Institute, Cary, NC, USA). A difference in textural attributes was found between the rice flours, together with a reduction in sweetness perception and a flouriness at the (higher) 4% usage level. Addition of rice flour did not prevent iciness developing on temperature abuse.

Replacement of sugars by polyols may both reduce the energy content of frozen desserts and minimise their contribution to elevated blood sugar levels. Bordi et al. (2004) replaced sucrose by maltitol in ice creams containing 12% milk fat and the ice creams were tasted by a randomly selected panel ranging in age from 20 to 26 years old and with 12 females plus 10 males. Water was sipped between samples and 9-point hedonic scales were used for taste, texture and appearance categories plus an overall impression category. The overall scores indicated no overall differences between the maltitol and sucrose sweetened ice creams, but the maltitol-sweetened ice cream was perceived as less sweet than that based on sucrose and the males indicated a preference for the sweeter product. Lee et al. (2005) monitored similar formulations over 6 months with a panel of 3 male and 16 female volunteers. Panels were held at monthly intervals and participants evaluated the samples within four categories: taste, texture and appearance on 6-point scales plus overall impression on a 7-point Likert scale. Gender was included in the analysis of covariance and no significant differences between the ice creams were found over the six months of the trial. However, a study by Stokols et al. (2005) used 200 participants drawn from both male and female students plus adults, comparing a control ice cream with one using maltitol as the sweetener. Questionnaires included taste, texture and appearance plus overall impression. The maltitol-based ice cream was found to be less sweet than the control, which overall was regarded as too sweet. Both male and female students preferred the less sweet tasting ice cream when compared to their adult equivalents. In a continuation of this work, three commercial sugar-substituted 'lite' vanilla ice creams (0–4.5% fat) were compared with a maltitol-sweetened control, using 200 participants evenly divided between the female and male university students and adults (Stokols et al. 2006). The questionnaire used 9-point hedonic scales and computers running Compusense software. The participants preferred the maltitol control though males rated it as not sweet enough, while the ice cream sweetened with lactitol was the least preferred.

Part of the sensory properties of fat globules is associated with the motion of the globules over the palate and this presents opportunities to use fat mimetics. Karaca et al. (2009) compared a microparticulated whey protein, a maltodextrin-based ingredient (DE not specified) and inulin, used at 4%, in ice cream with 0.1, 2 and 4% fat plus a control at 8% fat, the stabiliser blend being salep, guar gum and gelatin. Sensory tests for appearance (5 points), texture and mouthfeel (5 points) plus taste and aroma characteristics (10 points) used an experienced panel of seven panellists. Three replicates were examined, and ANOVA tests used SPSS v. 9 software (SPSS Inc., Chicago, IL, USA). Addition of inulin was associated with lower scores for appearance at all fat levels, whereas those ice creams with 0.1% fat appeared to exhibit the greatest differences in texture and mouthfeel from that of the control. Microparticulated protein at 4 and 2% fat levels gave better taste matches to the control while all substitutions at 0.1% fat were rated poorly, the latter performance probably reflecting the absence of a fat reservoir from which flavour compounds could be desorbed onto the palate.

Cadena and Bolini (2011) carried out time-intensity and acceptance tests on three pairs of traditional and light (minimum 25% fat reduced) commercial vanilla ice creams. Pre-selection started with 29 potential panellists with 19 passing to the training stage, from whom 10 were chosen for their discriminating ability and consistency. The computer-based scoring used a structured linear scale from 0 to 9 and samples were tested in triplicate. Timing was computer-based, each panellist taking the sample at the first signal and indicating the intensity of the attribute by mouse movement. After five seconds a second signal indicated that the panellist should swallow the sample while still recording intensity, while a third signal, 60s later, indicated the end of the test. Data collection and analysis used bespoke software. The three brands each exhibited different time-intensity plots with a common factor being greater intensity of the bitter flavour component in the reduced-fat variants, bitterness being absent in two of the three full-fat ice creams. One reduced-fat brand was noticeably less sweet and creamy than its full-fat partner whereas for the other two brands the reverse was found. Some of the differences were attributed to the use of cyclamate or sucralose in reduced-fat variants.

16.5.2 *Sucrose reduction*

Low-fat ice cream (4.1% fat) was used by Ozdemir et al. (2008) to compare different sweetener combinations with a sucrose-based control. High-fructose corn syrup (HFCS), glucose syrup (GS) and honey were used as partial or complete replacements. Eight trained panellists judged the seven ice creams, served at −10 °C, at 1 day and at 15 days from manufacture. Scoring was on the basis of colour, body and texture, taste and general acceptability, using a 1–9-point hedonic scale from very poor to excellent. Data was analysed using SAS v.6.12 (SAS Institute, Cary, IL, USA). Scores were similar or slightly higher after 15 days. The different sugars had significant effects on the quality parameters, with the 50 : 50 ratio sucrose : GS product scoring marginally higher than the sucrose control.

Wood (2011) investigated the partial substitution of sucrose by prebiotics, using either inulin or fructo-oligosaccharides (FOS) at 10, 20 or 30% substitution. The control contained 10% sucrose and all batches also contained 4.5% GS solids (36 DE). The 95 panellists were presented with four samples at a time, scoring each sample for sweetness, smoothness, taste and overall acceptability using a printed sheet for each sample by marking anchored 175 mm hedonic scales. Water was used as a palate cleanser between samples. Data was analysed using SAS v. 9.1. There were no significant differences between the control and samples with 10 or 20% substitution with inulin but at 30% substitution the ice cream was significantly less sweet than the control, also less smooth, less well flavoured and less acceptable. The effects of substitution by FOS were less clear cut though substitution at 30% was rated as less sweet than at 10% substitution but no less sweet than the control or at 20% substitution. Smoothness was rated less at 30% substitution, as was overall acceptability when compared to the control and at 10% substitution. Overall, the results supported the use of up to 2% of either inulin or FOS in a formulation.

When replacing sucrose, the challenge is to match both the perceived sweetness and the freezing depression, usually requiring two or more ingredients. Wheelan et al. (2008) looked at combinations of erythritol, tagatose, maltitol, polydextrose and trehalose plus sweetness supplementation by sucralose. A five-member expert panel was used for characterising physical properties of the first series of mixes, then a larger panel of 70 for flavour intensity and liking, with degree of liking marked on a 140 mm scale using the control as a centre anchor. Two potential formulations that met physical and sweetness requirements included 6% tagatose, 6% polydextrose and 3% maltitol, or 15% maltitol plus 2.5% trehalose.

Frozen yoghurt formulations can provide refreshing acidity, but the lower pH is usually accompanied by poorer water binding and reduced stability to storage. Rezaei et al. (2011) assessed the effect of guar gum and gum arabic as stabilisers, using an expert sensory panel of seven and a 5-point scale for judging colour, texture and total acceptability. Data was analysed by SAS v. 9.1 software (SAS Institute, Cary, NC, USA). The best acceptability was achieved with either 0.2% guar gum or 0.5% gum arabic, with gum arabic providing better physical properties.

Di Criscio et al. (2010) produced probiotic (fermented *with Lactobacillus casei* and *Lactobacillus rhamnosus*), prebiotic (added inulin) and synbiotic (containing inulin then fermented with *L. casei* and *L. rhamnosus*) ice creams. Unflavoured, vanilla and fruit-flavoured ice creams were produced. The samples were assessed by a semi-trained panel of 10 against 5 parameters: Flavour (scoring 1–3), homogeneity (1–3), consistency (1–4), taste (1–5) and colour (1–5). Such narrow and variable ranges, though easy to administer, do not make it easy for the panellists to log representative scores.

El-Kholy (2018) reported the inclusion of a powdered palm fruit in a probiotic frozen yoghurt. A panel of 10 assessed flavour, body and texture, appearance, melting resistance and overall acceptability. Data was analysed by ANOVA (SAS v9., SAS Institute, Cary, NC, USA). Substitution of milk fat by doum palm solids at 33 and 66% gave improved acceptability, though this was a small panel.

16.6 Consumer preference and acceptance testing

While trained panellists can provide detailed and reproducible information on the samples provided to them, the small number of people and the very nature of their training will result in their preferences being unlikely to reflect that of the general population. Thus, any preference testing, which may be of great commercial importance, should employ much larger and untrained panels.

Guinard et al. (1997) used a second panel of 146 untrained students split evenly between the sexes, to evaluate the same set of samples as used in the QDA investigation. These students tasted all 9 samples in one sitting but were asked to indicate their degree of liking for texture/mouthfeel, flavour and overall liking on a 9-point hedonic scale from 1 = 'dislike extremely' and 5 = 'neither like nor dislike' to 9 = 'like extremely'. RSM was applied to the ratings and indicated dome-shaped responses, the optimal values for fat and sucrose varying slightly between parameters. Optimum overall liking occurred at 14.3% sucrose and 14.8% fat.

The comparison between the three pairs of full and reduced-fat ice creams by Cadena and Bolini (2011) was continued with acceptance analysis, using 117 consumers with a mean age of 32 years (61% were females). Samples were served at -10 ± 2 °C and mouth washing with water between samples. Scoring used a 90 mm unstructured scale, anchored with 'dislike extremely' and 'like extremely'. The panellists were also questioned on their purchasing intentions in respect of each sample, using a 5-point scale. Data were analysed following the method of MacFie and Thomson (1994). All the samples of ice cream were accepted or appearance, aroma and texture though one brand had lower scores for texture. Two high and one reduced-fat ice creams were rated above the others and this separation was maintained in the bidirectional plot from internal preference mapping. The scores for purchase intention indicated a similar distribution.

Likewise, Varela et al. (2014) also evaluated the same six ice creams as for profiling, using 85 untrained vanilla ice cream consumers aged 23–68 years, with 30% males. Each consumer was invited to undertake two tests; the first being to score each sample for liking the appearance, texture and flavour, using a 9-point scale from 'dislike extremely' to 'like extremely'. The second test invited each consumer to tick any relevant attribute from the supplied list of 26: this check-any-that-apply (CATA) approach may be an alternative to trained profiling while giving an insight into consumer perceptions. Liking scores were subjected to ANOVA while the CATA data was analysed by multiple factor analysis, using XLStat 2009 software (Addinsoft, Paris, France).

Herald et al. (2008) carried out a comparison between egg yolks and alternatives in French-style ice creams using 102 untrained panellists. At each session, a panellist was supplied with four samples, one being the control. Unsalted crackers and distilled water were used as palate cleansers between samples. Degree of liking was evaluated on a 9-point hedonic scale, from 1 = dislike extremely to 9 = like extremely (Mielgaard et al. 1991) and ANOVA used Statistical Analysis System software (SAS Institute, Cary, NC, USA). The control with egg yolk was the preferred product.

In a similar exercise, Fragoso et al. (2016) investigated the sensory properties of ice cream inoculated with *Pediococcus pentosaceus*, using 54 untrained panellists in two sessions. In the first session, colour, creaminess and acidity were scored on a 5-point hedonic scale while in the second session acceptance on a 7-point scale was used. No difference was found between the control and inoculated samples

Difference tests can be useful where a small change has been made to a formulation or process and confirmation is sought whether this has had an effect on the product. There are a number of variations, including:

- Triangle tests where three samples are presented, one being different and the panellist is asked to choose the odd one out.
- Duo-trio tests where three samples including two reference samples are presented, one being labelled as the reference sample and the panellist being asked which of the other two are most like the reference sample.
- Paired tests – these may be used where looking for a known or unknown difference in a parameter, e.g. sweetness, or for preference between two batches.

With trained panels, it may be possible to obtain reliable data with less than 10 panellists but with untrained panellists expect to use ≥60 while consumer-based work would be better with ≥120. Where there are periodic changes in a product over a period of time then beware of drift; one change might not be detected by most consumers but after several changes there may be a noticeable change from the original product, creating an opportunity for a competitor.

Pintor et al. (2014) sought to optimise the use of inulin, a fructose polymer with prebiotic properties, as a fat replacer in ice creams. A 20-member trained panel, 75% female, was used for a series of duo-trio tests where the reference sample was marked 'R' and other samples carried random numbers. Each panellist was invited to indicate which attribute was different to that of the reference sample, following ISO (2004). At 3% inulin in the fat-reduced ice cream, there was not a significant ($P<0.05$) overall difference from commercial vanilla ice cream though a small effect on flavour was noticed.

References

Alvarez, V.B. (2009). Ice cream and related products. In: *The Sensory Evaluation of Dairy Products*, 2e (ed. S. Clark, M. Costello, M. Drake and F. Bodyfelt), 271–331. New York: Springer Science.

Bodyfelt, F.W., Tobias, J., and Trout, G.M. (1988). Sensory evaluation of ice cream and related products. In: *The Sensory Evaluation of Dairy Products*, 166–266. New York: Van Nostrad Reinhold.

Bordi, P., Cranage, D., Stokols, J. et al. (2004). Effect of polyols versus sugar on the acceptability of ice cream among a student and adult population. *Foodservice Research International* 15: 41–50.

Cadena, R.S. and Bolini, H.M.A. (2011). Time-intensity analysis and acceptance test for traditional and light vanilla ice cream. *Food Research International* 44: 677–683.

Cody, T.L., Olabi, A., Pettingell, A.G. et al. (2007). Evaluation of rice flour for use in vanilla ice cream. *Journal of Dairy Science* 90 (10): 4575–4585.

Di Criscio, T., Fratenelli, A., Mignogna, L. et al. (2010). Production of functional probiotic, prebiotic and symbiotic ice creams. *Journal of Dairy Science* 93: 4555–4564.

El-Kholy, W.M. (2018). Preparation and properties of probiotic low fat frozen yoghurt supplemented with powdered doum (*Hyphaene thebaica*) fruit. *Egyptian Journal of Dairy Science* 46: 67–78.

Fragoso, M., Pérez-Chaleba, M.L., Hernandez-Alcantara, A.M. et al. (2016). Sensory, melting and textural properties of fat-reduced ice cream inoculated with thermotolerant lactic acid bacteria. *Carpathian Journal of Food Science and Technology* 8 (2): 11–21.

Guinard, J.-X., Zoumas-Morse, Z., More, L. et al. (1997). Sugar and fat effects on sensory properties of ice cream. *Journal of Food Science* 62 (5): 1087–1094.

Herald, T.J., Aramouni, F.M., and Abu-Ghoush, M.H. (2008). Comparison study of egg yolks and alternatives in French vanilla ice cream. *Journal of Texture Studies* 39: 284–295.

Hyvönen, L., Linna, M., Tuorila, H., and Dijksterhuist, G. (2003). Perception of melting and flavour release of ice cream containing different types and contents of fat. *Journal of Dairy Science* 86: 1130–1138.

Imeson, A.P. (ed.) (1997). *Thickening and Gelling Agents for Food*, 2e. London: Blackie Academic & Professional.

ISO (2004). *ISO 10399:2004 (en) Sensory Analysis – Methodology – Duo-Trio Test*. Geneva, Switzerland: International Organization for Standards.

Karaca, O.B., Güven, M., Yasar, K. et al. (2009). The functional, rheological and sensory characteristics of ice creams with various fat replacers. *International Journal of Dairy Technology* 62: 93–99.

Lee, H., Stokols, J., Palchak, T., and Bordi, P. (2005). Effect of polyols versus sugar on the shelf life of vanilla ice cream. *Foodservice Research International* 16: 44–52.

Liou, B.K. and Grün, I.U. (2007). Effect of fat level on the perception of five flavour chemicals in ice cream with or without fat mimetics by using a descriptive test. *Journal of Food Science* 72 (8): S595–S604.

MacFie, H.J.H. and Thomson, D.M.H. (1994). *Measurement of Food Preferences*. London: Blackie Academic & Professional.

Marshall, R.T., Goff, H.D., and Hartell, R.W. (2003). *Ice Cream*, 6e. New York: Kluwer Academic.

McGee, C.E., Jones, J.O., and Park, Y.W. (2014). Evaluation of textural and sensory characteristics of three types of low-fat goat milk ice cream. *Small Ruminant Research* http://dx.doi.org/10.1016/j.smallrumres.2014.12.002.

Mielgaard, M., Civille, G.V., and Carr, B.T. (1991). *Sensory Evaluation Techniques*, 2e. Boca Raton FL: CRC Press.

Ozdemir, C., Dagdemir, E., Ozedemir, S., and Sagdic, O. (2008). The effects of using alternative sweeteners to sucrose on ice cream quality. *Journal of Food Quality* 31: 415–428.

Patel, S.S. and Wilbey, R.A. (1994). Thermal inactivation of g-glutamyltranspeptidase and *Enterococcus faecium* in milk-based systems. *Journal of Dairy Research* 61: 263–270.

Pintor, A., Severiano-Pérez, P., and Totosaus, A. (2014). Optimization of fat-reduced ice cream formulation employing inulin as fat replacer via response surface methodology. *Food Science and Technology International* 20 (7): 489–500.

Rezaei, R., Khomriri, M., Kashaniaejad, M., and Aalami, M. (2011). Effects of guar gum and Arabic gum on the physicochemical, sensory and flow behaviour characteristics of frozen yoghurt. *International Journal of Dairy Technology* 64 (4): 563–568.

Salem, S.A., Hamad, E.M., and Ashoush, I.S. (2016). Effect of partial fat replacement by whey protein, oat, wheat germ and modified starch on sensory properties, viscosity and antioxidant activity of reduced fat ice cream. *Food and Nutrition Sciences* 7: 397–404.

Soukoulis, C. and Tzia, C. (2008). Impact of the acidification process, hydrocolloids and protein fortifiers on the physical and sensory properties of frozen yoghurt. *International Journal of Dairy Technology* 61 (2): 170–177.

Soukoulis, C., Chandrinos, I., and Tzia, C. (2008). Study of the functionality of selected hydrocolloids and their blends with κ-carrageenan on storage quality of vanilla ice cream. *LWT – Food Science and Technology* 41: 1816–1827.

Soukoulis, C., Lyroni, E., and Tzia, C. (2010a). Sensory profiling and hedonic judgement of probiotic ice cream as a function of hydrocolloids, yoghurt and milk fat content. *LWT – Food Science and Technology* 43: 1351–1358.

Soukoulis, C., Rontogianni, E., and Tzia, C. (2010b). Contribution of thermal, rheological and physical measurements to the determination of sensorially perceived quality of ice cream containing bulk sweeteners. *Journal of Food Engineering* 100: 634–641.

Stokols, J., Bordi, P., Palchak, T., and Lee, H. (2005). Profiling of sensory evaluation of a no-sugar-added vanilla ice cream among specific age gender populations. *Foodservice Research International* 16: 86–92.

Stokols, J., Bordi, P., Palchak, T., and Lee, H. (2006). Sensory evaluation: age and gender profiling of lite, no-sugar-added vanilla ice creams. *Foodservice Research International* 17: 41–48.

Stone, H. and Sidel, J. (1993). *Sensory Evaluation Practices*, 2e. San Diego CA, USA: Academic Press Inc.

Tharp, B.W. and Young, L.S. (2013). *Tharp and Young on Ice Cream*. Lancaster PA, USA: DEStech Publications.

Thompson, K.R., Chambers, D.H., and Chambers, E.I.V. (2009). Sensory characteristics of ice cream produced in the USA and Italy. *Journal of Sensory Studies* 24: 396–414.

Varela, P., Pintor, A., and Fiszman, S. (2014). How hydrocolloids affect the temporal oral perception of ice cream. *Food Hydrocolloids* 36: 220–228.

Watts, B.M., Ylimaki, G.I., Jeffery, L.E., and Elias, L.G. (1989). *Basic Sensory Methods for Food Examination*. Ottawa: International Development Research Centre.

Wheelan, A.P., Vega, C., Kerry, J.P., and Goff, H.D. (2008). Physicochemical and sensory optimisation of a low glycemic index ice cream formulation. *International Journal of Food Science and Technology* 43: 1520–1527.

Wilbey R.A., Cooke T. and Dimos G. (1998). Effects of solute concentration, overrun and storage on the hardness of ice cream. *Ice Cream, SI 9803,* 186–187. Brussels: International Dairy Federation.

Wood J.M. (2011). Sensory evaluation of ice cream made with prebiotic ingredients substituted for sugar. Nutrition & Health Sciences Dissertations & Theses. 18. http://digitalcommons.unl.edu/nutritiondiss/18 (accessed 8 November 2022).

Index

Sensory Profiling of Dairy Products, First Edition. Edited by John J. Tuohy.
© 2023 John Wiley & Sons Ltd. Published 2023 by John Wiley & Sons Ltd.

www.ingramcontent.com/pod-product-compliance
Lightning Source LLC
LaVergne TN
LVHW062205200926
848484LV00006B/1324
* 9 7 8 1 1 1 9 6 1 9 2 1 5 *